D0909354

GENETICS AND
HORSE BREEDING

GENETICS AND HORSE BREEDING

WILLIAM E. JONES
Lake Elsinore, California

Lea & Febiger  *Philadelphia   1982*

Lea & Febiger
600 Washington Square
Philadelphia, PA   19106
U.S.A.

Library of Congress Cataloging in Publication Data

Jones, William Elvin.
  Genetics and horse breeding.

  Previous ed. published as: Genetics of the horse. 1971
  Bibliography: p.
  Includes index.
  1. Horse breeding. 2. Horses—Genetics. I. Title.
SF291.J75   1982   636.1'082   81-6059
ISBN 0–8121–0721–7   AACR2

Published in Great Britain by Bailliere Tindall, London

PRINTED IN THE UNITED STATES OF AMERICA

Print No: 5   4   3   2   1

preface

The foundation for this book was laid in 1968 when I was completing my work toward a Ph.D. in animal genetics under Dr. Ralph Bogart at Oregon State University. Dr. Bogart suggested that I put together a book on equine genetics, and he was a tremendous help in that endeavor, which resulted in the publishing of *Genetics of the Horse* in 1971. Two years later I completely revised the book for a second edition, which is now out of print. Although this new book does not carry Dr. Bogart's name, I am deeply indebted to this grand old livestock geneticist for directing me toward the work that has resulted in this volume.

Included in this text is some of the pertinent material originally found in *Genetics of the Horse*. The genetic material has been extensively updated. In fact, most of the former book's topics have been expanded, with a much greater emphasis on practical horse breeding.

Information in the area of horse breeding is expanding so rapidly that it is difficult to conclude a manuscript of this length without having new, even extremely vital, information come to light. Thus, revisions were made until the pages went to press.

During this past decade, the horseman's interest and general knowledge in equine genetics has boomed. Ten years ago reference material was scarce—mainly confined to old pamphlets and a few scientific articles. Today, dozens of horse magazines publish articles concerning sophisticated genetic concepts—often reprinted from a scientific journal.

Because of this great increase in written material relating to genetics and horse breeding, it is almost impossible to have a complete list of references. I have tried to include most of the important ones.

I hope that this volume will be useful to horse breeders in all stages of expertise. As the typical breeder is interested in only specific types and uses of horses, he will find some information that does not relate to his particular needs. Still it is surprising how much one learns when extrapolating knowledge from one area to another. Therefore the reader is encouraged to study all parts of the book, especially since many chapters build on the knowledge obtained from previous chapters.

Lake Elsinore, California   William E. Jones

contents

PART II. EQUINE GENETICS

PART IV. BREEDING MANAGEMENT

PART V. PRACTICAL HORSE BREEDING

part I

equine heritage

1

ancient equine history

The ice-age man watched from afar the herds of horses he found in his territory. Even then these gallant creatures were admired, and no doubt, the more enterprising men longed to be astride a prancing steed. Men painted their cave walls with pictures of the horse; they carved equine statues from mammoth tusks (Figure 1–1); and they made ornaments from stone depicting the horse (Figure 1–2).

The prominent place that the horse has come to play in man's life bears witness that the horse, even in the beginning, was more than a prey, decorative art, or some kind of symbol. The horse and man were destined to be partners in history. Perhaps man began to realize this connection, even during the ice age.

EVOLUTION

The horse evolved at a much slower pace than man. The first record of a horse-like creature is about 50 million years old. This primitive horse, eohippus, was no more than 12 inches in height and had four toes on each foot. Over the epochs this creature has increased in size and lost the use of all but one toe. The equine family seems to have remained much the same over the past several million years. Few other mammals have continued to exist with such little change over such a long period. The genetic makeup of the horse most certainly has been well suited for the niche it occupies.

The cavemen painted pictures of horses on their cave walls, indicating what appear to be both light and heavy horses. Since that time, remains of both have been found in the forested areas of central Europe and in Asia.

3

Figure 1–1.   A 32,000-year-old carving of a horse, from a mammoth tusk. (From American Museum of Natural History.)

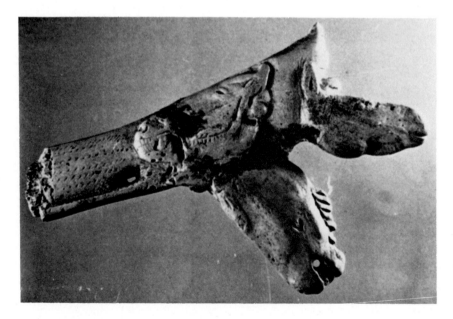

Figure 1–2.   A Cro-Magnon stone carving showing several horse heads.

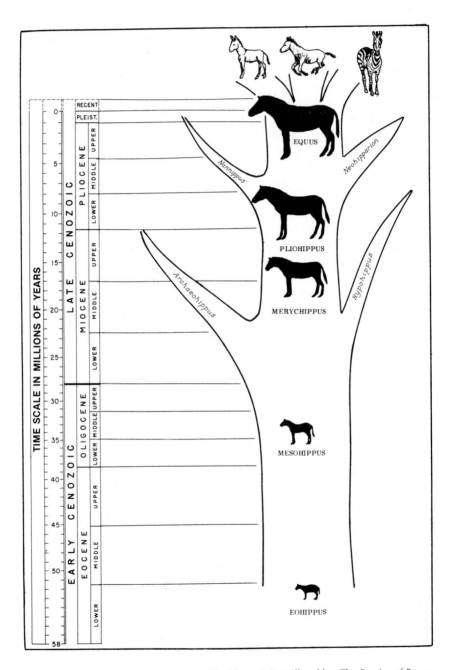

Figure 1–3.   The equine family tree. (Modified from D.P. Willoughby: The Empire of Equus. Cranbury, NJ, A.S. Barnes, 1978.)

Today probably eight distinct species of the genus Equus appear in the family equidae. Some of these species are known to interbreed and produce fertile offspring, an example being the Mongolian wild horse crossed with the domestic horse.

The distinction between a breed and a species is not always clear-cut. Usually groups of animals that cannot interbreed are classified as distinct species, while the various breed crosses are quite fertile. In the case of equine species, the lack of hybrid fertility stems from differences in chromosome number. Usually three divergent types of animals are considered to be within the equine family: the horse, the ass, and the zebra. Some of the chromosome differences among members of each of these types are explained in Chapter 8.

It is difficult to determine whether all of the present-day species evolved from a common ancestor or whether some represent stages of recent equine evolution. In all probability neither situation is entirely true. The fact that all present species hybridize, even though most are infertile, indicates that they are not too distant from a common ancestor. Simpson, a well-known horse expert, suggests that the zebra is the most primitive of the three types mentioned, and that the horse and ass may have evolved from zebra-like ancestors.

The early evolution of the equine occurred on the American continent and from there spread periodically to other parts of the world through various land bridges that have come and gone. Somewhat longer than a million years ago, the most recent ancestor of Equus migrated across the land bridge that existed in the area of the Bering Sea. This animal, now known as pliohippus, greatly resembled the modern zebra, probably stripes and all.

The evolution of the horse is illustrated in Figure 1–3. About 50 million years ago the great-grandfather of all horses, the tiny eohippus, grazed over the area which is now Wyoming, New Mexico, and Utah. This creature was rather stupid as judged by the size of its brain and the lack of a well-developed cerebrum.

Many species of eohippus (more accurately termed Hyracotherium) developed during the millions of years of the Eocene epoch. Some died out; others developed modifications that made them more successful. By the Oligocene epoch, some had lost one toe and had begun to depend upon the center digit more than the outer two; in addition, the facial portion of the skull had become slightly longer than the cranium. The brain had become more complex, which probably made a much more cunning animal. These creatures, known as mesohippus, were 6 to 8 inches taller than the eohippus. The changes in the equine skull as it evolved are shown in Figure 1–4.

During the Miocene epoch, some species that developed from mesohippus were better adapted to open grazing than to a forest habitat. Dependence upon the middle toe increased from generation to generation. Grazing rather than browsing created a problem, with teeth wearing away from the

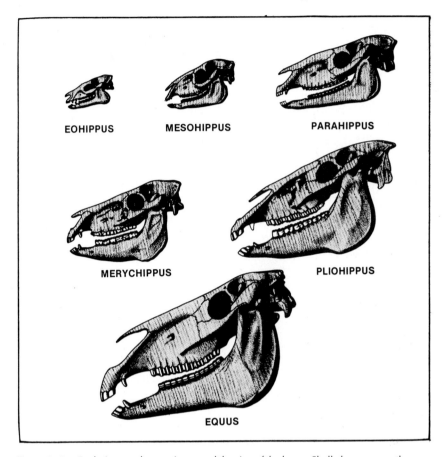

EOHIPPUS   MESOHIPPUS   PARAHIPPUS

MERYCHIPPUS   PLIOHIPPUS

EQUUS

Figure 1–4.   Evolutionary changes increased the size of the horse. Skull changes over the eons.

abrasive action of sand in the grass, and during the millions of years of the Miocene epoch the more successful grazers developed teeth with longer crowns and enamel ridges with cement between, much like the horse teeth of today. The skull length also increased to facilitate grazing. Ultimately the most successful equine species of this epoch turned out to be merychippus.

By the middle of the Pliocene epoch some equine one-toed species had evolved. Throughout the years, generation after generation increased in size so that pliohippus eventually resembled some zebras of today.

Equus was the typical equine of the Pleistocene epoch. No one knows how many species of this genus have evolved and died out in the thousands of years since then. The ancestors of the contemporary equine moved across the land of the Bering Strait and into Europe and Asia. They eventually migrated as far as the southern tip of Africa. Later the Bering Strait became covered with water, and still later, for unknown reasons the equine of the American continent died out completely.

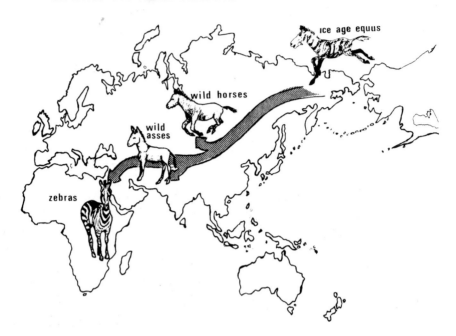

Figure 1–5.   Geographic development of the equine species from the ice age to the present. The three equine types, horse, ass, and zebra, each developed in a distinct geographic area.

Figure 1–6.   European rock drawings of pony types, approximately 5000 B.C.

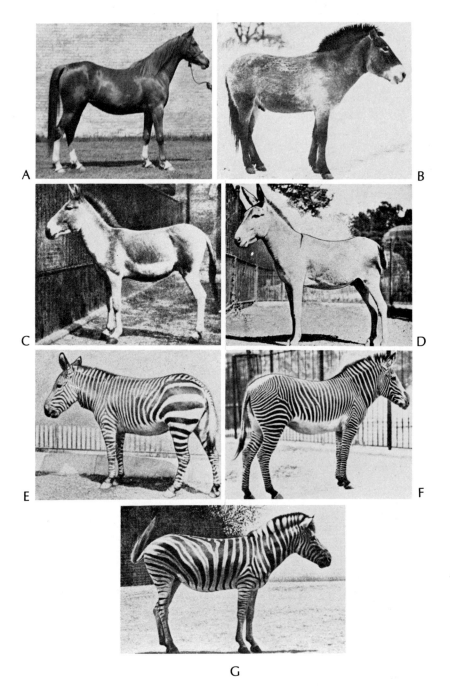

Figure 1–7.   The equidae species: A. *Equus caballus*, B. *Equus przewalski*, C. *Equus hemionus*, D. *Equus asinus*, E. *Equus zebra*, F. *Equus grevyi*, and G. *Equus burchelli*.

Three types of equines that are of special interest today developed in the old world. In the southern and eastern part of Africa the zebra species evolved. The various ass or "half-ass" species developed in northern Africa and Asia Minor, while the horse species developed in Mongolia and Russia (Figure 1–5). Several species of wild horse are known to have existed, all of which have become extinct except *Equus przewalski* and *Equus caballus*.

There is much evidence that *Equus caballus* developed along two lines often referred to as heavy horses and light horses, recognizable by their size and proportions. The light horse had a short face, a narrow snout, and slender legs and body. The heavy horse had a long, narrow skull with a prominent face and massive bones of the body and legs. Another difference is the enamel patterns of the teeth of the two. Knowledge of these early horses comes from cave paintings. These animals, shown in Figure 1–6, were of the pony type.

Modern members of the equine family have been divided into seven species (Figure 1–7). The two species of horse, as mentioned, are *Equus caballus* and *Equus przewalski*. There are probably two species of the ass type. *E. kiang* has still not been determined by karyotype (discussed in Chapter 8) to be a separate species, and there is doubt about whether *E. onager* and *E. hemionus* are separate species because of their identical chromosome number. Little doubt exists that there are three species of zebras: *E. zebra, E. grevyi,* and *E. burchelli*.

Characteristics other than chromosome number are more often used in differentiating species. One that is of value is the measurement or count of various bones of the body. A study by R.M. Stetcher revealed much variation in vertebrae number among the equine species as well as some variety among the domestic breeds (Table 1–1).

In Stetcher's study, most domestic horses were found to have 18 thoracic, six lumbar, and five sacral vertebrae. The Hemione was found to have one

Table 1–1. Varying Numbers of Vertebrae in the Equine Species. The numbers in parentheses represent the number of specimens examined by Stetcher.

Animal		Thoracic Vertebrae				Lumbar Vertebrae			
		17	18	18½	19	5	5½	6	7
Domestic horse	(55)	4	50	1		6	1	48	
Shetland Pony	(7)		7					7	
Arabian	(7)	4	3			1		6	
Przewalski's horse	(32)		23		9	16		16	
Ass	(11)	1	10			9		2	
Mule	(6)	2	4			1		5	
Hemione	(9)		8		1	9			
Zebra	(49)	7	41		1	5		44	
Grevy zebra	(14)		14			1		12	1

less lumbar vertebra in all cases. Most asses showed the same number of vertebrae as the horse. Przewalski's horse showed equal variation between five and six lumbar vertebrae. The Arabian horse, thought by many to have only five lumbar vertebrae, was found usually to have six, but the majority of Arabian horses have one less thoracic vertebra.

THE FOUR ANCIENT HORSE TYPES

Man's interest in the horse seems to have paralleled the development of culture. Ice-age men decorated the walls of their caves with portraits of the horse, although there is no evidence that the horse was domesticated at this early date. Rather, domestication probably began 5 or 6 thousand years ago.

The many horse breeds of today have all developed from a few ancient types of about 10,000 years ago (Figure 1–8). Controversy exists among the experts as to just which types were the primitive building blocks of the modern horse population. It is likely that the tarpan, the "cold-blooded" European horse, the "hot-blooded" Oriental horse, and Equus przewalski have all made valuable contributions to the heritage of the modern horse.

1.The tarpan, known as the wild horse of Europe, became extinct during the last century, but rather accurate descriptions of these animals are available. They were small and buckskin-colored. Either these horses developed from ancient pony types, as described by Anthony Dent in The Encyclopedia of the Horse, or the tarpan superseded the pony type and contributed heavily to its development as described by other authors.

Another author, T. Vetulani, claimed (in 1933) that the forest tarpan was an ancestor of the European pony types. An earlier author, Salensky (1907), observed that the 1866 tarpan was more horse-like than the Mongolian wild horse (E. przewalski). These "old" tarpans had no hock callosities, which is also true of some horses today. Still another writer, Antonius (1922), emphasizes the historical importance of the tarpan in The Empire of Equus:

> I believe that I was the first to repeatedly point out that, despite its probable admixture of blood from domestic horses, the Tarpan, as above mentioned, was a genuine type of wild horse, and as such, of greatest importance in connection with the origin of the domestic horse.

David Willoughby, in the previously quoted book, gives detailed analysis of varying scientific arguments about the tarpan. As to the geographical range of the creature he states:

> Hilzheimer. . . put the probable eastern limit of the tarpan's range (during historic times, presumably) as the 40th degree of longitude—roughly, the longitude of Archangel on the north and Rostov, southward, on the eastern shore of the Sea of Azov. Antonius's estimate would extend it five degrees farther to the east, to the longitude of the present city of Volgograd (Stalingrad). Just how far the tarpan spread westwards is not exactly known, although

Figure 1–8.   The four ancient horse types: A. the ancient Persian horse, B. the tarpan, C. the Mongolian wild horse, and D. the heavy cold-blooded forest horse.

Aurignacian, Solutrean, and Magdalenian paintings and carvings showing this type of wild horse have been found in various caverns in France and Spain. Indeed, it is said that the fierce, wild horses that the Roman legionnaires encountered in Spain were tarpans.

2. The cold-blooded European horse was an ancient inhabitant of European forests. Nearly pure strains of this type created the "Great Horse" of medieval times. These horses in turn were used to develop assorted draft breeds, and contributed blood to some of the light horse breeds as well.

Many authorities believe that a variety of large horse types may have been evolving during the Pleistocene period, a million years ago. These large horses had a limb-bone proportion different from the light horses. This difference included a relatively long upper arm and leg with a much shorter lower leg and forearm, which gave rise to a deeper chest and thicker body than that of a lighter horse of the same wither height. Most of the ancient heavy horse types had large rounded hooves, which suggests a soft forest or meadow habitat.

Fossil remains discovered so far do not show exceptional wither height. Measurements from several specimens of *E. caballus germanicus* give a wither height of about 15 hands and a live weight of about 1200 pounds. It is conceivable that the ice age encouraged development of more size in the heavy forest horse, resulting in greater wither height and greater weight.

Peter Willoughby, in *The Empire of Equus,* explains his views of the probable origin of the heavy horse:

> Of the heavy-built "coldblood" forms of caballus, probably the best-documented are *E.c. mosbachensis, E.c. abeli,* and *E.c. germanicus.* Since *abeli,* which came later than *mosbachensis,* presumably died out sometime during the Late Pleistocene, the most logical of these three "coldblood" types to have continued into post-glacial times and so possibly to have given rise to the modern draft horse is Nehring's *E.c. germanicus.* Finally, it should be borne in mind that fossil, or even sub-fossil, remains of members of the horse family are of comparatively rare occurrence, and that for every such species or subspecies unearthed and described there may have been ten, twenty, or even more forms that once existed but which were not preserved (or found).

3. The hot-blooded Oriental horse may be a type older than the history of horse breeding. Some experts have claimed that the Arabian horse originated from crosses of pony types and cold-blooded horses. The Arabian type of 3000 years ago, however, had many characteristics not present in any of the other ancient types of horse, raising the question, "How could a few centuries of horse breeding create the potent hotblood?" A more plausible explanation was originally proposed by Lady Ann Blunt, that the ancient hotblood was a type of its own in the wild, existing perhaps millenniums before man learned of domestication.

Scientists have always been extremely skeptical of any assumption regarding ancient horse types that cannot directly relate to fossil evidence. Yet they generally admit that there have been a great many horse types of which we have no scientific documentation.

Quite probably no horse alive is similar to the ancient hot-blooded oriental horse in all aspects. This hypothetical horse, of approximately 10,000 years ago, was small, probably between 13 and 14 hands, and may have been similar to the tarpan. Unlike pony types, he was very refined, with thin hair and a delicate bone structure. In all probability this ancient horse had a rather dished face, with a small muzzle. He may have had great speed, with anatomical and physiological traits which gave exceptional endurance. Lady Anne Blunt chose to name him *Equus agilus*.

The fact that little scientific evidence supports the theory of such a horse could be explained by their limited numbers for thousands of years. Small bands may have lived in isolated pockets of pasture and water throughout the arid Mideast.

Horse domestication began in the Near East. The earliest riding types eventually developed into three "old" lines: (1) a tall racy type which changed considerably to become known as the Turkmene; (2) a coarse, but speedy horse which changed over many centuries to be known eventually as the barb; and (3) a horse which over many centuries developed a high tail set and eventually became known as the Arabian.

4. *Equus przewalski* represents the ancient horse type, which links closely to the ass and zebra. From all indications the modern horse carries little of this type of blood.

Willoughby describes the Przewalski horse as more the cold-blood than the warm-blood type, and believes it may have descended from a larger-sized ancestor in which the percentage of "coldblood" was higher than it is today. He also believes that the Przewalski horse has made little, if any, contribution to the blood of our modern horses.

ANCIENT HORSE BREEDING

The earliest record of horse breeding was found a number of years ago in the highlands of southwestern Persia. An engraved stone (Figure 1–9) was

Figure 1–9.   A diagram of the 5000-year-old pedigree, seal 105F, found in Persia, carved on the surface of a stone among artifacts known to be used by man about 3000 B.C.

found among other material, dated about 3000 B.C. Amschler devoted years of study to the early history of equine domestication. He proposed that the engravings on this stone are a primitive pedigree. In support of his argument he pointed out that the horse heads are situated in four horizontal rows, and that the horses are of three categories with regard to their manes: upright manes, pendant manes, and no manes. In his interpretation the heads with upright manes represent wild stallions, those with pendant manes domesticated stallions, and those without manes, mares. The head forms were evidently carved carefully so as to represent specific types. The concave profile is a characteristic of the modern Arabian head, and the "Roman nose" is a familiar inherited characteristic of other breeds today.

Just what the entire pedigree records has not been determined, although it is quite evident that at this early date man was actively involved in horse breeding. If it could be deciphered, this stone might tell part of the early development of the hot-blooded horses.

Breeding and selection for the next 1500 years is probably what brought about development of the three basic light horse breeds: Arabian, barb, and Turk. Evidence suggests that the domestication of the heavy horse and interbreeding with the light horse did not occur until about 700 B.C. Prior to that time the light horse in Europe was principally the tarpan (Figure 1–10). A detailed analysis of the history of horse breeding is to be found in *A History of Horse Breeding*, by Daphne Machin Goodall.

Figure 1–10.   Modern "wild" horses of central Europe. (Courtesy of Daphne Goodall.)

Figure 1–11. European rock drawing of heavy horse type.

Early records indicate that the light horse was used mostly for pulling chariots rather than for riding. Stock and Howard cite evidence that it was almost 750 B.C. before horses were developed which were large enough to be ridden; other authorities set the date much earlier. Selective breeding for larger horses was probably quite effective in early Persia, as evidenced by horse racing and polo playing. Caesar's large war horses, on the other hand, were probably produced by crosses with the heavy horse or the coldbloods. These horses had probably been in Europe about as long as man. Cave drawings, like the one in Figure 1–11, can be cited as evidence.

DEVELOPMENT OF THE BREEDS

In some areas the heavy horse was developed into specific breeds without much crossing to the light horse, and vice versa. The heavy horse is exemplified by a 15th century engraving depicting the "Great Horse" (Figure 1–12). It shows a massive body capable of carrying a knight and his heavy armor. The lineage of many of the early breeds of both light and heavy horses is shown in Figure 1–13. The dispute continues as to whether the barb or the Arabian breed is the oldest. In all probability, both were in the process of development in 1500 B.C.

In 1680 B.C., a group of people known as Hyksos infiltrated Egypt with the use of horses of the light type. Before long these horses had spread along the northern coast of Africa, and both environmental and breeding selection began to develop a type later to be known as the barb. Today the native home of the barb is Algeria. It is a tough horse and able to exist on poor feed; it has a relatively long head and a low-set tail and is rather docile. Few pure strains exist today, as most have been interbred with Arabians. This

Figure 1–12.   The "Great Horse."

interbreeding occurred especially in the Libyan barb, the so-called "North African horse."

Throughout the history of horse domestication, various groups of people at various places and times have bred the Arabian horse, so that the Bedouins of the Arabian desert have called theirs the Elite Arab or the Original Arab. No doubt they brought about the finishing touches of the Arabian breed as we know it today.

About 750 B.C., Armenia and Media became breeding centers for what was evidently some of the ancient Persian stock; they developed what have since been called the Nisean horses. The so-called Turkmene race of horses was developed from the Nisean horses, according to Goodall. A number of Asiatic breeds have developed from this Turk blood.

Even before the Persians began horse breeding in the fourth millenium B.C., some of the Asian wild horses from which they had obtained their stock migrated into western and northern Europe. There, because of the severe environment, they evolved over hundreds of generations into somewhat smaller horses, now referred to as Celtic ponies. The Exmoor breed of today

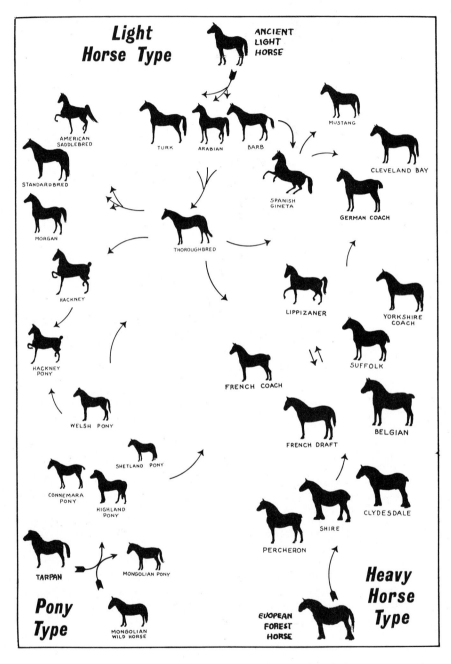

Figure 1–13.   Diagram showing the development of modern breeds from ancient stock.

most closely resembles the pony of the Celtic period. These animals served as foundation stock for most of the pony breeds in that area today. Some of the ponies were later interbred with heavy horses producing still other breeds.

The blood of all of these Old World breeds has been brought to America at one time or another, first from Spain with the explorers and the settlers. Evidently the American mustangs got their start from horses that were scattered by Indians when they raided the first settlement at Sante Fe, New Mexico. The early settlers on the East Coast brought in well-bred horses from England as well as those of barb, Arabian, and Turk breeding. The cold-blooded breeds were not brought to America in any significant numbers until the latter part of the 19th century.

The Ardennes breed of heavy horses, found today principally in France, is probably quite similar to the original wild heavy horse. These horses derive their name from the mountains of Ardennes in France and Belgium. The "Great Horse" was probably developed from these or similar coldbloods known as the Flemish type.

REFERENCES

Amschler, Wolfgang: The oldest pedigree chart. A genealogical table of the horse and pictures of horsemen dating back 5000 years. J. Hered. 26:233–238, 1952.

Dent, Anthony: The Horse: Through Centuries of Civilization. London, Phaidon Press.

Ewart, J. Cossar: The Smithsonian Report, 1904.

Goodall, Daphne: A History of Horse Breeding. London, Robert Hale, 1978.

Goodall, Daphne (translator): The Asiatic Wild Horse. London, J.A. Allen & Co., 1971.

Hope and Steinkraus (Eds.): The Encyclopedia of the Horse. New York, The Viking Press, 1978.

Riggs, Elmer S.: The Geological History and Evolution of the Horse. Geology Leaflet 13. Chicago, Field Museum of Natural History, 1932.

Salensky, W.: Prjevalsky's Horse. London, 1907.

Simpson, G.G.: Horses. Garden City, NY, Doubleday and Company, 1961.

Stetcher, R.M.: Anatomical variations of the spine in the horse. J. Mammalogy, 43(2):205–219, 1962.

Stock, C., and Howard, H.: The Ascent of Equus. Los Angeles County Museum Science Series. No. 8. Paleontology Publication, No. 5, July, 1963.

Vetulani, T.: Zwei Weitere Quellen zur Frage des Europaischen Waldtapons. Zeitschrift fur Sangetierkunde, Band 8, 1933.

Wentworth, Lady: Thoroughbred Racing Stock. London, George Allen & Unwin, 1960.

Willoughby, David P.: The Empire of Equus. Cranbury, NJ, A.S. Barnes, 1978.

2

ancestry of the breeds

A few horses found throughout the world can be classified as foundation breeds because of their ancient origins and their important contributions to the development of more recent breeds. It is difficult to sift out the truth about a particular breed because of the biases of the breeders. Usually they claim a much longer and purer lineage than can be substantiated. A registry can be helpful in this respect; a breed that has an old registry is generally less apt to contain recent crosses to other breeds.

There are many more horses of mixed blood than there are "purebreds;" however, "purebred" is a relative term. While most horses are predominantly one type, there is no such thing as an absolute purebred. Aside from recognized breeds of horses, many specific types of horses have developed because of various environmental factors and the needs of man in a particular locale.

ARABIAN

Arabian horses of quite diversified types can be found throughout the world. The people of many countries, outside of Arabia, have bred their own distinct type of Arabian horse for years. Such Arabians have been bred extensively in England, Egypt, Poland, the United States, Spain, Germany, and elsewhere.

Arabian breeding today, especially in the United States, is often conducted according to strain fads. Today's strains are quite different from the strains developed by the Bedouins of the desert. They are more like nationalities, developed within the past 100 years.

Figure 2–1.   Ansata Ibn Sudan, a straight Egyptian Arabian—1958 National U.S. Champion Stallion.

Figure 2–2.   Druzba, a Polish Arabian stallion.

The Egyptian type comprises horses tracing in all lines to horses bred in Egypt. There are several thousand Arabians of this type now in the United States, considerably outnumbering the Egyptian Arabian horse population (Figure 2–1).

The Polish type is made up of horses bred in Poland. In that country a definite type has developed through selective breeding over the past century. Polish imports to the United States have been well received, and have commanded high prices (Figure 2–2).

The American-bred Arabian horse can be found in distinct types also. Some of the larger breeding farms in the United States have produced as many horses as have entire countries elsewhere. Al-Marah, a large Arabian breeding farm, has developed the Al-Marah type. Other American types have developed from concentrated breeding to certain horses such as Raffles, Ferzon, and Fadjur (Figure 2–3).

The Persians claim that their breed is not only the original Arabian but absolutely the oldest domesticated breed. The discovery of seal 105 F has lent support to their claim. Since the time of Mohammed, however, the

Figure 2–3. 1977 U.S. National Champion Arabian stallion, Gai Parada, an American-bred Arabian.

desert Arabs have most zealously bred the Arabian horse, now often called Elite Arabs or Original Arabs. These horses became a part of their social and religious world. The desert Arabs placed more importance on the mare than on the stallion. Because of this priority, families have developed which can be traced back to certain mares. A number of families appear to be especially pure and are known as *asil*; all other Arabian horses are called *kudsh* by the desert Arab.

The Arabian horse is characterized by its distinctive head. The bulging forehead and concave face give a dished appearance. The eyes are large and set low in the head. A slight bulge at the back of the head and a light muzzle give the head a wedge-shaped appearance. The neck is arched, large, and rounded. The back is short, straight, and rather level at the hip, reflecting the long, somewhat horizontally set pelvic bones. As shown by Stetcher, the short back usually reflects a missing thoracic vertebra rather than a missing lumbar vertebra.

The Arabian horse has contributed to the development of every warm-blooded breed of horse and has even been crossed with the ancient barb and Turk horses to give our modern barb and Turkmene breeds. In this respect the Arabian could be considered to be the original source of the hotblood or, as the Arabian enthusiast would put it, the only purebred.

It might be pointed out, however, that from about 3000 B.C. until the time of Mohammed, in about A.D. 650, there was throughout Asia only one general type of good domesticated horse. Each locale developed its own name for this type. Over time this general type began to be referred to as the Arabian. Therefore, someone looking back on history tends to pick out the good horses and by definition call them Arabian even though they would not be accepted as Arabians by modern standards.

The modern Arabian breed is the product of the horse breeders of the Arabian desert during the last 1000 years (Figure 2–4). The blood of these horses was exported from time to time, principally to countries that exerted some political influence over the desert breeders. The Arabian breed in Persia has, therefore, been more or less continually infused with blood from the desert Arabian horse.

In A.D. 1700, the civilized world was becoming aware of the equine treasure to be found in the Arabian desert wasteland. During that year the Darley Arabian was foaled. The colt was discovered by Thomas Darley and shipped to England where there was great demand for Arabian horses. The British noblemen were interested in fast horses and were developing what has become the thoroughbred breed.

Before long Arabian breeding stud farms were developed in various countries. A German stud was established in 1760 and produced the famous sire, Turkmainatti. Horses were obtained from both Syria and Arabia. A stud was established in Babolna, Hungary in 1789 with horses originally imported from Arabia; later imports were mostly from Syria. This stud produced the Arabian sire, Shagya.

Figure 2–4. Modern Arabian breeders are proud of the ancient heritage of their horses. Arabian horse shows feature Arab costume classes.

In 1810, the King of Württemberg established an Arabian stud with horses from the German stud and from the desert. Count Wazlaw Rzewuski of Poland established a stud in 1828 with Arabian horses that he captured from Bagdad to Muscat (89 stallions and 45 mares).

A large number of good desert horses were collected at Cairo, Egypt, in the 1850s, by Abbas Pasha. These horses were dispersed by auction to the rest of the world in 1860. This auction was one of the early major sources of

Arabian blood available to European breeders. Shortly thereafter, the breeding of Arabian horses flourished in England.

Arabian breeding in the United States began to develop about the turn-of-the-century with sires such as Khaled, owned by Randolph Huntington. Khaled was by the famous running Arabian, Kismet from India, and out of the imported mare, Naomi. In 1906, the Hingham stud was established with horses from Arabia, and in 1912, the Maynesboro stud began principally with English imports.

BARB

Coming out of Persia at the time of the early Arabian and Turkmene horses was another horse called the Numidian. These horses were known to have been in Egypt in 1600 B.C. The breed was later modified by the desert environment and the needs of the people as it moved across North Africa during the following centuries. This horse became known as the barb from the Barbary states, now Ethiopia, Somaliland, and other parts of North Africa. According to Stock and Howard these horses reached Spain by 1000 B.C.

The late Lady Wentworth, of the Crabbet Arabian Stud in England and author of *Thoroughbred Racing Stock*, has described the old barb of 1000 years ago as a coarse-headed horse from North Africa. She maintained that 200 years ago the term barb was used interchangeably with Oriental and Arabian as most so-called barbs of the seventeenth and eighteenth centuries

Figure 2–5. A Lady Blunt painting of a barb.

came from Egypt, Arabia, and Syria, and according to modern definition would have been Arabian horses. Lady Wentworth's mother, Lady Ann Blunt, traveled the Arabian and North African deserts for many years with her husband buying horses. She described distinct differences between barb and Arabian horses. A painting of a barb by Lady Ann Blunt, about 1895, is shown in Figure 2–5.

The native country of the barb is northern Africa. The Libyan barb contains a strong mixture of Arabian blood, and various local strains contain other mixtures as well. A relatively pure strain of the barb is found in Algeria. Several types among these strains indicate infusion of some outside blood even in the Algerian barb. This outside blood is mostly Arabian also.

The barb has contributed to the development of other important breeds. Offspring of the famous Godolphin barb (1689) were used as the original stock for development of the English thoroughbred. The barb stallion, Guzman, founded the Spanish breed, Guzmanes. The horses brought to America by the Spanish explorers were strong in barb blood. These horses had an influence on the development of the wild horses of North America and some of the modern South American breeds.

In his book, *El Caballo de Paso y su Equitacion*, Luis de Ascasubi writes,

> In North Africa there was an undryable fountain of good horses. The Arab horses, because of the condition of the Arabian desert

Figure 2–6. This breed is being restored today through the careful breeding and selective linebreeding of the five Spanish-barb bloodlines that have been kept intact for 50 to 130 years by a few dedicated individuals. O-kum Ho-gahis is shown.

compared with the greater fruitfulness of North Africa, have always been a great deal more scarce than the barb. The barb, "Equus Caballus Africanus," is as excellent a horse as is the Arab horse, "Equus Caballus Asiaticus." But the two horses pertain to two completely different families of the horse kingdom.

A small group of breeders, headquartered at Colorado Springs, Colorado, is registering American horses that they consider to be of heavy Spanish barb blood (Figure 2–6). They contend that the early Spanish explorers in America were riding barb horses and that many of these horses were lost or left on the plains of the western United States. These horses became the nucleus for the mustang type. Some quarter horse historians attribute many characteristics of the "working" quarter horse to the barb blood, which comes to the quarter horse through the mustang.

TURKMENE

The Turkmene horses were well known in 1000 B.C. as race horses. They probably originated from the same Syrian stock from which the early Arabian breed was developed. King Darius of Persia reportedly had 30,000 horsemen mounted on Turkmene horses. About 500 B.C. the game of polo

Figure 2–7.   A Turk horse from Syria.

Figure 2–8.   The Byerly Turk, from an old painting.

originated in Persia and was played on these horses. Figure 2–7 shows a modern Turk horse from Syria. The breed spread into Europe and Asia, where it contributed to the development of others, such as the thoroughbred—principally by Byerly Turk (Figure 2–8) in 1724—and several Russian breeds, the Karabiar, the Jornad, and the Achal-Teke.

The modern Turkmene breed is found in Russia. These horses are of exceptional stamina and endurance, as well as extremely fast. They have a long neck with a medium-size head carried high, fairly prominent withers, long back, pronounced croup, and a sloping hip. It is interesting to note that the Turkmene, barb, and Arabian breeders of today claim the early Persian horses as the ancestry of their breeds.

TARPAN

Work on an American tarpan stud book began in 1970 with a census conducted among zoos exhibiting tarpans (Figures 2–9, 2–10). Ellen J. Thrall compiled information for the publishing of the book in 1975. A portion of her description of the history of the breed is quoted here.

> The story of the Tarpan horse might have ended in 1876 had it not been for two German zoologists with an interesting theory. Heinz Heck and his brother Lutz came from a family dedicated to zoology. During the early 1930's they began experimenting in the hope of genetically recreating the vanished Tarpan horse. They felt that every living creature was the product of its genetic make-up, and that genes, like pieces of a puzzle, might be reassembled resulting

Figure 2–9.   Duke, the first horse registered in the *American Tarpan Stud Book*.

in the recreation of a vanished species, if a source of similar genes could be found.

These hardy little animals had managed to survive the advances of human civilization upon their natural environment until the late 19th Century. They were found in the thickly forested regions of Germany, Russia, and Poland, but the human inhabitants who shared the region with them did not know the ancient history of the breed and believed that the Tarpans were feral ponies that had escaped from captivity and were living wild in much the same manner as the American Mustang.

By the mid-1800's the Tarpan herds had almost vanished because farmers would regularly shoot them if they strayed on to their

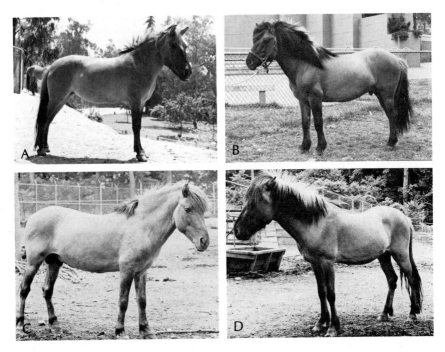

Figure 2–10.   Horses registered in the *American Tarpan Stud Book*.

property to steal domestic mares or eat crops. . . . The Polish government initiated a program to collect all animals of the Tarpan type and place them on preserves managed by the government. This program, although a worthy one, was too late to save the breed and they were not sure that they had found even one pure Tarpan. The last known true Tarpan is believed to have died in a Russian zoo in 1876.

Before they (the Heck brothers) could begin the actual work on their theory they had to study the history of the breed, and determine the exact conformation found on the extinct Tarpan. They had access to Tarpan skeletons on exhibit in German museums, so they set about cataloging the physical traits of the Tarpan, then they began a search for living horses that might possess one or more of the characteristics that had been found in the Tarpan. . . .. The first Tarpan horses were brought to the United States in 1954 when Hienz Heck, the son of Lutz Heck of Germany, brought the foundation animals for America's future herds from Tierpark Hellabrunn (The Munich Zoo) where they had been created, to the Chicago Zoological Park at Brookfield, Illinois.

BRITISH PONIES

Anthony Dent and Daphne Goodall in their book, *The Foals of Epona*, describe development of the early ponies of the British Isles. They tell of the islands being connected to the European mainland until 15,000 B.C. Then for at least 10,000 years these horses were isolated and took a course of development of their own. Dent and Goodall, considered to be the primary experts on British ponies, feel that Arabian blood was not introduced into British ponies until the Roman Conquest. These ancient ponies are called by some "Celtic" stock. The term is perhaps a misnomer but designates a primitive horse variety of the old tarpan.

The Welsh Mountain pony is a small, refined, well-bred riding pony (Figure 2–11). It has a recorded history dating back to the time of Julius Caesar. A current registry, which is maintained in Great Britain, contains four sections: A, B, C, and D. Section A contains the purest line, while the other sections include various types of crossbreeding. The Welsh pony has served as foundation stock for several other breeds such as the Highland pony.

The Welsh Pony Society of America is extremely active. The Society recognizes some variation in size of Welsh ponies and the registry is divided into two sections. Section A includes those ponies that are less than 12 hands high. Section B includes those over 12 hands. The ideal type of Welsh

Figure 2–11. A Welsh pony.

pony has a small, clean-cut head which is lightly dished; somewhat pointed ears set well upon the head; a lengthy neck set on long, sloping shoulders; and a well-muscled, deep body.

Lady Wentworth in her book, *Horses of Britain,* describes the mixed improved pony breeds of Great Britain. Many of these breeds were mixtures of Welsh and Exmoor. Other breeds described are the New Forest pony, the Dartmoor, the Dales, the Fell, the Connemara, and the Shetland.

The Connemara ponies, reaching 13 to 14 hands, originated from Celtic stock. During the Middle Ages both barb and Arabian blood was added to the breed to give refinement. This combination resulted in a pony with a deep body, medium legs, and a neat head and neck on good shoulders. A hardy breed, these ponies are also very docile. Their original color was dun with black mane and tail and a dorsal stripe. Gray, bay, brown, and black horses are found in the breed today because of crossbreeding.

The Exmoor pony is an ancient breed of pony from Great Britain (Figure 2–12). It is thought by some to resemble the Celtic pony. Some zoos of Europe breed these horses because of their scarcity and antiquity. These pones have remained relatively pure (compared to other pony breeds) over the centuries and probably more closely resemble the original wild pony of the Isles than does any other breed alive today. The Exmoor stands just over 12 hands high. A wiry coat is peculiar to the breed in winter, and bay, brown, and dun are the most common colors.

Figure 2–12.   An Exmoor pony.

Figure 2–13. The Shetland pony stallion, Patton.

Figure 2–14. A herd of miniature ponies, C.M. Bond owner.

The ancestors of the Shetland pony probably contained heavy horse coldblood (Figure 2–13). Because of back breeding to native British ponies, the average height of the Shetland breed today is no more than 9 to 10 hands high. Over the centuries breeding has been selective for other traits also, such as hardiness, prolific reproductive capacity, and a thick, heavy coat. A registry for the Shetland has been established in the United States. Breeding and selection by man under different standards have brought considerable changes in the breed. They have become more refined and have lost much of their hardiness.

The American Shetland Pony Club opened a registry for miniature ponies in 1972, called the American Miniature Horse Registry. The stud book records only those animals that measure 34 inches or less at maturity (Figure 2–14).

ARDENNES

The Ardennes breed is of ancient cold-blood origin. It is probably the modern horse that is closest to the original primitive heavy horse of the

Figure 2–15.   The Ardennes, probably the modern horse closest to the original primitive heavy horse of the European forests. (Courtesy of Daphne Goodall.)

European forests. It has contributed considerably to most of the heavy breeds of France and Belgium.

According to Goodall, the French Ardennes is the most direct descendant of the primitive horse. He has a deep, muscular body, coarse hair, and feathered legs, but has easy quick action and a gentle temperament (Figure 2–15).

KONIK

The Konik breed is quite popular in Poland and is said to be the nearest domesticated relative of the tarpan, the wild horse of Europe. It has participated in the development of most other Polish and Russian breeds. Seldom over 13 hands, the Konik is a stocky thick-necked horse. Extensive crossbreeding has brought much variability to the breed. The modern Konik, therefore, is perhaps more of a type than a breed (Figure 2–16).

HOLSTEIN

Breeding of the Holstein horse dates back to 1225 A.D. At that time primarily coldblood was found in the breed. Some 300 years later King Felipe II of Spain established his famous Royal Stud at Cordoba, principally using Holstein horses. Subsequently much Spanish blood, Spanish Gineta, infused the breed and the Holstein became known as a warm-blooded horse. In the recent past, much English thoroughbred blood has also been introduced, bringing more elegance to the breed (Figure 2–17).

Today there are just over 3000 purebred Holsteiners in Germany, the breed's homeland (Figure 2–18). A registry, established in the United States in 1978, records purebreds from other countries as well as half thoroughbred and Holstein crosses. This registry requires inspections for physical and

Figure 2–16.   A Konik horse from Poland.

Figure 2–17. The Holstein stallion Casanova, owned by Mr. and Mrs. Frank Magadance, Shakopee, MN.

Figure 2–18. A team of Holstein horses from West Germany. (Courtesy of Daphne Goodall.)

genetic traits that are necessary to improve the breed. The Registry requires purebred Holsteins to be a minimum of 16 hands high.

ANDALUSIAN

The Andalusian is an old Spanish breed of horse (Figure 2–19) dating back to the Moorish invasions in the 8th century A.D., when barbs and Arabians were crossed with the local Spanish ponies and later with larger horses. The Andalusian has been important in the development of many other breeds, partly because of the strong influence Spain had in the past on other cultures. Many of the horses brought to the New World were Andalusians, which played a role in establishing the criollo and the Campolino breeds, as well as the North American mustang. Evidently Andalusians were often crossed with the Lipizzaner, the Holstein, the hackney, and the Kladruber.

The Andalusian today is usually about 15 hands high with a Roman nose and a well-proportioned body. He is an extremely showy and active horse, often used as a jumper (Figure 2–20).

PERCHERON

A few hundred years A.D., in central Europe the development of heavy horses was moving in three directions: toward the Lipizzaner, from Andalusian crosses; toward the "Great Horse;" and toward the Flemish type. The latter served as a foundation for the Percheron breed, which was refined with some infusion of barb, Turk, and Arabian horse blood. Percherons were developed in France and have been exported to other countries, where they

Figure 2–19. A Spanish Andalusian.

Figure 2–20.   A modern Andalusian stallion, Legionario III, ridden by Albert Ostermaier.

Figure 2–21.   The Percheron stallion Calypso.

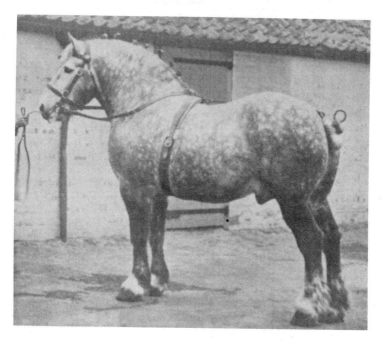

Figure 2–22.   A Percheron from England.

have occasionally been used in the development of other breeds, the heavy draft horse of Russia as an example.

At the turn-of-the-century there were more than 10,000 registered Percherons in the United States. One of the most famous sires of this time was Calypso (Figure 2–21). Imported from France in 1900, Calypso became reserve champion at the International Livestock Exposition. He was a linebred Brilliant 111, standing about 16¾ hands. Not of the largest type, he weighed about 1900 pounds in breeding condition. Percherons are the only breed of horses that are always born black. A large proportion of them turn gray, however (Figure 2–22). The head of the Percheron is clean-cut, although often fairly long. The legs are free from the feathering characteristics of the shire. Height varies from about 15½ hands to 17 hands.

LIPIZZANER

The Lipizzaner is one of the oldest breeds of heavy horses. His origin may have been somewhere in the Dark Ages, even before the development of the "Great Horse." He now shows more refinement than most of the other coldbloods because of crossing with Andalusian and Kladruber horses and centuries of selective breeding (Figures 2–23, 2–24).

Figure 2–23.   A Lipizzaner.

Figure 2–24.   A Lipizzaner stallion exhibited in the United States. (Courtesy of Otto Herrmann.)

Figure 2–25.   A Lipizzaner trained after the fashion of the famous Spanish Riding School.

The breed name was derived from the small Yugoslavian town of Lipizza, near the Italian border where in 1580, Maximilian II established a royal stable and stud farm. The original stock consisted of nine Spanish stallions and 24 broodmares. Later, some other blood including Arabian horse was introduced. The breed spread throughout Austria and Czechoslovakia. Six principal lines make up the breed today. These lines go back to the sires, Pluto (1766), Concersano (1767), Maestoso (1773), Favor (1779), Neopolitano (1790), and Siglavy (1810).

In the 16th century, the now famous Spanish Riding School of Vienna, Austria, was established. The Spanish horses and style of riding were supreme at that time, and horsemanship became a fine art for noblemen (Figure 2–25). The school today uses only Lipizzaners which are obtained from a breeding farm at Piber, Austria.

Lipizzaner blood contributed to other breeds of Europe such as the old Frederiksborg horse admired by the cavaliers of the Middle Ages, and the Holstein and Knabstrup breeds.

Lipizzaner horses are predominantly gray, but sometimes are bay or chestnut. They have a large head with expressive eyes; refinement is shown in the ears, mane, tail, and the small feet. The horses generally stand just over 15 hands and are heavily muscled.

KLADRUBER

The Kladruber breed almost died out at the end of the Second World War. Those horses of the breed that remained were crossed extensively with Oldenburg and Hanoverian horses. The modern Kladruber breed can now be found in Czechoslovakia (Figure 2–26).

This breed consists of horses with a large semi-draft horse build. They are nearly always gray and range from 17 to 19 hands. They are well proportioned throughout.

Originally the breed was developed from heavy native horses of the Alps, infused with barb and Turk horse blood. Earliest records of the breed indicated that it came into being in about the 15th century A.D. Even in the early days, the Kladruber were crossed with other breeds such as the Lipizzaner. The breed name was derived from the Stud Kladrub in Bohemia.

SHIRE

The beginning of the shire breed is obscured in the Dark Ages. Most authorities agree that the "Great Horse" had a strong influence in the founding of the breed. In early English days extremely heavy horses were desirable for pulling great loads (Figure 2–27). Selection over the centuries has been for larger horses.

It was once fashionable to have much feathering on the legs, and horses were selected for extremes of this trait. This characteristic is not considered

Figure 2–26.   A Kladruber horse from Czechoslovakia.

Figure 2–27.   The shire horse, Elton Lad, of the Big Foot Ranch.

to be advantageous now and has decreased somewhat. The shire has played an important part in the development of other heavy breeds, particularly the Clydesdale, Cleveland bay, and Suffolk.

CRIOLLO

The criollo is a popular breed of horse in South America. The breed developed from the Spanish horses left by Pizarro in 1532. The changes which have occurred in the breed since that time have been due mainly to selective environmental factors such as terrain, climate, and nutrition (Figure 2–28).

In 1930, Solanet made an extensive study of criollo horses in South America and found the breed present in Uruguay, Argentina, Brazil, Paraguay, and Chile. The breed is also quite popular in Peru.

There is some variation in size of the criollos; they range from 13½ to 15 hands. The head is short and pyramid shaped. The body is fairly heavily muscled. There is also a much more refined and smaller type of horse known as Costino. Evidently it is a purer line from the Spanish horses of Pizarro.

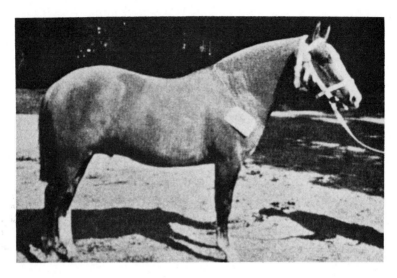

Figure 2–28.   The criollo horse. (Courtesy of David Willoughby.)

CLEVELAND BAY

When coaches first came into use in the early days of England, the Cleveland bay was developed as a strong coach horse. Originally a heavy type, strictly cold-blooded horse, the Cleveland bay was developed from the Chapman pack horse (Figure 2–29). Today, the Cleveland bay is crossed with the thoroughbred to produce good hunting horses; however, the purebred contains very little thoroughbred blood.

Almost always bay in color, the modern Cleveland bay has a large head and a fairly long neck, with a strong, well-made body which shows much quality (Figure 2–30). A Cleveland bay registry has been recently established in the United States. The breed has contributed to the development of a number of other breeds such as the Yorkshire coach horse, the Clydesdale, and the Russian Uladimir, a heavy draft horse.

THOROUGHBRED

During the 16th century, the English were avid horse breeders and the sport of horse racing developed. At this time the English ponies were being crossed with the Arabian and barb horses. This crossing increased the size and speed of the English horses and brought about a new type. Perhaps some Andalusian blood was introduced to give even more height to the horse.

Many types of crosses were tried during the first few decades of the development of the breed. The imported stallion Godolphin barb (1689) was crossed successfully with these horses. Seventeen years later, the Darley

Figure 2–29.   A Cleveland bay of the old type. (Courtesy of Mrs. William Dorman.)

Figure 2–30.   A modern Cleveland bay.

Arabian was bred to mares of the developing breed (Figure 2–31). Finally in 1724, the imported Byerly Turk was extensively bred to the better mares. Within a few decades, English breeders began to stabilize the developing breed by line breeding. *The Racing Calendar,* first published in 1752, helped to center interest around the developing breed. By 1786, a list of distinguished race sires was published, and in 1791, James Weatherby, Jr., published *An Introduction to a General Stud Book.* This stud book listed foals chronologically under the names of their dams, indicating that private records had been kept on "pedigreed" horses. Finally in 1793, Volume I of the *General Stud Book* was published, and has been continued to today.

Some of the early thoroughbred horses were exported to America. As early as 1745, in Maryland, racing was held "between pedigreed horses, in the English style." The sport spread throughout Virginia and the Carolinas during the Colonial period.

Three of the most important early thoroughbred sires were Matchum, 1748; Herod, 1758; and Eclipse, 1764 (Figure 2–32). According to a genetic analysis conducted by Steele in 1944, these three foundation sires contributed the following proportion of genes to the modern American thoroughbred; Matchem, 5 or 6 percent; Herod, 17 or 18 percent; and Eclipse, 11 or 12 percent. Later sires who contributed significantly to the gene pool of the American thoroughbred were: Hermit, 1864; Hampton, 1872; Galopin, 1872; Isonomy, 1875; Bend or, 1877; St. Simon, 1881; and Domino, 1891.

The *American Stud Book* was published in 1873 after an abortive publication of Volume I in 1868. The Jockey Club was organized by influential sportsmen in 1894, and this group purchased the *American Stud*

Figure 2–31.   The Darley Arabian.

MATCHUM

HEROD

ECLIPSE

Figure 2–32.   Early drawings of Matchum, Herod, and Eclipse.

Book in 1896. Because American thoroughbreds could be registered in the General Stud Book of England, the American Jockey Club was slow in developing. In 1913, however, the British Jockey Club passed a recommendation known as the Jersey Act, which prohibited the registration of thoroughbreds who could not be traced in all lines to ancestors already recognized by the British Jockey Club.

Two hundred years of intensive breeding of thoroughbreds has concentrated the hotblood and developed the thoroughbred horse to a very prepotent breeder. Thoroughbred crosses have been used to refine many other breeds. The Cleveland bay, as we have seen, although developed from cold-blooded horses, is now classified by some authorities as a warm-blooded breed due to extensive crossing with the thoroughbred. The quarter horse and other North American breeds have been developed and improved by much thoroughbred blood.

Since the turn-of-the-century, thoroughbred sires have gained popularity in the United States by the number of stakes winners they have sired. The Top Ten producers in descending order are Nasrullah, Bold Ruler, Court Martial, Broomstick, Mahmoud, Teddy, Sir Gallahad III, Man O'War, Princequillo, and Khaled. In the last few decades breeding popularity has centered around the Triple Crown Winners, the 3-year-olds who have won the Kentucky Derby, the Preakness Stakes, and the Belmont Stakes (Figure 2–33, Table 2–1).

Figure 2–33.   The Triple Crown winner, Secretariat.

Table 2–1.   Triple-Crown Winners.

1919	Sir Barton
1930	Gallant Fox
1935	Omaha
1937	War Admiral
1941	Whirlaway
1943	Count Fleet
1946	Assault
1948	Citation
1973	Secretariat
1977	Seattle Slew
1978	Affirmed

HACKNEY

Records in the *Hackney Stud Book* date back to 1755. The breed was developed from native ponies of the British Isles, the Norfolk trotter, and the early thoroughbred. Intensive selection in breeding has made the hackney a very showy horse with high action (Figure 2–34). In the days before the automobile, a fast harness horse was very important; and it is said that the hackney could trot a distance of 17 miles in one hour. The modern hackney breed is found in many countries outside of its native England. It is usually

Figure 2–34.   An English hackney at the turn of the century.

maintained as a show horse where exaggerated knee and hock action is much admired.

Hackneys range in size from 14⅓ hands to 15⅓ hands. They have a small but relatively long head, a crested neck, strong, rather straight shoulders, and exceptionally strong hocks. Most of the hackney horse characteristics have been passed to the hackney pony, which was developed from the hackney horse.

SUFFOLK

The Suffolk breed has evolved from the time of the "Great Horse," but has been established as a fixed type only in the last 250 years (Figure 2–35), owing to intensive selection in the eastern counties of England, particularly Suffolk County. The old-type Suffolk horse was at times crossed with the hardy native mares. This cross resulted in a horse who could do a large amount of work on comparatively little feed, but who was somewhat smaller than other draft animals.

TRAKEHNER

The Trakehner breed originated in East Prussia with written pedigree records spanning the past 230 years (Figure 2–36). In 1732, Friedrick

Figure 2–35.   Suffolk stallion, Ironside Jack, bred by Marble Ranches, Deeth, NV. (Courtesy of the American Suffolk Horse Association.)

Figure 2–36.   The Trakehner stallion Carajan II.

Wilhelm selected outstanding horses from seven royal breeding locations and founded the breeding farm of Trakehner. Highly selective breeding has been conducted with these horses since that time. In 1817, a few English thoroughbreds and Arabian horse sires were brought in. A few more were brought in during 1826 and again in 1837. Selection and culling have been extremely strict over the years, resulting in a very refined horse.

Before World War II, there were over 20,000 head of Trakehner breeding stock, but only a few hundred survived the war. Today the West German Trakehner is bred by a government-controlled breed association. Recently a Trakehner Breed Association was formed in the United States.

The Trakehner horse is used mainly for dressage, jumping, hunting, steeplechasing, and half-bred thoroughbred racing. The breed is characterized by a strong refined type head, sloping shoulders, a long graceful neck, a moderately short back, and strong, well-developed bones and muscles. The height is generally just over 16 hands.

The Trakehner registry in West Germany is known as Verbond. Membership in 1973 stood at 2626, with a total of 3036 mares and 273 stallions registered. The American Trakehner Association was formed in 1974. The organization of several hundred members published an English version of the classic German book, *Trakehner Horses, Then and Now.*

BIBLIOGRAPHY

Albright, Verne: The Peruvian Paso and His Classic Equitation. Ft. Collins, CO, Caballus Publishers, 1976.

Chivers, Keith: The Shire Horse. London, J.A. Allen & Co., 1976.

Dent, A.A., and Goodall, D.M.: The Foals of Epona. London, London Galley Press, 1962.

Edwards, Gladys Brown: The Arabian: War Horse to Show Horse. Temecula, CA, Rich Publishing Co., 1973.

Emsinger, M.E.: The Complete Encyclopedia of the Horse. Cranbury, NJ, A.S. Barnes & Co., 1977.

Goodall, D.M.: Horses of the World. London, Country Life Ltd., 1965.

Heck, H.: The Breeding of the Tarpan. Oryx *1*:338, 1952.

Mohr, Erna: The Asiatic Wild Horse. Translated by Daphne Machin Goodall. London, J.A. Allen & Co., 1971.

Pines, Philip A: The Complete Book of Harness Racing. New York, Grosset & Dunlap, 1970.

Robertson, William H.: The History of Thoroughbred Racing in America. New York, Bonanza Press, 1964.

Sanders, A.H., and Dinsmore, Wayne: A History of the Percheron Horse. Chicago, The Breeders Gazette, 1917.

Schilke, Dr. Fritz: Trakehner Horses, Then and Now. Translated by Helen K. Gibble. Norman, OK, American Trakehner Association, 1977.

Solanet, E.: The criollo horse of South America. J. Hered. *30*:548, 1930.

Thrall, Helen J.: The Tarpan Studbook. Ft. Collins, CO, Caballus Publishers, 1965.

Tinker, Edward Larocque: The Horsemen of the Americas. Austin, University of Texas Press, 1967.

Wentworth, Lady Ann: Horses of Britain. London, George Allen and Unwin, 1947.

Wentworth, Lady Ann: Thoroughbred Racing Stock. London, George Allen and Unwin, 1938.

Willett, Peter: The Thoroughbred. New York, G.P. Putnam's Sons, 1970.

Zaher, A.: The Genetic History of the Arabian Horse in America. PhD Thesis, Michigan State University, 1948.

3

modern breeds and registries

Horse breeds that have been developed principally within the past 200 years have been formed from the foundation breeds. They generally exhibit few new or unique traits, but rather different combinations of traits from other breeds, because it usually takes longer than 200 years to bring out a unique trait and fix it in a breed by inbreeding.

The more genes involved in the uniqueness of the new breed, the longer it will take to establish a true breeding group of animals. Some of the modern breeds have originated from stock with one or two unique traits, but with too many other questionable traits. Other modern registries seem to be suffering from lack of a permanent, clear-cut set of goals. These pitfalls are discussed in more detail in Chapter 4.

Throughout the history of the domestication of the horse the predominant factors in selection have been centered around the ability of the horse to do a particular job for man. Recently, with machines taking over the duties of the horse, most of the older breeding and selection programs have been revamped in their goals and ideals.

Out of this modern upheaval of selection goals has come increased interest in the color of a horse. Horses are kept now for enjoyment, at least in the United States, and a large part of the enjoyment is in the appearance of the horse. Of course, people have various ideas about which color looks the best. The oft-quoted statement from the old-timer, "I never saw a good horse with a bad color," does not mean as much today as formerly. Some people prefer blacks, or the diluted grullo (smokey) color, while others are partial to palominos. The solid white is sought by some, although many seem to prefer dark spots on the white horse. Some color patterns can be very striking, such as the white hip blanket on a dark horse, decorated with a few large spots. The genetics of coloring is discussed in Chapter 9.

A number of color registries have developed from this interest in color. Horses of a specific color are often referred to as breeds; however, until their characteristics are more firmly fixed and the horse breeds true, in both color and conformation, it is referred to as a color registry and not a breed registry. Many color registries have conflicting ideals regarding color. Many times a desirable color pattern is the product of a heterozygous condition, thus preventing it from ever being established as a true-breeding trait.

There are approximately 50 horse breed registries in the United States. Most of them are included in the chart at the end of the chapter. A few of these registries date back into the last century. One of the oldest registries is the American Shetland Pony Club, along with the American Hackney Association, the American Saddle Horse Breeders Association, the American Shire Horse Association, and the American Jockey Club. Paralleling the growth in horse numbers has been an increase in the number of breed registries. Quite a few have been formed within the last decade.

The American Quarter Horse Association has registered over 1.4 million horses. Covering a longer period of time, 613,000 thoroughbreds have been registered by the American Jockey Club. Standardbreds are in third place in total number of horses registered, with 392,000 through 1978. Next in number are Appaloosa horses with 289,000 registrations through 1978.

The 1978 registration numbers give a clear picture of which breeds are currently growing at the most rapid rate. Again, the quarter horses lead. Surprisingly, the Arabians, including half-Arabians, were second in total numbers registered, with nearly 30,000 registered during the year. Thoroughbreds were in a close third place, with 28,000 registered, and Appaloosas followed with 17,767 registrations. (See table at end of chapter.)

THE ALBINO

Many mammalian color geneticists believe that no true albinos can be found among horses, in spite of the existence of albino deer, rhinoceros, cattle, rabbits, guinea pigs, beavers, and mice. Traditionally, true albinism has been described as a complete lack of pigment including the iris of the eyes, which are pink. Animals that are white in color with blue eyes are often known as mock albinos.

Albinism has been studied quite extensively in man, and is technically described as a lack of the enzyme necessary to change the biochemical called DOPA into melanin. Because of this rather specific definition, geneticists are generally cautious about describing a white animal as an albino unless it is known that the formation of pigment is stopped at this particular stage of the process. True albinism has always been shown to be a recessive trait, and it is necessary for both sire and dam to pass the gene to a foal before the white color is produced.

A number of other genes produce a white horse. Some of these are dominant, and others are semidominant. With a dominant gene it is necessary for only one parent to pass the gene to a foal in order to produce the white color. Semidominant genes when passed to a foal by only one parent bring about an intermediate effect in the color. For example, the palomino color is due to a semidominant gene; and when the foal receives that gene from both parents, the foal is a faded yellow or a near-white, often known as creamello.

Several years ago, an article by Castle and King, in *The Western Horseman*, described various grades of albinos caused by the gene which also produces the palomino color. The authors were correct in pointing out that this gene can produce an almost white horse, when homozygous (see Chapter 6), but studies since that time have shown that it is not related to the true albino gene.

Figure 3–1.   A registered white stallion, Virginia Hyde up.

The amount of white on pinto horses is influenced by a number of modifying genes, as explained in Chapter 9. Occasionally these modifications are such that an almost pure white horse is produced; however, a small telltale spot of pigment can generally be found somewhere on the body of the horse. These white horses are also of the dominant variety of pinto or paint called tobiano. The recessive or overo pattern produces a lethal abnormality of the digestive tract if the horse is born completely white.

The American Albino Horse Club was organized in 1937, to register a type of dominant white horse which sprang from a stallion known as Old King. Old King was bred to a group of Morgan horse mares on the White Horse Ranch at Naper, Nebraska, and the offspring are quite numerous in the United States today. A genetic study, by Pulos and Hutt, in *The Journal of Heredity,* of the offspring of Old King revealed that the gene responsible for the white color is actually lethal to an embryo when it receives one gene from each parent. The white color, therefore, is always a heterozygous condition (see Chapter 5). When two of this type of white horse are bred together, one-fourth of the developing embryos will die and be resorbed into the dam's system or be aborted, often with no evidence that the mare was ever pregnant. A white "albino" stallion from the White Horse Ranch is shown in Figure 3–1.

The Albino Registry has now changed its name to the Creme and White Registry. They register the solid white Old King descendants as "white," and the "diluted" white horses as "creme." They no longer claim that their horses are albinos.

THE APPALOOSA

The Appaloosa Horse Club of Moscow, Idaho, serves as a registry for what is generally termed the Appaloosa horse. It is by no means the first registry for horses with Appaloosa characteristics, as the Knabstrup registry of Denmark and the ranger horse registry in Colorado had both registered horses with Appaloosa type markings much earlier. The Appaloosa Horse Club was formed in 1938, and these registered horses descended from horses the Nez Percé Indians had bred and called Appaloosa. Currently there are over 308,767 registered Appaloosa horses (Figure 3–2).

A number of genes are involved in Appaloosa patterns (see Chapter 9), and their expression is altered in various ways by modifying genes. Some of these genes, under certain conditions, produce no clearly evident body spotting but rather are detected only by the presence of horizontal light stripes on the hooves, a noticeable white sclera of the eye, and a skin mottling about the nose or genitalia. These are the minimum characteristics necessary to register a horse, regardless of ancestry, with the Appaloosa Horse Club. There are no conformation requirements other than a height of 14 hands, or over, and no pony or draft horse characteristics.

Registration rules state "Appaloosas foaled in 1970 and after must have

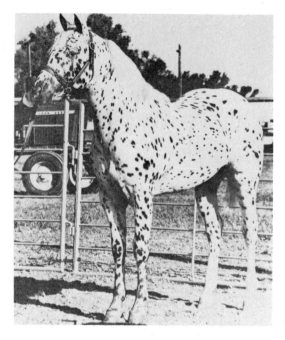

Figure 3–2.   The Appaloosa stallion Illusion.

both parents registered or identified with the Appaloosa Horse Club and/or a recognized breed association in order to be eligible for registration." In effect, this registry is an open registry, as it allows perpetual infusion of outside blood, and will result in a continually fluctuating breed type. The registry differentiates between permanent registration, for which horses must have a "typical Appaloosa coat pattern making them easily recognizable as Appaloosas," and breeding stock registration, which requires the minimum white sclera, striped hooves, and mottled skin (Figure 3–3).

THE BASHKIR CURLY

A registry was formed in 1971 for a small number of U.S. horses which had the unusual trait of curly hair. As the owners of these horses pooled their information, they discovered that most of the horses involved were of Russian origin and carried the characteristics of the Russian Bashkir breed.

These horses are well adapted for a cold climate. Their curly coat grows 4 to 6 inches long in the winter, and they carry an extra layer of fat under the skin. Their small round nostrils limit the cold air intake. The hair of the mane, and sometimes the tail, will shed out in the summer. The mane hair is very soft and kinky. The winter coat patterns range "from a crushed velvet effect, to a perfect marcel wave, to extremely tight ringlets over the entire

Figure 3–3. Photographs showing variations in the amount of white on Appaloosa horses.

Figure 3–4.   The Bashkir curly horse, Genghis Khan P-11, owned by Georgette Jessen.

body." The breeders claim that distinctive differences do exist in the basic structure and composition of the hair of these horses (Figure 3–4).

The breed is of medium size, gentle disposition, and various colors. Breeders describe them as a no-nonsense horse with an inclination to do all that is asked of them. These horses are used principally for gymkhana, eventing, competition, and endurance trail riding.

THE BELGIAN

The Belgian breed, which has been developing its present form since 1850, is descended from the ancient Flemish type of heavy horse in Belgium. In 1886, a stud book was begun for the breed. Belgians are compact horses, averaging from 16 to 17 hands, and weighing a ton. Their legs are free from feathering; they have a medium-sized head on a short, heavily crested neck.

THE BUCKSKIN

Two separate registries have developed for the buckskin horse in the United States: the International Buckskin Association and the American Buckskin Association. The buckskin pattern is a golden body color with

Figure 3–5.   The buckskin horse, many have a dorsal black stripe from mane to tail.

black on legs, mane, and tail. Generally a black strip runs down the center of the back connecting the mane and tail. The ancient tarpan horse was of buckskin color pattern. A modern buckskin horse is shown in Figure 3–5. The inheritance of this color pattern and its relationship to palomino, red dun, and creamello are covered in Chapter 9. Extravagant claims are made by breeders as to endurance and other favorable characteristics of these horses. The only scientific evidence for these claims, however, is the remote possibility of a genetic linkage of these favorable traits with color, as discussed in Chapter 7.

THE CHICKASAW

According to the Chickasaw Horse Association, the Chickasaw horse was developed by the Chickasaw Indians from stock brought to the American continent by DeSoto in 1539. In late 1959, a registry for the horse was developed by the Chickasaw Horse Association.

Standing from 13.1 to 14.2 hands high, the Chickasaw horse has a short head with unusual width between the eyes, a short neck, short ears, short back, square stocky hips, a low set dock, and a wide chest. He is a short coupled horse with quick action and can be found in all colors. Some authorities say that he is an early ancestor of the quarter horse and may be the same horse that inhabited the early prairies and was known as the mustang.

THE CLYDESDALE

The Clydesdale was bred in Scotland from principally shire ancestry. Cleveland bay blood has been introduced, giving the breed less size but more style and action.

The Clydesdale Horse Society of Great Britain and Ireland was organized in 1878, and shortly thereafter Clydesdales were exported in large numbers to Canada and the United States.

Clydesdales show the shire feathering of the legs and usually have white on the lower extremities and face. They stand about 16 hands and generally weigh 1600 to 1800 pounds (Figure 3–6).

THE DON

The Russian don breed is descended from an ''old don'' breed with much recent crossing to thoroughbred and Turkmene horses. The result is a breed

Figure 3–6. The Clydesdale horse.

with superior stamina possessing many of the original characteristics of the early Syrian horses. They average about 15.2 hands in height and make excellent saddle horses.

THE FJORD

The fjord is often known as the Norwegian breed or the Westlands pony (Figure 3–7). Heavily muscled and standing around 14 hands, the fjord is heavy enough to make an ideal harness pony suitable for many types of farm work. At the beginning of the 20th century, many fjords were imported into Denmark where they have had a marked effect on the general horse population there.

THE FRIESIAN

The old Friesian breed is a Dutch horse of ancient origin, which became practically extinct in the early part of the 20th century (Figure 3–8). At that time the few remaining horses were crossed with Oldenburg horses, producing the modern Friesian breed.

These horses are always black, with feathering of the legs, and they

Figure 3–7.   The fjord horse.

Figure 3-8.   An old photograph of a Friesian horse.

average about 15 hands high. Friesians are compact, muscular horses suitable for saddle or harness uses.

THE GALICENO

The Galiceno originated in Galicia, a province in northwestern Spain, and has been in existence on the American continent for several centuries. Bred in the coastal regions of Mexico, the horses are prized for their riding ease, endurance, and functional size (Figure 3-9).

The Galiceno was first imported to the United States in 1958. The Galiceno Horse Breeders Association was founded in 1959 to "record, perpetuate and preserve the breed." The Galiceno horse is believed to be an ancestor of the mustang that thrived on the plains of the southwestern United States.

The Galiceno horse is small, ranging from 12 to 13.2 hands at maturity, and weighs between 625 and 700 pounds. The head shows refinement with good width between the prominent eyes. Ears are pointed, of medium size, with the tips curved inward slightly. The neck is slightly arched with a clear-cut throat latch and runs well back into the moderate withers. The body of the Galiceno horse is muscled with a well-sprung rib cage, a short and extremely straight back. The croup slopes slightly with the tail set moderately high. The rear legs are slightly more under the horse than with other breeds.

THE GOTLAND

The modern Gotland breed is named after the Swedish Island of Gotland where ponies were isolated for 15,000 years. Descendants of these horses were instrumental in the Goth's overthrow of Rome in the 5th century A.D.

In the 1880s, the Swedish government, fearing the extinction of the small horse, placed the remaining Gotland horses under strict government protec-

Figure 3-9.   The Galiceno mare Canta.

Figure 3-10.   A Gotland horse.

tion and supervision. Some Oriental blood was used in the breeding program at this time to ensure a strong genetic base.

In the United States, the breed is a small horse standing between 12 and 14 hands. Many color varieties are produced. The Gotland Leeward Rurik is shown in Figure 3–10.

THE HACKNEY PONY

The hackney pony is of hackney horse breeding with infusion of Welsh and Fell pony blood (Figure 3–11). The pony has retained the characteristic traits of a hackney horse but is much smaller in size, ranging from 12.2 to 14.1 hands high. The hackney pony shows extremely high action in the knees and hocks.

THE HAFLINGER

The Haflinger is a robust Austrian mountain horse. A registry has recently been established in the United States and also in Great Britain. These horses have an ancestry in common with the Gotland, as the Goths left horses in

Figure 3–11.   A hackney pony mare with foal.

Austria as long ago as A.D. 555. Haflingers today are all chestnut in color with flaxen mane and tail. They are small in size with a heavy build.

THE HANOVERIAN

The Hanoverian was developed as a type in Germany in the 18th century. Prior to World War II, the breeding of Hanoverians was centered at the State Stud Farm at Celle where in the spring, stallions were sent out to about 60 service stations. The Hanoverian is now the largest in number of the light horse breeds in Europe (Figure 3–12). At the Munich games, the German equestrian team consisted of all Hanoverian horses, except one, and was considered the outstanding team in the Olympics. Of the combined teams represented at Munich, eight nations entered 31 Hanoverians.

There are four Hanoverian sales in Germany each year, one broodmare sale, one stallion sale, and two carefully selected performance horse sales. For the performance sales, 400 horses are selected, out of which 80 of the best are chosen for the sale. These horses come to Verden for 5 weeks of training and are examined on the basis of character, temperament, talent, and capacity. They are graded during the training period, and all results are published prior to the sale.

The breed exhibits enough variation of type so that choices can be made between heavy or light horses, elegant and intelligent horses with dressage talent, elastic cross-country horses, and strong, sturdy horses for more

Figure 3–12. A Hanoverian horse.

utilitarian purposes. People from all over the world are now purchasing the Hanoverian horse for international competition. You will see Hanoverians in all countries, including Canada and the United States, as dressage and jumper horses. Wherever Hanoverians are found, they are making an outstanding name for themselves as performance horses.

Twice a year, in April and October, there are the world famous "Elite" shows. These events, as well as the sales, serve as the showcase for the Hanoverian breed. You will always see international riders entering the shows for the purpose of purchasing Hanoverians.

The dressage horse with the highest sum of prizes in the world is Doubette, a Hanoverian which was originally sold at the Verden auction sale. At recent sales, Hanoverian horses were sold to such countries as Austria, Belgium, Luxemburg, Denmark, Czechoslovakia, England, France, the Netherlands, Greece, Southwest Africa, Italy, Canada, Mexico, the USSR, Portugal, Sweden, Switzerland, Turkey, and the United States.

THE KNABSTRUP

The Danish Knabstrup registry originated with a spotted mare called Flaebehoppen. The color pattern, which is similar to some of the "leopard" Appaloosa patterns, has been perpetuated by using offspring of this foundation mare (Figure 3–13). Breeding records have been maintained since about 1808. Recently, crossing with other spotted horses has diversified the

Figure 3–13. The Knabstrup, a Danish spotted horse.

conformation type, so that what might originally have been classified as a breed is now simply a registry.

The color pattern of the Knabstrup is a white background with pigmented spots over the entire body. There is some indication that crossing with non-spotted horses has a greater tendency to reduce the white background on the head and shoulders than on the posterior parts of the horse. There is little doubt that the same genes are involved in the Knabstrup and the Appaloosa color patterns.

THE MORGAN

The horse, Justin Morgan, was foaled in 1793. He was reportedly by Briton and out of a daughter of Diamond. The sire and dam were strong in the blood of thoroughbreds and Arabians. Justin Morgan founded the Morgan breed principally because of his tremendous prepotency.

Although breeding records date back more than 100 years, the type of early ancestry, other than Justin Morgan, seems to have been forgotten by all concerned. The concentrated line breeding to this sire eventually produced a type which, according to W.H. Gocher, "in time became the trademark of the first great family of horses, and it is so different from all others that when an Australian or an Englishman refers to a 'Vermont Morgan,' an idea of his general appearance is conveyed lucidly."

Figure 3–14.   The Morgan stallion Flyhawk 7526. (Courtesy of Nancy Buckles.)

Figure 3–15.   Senator Graham, a Morgan horse of the late 40s. (Courtesy of Nancy Buckles.)

The Morgan has the general body conformation of a thoroughbred, with a short back and sloping pasterns (Figures 3–14, 3–15). The head is held high on the neck which is very wide where it joins the body. Morgan breeders emphasize the great endurance of the horse. Morgans have been instrumental in the establishing of other American breeds.

THE NONIUS

The Nonius breed originated with a Hungarian stallion (1810) known as Nonius (Figure 3–16). Over the years since then, a large and small type have been developed. The large Nonius is a middle-weight horse, standing over 15.3 hands and suited for riding or driving. The small Nonius was produced by more crossing with Arabian horses. The result is a finer, shorter-legged horse standing no more than 15.3 hands high.

THE ORLOV TROTTER

The Orlov trotter was originated in 1777 by Count Alexis Grigoroevich Orlov. Various types of horses were used in developing this breed, including the Arabian and the thoroughbred. Selection for trotting ability made the breed world famous. Orlov trotters were used as a basis for the American trotting horse. Orlovs resemble the Arabian horse but are much larger, ranging in height from 16 to 17 hands (Figure 3–17).

Figure 3–16.   An old photograph of a Nonius horse.

Figure 3–17.   An old photograph of an Orlov trotter.

THE PAINT

The American Paint Horse Association of Fort Worth, Texas, registers paint horses which are sired by either registered paints, thoroughbreds, or quarter horses. The registry was organized in 1962 and had registered 15,000 paint horses by 1970.

The registry recognizes two distinct genes, both of which produce the paint color pattern, and both of which have existed for centuries in various breeds of horses. The tobiano (to-bi-an-o) gene is dominant and causes large patches of white to occur at various locations over the body or limbs of the horse (Figure 3–18). Generally the head is marked like that of a solid-colored horse, and the legs are almost always white below the knees and hocks. Occasionally a modifying gene may cause the borders between the white and pigmented areas to be roan or somewhat speckled in appearance.

The second type of paint is known as overo (o-ver-o), a recessive gene commonly found in many other breeds such as the hackney and Clydesdale (Figure 3–19). Modifying genes determine the extent of white over the legs and body, as explained in Chapter 8 (Figure 3–20).

Figure 3–18.   The tobiano paint horse, a dominant spotting pattern.

Figure 3–19.   The overo paint horse, a recessive spotting pattern.

Figure 3–20.   The overo and tobiano paint mixture.

THE PALOMINO

The Palomino Horse Breeders of America, Inc., was established in 1941 and since then has served as a registry for palomino horses. Palomino horses are defined as having "a body color near that of an untarnished U.S. gold coin, and a white mane and tail" (Figure 3–21). The registry maintains some conformation standards, excluding horses with draft horse or pony characteristics. To be registered, a horse must be over 14 hands high and weigh 900 pounds. In addition, the sire or dam must be a registered palomino, thoroughbred, quarter horse, American saddle horse, Arabian horse, Morgan horse, Tennessee walking horse, American remount, or a trotting horse.

Figure 3–21.   The palomino horse.

Research has shown that the palomino color is due to a heterozygous condition and therefore could never become fixed in a population to the extent that it would breed true from generation to generation. Some evidence, however, indicates that with a dark chestnut genetic background, a light shade of palomino may be homozygous, and could therefore be true breeding. Definite proof either way remains to be brought forth.

THE PERUVIAN PASO

The Peruvian Paso horse comes only from the South American nation of Peru (Figure 3–22). The horses which are referred to as Paso Finos have a number of origins: Colombia, Puerto Rico, the Dominican Republic, and Cuba, to name the foremost. The Peruvian breed has been without the introduction of outside blood for many centuries. One of the results of this is

Figure 3–22.   Cascabel, Peru's National Champion of Champions in 1975–76, was one of the Peruvian Paso horses shown to President Carter's wife during her visit to Peru.

that the Peruvian Paso horse can guarantee 100% transmission of his gait. The Paso Finos cannot.

The Peruvian Paso breed has been popularized in the United States, with the formation of a registry and the importation of many horses. Verne Albright, in his book, *The Peruvian Paso and His Classic Equitation,* writes,

> The Peruvian Paso horse has been developed in Peru during the last 400 years. The fact that such a breed has been brought into being is nothing less than a minor miracle. The Peruvian horse combines qualities which are exceedingly difficult to combine. Some of his attributes could almost be called enemies of one another: He is spirited, yet easy to handle; he has sparkling, brilliant action, yet he is unbelievably smooth and enduring; he has termino, yet he is sure-footed; he has fitness of build, yet he is powerful.

THE PINTO

The Pinto Association of America, Inc., located at San Diego, was founded on the stud book of the Pinto Horse Society, oldest of pinto registries. The Pinto Association was founded in 1956, about 6 years prior to the establishment of the Paint Horse Association of Fort Worth. The Pinto Association covers many fields of activities, organized "for the improvement of the Pinto breed by encouraging Pinto horse classes in all horse shows, and to promote good sportsmanship."

The directors of the Pinto Horse Association are very active and knowledgeable. Horses registered with them are placed in one of a number of divisions reflecting the genetic diversity of the color and extensiveness of the pedigree. These divisions are: 1, tentative; 2, permanent; 3, approved breed; and 4, premium division. They also maintain a pony registry.

The Pinto Association recognizes both a tobiano and overo color pattern; however, many pinto breeders feel that there are both dominant and recessive types of overos. A 50:50 ratio of white to color is the most desirable in both color patterns.

A Pinto Horse Association Champion and Register of Merit program has helped tremendously in recent years to improve the caliber of pintos. Register of Merit certificates are earned in individual performance classes and racing. Performance events include western, hunter, and saddle horse classes. Register of Merit points combined with halter points may lead to the title of Pinto Horse Association Champion.

Pinto coloring, which is produced by the same two genes of the paint pattern, can be bred into any breed; and by means of continual crossing back to horses of the original breed, a relatively pure strain of pinto horses with almost 100 percent of the original breed characteristics can be retained. Therefore, the Association recognizes a number of halter divisions reflecting these breed variations within their color registry. The most popular division is the stock-type pinto; others are pleasure-type pintos and saddle-type pintos.

The Pinto Association claims that the pinto horse is a breed, albeit their encouragement and recognition of distinct types within their registry will never allow development of a distinct pinto breed. The registry is to be commended, however, on its efforts to register all pintos and to keep complete records of all ancestry.

THE PONY OF AMERICA

The Pony of America Club was organized in 1954 as a registry for ponies with Appaloosa spotting. To be eligible, the pony must be between 12.3 and 13.2 hands high and have the Appaloosa characteristics (Figure 3–23) of horizontal light stripes on the hooves, a noticeable white sclera of the eye, and a skin mottling about the nose or genitalia and Appaloosa color. Registration is open to any pony that qualifies in height and color characteristics and whose sire or dam is a registered P.O.A.

THE QUARTER HORSE

The quarter horse may have originated as long ago as the 18th century in the United States. At least the name "quarter horse" was used at that time to

Figure 3–23. The Pony of America Supreme Champion, Salty Little Britches.

denote a horse that was extremely fast at one-quarter of a mile. Of course, these horses were mainly of thoroughbred extraction. During the 19th century, match racing for a quarter of a mile was very popular, and horses were bred particularly for this ability.

Several famous sires established themselves at this time and are now known as the foundation horses of the quarter horse breed. Printer, by the famous Janus, foaled in 1800, was one of the first, followed by Tiger, 1812; Copperbottom, 1832; Shilow, 1840; Steeldust, 1845; Old Billy, 1858; Cold Deck, 1860; and Lock's Rondo, 1880, all established families which were well known in the quarter-mile racing world of the last century. About the turn-of-the-century, principally in Oklahoma and Texas, quarter horse people began thinking in terms of a breed. Many of these breeders referred to the horse as a quarter horse. Several important sires came into prominence at about this time. Traveler, 1885; Peter McCue, 1895; Little Joe, 1904; Possum, 1910; and Old Sorrel, 1915, are known as the most important founders of the breed. Many of the better horses in the first quarter horse stud books trace to these sires.

In 1941, The American Quarter Horse Association was established, with the grand champion stallion at the Fort Worth Stock Show being given the honor of registration number one. The American Quarter Horse Association is now the largest horse registry in the world. An analysis of the first 10 horses registered in 1941 gives a fairly representative sample of the kind of blood with which the breed was established. Wimpy, P-1, was a King Ranch horse by their famous sire Old Sorrel, who traced to Peter McCue. Rialto, P-2, was sired by a well-known quarter running horse of the day, known as Billey Sunday, and was out of a mare by Little Joe. Joe Reed, P-3, was principally of thoroughbred blood. Joe Bailey, P-4, traced to Traveler on his sire's side and to Old Joe Bailey on his dam's side. Chief, P-5, was a son of Peter McCue. Oklahoma Star, P-6, was a thoroughbred. Columbus, P-7, was sired by a horse of mixed blood, while his dam was by Old Billy. Colonel, P-8, traced to Peter McCue through both his sire and dam. Old Red Buck, P-9, traced to Printer on his sire's side and to Cold Deck on his dam's side. Old Jim, P-10, was sired by a horse of Traveler blood.

Rhoad analyzed the ancestry of the modern quarter horse and found an extremely heavy concentration of thoroughbred blood. Denhardt, who was instrumental in establishing the quarter horse registry, took issue with him, however, maintaining that early breeders were quite selective and tended to breed more to the horses from the good quarter horse families. At the present time the breed seems to be basically of three types of ancestry: thoroughbred, old established quarter horse families, and nondescript (Figure 3–24).

Much of the nondescript blood was introduced into the breed during the 1950s when the registry was still "open" by inspection and prices were relatively high for registered quarter horses. Since that time, the registry has closed to all but those horses with sire and dam being permanently registered quarter horses, and under certain circumstances to a horse with

Figure 3–24.   A working-type quarter horse, Sonora Sorrel. (Photo by Darol Dickinson.)

one parent a permanently registered quarter horse and the other parent a registered thoroughbred. In the latter case, the applicant must become Register of Merit (R.O.M.), which is a registry within the Association for horses that excel in performance, and must also be inspected for conformation before becoming permanently registered.

Racing of Quarter Horses

Quarter horse racing has become extremely popular and highly competitive. The more successful breeders seem to be crossing certain lines of thoroughbreds with some of the old established quarter horse lines to obtain the best results. The thoroughbred, Three Bars, has, without a doubt, contributed more recently to the quarter horse breed than any other horse from outside the Registry.

Quarter horse racing has become big business, and as a result has changed the nature of the quarter horse breed in the past decade. The advent of the All American Futurity, billed as the "World's Richest Horse Race," and other futurities offering large purses have caused many breeders to infuse large doses of thoroughbred blood into quarter horse stock. The popular lines have developed from the horses that have been the greatest winners (Figure 3–25). Following is a list of quarter horses that have been named "World Champion Quarter Running Horses" since 1960.

Vandys Flash
Pap
No Butt
Jet Deck
Goetta
Go Josie Go
Laico Bird
Kaweah Bar
Easy Jet
Charger Bar
Mr. Jet Moore
Truckle Feature
Tiny's Gay
Easy Date
Dash for Cash (Figure 3–26)

Figure 3–25. Easy Jet, one of the top sires of quarter running horses. (Courtesy of Walt Wiggins.)

Figure 3–26.   Dash for Cash, 1976–77 Champion Quarter Running Horse.

Today there seems to be somewhat of a split in the American Quarter Horse Association as to goals for the breed. Some racing interests are breeding for nothing but speed, excessively diluting the original quarter horse blood with that of the thoroughbred. On the other hand, some breeders cling to the old bloodlines to the point that they are neglecting some of the recently developed characteristics which have greatly improved the breed. Although there are several types among registered quarter horses, breeders are rapidly establishing a prepotent breed and have a tremendous future ahead.

THE AMERICAN SADDLE HORSE

The American saddle horse originated from the thoroughbred, initially from the famous stallion Denmark (1839). Through the years, Morgan horse, Narragansett pacer, and trotter blood was introduced. A registry was established in 1891, which until recently was maintained on a restricted open basis. Originally this allowed for the incorporation of good traits from

Figure 3–27.   Plainview's Julia, an American saddle horse.

other breeds, but according to Steele, the breed has become severely hybridized. Diversity of objectives has led to minor breeds within the breed. Recently the American Saddle Horse Breeders Association billed their horse as "popular for pleasure riding, driving, as a hunter or jumper, parade horse, cow horse as well as a show horse" (Figure 3–27).

Although the breed originated from two inbred lines, Chief and Denmark, later much outbreeding was practiced. Closing of the registry to all horses not strictly of purebred parents may eventually stabilize the breed. The ideal American saddle horse averages 15 to 16 hands in height and is from 1000 to 1200 pounds in weight.

SPANISH BARB

A Spanish barb registry has been established in the United States, "to faithfully restore the historic Spanish-Barb horse." The breed standard is based on "horses out of proven bloodlines only" and numbered in the permanent division of the registry, as well as documented descriptions given of the Spanish-Barb horses which were bred and used in North America

Figure 3–28.   A Spanish-barb horse, with Walt Banner up.

during the 15th through the 17th Centuries. A Spanish barb horse is shown in Figure 3–28.

SPOTTED HORSE

One of the oldest spotted horse registries is the Morocco Spotted Horse Co-operative Association of America. The organization was begun in 1935 and was given its unusual name after pedigree studies showed that many of the foundation horses in the registry traced back to imported English hackney and French coach horses and through these to barb horses of Morocco (Figure 3–29).

The organization has a total registry of about 2000 horses. It is one of the few organizations which refuses to register animals because of hereditary faults. Listed as faults are cocked ankles, straight pasterns, bog or bone spavin, parrot mouth, chorea, sickled hocks, blindness, navicular disease, cryptorchidism, side bones, ring bone, a large head, long ears, jack, curb, club foot, and cloven hoof.

Spotted horses have been bred off and on for centuries in countries throughout the world.

In 1963, the Spotted Horse Society was established in Great Britain as a

Figure 3–29.   Starbruk Leopard, from Barbary States.

registry for horses of the Appaloosa color pattern. The society has registered horses of all the typical patterns found in the United States in the Appaloosa Registry.

STANDARDBRED

The standardbred is often known as the "American trotter and pacer" and sometimes as the "harness horse." The registry, established in 1876, is known as the U.S. Trotting Register. Although trotting horses were first bred in England, in such breeds as the Norfolk trotter, harness racing has never been as popular there as in the United States. The standardbred was developed along the Atlantic seaboard principally from thoroughbred ancestry such as Messenger, 1780, and Diomed, 1777. Bellfounder, imported in 1822 (reportedly a Norfolk trotter), was also an important ancestor.

One of the early popular standardbreds was Boston Blue, who trotted a mile in 3 minutes in 1818. The world trotting record was held for many years by Greyhound, 1 mile in 1 minute, 55¼ seconds made in 1938. In 1969, Nevele Pride (Figure 3–30) erased this long-standing speed record. Interest in harness racing is on the rise with more than 30,000 members in the United States Trotting Association and about 450 trotting tracks.

The standardbred is of general thoroughbred type, averaging 15.2 hands and weighing 820 to 1180 pounds. These horses generally have good action as either pacers or trotters. Philip Pines in *The Complete Book of Harness Racing* says that the tendency to trot or to pace is inherited, but that the ability to maintain that gait at high speeds over long distances is the result of training.

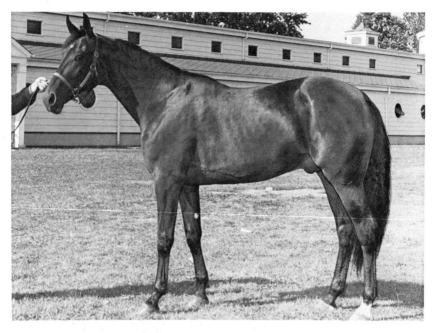

Figure 3–30. The standardbred trotter, Nevele Pride. He erased the long-standing speed record, over a mile, of Greyhound late in 1969.

Figure 3–31. Whata Baron, a standardbred pacer.

Figure 3–32.   Speedy Somalli, a standardbred trotter.

Pines gives examples of how a sire passes on his gait:
> Out of seventy-six two-minute performers sired by Adios, seventy five of them were pacers. The pacer, Tar Heel, produced thirty-five two minute performers, all of them pacers. On the trotting side there was Rodney, whose sixteen two-minute offspring included fourteen trotters. Star's Pride sired fifteen Standardbreds who could claim two-minute records, thirteen of them at the trotting gait.

In the trot a horse moves his legs diagonally: Front right, rear left; front left, rear right. In pacing the action is the opposite. The right front and rear legs move in the same direction at the same time. Whata Baron, a pacer, is shown in Figure 3–31. Speedy Somalli, a trotter, is shown in Figure 3–32.

THE TENNESSEE WALKING HORSE

The Tennessee walking horse was developed from a wide variety of types including standardbreds, thoroughbreds, and Morgan horses. The Tennessee

Walking Horse Breeders' Association of America was formed in 1935, establishing pedigrees tracing to such horses as Copperbottom, Hals, Slashers, Grey Johns, Bullets, Whips, Blue Jeans, Brooks, Stonewalls, Pilot, and Denmark.

Allen, F-1, was picked to head the registry because of his outstanding contribution to the breed. He was by Allendorf, by Onward, by George Wilkes, R-54, by Hambletonian. His dam was Maggie Marshall, by Bradford's Telegraph, by Black Hawk, by Sherman Morgan. He was a black horse foaled in 1886, a double-gaited horse, trotter and pacer. His sire, Allendorf, was a fashionable harness horse of the day, while his dam was of the best Narragansett pacing blood available.

The breed is famous for its natural overstride at a running walk, that is, placing of the back hoof ahead of the print of the fore hoof (Figure 3–33). Some of these horses are capable of overstriding by four feet or more. They generally have a neat head with well-shaped, pointed ears, and a long, graceful neck set on a well-muscled sloping shoulder. They average 15½ hands in height and weigh 1000 to 2000 pounds.

Figure 3–33. Sensational Shadow, a Tennessee walking horse, demonstrates the natural overstride at a running walk.

Breed Association	Date Registration Began	Total Number of Horses Registered Since Registry Began (through 1978)	Number of Horses Registered in 1978	Total Number of Registry Members (through 1978)	Registration Fee for Foals	
					Members	Non-Members
American *Andalusian* Association	1963	390	*40	****	$15.00	****
The *Appaloosa* Horse Club	1939	*289,000	17,767	54,408	20.00†	$40.00†
National *Appaloosa* Pony, Inc.	1963	*5,264	83	3,050	10.00†	15.00†
Half-Arabian & Anglo-Arabian Horse Registry	1930	*195,000	14,696	28,000 (IAHA)	10.00	20.00
Arabian Horse Registry of America, Inc.	1908	182,000	16,669	11,500 (60,089 owners)	15.00†	40.00†
American *Bashkir* Curly Registry	1971	127	27	66	10.00	****
Belgian Draft Horse Corporation of America	1887	*85,000	2,416	****	10.00	15.00
International *Buckskin* Horse Association	1971	5,011	980	1,500	5.00†	****
Chickasaw Horse Association, Inc.	1959	632	0	632	7.50‡	****
Cleveland Bay Society of America	1886	3,673§	****	117§	10.00	****
Clydesdale Breeders of the U.S.	****	****	178	400	10.00	****
American *Connemara* Pony Society	1956	*2,200	98	*350	10.00†	****
American *Crossbred* Pony Registry	1960	*400	16	††	7.50	****
Creme & White Horse Registry (formerly the American Albino Association	1937	3,000	80	65	5.00†	****
Missouri Fox Trotting Horse Breed Association Inc.	1948	15,231	1,729	1,966	15.00	****
Galiceno Horse Breeders Association	1959	2,234	36	70	10.00‡	****
American *Gotland* Horse Association	1960	366	38	120	2.50	7.50
American *Hackney* Horse Society	1891	*22,654	*590	435	20.00†	****
Hafflinger Association of America	1976	221	112	61	10.00	15.00
The American *Hanoverian* Society	1973	*300	100	****	15.00	25.00

Breed Association	Date Registration Began	Total Number of Horses Registered Since Registry Began (through 1978)	Number of Horses Registered in 1978	Total Number of Registry Members (through 1978)	Registration Fee for Foals	
					Members	Non-Members
The American *Holstein* Horse Association	1978	2	2	10	10.00	****
The *Horse of the Americas*	1973	310	22	64	10.00	****
Royal *International Lipizzaner* Club						
International *Miniature* Horse Registry	1976	*2000	1,057	****	20.00‡	****
Morab Horse Registry of America	1973	249	67	40	25.00†	35.00†
The American *Morgan* Horse Association	1891	71,389	3,519	4,611	25.00†	35.00†
Spanish Mustang Registry Inc.	1957	735	35	125	15.00†	****
Norwegian Fjord Horse Association	1978	105	90	20	10.00	****
American *Paint* Horse Association	1962	51,000	8,363	7,923	15.00†	40.00†
The *Palomino* Horse Association	1936	9,120	115	250	15.00	20.00
Palomino Horse Breeders of America	1941	38,200	2,138	2,700	5.00†	20.00†
American *Part-Blooded* Horse Registry	1939	12,000	450	****	12.00	****
Paso Fino Owners and Breeders Association, Inc.	1973	3,090	621	800	8.00†	****
Percheron Horse Association of America	1876	253,982	580	550	10.00	12.50
American *Peruvian Paso* Horse Registry	1978	−100	−50	−50	15.00	****
American Association of Owners and Breeders of *Peruvian Paso* Horses	1967	*1,662	151	248	25.00†	****
Peruvian Paso Half-Blood Association	1969	275	31	****	15.00‡	****
Peruvian Paso Horse Registry of North America	1970	1,100	171	220	25.00	****
The *Pinto* Horse Association of America Inc.	1956	31,832	1,870	12,000	30.00†	50.00†
Pony of the Americas Club	1955	23,000	1,000	1,500	10.00†	****

Breed Association	Date Regis- tration Began	Total Number of Horses Registered Since Registry Began (through 1978)	Number of Horses Registered in 1978	Total Number of Registry Members (through 1978)	Registration Fee for Foals	
					Members	Non- Members
American *Quarter Horse* Association	1941	*1,400,000	119,287	98,000	15.00	25.00
Half-Quarter Horse Registry of America						
National *Quarter Horse* Registry	1956	****	****	****	15.00	20.00
Standard *Quarter Horse* Association	1963	7,146	400	500	10.00	****
National *Quarter Pony* Association, Inc.	1976	415	104	142	15.00	25.00
Racing Horse Breeders Association of America	1971	10,000	1,725	2,300	20.00	****
Colorado Ranger Horse Association	1938	2,053	138	657	10.00†‡	15.00†‡
American Remount Association, Inc. *(Half-Thoroughbred)*	1916	*25,000	127	****	25.00	****
American *Saddle Horse* Breeders Association	1891	168,643	3,803	611	25.00	****
The *Half Saddlebred* Registry of America	1971	*500	103	*400	10 00†	****
American *Shetland Pony* Club	1888	*132,600	*560	500	11.00‡	****
American *Shire* Horse Association	1885	****	77	125	15.00†	30.00†
Spanish Barb Breeders Association	1972	102	20	78	12.00	****
The United States Trotting Association *(Standardbred)*	1876	446,303	13,991	43,362	15.00†	****
American *Suffolk* Horse Association	1907	2,881	29	59	7.50†	15.00†
American *Tarpan* Stud Book Association	1971	99	6	12	no fee	
Tennessee Walking Horse Breeders & Exhibitors Association	1935	*171,845	*6,500	4,500	15.00†	30.00†
Thorcheron Hunter Association	1974	36	5	43	no fee	
The Jockey Club *(Thoroughbred)*	1894	*790,000	31,000	75 (it is a club)	60.00	****
The American *Trakehner* Association	1975	1,011	208	*300	10.00†	****

Breed Association	Date Registration Began	Total Number of Horses Registered Since Registry Began (through 1978)	Number of Horses Registered in 1978	Total Number of Registry Members (through 1978)	Registration Fee for Foals	
					Members	Non-Members
The North American *Trakehner* Association	1978	104	52	77	15.00	****
International Trotting & Pacing Association (*Trottingbred*)	1964	6,000	500	1,500	12.00	****
American *Walking* Pony Association	1968	189	18	40	15.00	****
Welsh Pony Society of America	1906	26,880	*450	650	10.00†	****
Wild Horses of America Registry, Inc.	****	****	****	****	****	

*Approximate figure.
†Registration fee varies with the age of the animal.
‡Registration fee varies with the sex of the animal.
§Information is based on 1972 figures.
****The information was not submitted.

BIBLIOGRAPHY

Albright, Verne: The Peruvian Paso: and his classic equitation. Ft. Collins, CO, Caballus Publishers, 1976.

Denhardt, Robert: Quarter Horses: A Story of Two Centuries. Norman, OK, The University of Oklahoma Press, 1967.

Dent, Anthony: Cleveland Bay Horses. London, J.A. Allen & Co., 1978.

Goodall, Daphne Machin: Horses and Their World. New York, Mason/Charter Publishers, 1976.

Haines, Francis: Appaloosa: The Spotted Horse in Art and History. Ft. Collins, CO, Caballus Publishers, 1973.

Harrison, James C.: Care and Training of the Trotter and Pacer. Columbus, OH, U.S. Trotting Assoc., 1968.

Laune, Paul: America's Quarter Horses. Garden City, NY, Doubleday & Co., 1973.

Mellin J.: The Morgan Horse. Brattleboro, VT, Stephen Greene Press, 1961.

Nagler, Barney: The American Horse. New York, The Macmillan Co., 1966.

Pulos, W.L., and Hutt, F.B.: Lethal Dominant White in Horses. J. Hered. 60:59–63, 1969.

Rhoad, A.O.: The American Quarter Horse. Quarter Horse J., March 1961, p. 33.

Schilke, Dr. Fritz: Trakehner Horses: Then and Now. Norman, OK, American Trakehner Association, 1977.

Self, Margaret Cabell: Horses of Today. New York, Meridith Press, 1964.

Taylor, Louis: The Horse America Made: The Story of the American Saddle Horse. New York, Harper & Row, 1961.

Welsh, Peter C.: Track and Road: The American Trotting Horse. Washington, D.C. Smithsonian Institute Press, 1969.

4

breeding goals

It is common practice among registries to list the breeder of a horse as the one who owned the mare at the time of service. This practice implies that this individual gave careful thought and planning to the production of the foal; that is, the breeder decided what stallion to breed to the mare; in other words, he had a purpose in mind. Regrettably, this is not usually the case, but the designation "breeder" implies such forethought.

All too often the owner of a mare simply breeds her to the closest stallion, or the cheapest, with no regard to what kind of a foal will be produced. This approach could be regarded as horse production, but certainly not as horse breeding.

John Hislop, a well-known horseman, once wrote,

> When those of my generation come to be asked by their grandchildren, "Who was the greatest breeder?. . ." the answer will probably be, "Tesio. . ." . . . (he) produced his many great racehorses, sires and brood mares from comparatively limited resources. He never had a large number of mares in his stud at any given time, did not spend a great deal of money on bloodstock and sold most of the best horses he bred. Senator Tesio was a genius.

Such can be the tribute to the best horse breeders.

Horse breeding has been mostly an endeavor of wealthy families or governments, because it is a long-time effort and takes vast resources. The average man can indulge in the enterprise, as Federico Tesio did, only if he takes advantage of the previous progress made by others. The beginner will generally find, after a thorough survey of the situation, that someone else has already been working a lifetime to do what he has in mind. It would be fruitless to cover the same ground again.

Long-term horse breeding goals can take one of three general directions. The first is the overall goal of recreating a breed or type that has vanished. Here the breeder attempts to select certain traits that are scattered throughout a large number of animals, and put them together into a final product which breeds true. An example of this type of endeavor is the recreation of the tarpan horse. Breeders involved in this effort have the scientific description of the wild tarpan, which once roamed Europe, as their ideal. The Heck family has worked for years to bring the extinct tarpan back into existence. The original type is illustrated by cave paintings (Figure 4–1), and further information about size and weight is obtained from skeletons.

The second general direction that horse breeding can take is the preservation of a breed or type as it is, without allowing genetic drift. In this instance a breeder must be aware of the history and development of the type he wishes to preserve. He must be committed to the premise that the type is already perfect as it is. An example of this type of breeding was the undertaking of Wilfred and Ann Blunt in the last century to collect the best examples of the remaining Arabian horses in their homeland, and breed them to perpetuate a type of horse that was rapidly disappearing (Figure 4–2). Arabian horse historians now recognize these dedicated horse breeders as the ones who saved the Arabian horse from extinction, or at least from drastic change that would have undone what had been accomplished over hundreds of years.

Figure 4–1.   This ancient cave painting serves as a model for breeders to recreate the tarpan horse today.

Figure 4–2. Wilfred and Lady Ann Blunt, saviors of the Arabian horse.

A third direction that horse breeders can take is an attempt to bring a completely new type of horse into existence. This track has been pursued time and time again with the development of new breeds. The cold-blooded horses of Europe were refined with Oriental blood to produce the Lipizzaner and the Holstein, as examples. Also the thoroughbred was developed over several hundred years by the continual improvement of existing stock. The quarter horse breed has been developed through the dedicated efforts of many good horse breeders. Robert Denhardt wrote of Ott Adams, in *Quarter Horses,*

Figure 4–3. Ott Adams, great quarter horse breeder of South Texas.

Ott Adams certainly rates as one of the best Quarter Horse breeders who ever lived. Two of Ott's best-known sayings were that the only way to get speed is to breed speed to speed and a Quarter Horse can run his best for as long as he can hold his breath. Certainly he had this first axiom in mind when he bought Della Moore to breed to Little Joe.

During his lifetime, Ott Adams (Figure 4–3) made much progress toward improvement and development of the quarter horse breed.

GOALS AND BREEDING PROGRESS

Raising a few foals from time to time can be very enjoyable. Producing an annual crop of horses for sale can be profitable. But, a "true horse breeder" is driven by a lifelong goal to create or preserve that "perfect"horse; enjoyment and profit are merely fringe benefits to him. The real satisfaction comes as he sees progress toward his goal.

It is not easy to improve a horse breed. It does not happen in a year, or even 10 years; and without intelligent, diligent, and even ruthless selection over a period of many, many years, a breeder may finish his lifetime work without an inch of progress. A true horse breeder takes his charge seriously. He begins his endeavor by clearly defining what type of horse he wants. This initial definition or description of the goal must be put down in extreme detail. Otherwise, the tendency is to change his goal with every whim that appears.

No progress can be made toward an ideal type unless a breeder maintains his predetermined goal without faltering throughout the many years that will be required to bring about significant results. From the very beginning the successful horse breeder must look years ahead and accurately project beyond the short-term fads, deciding what is really important in the way of horseflesh. Many horsemen have tried to be true horse breeders but have missed the mark because they have vacillated. Doubts and mind-changing come to those who do not think deeply about the matter initially and to those who let others make the choices for them. Goals will shift and fade for those who are overly concerned about the show ring winners each year and how these horses compare with their breeding stock.

Over the years the standards of judging halter classes at shows have continued to evolve. Breeders engaged in long-term programs should not adjust their goals to reflect the changing whims of the show ring. Following such a program leads in circles with the breeder never reaching a prescribed goal.

To be successful at breeding and showing over the long haul requires considerable foresight. The relatively simple process of breeding a high-priced stallion to high-priced mares and then selling high-priced foals does not necessarily constitute a sound breeding program. At any rate, it does not generally command deep respect from fellow breeders. A program that begins with stock chosen on the basis of specific traits, rather than prices, and continues on the basis of a predetermined system, rather than the dictates of fickle buyers, will prove to be much more profound.

THE ROLE OF THE REGISTRY

The concept of a breed registry is one that has evolved in the last century. The forerunner was the Jockey Club, established in 1750 in England. The Jockey Club began as a social club open to all who entered horses in races. Eventually it developed the authority to regulate the sport and was limited to a small influential group.

Even before the Jockey Club was established, there were attempts to publish records of horses, principally racing records. John Cheny was first with his *An Historical List of All Horse-matches Run*. In 1773, the Jockey Club authorized James Weatherby to publish what was then called a Calendar. For a decade Wetherby struggled with thoughts as to how pedigrees should be recorded. Then in 1790, he announced his *Stud Book*. This was revised in 1793, and the first supplement was published in 1800.

Today those involved in horse breeding recognize the importance of pedigrees and accurate breeding records. If nothing else, accurate breeding records allow breeders to choose intelligently the ancestry of their stock. Even Weatherby, at that early date, recognized a need to correct the "increasing evil of false and inaccurate pedigrees." The term thoroughbred was not used in Volume 1, but the concept was there in the reference to "half-bred" horses, which were not admitted to the *Stud Book*.

Peter Willett in *The Thoroughbred* writes,

It seems that what Weatherby had in mind as the condition for eligibility for the *Stud Book* was direct descent from a small number of original mares, less than a hundred, who were regarded as the foundation stock of the thoroughbred. The pedigrees of these mares could be traced back no further, and had their productive careers mostly in the second half of the Seventeenth Century.

The matings of these mares and their offspring and descendants, with the Eastern stallions and their offspring and descendants who were imported during a period of approximately a century following the Stuart Restoration, formed the breed which Weatherby accepted as thoroughbred and eligible for his *Stud Book*. This interpretation is corroborated by a later editor, who stated in the Advertisement to Volume 14 "that a recent importation of Arabians from the best Desert strains will, it is hoped, when the increase of size has been gained by training, feeding and acclimatisation, give a valuable new line of blood from the original source of the English thoroughbred."

It may be noted in passing that this hope was to be disappointed, for events proved that the genetic pool from which successful racehorses could be bred was already filled and, as Admiral Rous pointed out so trenchantly, the Arabians were so inferior as racehorses that they could not be assimilated with the thoroughbreds of the Nineteenth Century. Two centuries of selective breeding in a favorable environment had set an unbridgable gulf between the English racehorse and his desert ancestors.

James Weatherby (Figure 4–4) and his descendants had set a basic premise for the future definition of a purebred. This culminated in what was later known as "The Jersey Act." Willett describes it:

At the Jockey Club meeting in the spring of 1913, Lord Jersey, the Senior Steward of the previous year, put the case for making the regulations even stricter and more precise. His motion was passed by the Club and acted on by Weatherby, who inserted the following rule for admission in Volume 22, published the same year: "No horse or mare can, after this date, be considered as eligible for admission unless it can be traced without flaw on both sire's and dam's side of its pedigree to horses and mares themselves already accepted in the earlier volumes of the Book."

At the end of the last century, Americans were a bit too independent to believe strongly in a closed stud book. There was, as a result, no central authority or record. William Robertson described the situation in *The History of Thoroughbred Racing in America*,

There was the appalling confusion in the matter of names. Every horse of any pretension to quality had a regiment of namesakes tagging along after him—there were "Januses," "Eclipses," "Partners," "Messengers" and "Travellers" in profusion, of both sexes. The same lack of ingenuity was evident in the selection of more pedestrian titles, and historian Hervey counted 102 different females named "Fanny" this or that, and 139 "Johns," in the first two volumes of the *American Stud Book*.

Figure 4–4.   James Weatherby.

The American Jockey Club took over the publication of the *American Stud Book* in 1896. The concept of "purebred" had developed in the average horseman's mind, and it was generally accepted that a breed had need of a registry. Nearly a dozen breed registries had been initiated in England, including the Clydesdales and the Suffolk. A Holstein registry had been started in Germany in 1886.

Arab horse breeders had for centuries lived with the concept of purebred horses. They called these horses *asile,* which meant that such horses were pure in bloodlines as far back as records were kept. A book published in 1888, *The Horse and His Rider,* tells of the Arab concept of purebred. "The horses of Arabia are divided into two classes, ignoble and noble; the former they call by a name which signifies 'without pedigree;' the latter by another name, which means 'known for two thousand years.' " When Europeans bought horses from the desert, they demanded more proof than the Arab's word that the horse was what it was represented to be. The Arabs would then produce a document similar to the one shown in Figure 4–5—the earliest form of a registration certificate.

As the Arabian horse became more popular in the United States, the early 20th century breeders began to think in terms of an Arabian registry in this country. They struggled with the questions, "What is the role of a registry? Is it to be a club, or society? Should it be a service organization for breeders?"

In those days the common man had little interest in such things; consequently the development of a registry or breed society involved a chosen few breeders, and was set up to be rather dictatorial in its policies.

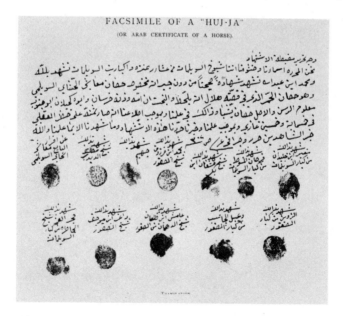

Figure 4–5.   An Arabian horse registration certificate, approximately 100 years old.

The Arabian Horse Club Registry of America, established in 1908, had only a small group as members and had no democratic process whereby future owners of Arabian horses would have a voice in the policies of the club or registry. The style and procedures of the Jockey Club prevailed. It was not until a half-a-century later that registries were conceived as working democracies.

In the past few decades, as registries have matured, the autocratic types have become a bit more democratic in their methods, and many democratic registries have grown so cumbersome that individual members have lost much of their voice in control of the organization. Today several registries are large corporate entities involved in not only registering horses, but also controlling and keeping records on the many activities related to the breed and actively publicizing the breed (Figure 4–6).

The American Holstein Horse Association, Inc., organized in 1978, established a definition of purpose which is fairly characteristic of what registries today consider as their role:

> To establish, maintain, and operate a non-profit association, as specified under the laws of the State of New York, of breeders,

Figure 4–6. The headquarters building for the American Quarter Horse Association, in Amarillo, Texas, the world's largest registry.

owners and friends for the promotion and preservation in the Western Hemisphere of the Holstein Horse; to maintain a public registry of Holstein Horses; to mark or brand approved stallions and mares with the association's Branding Seal; to disseminate information to breeders, owners and friends pertaining to the breeding and raising of Holstein Horses; to promote the performance of the Holstein Horse in Dressage, Three-Day Eventing and Hunting and Jumping, as well as Driving, and generally to do all things appropriate to encourage a public understanding of the Holstein Horse, its breeding and performance.

In the past, registries have generally failed to recognize their all-important influence on the direction of development of the breed. Regrettably, some registries have, for a time, been dominated by a few influential breeders who sought to profit from the way registry policies related to their own breeding program. Some registries have begun with a very noncommittal approach to breed goals. They took the attitude that their role was simply to record; they left it up to each individual breeder to produce whatever he thought was best. As time passed, breeders' efforts were frustrated because of the wide diversity of such goals. This happened, in a way, with the Appaloosa Horse Club. The original concern was only with color. Fortunately today, because of breed standards set by the registry, the Appaloosa is becoming a type rather than just a color. It could not happen without the registry becoming a driving force. Today the Appaloosa Registry has tightened limitations of color patterns; for example, they are ruling out roans.

A registry should serve the breed—the horse. Those who are in control of registry policies have an awesome responsibility to decide first whether the overall goal is to preserve the breed as it is, or to improve it. If improvement is the goal, then a breed standard must be carefully defined, with policies set which will move succeeding generations of the breed toward the standard.

Regardless of whether the goal is to recreate, preserve, or improve, it is important for a registry to establish policies that will guard against genetic deterioration of the breed. A prime example is the Arabian breed. The general overall goal of breeders is to preserve the "as-is" situation. But concentrated inbreeding has increased the frequency of genetic defects in the breed. Of principal concern is C.I.D., which is explained in Chapter 10. As of this writing the Arabian Registry has not responded to the challenge of establishing policies which will reduce and eventually eliminate this genetic defect in the breed. But should this not be one of the prime roles of a registry?

The older registries struggle under archaic organizations and bylaws. Most of these registries began with limited purposes, and today find it difficult to adapt to an expanding role. New registries have the advantage of beginning with more comprehensive purposes and as a result can more effectively direct the breeding toward established goals.

TYPE AND USE VERSUS GOALS

Horses are used for many activities related to both recreation and work. It is logical, therefore, to establish breeding goals in relationship to the type and use of the horses involved. Breeding for speed is quite different from breeding for cutting ability. A breeder seeking to develop the ultimate pleasure-riding horse will be using a different set of criteria in selection than will a breeder wanting to improve the pulling power in his draft horses.

Regardless of the type and use of the horses in a breeding program, some goals will remain the same for all breeders. A breeder must always be on guard against defects of soundness which tend to assert themselves in each generation. Continual elimination of these defects should be included in the goals of every breeder. (For more information, see Chapter 10.)

Goals for Breeding Racehorses

Mother Nature has been selecting horses on the basis of speed throughout the evolution of the creature. Speed has always been the principal mechanism of defense for the horse. During this evolution, feet changed to enhance speed, size increased to improve speed, brain capacity became larger and thus increased coordination for greater speed, and many other subtle changes occurred which made for a faster horse.

Beginning several thousand years ago, man has attempted to bring about more speed in the horse by selective breeding. The Arabs and Turks did a fine job in their day. For the past 300 years breeders have been improving the thoroughbred, but there may be a certain physiological limit to how fast a horse can run. E.P. Cunningham in his paper, "Genetic Studies in Horse Populations," at the 1975 International Symposium on Genetics and Horse Breeding, proposed that speed improvement in the thoroughbred had reached a plateau since about 1910. He charted speed since 1850 by averaging each decade's winning time at the St. Leger, Oaks, and Derby (Figure 4–7). Improvement is evident in the first 60 years, and thereafter winning times have remained relatively constant. Others at the symposium suggested that preoccupation with speed, as such, may have outlived its usefulness, and now in order to make progress, breeders must set their sights specifically on complementary factors.

Speed ability comes from a favorable combination of many traits. With a closed stud book there is no chance of bringing in new traits; thus, developing the ultimate in speed means finding the best combination of all the existing traits that contribute to speed.

When it comes to the racehorse, the term "fast" must be qualified to have meaning. Comparing one horse to another to see which is fastest is not easy. One horse may be fastest at 6 furlongs, while another is fastest at 10 furlongs. One horse may be fastest at 6 furlongs at 2 years of age and slow at 5 years of age. Even with the same age and sex of horses in the same race, a good

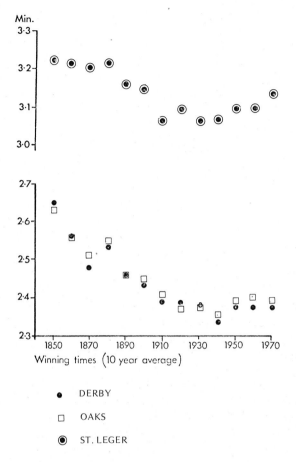

Figure 4–7. Winning times of the British Triple Crown, each point representing a 10-year average. (Data from E.P. Cunningham: Genetic Studies in Horse Populations.)

comparison of ability may not be possible because of the various handicapping weights carried.

Speed has been defined in many ways. Some choose to speak in terms of a certain time over a certain distance, neglecting such things as track conditions and competition. Others speak of speed in terms of a combination of time and handicapping weight carried. Others feel that overall earnings are an accurate measure of speed. Nat Kieffer has described a system called "the performance rate," in his paper "Heritability of Racing Capacity in the Thoroughbred."

In summarizing his feelings about goals, Nat Kieffer recommends:

> a breeding program in which the basic philosophy is breed the "best" to the "best"! This is not to say that conformation of the potential parents should not also be considered, but performance of

the parents should be the dominant principle on which breeding programs are based.

Racehorse breeders have each developed their own formulas and goals. As an example, Greg Fox, of Sparta, Kentucky, once advertised in *The Thoroughbred Record* to assist breeders in setting their goals. He proposed a complicated system composed of (1) conformation factors, (2) pedigree factors, and (3) performance factors. These factors are all covered in detail in Part III of this book.

A racehorse breeder must never forget the importance of environment in producing a fast horse. Early training, proper nutrition, the ability of the jockey, and other factors, all play a part. The 1978 U.S. Triple Crown was captured by Affirmed over his close rival Alydar (Figure 4–8) because of superior heart and determination: just how much of that was due to heredity is impossible to determine.

Federico Tesio, one of the most successful racehorse breeders in history, and author of *Breeding the Racehorse,* described his original goal in breeding: "My aim was to breed and raise a race-horse which, over any distance, could carry the heaviest weight in the shortest time."

Figure 4–8. Affirmed wins the Triple Crown title.

Goals for Breeding Endurance Horses

The biggest problem in breeding for endurance is identifying those traits which bring about endurance ability in a horse. One can guess at some primary traits that contribute to good endurance: a high red blood cell count, a high hemoglobin content of the cells, a large lung capacity, large-sized air passages, proper endocrine balance, strong musculature, fine bone structure, good temperament, adequate size, proper nervous coordination, good tendon strength, and proper functioning of the heart.

Dr. Bill Throgmorton, who is well known in competitive trail riding circles, outlines four general areas of concern when setting goals in breeding endurance horses. The first is movement. It should be free-flowing without wasted action. A horse should not have excessive muscle, or a short, thick muscular body, as this type expends more energy per mile than do lightly muscled horses. The second area of concern is stride. A long stride is very important, allowing a horse to cover more distance in less time. Dr. Throgmorton says that the best competitive trail horses can walk 5 miles an hour.

Conformation is the third area of concern. A good sloping shoulder, a long underline, and a proper slope of the pastern all help bring about a good stride. Straight legs are essential so that there is no interference, forging, cross-firing, or overreaching. The withers and back must be built properly to keep away from soreness. Dr. Throgmorton also stresses the importance of large air passages when analyzing conformation. He says,

> The ability of a horse to take in an adequate amount of air can be evaluated by checking the nostrils and throatlatch and listening to him breathe after exercise. The nostrils should be large and open so as not to restrict air flows. The nose should not be too narrow as this can result in reduction of the internal air passage volume.

He adds,

> The area between the jaws and the throatlatch must not be narrow and restricted. Any restriction of the air passages can be heard after a short, vigorous exercise. If there is any doubt about restriction, then listen to the nose, throat, trachea, and lungs with a stethoscope. Any restriction will cause air turbulence that can be heard.

Temperament is the fourth consideration in breeding a good endurance horse, according to Dr. Throgmorton. A horse must not be the type to fret over being without his buddy, as he will expend extra energy and can seriously hamper efficient travel over rough country.

An interesting genetic question is "Does a good performance record equal good breeding potential?" The general feeling among horsemen is yes, if all else is equal. However, sometimes a good breeding prospect cannot compete well himself. In other words, aside from the goal of producing winning endurance horses, such as the great Kamazar (Figure 4–9), the goal

Figure 4–9.   Bill Chambers and Kamazar, first annual National Champion Competitive Trail Horse.

of the breeder should center around the production of good movement, good conformation, and the best temperament.

Goals for Breeding Stock Horses

As for most types of performance horses the best cutting horses often come from a long line of good cutting horses. As an example, Sonita's Last was sired by World Champion cutting horse, Peppy San. Her dam, Sonita Queen, is the producer of another great cutting mare, Moria Girl.

A stock horse must be intelligent and agile, with endurance and stamina. The relative emphasis to give each trait in setting a breeding goal is debatable, but "cow sense" is of extreme value. It will be necessary to rate intelligence subjectively, perhaps by a score of 1 to 10, with the higher figures denoting superiority. Horses can be put through workouts early to estimate agility and endurance. In general, a horse must have sound conformation to hold up over years of activity in working with livestock. Temperament will be important because the horse cooperates closely with man in working livestock. Size can be an important consideration if a stock horse is to be used for heavy roping.

Breeding goals for the stock horse, then, must center around those traits which the breeder considers most important. Temperament, cow sense, agility, and size should all be important.

The author of *Training and Riding the Cutting Horse,* Dean Sage, described goals in breeding cutting horses:

> He should be a horse which is not too large and, above all, one which moves easily and does not show a tendency toward clumsiness. A short back and fairly close coupling will indicate agility. Hind legs that set just a little underneath the body are desirable for a cutting horse, since they will help him work off his hocks. Forelegs that set close together will make for limber supple action, while forelegs set apart will generally mean a shorter choppier type of action. . . both types have their merit. A good neck is important, for a horse uses his head and neck to aid his balance. His neck should have length and should be set on an angle with his shoulders that permits an easy, low head carriage. Too short a neck will make for awkward balance, and too upright a one will produce high head carriage.
>
> Action is the key to performance in a cutting horse. A horse can be taught what to do, but he must have the physical ability to do it. The points of conformation given above are merely indicative of possession of this ability. They are not absolute criteria, and they will most certainly have exceptions, for if there is one thing certain about horses, it is their uncertainty and the futility of trying to lay down hard and fast rules about them. The inclination to play is an important element in the make-up of a cutting horse. Cow sense it is often called, which implies that the cutting horse has an intuitive understanding of what cattle are going to do. . . (but) the trained cutting horse will demonstrate the same so-called instinct with almost every kind of moving object. It is more reasonable to attribute his propensity to a willingness or inclination to play or move with an animal, once he has been shown what is expected of him.

Goals for Breeding Pleasure Horses

The well-known western horse trainer, Dave Jones, described his idea of the best pleasure horse in his book, *The Western Horse:*

> I think the personal pleasure horse should be a gentle gelding and a good representative of his breed. Conformation of the western horse can be generally described as a pleasing appearance. The head should be refined, with plenty of width between the eyes. The eyes should be large and tranquil, and the ears should be small and sharply pointed. I prefer a neck that is refined and fairly long. The withers should be prominent, but not so high as to make the

purchase of a special saddle necessary. The horse should have a fairly deep girth, and the underline shouldn't show a steep cut up to the flank; that is, the horse should be deep through the loin. Legs should appear sturdy with enough bone to support the horse's weight.

In breeding horses strictly for pleasure, soundness will come a bit further down the list of goals. A new concept in appraising horses is serviceability, in which the uses of a horse are one of the major considerations. Thus a horse which will be used for walk-trot and occasional canter can have a great many more conformation faults than a horse which is to be used for competitive sports. Sometimes the production of traits antagonistic to soundness will be the goal. The Peruvian Paso, for example, has been bred to be "the Cadillac of pleasure horses," and in the course of developing the breed, the trait of winging out with the front feet (Figure 4–10) has been ingrained. Peruvian Paso breeders find this trait desirable, in spite of the fact that winging can lead to leg damage. In *The Peruvian Paso,* Vern Albright describes this quality:

> . . . throwing his forefeet to the outside, swimming exuberantly down the road, wasting his energy. . . Wasting? Well not exactly. There are times in life when one longs for the superfluous. And in this world, dominated by statistics, efficiency, and materialism, it is a joy to see a "Zanero" come dancing down the road, rhythmically

Figure 4–10.   The trait of "termino," or winging out, is desired by Peruvian Paso breeders.

tapping out his beat, glowing with good humor, elegantly performing his useless termino.

In the same respect, the exaggerated knee action of the Tennessee walking horse can be desirable to the pleasure-horse rider. Beauty is in the eye of the beholder. Anything goes as far as goals are concerned in pleasure horses, from the gangly saddle horse to the pony. A good disposition is perhaps the one trait that is important in all types of pleasure saddle horses.

Goals for Breeding Jumping Horses

Jumping competition, 3-day eventing, and over jump races have increased dramatically in worldwide popularity. Horses with good jumping ability are more and more in demand. Many breeders are seriously trying to fill the gap by producing better jumping horses. The question, "How can we measure jumping ability?" was answered by C. Legault in his paper, "Can We Select Systematically for Jumping Ability?" at the International Symposium on Genetics and Horse Breeding. Legault said:

> Jumping is a complex gesture in which the morphology (development and shape of skeletal muscles), nervous influx, and the degree of training of the animals are involved. Moreover, jumping ability can be characterized by notions such as "strength," "style," and especially "regularity." In fact, the horse is essentially supposed to be the faithful, agreeable, and strong partner of the rider.

In his book, *Steeplechasing,* John Hislop writes:

> In the breeding of jumpers, conformation, size and substance is a far more important aspect than pedigree. A badly made weed will

Figure 4–11.   Seamus Hayes, a believer in the "Italian System" of show jumping.

sometimes win a good race on the flat, but it will never carry 12 stone over fences—or even hurdles—and it is a waste of time and money to try and produce good jumpers from anything except mares and sires of reasonably good bone and build.

The well-known jumping competitor, Seamus Hayes (Figure 4–11), is quoted as saying:

Intelligence is very important in a horse that is going to make a show jumper. If you are going to put your trust in him, and let him make up his own mind when he's jumping, you have to believe he has some sort of a brain in his head. Horses must like jumping—they can't be forced.

Goals for Breeding Draft Horses

Breeding draft horses has made a comeback in recent years. Draft horse numbers and prices are increasing substantially. As always, some breeders look to the show ring to point the direction of their breeding program, but those who can take a lesson from the past realize the folly in this action and concentrate on producing horses with the best pulling power. A British breeder of the 1890s, Archibald M'Meilage, is quoted in *The Shire Horse:*

Aiming at the development of qualities which are enduring rather than temporary, in some localities too much anxiety to produce fancy animals with exaggerated showyard points may have led to the neglect of more solid and enduring excellencies.

Back in the days when the tractor was just beginning to overtake the horse in pulling supremacy, Iowa A&M College conducted some research and published a pamphlet, *Testing Draft Horses*. A quote from that publication should help the modern breeder to set some goals:

The ability of horses to pull depends upon their strength, and the available footing. The ability of the driver, however, the disposition, and training of the horses are all important in getting them to put forth a maximum effort. It is possible for a horse well trained to exert a tractive pull equal to his weight. Sore shoulders, more than fatigue or loss of weight, limited the tractive pull of the teams in these experiments. The conformation of the shoulder, the way the collar fits and the character of the work may be the limiting factors in a horse's ability to work.

Figure 4–12 shows a team of horses weighing 2305 pounds starting a load of 2000 pounds. The tremendous strain on the hocks suggests that strong hocks should be an area of special concern to the breeder. The book, *Horse Power,* by Frank Lessiter tells an interesting account of Darwin Stark's attempt to breed the world's heaviest horse. Upon locating a large team, Dick and Tom, in the mountains of West Virginia, Darwin gathered together $10,500 in $100 bills to persuade the logger to sell the team (Figure 4–13). This event took place in 1976. The team weighed a total of 5700 pounds.

Figure 4–12.   Draft horses require strong hocks.

Figure 4–13.   Dick and Tom, a Belgian team, with owner Darwin Starks.

The largest, Tom, stood 20 hands high. After doing some exhibiting of the Belgian team, Darwin eventually sold the smaller horse for $10,000.

BIBLIOGRAPHY

Albright, Verne: The Peruvian Paso: Its Classic Equitation. Ft. Collins, CO, Caballus Publishers, 1974.
Blake, Neil French: The World of Show Jumping. Fort Collins, CO, Caballus Publishers, 1975.
Chivers, Keith: The Shire Horse. London, J.A. Allen & Co., 1976.
Denhardt, Robert: Quarter Horses: A Story of Two Centuries. Norman, OK, University of Oklahoma Press, 1967.
Denhardt, Robert: The King Ranch Quarter Horses. Norman, OK, The University of Oklahoma Press, 1970.
Hislop, John: Steeplechasing. London, J.A. Allen & Co., 1970.
Jones, Dave: The Western Horse. Norman, OK, University of Oklahoma Press, 1974.
Kieffer, Nat: Heritability of Racing Capacity in the Thoroughbred. Proc. Int. Symposium on Genetics and Horse-Breeding, Dublin, Ireland, 1975.
Lessiter, Frank: Horse Power. Milwaukee, Reiman Publications, 1955.
Mortimer, Roger: The Jockey Club. London, Cassell, 1958.
Napier, Miles: Breeding a Racehorse. London, J.A. Allen & Co., 1975.
Proc. Symposium on Genetics and Horse Breeding, 1975.
Robertson, William: The History of Thoroughbred Racing in America. New York, Bonanza Press, 1964.
Sage, Dean: Training and Riding the Cutting Horse. Colorado Springs, Western Horsemen, 1965.
Solá, Gloria A.: Environmental Factors Affecting the Speed of Pacing Horses. M.S. Thesis, Ohio State University, Columbus, OH, 1969.
Tesio, Federico: Breeding the Racehorse. London, J.A. Allen & Co., 1958.
Tweedie, W.: Major-General (reprinted from 1894 edition). Alhambra, CA, Bordon Publishing Co., 1961.
Wall, John F.: Breeding Thoroughbreds. New York, Scribners Sons, 1946.
Willett, Peter: The Thoroughbred. New York, G.P. Putnam's Sons, 1970.
Wynmaker, Henry: Breeding and Stud Management. London, J.A. Allen & Co., 1950.

5

heredity versus environment

Heredity does not work in a vacuum. It expresses itself only within the proper environment. Under varying environmental conditions, succeeding equine generations tend to adapt. Horses have evolved and continue to proliferate upon this planet because of their adaptability. It allows them to survive the rigors of a severe winter as well as the extreme temperatures of a hot summer.

A simplified way of explaining adaptability in terms of genetics is to say that not all of the genes are functioning at any one time in the life of an animal. Most genes have a biochemical switch which may turn them on or off. It is known that other genes may act as the trigger for the switch, which is known as gene interaction. Often, certain factors in the environment, such as sunlight or temperature, may act as the trigger for the switch, leading to adaptability.

This interaction produces a winner of the Triple Crown or a World Champion cutting horse. Heredity provides the propensity to perform at a maximum level, and an excellent trainer provides all the necessary environmental factors, including training. The combination makes a champion.

This blend is one reason why horse breeding is more an art than a science. It takes extreme insight to maintain the proper balance of all factors to produce the best product. Horse breeding is much like cooking. Heredity is the flour, lard, salt, and other ingredients. The cook puts these together in the proper amounts, mixes them in the right way, and cooks them in a correct manner. The result can be terrible, mediocre, or excellent. This lack of predictability is also true of the production of champion horses.

NATURAL SELECTION

Mutation is the basis of evolution and of progress in horse breeding. Details of the nature of genes and mutation are found in Chapter 6. At this point it is adequate to understand that a gene is a large molecule within the cells of the body. The structure of the molecule will occasionally change, resulting in a mutation.

Mutations bring about subtle physiological differences, or drastic outward

Figure 5–1.   A mutant causing excessive mane and tail growth (above); a mutant causing excessive hoof growth (below).

change (Figure 5–1). Over 99 percent of the mutations which have been studied from all forms of life have been found to be harmful, that is, harmful to the animal in the environment in which it finds itself. As the environment changes, or varies, these "harmful" mutations may, in some instances, foster adaptability to the new environment.

There is probably more chance, in the equine, for adaptation than can ever be realized. When the species remains a free interbreeding unit, a somewhat uniform direction of evolutionary adaptation occurs along the lines of one of the potentialities. When a small group becomes isolated from the others, some of the potentialities are unavailable in their pool of genes. Adaptation will then proceed along a line of potentiality which is different from that of the parent group. This probably happened during the ice age when the then-existing equine type became divided into isolated groups. As a consequence the tarpan developed in one area, the cold-blooded, heavy horse developed in another area, and the hot-blooded horse in still another area. In more recent times (about 15,000 B.C.) a small group of tarpan horses were isolated as the sea rose and created the present British Isles. As the environment began to differ for the isolated group and the parent group, the

Table 5–1.   Rate of Mutation of Genes of Fruit Flies and Man.

Gene	Rate of Mutation
Fruit Flies	
Y	1/3,873 (irradiated)
sc	1/1,936 (irradiated)
w^a	0/11,620 (irradiated)
ec	1/645 (irradiated)
ct	0/11,620 (irradiated)
v	1/1,936 (irradiated)
m	1/3,873 (irradiated)
q	1/1,936 (irradiated)
f	1/1,936 (irradiated)
car	0/11,620 (irradiated)
Human	
Pelgar	1/12,500
CS Dwarf	1/14,300
Ret Blast	1/43,500
Aniridia	1/200,000
Epiloia	1/83,333
Albinism	1/35,700
A. idiocy	1/90,909
Ichthyosis	1/90,909
Color blind	1/35,700
Hemophilia	1/31,250

Modified from A.M. Winchester: Genetics. Boston, Houghton Mifflin Co., 1966 genetic and physical makeup of the two groups became different. This process is known as natural selection.

The rate at which natural selection can occur is determined by the rate of mutation. There is no information available regarding the mutation rate of various genes in the horse. However, it is known that genes are much alike, from fruit flies to man, and often generalities can apply to all animal types. Research reported by Winchester gives the mutation rate for 10 genes of the fruit fly as well as 10 genes of man (Table 5–1). Some genes mutate more frequently than others, and the mutation rate can be increased with the use of irradiation, as was the case with the fruit flies. The 10 fruit fly genes mutate, on the average, at the rate of 1 in 2420 flies. Without radiation the 10 human genes would mutate at a slower rate. On the average, this was 1 in 63,810 individuals. If the human average can be applied to the horse, each horse would have one mutation for each 63,810 genes in his genetic makeup. Given a specific gene, a mutation in that gene will occur, on the average, once in 63,810 horses. It becomes evident that the greater the horse population, the greater will be the effects of natural selection.

NUTRITION AND HEREDITY

Nutrition is one of the major factors in the environment of a horse. It has a drastic effect on phenotype, or physical appearance, which is brought about by a physiological interaction with various genetic factors. Nutrition does not affect the genotype, or the genetic nature, of a horse, but certainly plays an important role in bringing out a particular genetic potential.

The drastic effect that nutrition has on phenotype may lead one to assume that poor nutrition for long periods of time over many generations brings about eventual changes in genotype. Often, this is true, but it occurs in an indirect manner. Poor nutrition alters selection but in no way changes the genes that will be passed on to the next generation.

Many inherited characteristics express themselves only in an environment of good nutrition. For example, it is difficult for a horse to become interested in cutting or any other performance feat when his whole system is under stress from poor nutrition. The performance of a horse is closely related to his mood or temperament, which in turn is closely related to how well he feels.

Poor nutrition can mask many outstanding abilities of a horse. In racing, for instance, speed and endurance depend upon a number of physiological factors in the body. One of these is a large supply of red blood cells to carry oxygen to the tissues. An undernourished horse is not capable of producing all of the cells necessary, and often nutritional anemias occur.

A deficiency or imbalance of minerals in the diet of a racehorse can cause him to break down with pulled ligaments, fractured sesamoids, and other difficulties. As a result these horses never make a name for themselves even though they may have had the potential of Man-O-War. When selection is

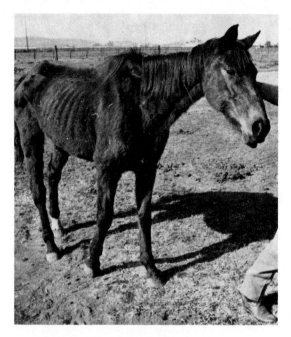

Figure 5–2.   A worm-infested horse of good breeding may be culled from a breeding program because his true potential cannot be predicted.

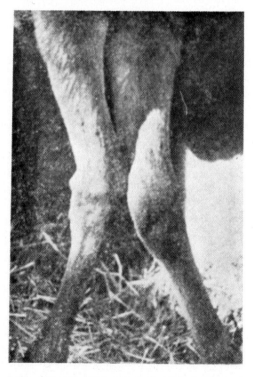

Figure 5–3.   Poor nutrition during gestation may produce a foal with crooked legs. (Courtesy of American Quarter Horse Association.)

for speed, the criterion is winning; and a horse with a cracked sesamoid cannot win. He is, therefore, effectively culled from the breeding program regardless of how great his genetic makeup is.

Internal parasites and chronic disease can drastically affect the performance of a foal. A potbellied runt has a poor chance of ever showing his genetic potential and as a result will not be selected as breeding stock. A colt that is well bred but heavily infested with parasites (Figure 5–2) may be culled unnecessarily.

Since a horse is no better than the legs that are under him, straight legs and feet are usually an important selection criterion. The nutrition of the mare during gestation can have a tremendous effect on straightness of the legs and feet of the foal (Figure 5–3). It is certainly true that much crookedness is genetic in origin and good nutrition would not have improved the situation. Still, in order to determine which kind of crookedness one is dealing with, good nutrition for all of the horses in the breeding program is essential.

On the other side of the coin, poor nutrition may sometimes bring out the expression of a partial lethal gene, which might never have been discovered under ideal nutritional conditions. For example, some horses with an excessive amount of white about the eyes, particularly the sclera, are susceptible to many types of eye diseases. This condition becomes even more evident when the amount of vitamin A in the diet is low. Under conditions of ideal nutrition, however, no eye problems are encountered, and this trait could not therefore be selected against.

It is best to gear the level of nutrition to approximate what it would be under field conditions. If one were breeding cow horses for general ranch use, he should feed his breeding stock at the same level that they would be fed on the ranch. With this procedure, those partial lethals that will express themselves under field conditions can be discovered and selected against in the breeding program.

The way a horse utilizes his nutrition is governed by the anatomy of his gastrointestinal tract. The typical tract is a tube over 100 feet long (Figure 5–4). Modifications of the tube along its length form the esophagus, stomach, small intestine, cecum, large intestine, and rectum. The gut of the horse is intermediate in structure between man and the cow, so it is improper to feed a horse the same type of concentrated diet that humans eat, and by the same token, a horse cannot utilize such a high fiber diet as the cow. The equine has evolved as a grazing animal, functioning best by taking in small amounts of forage throughout the day. Digestion of the food in the stomach takes 10 to 15 minutes, and maximum capacity is 3 to 5 gallons. Understandably he is not able to take in large amounts of food or water at one time. The major amount of digestion occurs as the food passes through the small intestine, which varies in length from 70 to 90 feet.

The modern practice of pushing large amounts of concentrates into our horses brings about many colic problems. More study is needed concerning variations in intestinal structure among the breeds in an effort to explain why

Figure 5–4.   The intestinal tract of the horse. The stomach (st) holds 3 to 5 gallons; the small intestine (d) is 70 to 90 feet, and not all is shown here; the cecum (a) holds 7 to 8 gallons; the large colon (e,f,g) is 20 to 24 feet.

some horses are prone to colic and why some are more efficient in utilization of feed. Proper feeding techniques for the horse are explained in Chapter 20.

Willoughby, in *Growth and Nutrition in the Horse,* provides the chart shown in Table 5–2, which gives optimum chest girth and weight of a growing foal. The diligent horseman, wanting to provide the best nutrition for expression of his foal's true genetic nature, will keep proper growth in line with age.

Willoughby has given the guidelines for maintaining a proper weight to height ratio in growing foals. Dr. D.D. Kronfeld has presented evidence that overfeeding can cause a knuckling over at the fetlocks in foals (Figure 5–5). He also believes that the common practice of pushing foals too rapidly is detrimental to their best potential.

Young horses require 0.6 percent calcium and 0.4 percent phosphorus in their diet. Calcium and phosphorus deficiency, or imbalance, will cause poorly formed and soft bone and tendons, which often begin to bow. Excess dietary phosphorus decreases calcium utilization. Grains contain a higher ratio of phosphorus than calcium, and the subsequent release of calcium from bones causes many problems including lameness.

Vitamin supplements are not necessary as a routine practice if good quality feed is available. Their effect can be dramatic when given to

Table 5–2. Optimum Chest Girth and Weight for a Growing Foal.

Optimum Weekly Gains in Chest Girth (top rows) and Body Weight (bottom rows)
Figures apply to all breeds and both sexes.

Age Interval	Chest Girth, in., at Birth (top row) and Body Weight, lbs., at Birth (bottom row)										Relative Increase	
	22.00 / 40	23.38 / 47	26.20 / 64	28.55 / 81	30.72 / 99	32.54 / 116	34.18 / 133	35.69 / 150	37.09 / 167	38.39 / 184	Chest Girth	Body Weight
Birth– 1 mo.	0.847 / 4.11	0.973 / 5.58	1.146 / 8.48	1.283 / 11.37	1.386 / 14.26	1.476 / 17.15	1.556 / 20.04	1.627 / 22.94	1.693 / 25.83	1.754 / 28.72	1.0000	1.0000
1 mo.– 2 mo.	0.567 / 3.62	0.651 / 4.91	0.767 / 7.46	0.859 / 10.01	0.928 / 12.56	0.989 / 15.11	1.042 / 17.65	1.090 / 20.20	1.134 / 22.74	1.175 / 25.29	0.0086	0.8805
2 mo.– 3 mo.	0.427 / 3.13	0.491 / 4.25	0.577 / 6.45	0.647 / 8.65	0.699 / 10.85	0.745 / 13.05	0.785 / 15.25	0.822 / 17.45	0.855 / 19.65	0.886 / 21.85	0.5046	0.7609
3 mo.– 4 mo.	0.313 / 2.65	0.360 / 3.60	0.424 / 5.46	0.475 / 7.33	0.513 / 9.19	0.546 / 11.06	0.575 / 12.93	0.603 / 14.79	0.628 / 16.66	0.651 / 18.53	0.3693	0.6446
4 mo.– 5 mo.	0.255 / 2.35	0.294 / 3.19	0.345 / 4.84	0.387 / 6.50	0.418 / 8.16	0.445 / 9.82	0.468 / 11.47	0.491 / 13.12	0.511 / 14.77	0.529 / 16.42	0.3016	0.5719
5 mo.– 6 mo.	0.216 / 2.17	0.248 / 2.94	0.292 / 4.46	0.328 / 5.98	0.354 / 7.50	0.377 / 9.03	0.397 / 10.55	0.417 / 12.08	0.434 / 13.60	0.450 / 15.12	0.2557	0.5267
6 mo.– 7 mo.	0.197 / 2.06	0.226 / 2.79	0.267 / 4.23	0.299 / 5.67	0.323 / 7.12	0.344 / 8.56	0.362 / 10.01	0.380 / 11.45	0.396 / 12.90	0.412 / 14.35	0.2328	0.4992
7 mo.– 8 mo.	0.188 / 1.99	0.215 / 2.70	0.255 / 4.10	0.286 / 5.50	0.309 / 6.89	0.329 / 8.29	0.347 / 9.68	0.364 / 11.08	0.379 / 12.47	0.393 / 13.87	0.2220	0.4830
8 mo.– 9 mo.	0.180 / 1.94	0.206 / 2.63	0.244 / 4.00	0.273 / 5.36	0.295 / 6.73	0.314 / 8.09	0.331 / 9.46	0.348 / 10.82	0.362 / 12.19	0.377 / 13.55	0.2131	0.4717
9 mo.–10 mo.	0.174 / 1.90	0.199 / 2.57	0.236 / 3.91	0.264 / 5.25	0.285 / 6.59	0.303 / 7.92	0.319 / 9.26	0.335 / 10.60	0.350 / 11.94	0.364 / 13.27	0.2053	0.4620
10 mo.–11 mo.	0.169 / 1.86	0.193 / 2.52	0.229 / 3.83	0.256 / 5.14	0.276 / 6.45	0.294 / 7.76	0.310 / 9.07	0.325 / 10.38	0.339 / 11.69	0.352 / 13.00	0.1990	0.4523

	300	400	600	800	1000	1200	1400	1600	1800	2000		
11 mo.–12 mo.	0.164 / 1.82	0.188 / 2.47	0.222 / 3.75	0.249 / 5.03	0.269 / 6.31	0.287 / 7.59	0.302 / 8.87	0.316 / 10.15	0.329 / 11.43	0.341 / 12.71	0.1938	0.4425
12 mo.–15 mo.	0.155 / 1.68	0.178 / 2.28	0.210 / 3.46	0.235 / 4.64	0.254 / 5.82	0.271 / 7.00	0.286 / 8.18	0.299 / 9.36	0.311 / 10.54	0.322 / 11.72	0.1827	0.4082
15 mo.–18 mo.	0.125 / 1.37	0.145 / 1.87	0.169 / 2.84	0.189 / 3.81	0.204 / 4.78	0.217 / 5.75	0.229 / 6.72	0.239 / 7.68	0.249 / 8.65	0.258 / 9.62	0.1476	0.3349
18 mo.–21 mo.	0.076 / 1.17	0.087 / 1.58	0.103 / 2.40	0.115 / 3.22	0.124 / 4.04	0.132 / 4.86	0.139 / 5.68	0.145 / 6.50	0.151 / 7.32	0.156 / 8.14	0.0895	0.2833
21 mo.–24 mo.	0.037 / 1.11	0.043 / 1.50	0.050 / 2.27	0.056 / 3.05	0.060 / 3.82	0.064 / 4.60	0.067 / 5.37	0.070 / 6.15	0.073 / 6.93	0.076 / 7.71	0.0443	0.2682
24 mo.–30 mo.	0.030 / 0.93	0.034 / 1.26	0.041 / 1.92	0.045 / 2.57	0.049 / 3.23	0.052 / 3.88	0.055 / 4.53	0.057 / 5.18	0.059 / 5.83	0.061 / 6.48	0.0350	0.2256
30 mo.–36 mo.	0.024 / 0.61	0.027 / 0.84	0.033 / 1.27	0.036 / 1.71	0.039 / 2.14	0.042 / 2.57	0.045 / 3.00	0.047 / 3.43	0.049 / 3.86	0.051 / 4.29	0.0287	0.1497
36 mo.–42 mo.	0.019 / 0.42	0.021 / 0.57	0.026 / 0.87	0.029 / 1.16	0.031 / 1.46	0.033 / 1.75	0.035 / 2.05	0.037 / 2.34	0.039 / 2.63	0.040 / 2.92	0.0220	0.1020
42 mo.–48 mo.	0.014 / 0.24	0.016 / 0.32	0.019 / 0.49	0.021 / 0.66	0.023 / 0.83	0.024 / 1.00	0.025 / 1.17	0.026 / 1.34	0.027 / 1.51	0.028 / 1.68	0.0172	0.0582
48 mo.–54 mo.	0.007 / 0.12	0.008 / 0.16	0.009 / 0.25	0.010 / 0.33	0.011 / 0.42	0.012 / 0.50	0.013 / 0.59	0.014 / 0.67	0.014 / 0.76	0.015 / 0.84	0.0086	0.0293
54 mo.–60 mo.	0.003 / 0.09	0.004 / 0.12	0.004 / 0.18	0.005 / 0.24	0.005 / 0.30	0.005 / 0.36	0.006 / 0.42	0.006 / 0.47	0.006 / 0.53	0.006 / 0.59	0.0032	0.0210
Adult Chest Girth, In.	45.66	50.53	58.23	64.36	69.41	73.75	77.64	81.18	84.43	87.45		
Adult Body Weight, lbs.	300	400	600	800	1000	1200	1400	1600	1800	2000		

NOTE: To derive the total amount to be gained between birth and 60 months (in either chest girth or body weight), multiply each weekly increase from birth to 12 months inclusive by 4⅓; from 12 months to 24 months inclusive by 13; and from 24 months to 60 months inclusive by 26; then add the three products. For example, with a mature body weight of 600 pounds, the birth weight is approximately 64 pounds, which leaves a remainder of 600–64, or 536 pounds, to be gained from birth to 60 months. The 536 pounds is distributed according to the weekly increments listed in the bottom rows in the table under 64 pounds birth weight.

Modified from D.P. Willoughby: Growth and Nutrition in the Horse. A.S. Barnes, 1975.

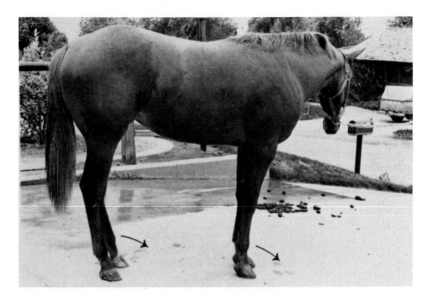

Figure 5–5.   A yearling knuckling over at the fetlocks because of overfeeding, which triggered a mineral imbalance.

debilitated animals. Vitamin A is perhaps one of the most critical. Good green hay will provide adequate vitamin A.

Horses have individual needs, some requiring much more feed to hold their weight than others, which can get by on very little and still remain fat. When temperatures dip, a bit more energy will be required. Those horses on a heavy working schedule need more nutrients than those standing in the paddock. Heavily parasitized horses may require much more feed before they have been wormed.

FERTILITY AS A SELECTIVE FACTOR

Although a good mare may be kept in the broodmare band, she is effectively culled from the breeding program if she does not produce foals. This elementary deduction is all too frequently overlooked. Carrying the same thought further, it can be concluded that poor mares producing at 100 percent capacity would have more effect on the breeding program than an excellent mare producing at 50 percent capacity. Fertility, therefore, is an important consideration in a breeding program, although environmental factors are the principal causes of infertility.

Fertility in horses is not an all-or-none condition; levels of fertility ranging from complete sterility to high fertility exist. In the horse, where the birth of only one young is the rule, expressions of fertility level are found in variations in length of heat and in the estrous cycle, ease of settling in the mare when bred, and semen quality and willingness and ability to serve in the stallion.

Genetic Fertility

Fertility may be affected by inheritance in three general ways. First, lethals can cause death of the embryo or make the germ cells inviable or nonfertilizable and cause reductions in fertility. Some mares will breed and show no heat periods for some time, after which they start cycling again. This history suggests that they became pregnant but that the embryo died and was resorbed, after which the mare came back into heat.

Second, inherited endocrine disturbances, particularly disturbances of the pituitary and thyroid glands, will usually result in impaired fertility. The testes of the male and ovaries of the female are under the control of gonadotropic (follicle-stimulating and luteinizing) hormones from the anterior pituitary gland. The response of the testes to hormones from the anterior pituitary gland is influenced by secretions of thyroxine from the thyroid gland.

Third, fertility is affected by heritable traits in which many genes (called polygenes), each with a small effect, influence fertility or the components of fertility. In general, fertility is a lowly heritable trait in all farm mammals, and horses are no exception.

It appears that multiple ovulations occur rather frequently, 3 to 15 percent in horses and even in the mare mule. Brelanski observed ovulation in both ovaries 4 percent of the time; however, twinning is much less frequent than would be expected from the number of ovulations. Generally, twinning in horses is less than 1 percent, although Blakeslee and Hudson observed 3.2 percent twinning in draft mares.

Rossdale believes that twinning is the leading cause of early embryonic death in the equine. The great majority of twins are either aborted or born prematurely. Shortly after the fourth month one twin of a pair often dies, and this situation may lead to the death of the other.

A Russian research report claimed 199 aborted fetuses from 266 twin pregnancies. Of these, only 16 produced two live foals, and 13 produced a single live foal (or at least only one that could be reared).

There is evidence that the twinning tendency is under genetic control because twinning appears to occur in certain families. Robertson studied a mare that foaled a horse and a mule from two breedings in the same heat, one to a stallion and the other to a jack. This mare produced two other sets of twins and her half sister also produced twins. That all the twins were dizygotic (result of fertilization of two eggs produced by the mares) suggests a genetic tendency for multiple ovulations.

Identical twins are different from fraternal twins. Identical twins are always of the same sex and show identical inherited characteristics (Figure 5–6). In formation of identical twins, cells of the very early embryo separate,

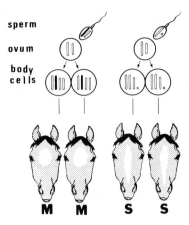

Figure 5–6.   Identical twins are formed from one sperm and one ovum. The sex is always identical. Body spotting varies only slightly.

forming a second identical embryo. Both of these come from the same sperm and same ovum. Fraternal twins may be of opposite sexes and show variations in inherited traits (Figure 5–7). Two separate ova come from the ovary to form fraternal twins, and they do not contain identical sets of genes. They are also fertilized by separate sperm, so they are no more alike than litter mates of a pig or dog.

Fertility, which is generally low in horses, is influenced by age, breed, and lactation status, according to a number of researchers. Speilman reported 60 percent of range mares bred (47 percent of services) foaled. Some stallions had much higher and others much lower than the average ability to settle mares. Very young and old mares were lower in fertility than mares in their

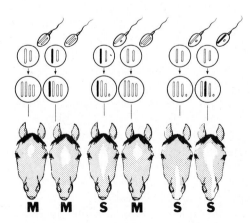

Figure 5–7.   Fraternal twins are a result of separate ovum and sperm. Genetically they are no different from other full sibs. Sexes of fraternal twins can be the same or different.

peak of reproduction. They observed only one set of twins among 567 foals. Higher fertility was observed by Hutoon and Meacham, whose study included 1876 mare years on 14 horse farms. They reported 80.1 percent conception, 73.8 percent foaling, and 70.8 percent weaned of mares bred. Conception was 79.5 percent in lactating, 72.5 percent in barren, and 78.6 percent in maiden mares. Fertility increased up to 9 years of age; then a plateau was established, after which fertility declined at 15 years of age. Twinning occurred in 1.1 percent of the births, and 0.6 percent of the mares never showed heat.

Environmental Infertility

It is generally recognized by good horse breeders that overfatness may interfere with fertility of both mares and stallions. Also, animals that are quite thin due to lack of feed, improper diet, or heavy parasitism are usually low in fertility. Since the heritability of fertility is low, one would not expect to make rapid improvement in fertility of horses by selection. Because fertility is very important, one should do everything possible to create an environment that will encourage a high level of fertility. Animals that are increasing from moderate state of flesh because of improvement in nutrition are usually higher in fertility than animals that are overfat, thin, or going down in condition.

A leading equine reproductive veterinarian, Dr. John P. Hughes of the University of California Davis, once listed 16 problems that can contribute to a mare's failure to conceive or produce a live foal. Half of these are largely genetic in nature, or at least have a genetic predisposition:

1. Selection of broodmares based on physical ability, rather than reproductive performance.
2. Failure of the corpus luteum to regress.
3. Behavioral anestrus (silent heat).
4. Pneumovagina (wind-sucking).
5. Chromosomal errors leading to reproductive failure.
6. Aging.
7. Immune phenomena which may be involved in infertility.
8. Stallions of questionable or low fertility.

The other eight problems he listed are primarily environmental:

9. The short breeding season.
10. The physiologic breeding season starting 2 to 3 months after man's imposed breeding season.
11. Failure to tease the mare properly and detect estrus.
12. Endometritis or metritis due to infection.
13. Abnormalities of the reproductive tract of a pathologic nature (adhesions, endometrial fibrosis, endometrial cysts, tumors).
14. Nutritional deficiencies.

Table 5–3.   A Detailed Foaling Record Maintained Over the Years Can be Helpful in Differentiating Genetic Problems from Environmental Problems.

Year	Age	Gestation, Days	Del. Time, Min.	Placenta Expelled	Remarks (Foal)	In Foal At	Remarks

15. Abortion.

16. Non-selective breeding of mares at foal heat.

When detailed foaling records are kept on a herd basis over a period of years, it becomes evident that certain mares are prone to specific problems year after year. Many of these can be attributed to genetic variations. Dr. Alan Purvis has kept such records, and reported his results to the A.A.E.P. convention in 1977. Some mares are prone to colic, from mild to severe. Some mares seem to have uterine inertia quite often at foaling. The length of gestation seems to be rather consistent from year to year for each mare, with generally no more than a week's variation. When a detailed record is kept over the years with use of a chart such as the one in Table 5–3, it becomes much easier to determine whether certain problems are idiosyncracies of the mare or related to environmental effects.

INHERITANCE OF GROWTH AND SIZE

Inheritance may affect growth rate in four ways: (1) inherited endocrine disturbances; (2) nutrition-genetic interrelations, (3) inherited enzyme deficiencies, and (4) additive genetic action. The two endocrine glands which are concerned with growth are the pituitary and the thyroid. If the pituitary gland does not produce normal quantities of the growth hormone (somatotropin), the animal will be a dwarf (much undersized). Dwarfism from a pituitary growth hormone deficiency is proportionate; that is, all the parts of the animal grow so slowly that it is small but not out of proportion. Sometimes an abnormal pituitary gland may not function in certain respects, and several of the hormones it normally produces may be either lacking or produced at such low levels as to influence other body functions. Some proportionate dwarfs are also sterile or low in fertility because both the growth hormone and the gonadotropins are not produced in normal amounts.

If the thyroid gland does not produce sufficient thyroxine in young animals, it causes a retardation in growth and results in a dwarf. A dwarf resulting from a deficiency of thyroxine is disproportionate; it has a large head, may be thicker than normal through the shoulders, and is usually shorter legged than normal. Disproportionate dwarfs are usually low in fertility or may even be sterile because thyroxine is needed in all body functions, including reproduction. Thyroid function in horses is much less well understood than that in humans. Therefore, it would be difficult to correct a hypothyroid condition in a foal, even though this result might be accomplished in a young horse by giving thyroxine. Disproportionate dwarfs usually have difficulty breathing and often show a potbellied condition because growth of the intestine is not retarded as much as growth of the body framework; consequently, this type of dwarf has more gut development than it has space. Dwarf horses are being bred as miniature horses and seem to be appreciated by some (Figure 5–8).

Stallions usually grow more rapidly in early life than mares, perhaps because of the stimulatory effect of testosterone (the male hormone produced by the testicles), which increases nitrogen anabolism (the retention of nitrogen by growth of muscular tissue). It has been shown that testosterone given to female cattle will increase their rate of gain and cause them to produce more lean and less fat in the carcass. No information has been found in the literature on the effects of testosterone in horses, but it might be conjectured that muscular development in geldings would increase. Testosterone could adversely affect reproduction if given to mares. If it were given to geldings, it would give them stallion characteristics of behavior and appearance which might be objectionable.

Genetic-nutrition interrelations are most noticeable in their effects on growth rate. Young animals that have an inherited condition in which certain nutrients are required in larger than normal amounts will not grow properly under ordinary conditions. If additional needed nutrients are given, the animal with the abnormal requirement will grow at the normal rate. Since an animal has a different nutritional requirement for maintenance after it is grown than it has for growth, an animal with abnormal nutritional requirements for growth in early life may be quite able to cope with ordinary conditions after it is grown.

Figure 5–8.   A miniature pony from the Bond Miniature Horse Farm.

The work that was done with pigs by McMeekan and Hammond showed that underfeeding in early life interfered with muscular growth; and if such animals were fed well later, they simply stored large amounts of fat tissue. Those who have studied growth in animals find that the order of the growing process is: (1) vital organs such as intestines, heart, lungs, and liver; (2) bones; (3) muscle; and (4) fat. Undernourishment which prevents much fattening is not undesirable; but when it reaches the point of preventing normal muscle growth, it is detrimental and will influence the subsequent usefulness of the animal.

Although the enzyme system of horses has not been well studied, extensive studies in man have found that some individuals inherit the lack of certain enzymes which are needed in the metabolic processes. For example, an enzyme needed for the metabolism of an amino acid, if absent, may lead to grave difficulties if the amino acid is partly metabolized to a harmful product and the metabolic process is discontinued because of the lack of an enzyme. Not only is growth inhibited by inherited enzyme deficiencies, but many abnormal conditions result. If the harmful products of partially metabolized amino acids affect the nervous system, mental disturbances, blindness, and other serious conditions may result.

Growth Heritability

Growth in most animals is one of the most highly heritable traits; that is, growth is generally under additive genetic control. With a high heritability for growth, increased growth rate by selection should be possible. The small Arabian horse, Raffles (Figure 5–9), has been line bred to for many years in America, and the progeny have increased in size through selection.

In many animals, growth of the young is associated with adult body size, so that larger animals produce more rapidly growing offspring. This statement does not necessarily mean that the only way that young animals can attain rapid growth is by increasing the size of the adults. It appears that genes may act at different phases of development; if true, one could select for rapid early growth and smaller adult size and obtain rapidly growing young without large adult size.

In horses, preferred size is extremely varied (Figure 5–10). Large size is not required in children's horses, whereas small size is not desirable in jumping horses. The offspring of small Shetland ponies bred to large draft horses

Figure 5–9.   The Arabian, Raffles, progenitor of many modern Arabians, was very short. Over the years, size in his descendants has been increased through selection.

Figure 5–10.   Variations in the size of horses.

show that horses have great genetic capacity for developing almost any desired size.

Growth can be considered in three stages: (1) prenatal, (2) preweaning, and (3) postweaning. In rabbits there are differences in the rate of cell division between large and small rabbits as early as 4 hours after fertilization, and at 40 hours after fertilization there is a marked increase in cell divisions of fertilized eggs of the large breed over that of the small breed. Two factors could be operative in this difference in cell divisions of large and small rabbits—genetic differences in the eggs and environmental differences in the uterus in which the eggs were developing. Walton and Hammon have demonstrated that uterine environmental influences may occur in horses (see Chapter 8).

The physical differences between breeds of horses (Table 5–4) are largely genetic in nature. The sex differences within a breed, however, are rather significant as these are undoubtedly hormonal in origin. The numbers in Table 5–4 are averages taken from many horses in each breed. These measurements were taken generally 5 to 80 years ago and sizes may have

Table 5–4. Breed Physique Variations.

BREED	Wither Height, inches		Chest Girth, inches		Cannon Girth, inches		Body Weight, pounds		Date of Survey
	Male	Female	Male	Female	Male	Female	Male	Female	
Arabian	59.69	58.96	67.51	66.93	7.35	7.12	901	952	1937
Anglo-Arab	60.02	59.29	69.10	68.46	7.44	7.13	967	996	1937
English T.B.	63.11	—	71.77	—	7.99	—	1131	—	1891
Standardbred	62.80	62.00	70.24	68.50	7.99	7.50	1130	1000	1894
Orloff Trotter	62.05	61.77	68.07	68.94	7.64	7.48	958	958	1930
Quarter Horse	60.00	58.50	73.00	73.10	7.90	7.40	1201	1172	1961
Lipizzan	60.02	59.17	71.26	70.16	8.03	7.64	1102	1015	1935
Gidran	62.09	61.50	71.77	71.40	8.03	7.72	1062	1017	1936
Large Nonius	65.20	65.00	78.35	76.86	8.62	8.24	1360	1372	1936
Small Nonius	62.44	62.32	74.88	73.96	8.23	7.87	1144	1162	1936
East Prussian	63.54	63.46	73.70	76.06	8.27	7.83	1222	1195	1891
Zemaitukai	56.30	55.12	66.54	65.75	7.68	6.93	902	910	1937
Coach	63.74	64.75	73.86	74.50	8.43	8.00	1277	1250	1894
Hanoverian	64.06	64.02	74.76	77.16	8.54	8.03	1296	1288	1912
Holstein	64.09	64.65	75.12	76.93	8.58	8.07	1305	1329	1891
Oldenburg	64.45	64.61	76.97	79.29	9.02	8.50	1398	1400	1891
Percheron	65.48	—	79.87	—	9.66	—	1523	—	1928
Belgian	63.70	64.37	80.47	79.92	9.80	9.21	1526	1634	1891
Clydesdale	64.41	64.37	80.12	79.92	10.47	9.68	1702	1561	1891

Modified from D.P. Willoughby: Growth and Nutrition in the Horse. A.S. Barnes, 1975.

increased in most breeds since then. Improved nutrition as well as selection in breeding may both contribute toward development of larger horses.

Growth Rate

If animals are held back by limitations on amount of feed during the rapidly growing phase but are not damaged by a nutritionally unbalanced diet, they will grow at a high rate when feed is provided. This rapid growth rate following a period in which growth has been retarded by restricted food intake is called compensatory growth. The animal is attempting to compensate for the previous slow growth it experienced. However, an animal that has been held back usually will not match in total growth the one making good continuous gains. Maximum total growth of a foal depends upon continuously rapid gains.

Preweaning gains are highly influenced by the milk production of the mare. Two important aspects to milk production in mares are:

1. Mares must give an adequate supply of milk for the foal at birth so that it can get a good start in life.
2. The mare must give sufficient milk for the foal until it reaches the stage where it can do well on pasture or hay and grain.

If a mare is late in coming to her milk, the foal may be so retarded in early life that it will never do well. In fact, foals that are undernourished in early life are more susceptible to an adverse environment and to disease organisms and are likely to die. Mares must also give sufficient milk for the proper growth of the goal until it is large enough to eat feed, because the foal is completely dependent upon the milk from its mother in early life.

Postnatal growth of horses depends on their nutrition and inheritance. Whichever is the weakest—nutrition or genetics—becomes the more important. When nutrition is adequate, variations in postnatal growth are largely genetic in nature. A show class of yearlings is generally a group of well-fed animals; therefore, differences in size reflect age and genetics. A halter class of a single breed generally comprises animals so similar in breeding that the variation in size can be almost entirely attributed to age differences.

Genetic variations do occur within each horse type (Table 5–5). Some breeds are slower maturing than others. The smaller breeds tend to have a foaling weight that is a higher percentage of the adult weight; that is, the larger breeds must grow faster. The weights in Table 5–5 are optimum weights at each age depending upon birth weight.

Rapid growth is not necessarily superior. Statistics indicate that rapid growth is often associated with a higher early mortality. Some factors of growth are rather constant regardless of breed size. For instance, the body weight at 3 months is approximately half of the 6 months' weight, and thereafter the rate of gain slows.

Table 5–5. Variations in Growth and Weight.

	Birth Weight		Body Weight, lbs., at age of:								
	Lbs.	Percent of Adult Weight	1 mo.	3 mos.	6 mos.	1 yr.	1½ yrs.	2 yrs.	3 yrs.	4 yrs.	5 yrs.
Shetlands	30	15.20	41	60	80	112	138	157	183	194	197
	35	13.60	50	75	101	143	177	202	237	251	256
	40	12.72	59	90	122	175	217	248	291	309	314
	45	12.06	68	105	144	206	256	294	344	366	373
	50	11.59	76	119	165	238	296	340	398	424	432
Larger Ponies	55	11.22	84	133	186	269	336	385	452	481	490
	60	10.94	93	148	207	301	376	431	506	539	549
	65	10.71	102	163	228	332	415	476	560	596	607
	70	10.52	111	178	249	364	455	522	614	653	666
	75	10.36	119	192	270	395	494	568	668	710	724
	80	10.22	128	207	291	426	534	614	722	768	783
	85	10.10	137	222	312	457	573	659	776	825	841
Light Horses	90	10.00	146	237	333	489	613	705	830	883	900
	95	9.91	154	251	354	520	652	750	884	940	958
	100	9.83	163	266	375	552	692	796	938	998	1017
	105	9.76	171	280	396	583	731	841	991	1055	1075
	110	9.70	180	295	418	615	771	887	1045	1113	1134
	115	9.64	189	310	439	646	810	933	1099	1170	1192
	120	9.59	198	325	460	678	850	979	1153	1228	1251
	125	9.54	206	339	481	709	890	1024	1207	1285	1309
	130	9.50	215	354	502	741	930	1070	1261	1343	1368
Medium to Heavy Draft Horses	135	9.46	223	369	523	772	969	1115	1315	1400	1426
	140	9.42	232	384	544	803	1009	1161	1369	1458	1485
	145	9.39	241	398	565	834	1048	1207	1423	1515	1543
	150	9.36	250	413	586	866	1088	1253	1477	1572	1602
	155	9.33	258	427	607	897	1127	1298	1531	1630	1661
	160	9.30	267	442	629	929	1167	1344	1585	1688	1720
	165	9.28	275	456	650	960	1206	1389	1638	1745	1778
	170	9.26	284	471	671	992	1246	1435	1692	1802	1837
	175	9.23	293	486	692	1023	1285	1481	1746	1859	1895
	180	9.21	302	501	713	1054	1325	1527	1800	1917	1954
	185	9.19	310	516	734	1084	1365	1573	1854	1974	2013

Modified from D.P. Willoughby: Growth and Nutrition in the Horse. A.S. Barnes, 1975.

The date for setting age requirements in all halter and performance contests is January 1. Thus, according to the record, a foal born on December 31 is a whole year older than a foal born the following day. In fact, the two might not have been foaled more than minutes apart; but, for contest purposes, they would be in different age groups.

As a result, most breeders try to breed for January foals. As 2-year-olds, these horses have a decided advantage over June foals, owing to the 6 months' difference in their ages. Breeders raising horses for racing as 2-year-olds especially realize the importance of this fact. A June foal is just not mature enough to race as a 2-year-old. For that matter, many horsemen have come to realize that even the January colt is not mature enough at 2 for any extensive amount of racing training.

Variation in the size of weanlings among genetically similar animals is basically a reflection of the milk-producing ability of the dam. As yearlings, the variation is principally a reflection of the level of postweaning nutrition. Those weanlings that are grained heavily will grow faster, while those that have a lower state of nutrition, but entirely adequate, will not generally catch up until they are 2 or 3 years old.

Growth may be considered as the development or production of muscular tissue. If horses are materially restricted in food during the rapidly growing phases, either preweaning or postweaning, muscular development will be impeded. Although the animals will show some compensatory growth when good feed conditions follow adverse periods, they never develop as they would have done had they been allowed normal continuous growth. Thus, restrictions in feed during the growing period can interfere with normal muscle development if the horse is severely restricted for a considerable period of time. Short periods of food restrictions or only minor restrictions in food intake do not materially interfere with muscle development.

MAN AND MOTHER NATURE

Without man, the horse evolved into a creature extremely fit to survive in the wild. Today the extremes of horse types developed by man could not survive in the wild. Man seems to have a different set of criteria for selection than Mother Nature. Man changes the environment and selects for features best suited to his needs. As an example, a pair of Shetland ponies as they appeared in the wild is shown in the upper portion of Figure 5–11. The change brought about in the Shetland ponies is shown in the lower portion of the same figure.

As discussed in Chapter 13, there is a continual selection pressure from Mother Nature. Animals that cannot survive in their environment are eliminated. Man comes along and changes the selection pressure, resulting in a new type that can only survive in the environment that man provides. An example is the increase of horses with parrot mouth (the upper teeth do not match the lower). This condition makes it impossible for a horse to

Figure 5–11.   Shetland ponies as they were in the wild (above) and as changed by man (below).

graze, and he and his progeny must forever be dependent upon a supply of hay and grain from man. In the natural selection, horses with this condition would be eliminated.

BIBLIOGRAPHY

Bengtsson, G., and Knudsen, O.: Feed and ovarian activity of trotting mares in training. Cornell Vet. 53:404, 1963.

Blakeslee, L.H., and Hudson, R.S.: Twinning in horses. J. Anim. Sci. 1:118–155, 1942.

Castle, W.E., and Gregory, P.W.: The effects of breed on growth of the embryo in fowls and rabbits. Science 73:680–681, 1931.

Cunningham, Kirby: A Study of Growth and Development in the Quarter Horse. Bull.. No. 546, Louisiana State Univ. Agric. Exper. Sta., Nov. 1961.

Evans, J.W., et al.: The Horse. San Francisco, W.H. Freeman & Co., 1977.

Gluecksohn-Schoenheimer, S.: The morphological manifestations of a dominant mutation in mice affecting tail and urogenital system. Genetics 28:341–348, 1943.

Green, D.A.: A study of growth rate in thoroughbred foals. Br. Vet. J. 125:539, 1969.

Hammond, J.: Farm Animals—Their Breeding, Growth and Inheritance. New York, Longmans, 1938.

Kibler, H.H., and Brody, S.: Growth and Development. LVII. An Index of Muscular-work Capacity. Mo. Agric. Exp. Sta. Res. Bull. 367, 1943.

Proceedings, American Veterinary Medical Association. Golden, CO, AAEP, 1977.

Rossdale, P.D.: The horse: from conception to maturity. Arcadia, CA, Calif. Thoroughbred Breeders' Association, 1972.

Salensky, W.: Prjevalsky's Horse. London, 1907.

Van Niekerk, C.H., and van Heerden, J.S.: Nutrition and ovarian activity of mares early in the breeding season. J. S. Afr. Vet. Med. Assoc. 43:351, 1972.

Vetulani, T.: Zwei Weitere Quellen zur Frage des Europaischen Waldtapons in Zeitschrift fur Sangetierkunde, Band 8, 1933.

Walton, A., and Hammond, J.: The maternal effects of growth and conformation in shire horses—Shetland pony crosses. Proc. R. Soc. Lond. 125:311–335, 1938.

Werscher, F.: Clarifying the hereditary and environmental relationships in equine mallenders. Tieraerztl. Umsch. 4:318–320, 1949.

Willoughby, David: Growth and Nutrition of the Horse. San Diego, CA, A.S. Barnes & Co., 1975.

part II

equine genetics

6

inheritance in horses

Understanding of inheritance in horses has come a long way in the past century, and especially in the last two decades. Horse breeders have for centuries made attempts to explain their observations about heredity. They often seized upon the first explanation that came to mind without real evidence of the truth. As a result, many false notions about inheritance have been passed by word of mouth among horsemen, and can still be found in older books about horse breeding.

One of the first explanations concerning inheritance was that blood was the genetic determiner. From this thesis came the concept of percentages of blood of certain ancestors, and terms such as "bloodline," "blue blood," "bad blood," and "blood will tell." The British still use the term "blood stock" to denote their purebred racehorses. Science has proven that blood has nothing to do with inheritance—every drop of blood could be drained from a horse and replaced with blood from another horse without changing the genetic nature of the animal.

Horse breeders also have developed the misconception that the age of the sire or dam influences the quality of the foals. Science has found not one shred of evidence that age of parents affects heredity. Another false idea is that if a purebred mare is bred to a mongrel stallion one year, she cannot then ever produce a purebred again. Breeders thought the production of a mongrel foal would somehow change the genetic makeup of the mare. It is known today that this concept is entirely false.

GENES

Today genetics is discussed in terms of individual units of heredity within cells, known as genes. No one knows how many genes are involved in the

137

total genetic makeup of a horse, though somewhere between 10,000 and 60,000 would be a good guess. Many of these genes are important to an animal only during a short period of embryonic development; others continue to play an important role throughout the life of the animal.

A gene is a biochemical formula. It does not itself change or develop into an important part of bodily structure; rather it serves as a genetic key or formula for the exact structure of a specific protein within the living organism. Genes reside within the chromosomes, which are found in the nuclei of cells. From this location they direct the production of the proteins so vital to cellular life. The body of a horse is composed of many types of proteins, and each protein has its corresponding gene. These relationships are explained in greater detail in Chapter 8.

Molecular genetics has almost outgrown the term gene. It originally was coined to signify an entity so basic that changes in its structure or function were not possible. Now we know that the genetic material of a cell is a long strand of chemical DNA (desoxyribonucleic acid). A certain segment along its course, now properly referred to as a locus, roughly corresponds to the old concept of a gene. Each locus directs the production of a specific type of protein (Figure 6–1). The chemical makeup of a locus occasionally is accidentally altered (mutation), producing another variety of the same gene, known as an allele.

DIPLOID TO HAPLOID

Normally the somatic cells or tissue cells of the body contain paired chromosomes, as described in Chapter 8. We speak of the paired condition of the chromosomes in body cells as the diploid condition. Cells are produced in the body by a process known as mitosis, in which one of the cells divides to produce two daughter cells; each of the chromosomes is reduplicated, and in the process each daughter cell gets a complete set of chromosomes.

In the ovaries of the female, and the testes of the male, another type of cell division can take place. Called meiosis, this process differs markedly from mitosis. In meiosis, first the two members of each pair of chromosomes come together in a close union called synapsis. These chromosomes are reduplicated to produce a four-parted body, called a tetrad, for each original pair of chromosomes. Then, two rapid cell divisions take place without further chromosome reduplication. The cells produced now have half the number of chromosomes present in the body cells (Figure 6–2). They have one member of each pair of chromosomes. Thus by meiosis the chromosome number is reduced in each sperm (the male germ cell) or each ovum (the female germ cell). This process of gamete production is schematically outlined for the male and female in Figure 6–3. For simplicity, only one pair of chromosomes is shown.

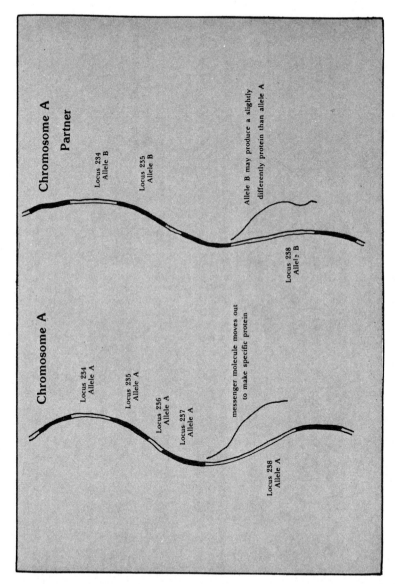

Figure 6–1. DNA makes up a chromosome, which is made up of units known as genes.

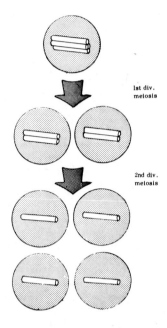

Figure 6–2.   A diagram of meiosis.

In the male each original primary spermatocyte gives rise to four sperm, whereas in the female each primary oocyte gives rise to only one ovum or egg. The reduction in chromosome number from the diploid to the haploid (half) condition in the production of the ovum is accomplished by the formation of the first and second polar bodies, which degenerate. Almost all of the food material goes into the formation of the egg. This added food contained in the egg is important for the survival of the new individual formed when the sperm unites with the egg.

When a mare is in heat, the egg is ovulated or extruded from the ovary; and if the mare is bred, sperm are deposited in the female tract. When a sperm unites with the egg, the paired condition of the chromosomes is re-established and a diploid individual results (Figure 6–4).

The reason for presenting material on what happens to the chromosomes in the production of eggs and sperms and in fertilization is that areas of activity (genes) in the chromosomes govern the development of traits. Thus, the principles of genetics are based on the knowledge that genes are in the chromosomes and that the distribution of the chromosomes in gamete production and fertilization governs the distribution of the genes.

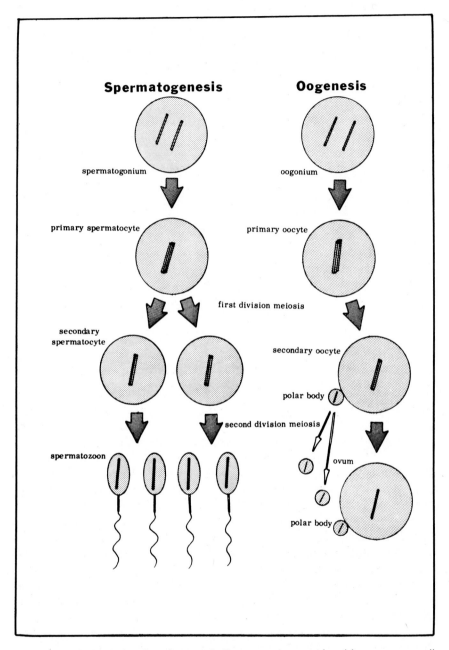

Figure 6–3.   Spermatogenesis and oogenesis. Four gametes are produced from one germ cell in the male, while only one is produced in the female.

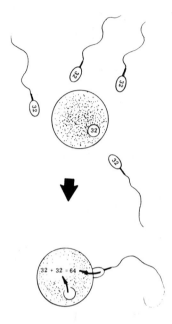

Figure 6–4.   Haploid to diploid in the horse.

SIX FUNDAMENTAL TYPES OF MATING

With only one pair of alleles taken into consideration, only three kinds of zygotes or individuals are possible. For example, when B equals black and b equals chestnut, three kinds of horses can be produced: BB (black), called homozygous black; Bb (black), called heterozygous black; and bb (chestnut [Figure 6–5]). "Zygous" refers to the zygote or individual and "homo" means alike, while "hetero" means unlike. Thus the homozygous individual has genes that are alike, while the heterozygous individual has genes that are unlike.

Six types of matings are possible when the homozygous dominant (BB), the heterozygous (Bb), and the recessive (bb) kinds of individuals are available. These six fundamental types of matings, in which only one pair of alleles is of concern, are called monohybrid (mono means one) matings. The same genes previously used serve for illustration in Figures 6–6 through 6–11. In this case, B, black, is dominant to b, chestnut, and it is assumed that no other genes interfere with their expression. What is presented by black and chestnut in the six types of mating applies to any pair of genes in any organism. The only difference that may be encountered is that in some cases there is no dominance; consequently, the heterozygote expresses an intermediate type between the two homozygotes.

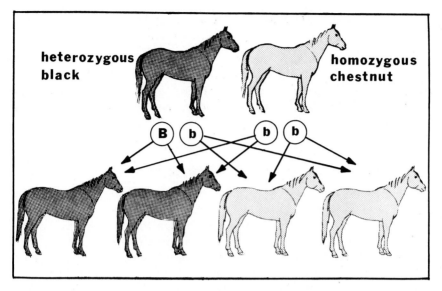

Figure 6–5.   Three types of zygotes are possible when one pair of alleles is considered. Here heterozygous black is mated to homozygous chestnut, producing only two of the types.

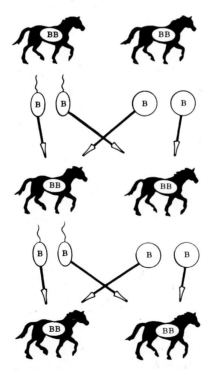

Figure 6–6.   Mating of homozygous dominant to homozygous.

Homozygous Dominant × Homozygous Dominant

Homozygous dominants crossed with homozygous dominants produce all homozygous dominants (Figure 6–6). Each of the homozygous dominants produces only one kind of gamete, the one carrying the dominant gene, whereas the heterozygote produces two kinds of approximately equal numbers. One carries the dominant gene, the other the recessive gene (Figure 6–5).

Homozygous Dominant × Heterozygote

There is equal chance for a gamete with the *B* allele, produced by the homozygous dominant individual, to unite with either of the gametes produced by the heterozygote. Although this type of mating gives homozygous individuals and heterozygous dominants in approximately equal numbers, all of them look alike (Figure 6–7). In this case they are all black. Because of chance in gametic union the exact 1 : 1 ratio would not be likely to occur in practice. The homozygous black horse bred to the heterozygous black mare will also produce all black offspring; however, half of them will be homozygous and half of them will be heterozygous.

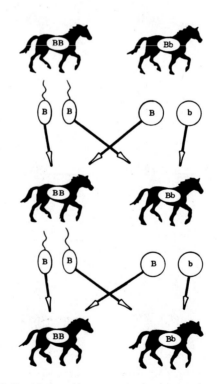

Figure 6–7.   Mating of homozygous dominant to heterozygous.

A herd of entirely black horses could perpetuate a few heterozygous individuals and continue to produce only black colts. As the number of heterozygous individuals in the herd increases, so will the chance increase that two heterozygotes will be bred together and thus possibly produce a color other than black.

Homozygous Dominant × Homozygous Recessive

Both the homozygous dominant and the homozygous recessive individual will produce only one kind of gamete, since both are pure, but naturally one produces gametes carrying the dominant gene, B, while the other produces gametes carrying the recessive gene, b (Figure 6–8). This kind of mating gives all heterozygotes, and all are dominant or black in appearance.

It can be seen that when a homozygous dominant is used for breeding, all its offspring will show the dominant character, regardless of the kind of animal with which it is mated, i.e., another homozygous dominant or a homozygous recessive. This occurrence must be contrasted with results obtained when a heterozygote is mated to another heterozygote, in which case the offspring are not all alike in appearance. This difference in kind of offspring distinguishes homozygous from heterozygous dominant individuals. Such knowledge is the basis of progeny testing, discussed later in this chapter.

Heterozygote × Heterozygote

Each of the heterozygotes produces approximately equal numbers of gametes carrying dominant genes and those carrying recessive genes. There is also equal chance for union of gametes produced by one of the

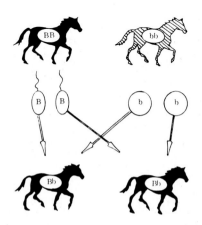

Figure 6–8.   Mating of homozygous dominant to homozygous recessive.

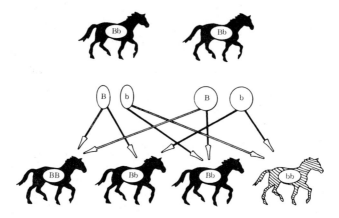

Figure 6–9.   Mating of heterozygous to heterozygous.

heterozygotes with the gametes produced by the other heterozygote (Figure 6–9).

A ratio of 3 black to 1 chestnut is shown by appearance, but genetically the ratio is 1 homozygous dominant (black) to 2 heterozygous dominant (black) to 1 recessive (chestnut). Any group of animals that look alike would be placed in the same phenotype. Any group of animals having the same genetic makeup would be in the same genotype. These ratios, both genotypic and phenotypic, will be only approximately correct with small numbers because of chance in gametic union.

It is possible to produce homozygous offspring, either dominant or recessive, from double heterozygote crosses. Only the recessive homozygote is visibly distinguishable.

Heterozygote × Homozygous Recessive

In this mating the phenotypic and genotypic ratios are both 1 to 1 (Figure 6–10). The phenotypic ratio is 1 black to 1 chestnut, and the genotypic ratio is 1 heterozygote to 1 recessive. This mating should be contrasted with the mating of a homozygous dominant to a recessive. It can be seen, by comparing these situations, why the progeny test of mating a desirable dominant to undesirable recessives can differentiate the homozygous from the heterozygous dominant desirable.

In determining whether a desirable animal is heterozygous or homozygous dominant, a progeny test in which the individual is mated to a known recessive is helpful. Only the heterozygous individual will produce recessive foals from such a mating.

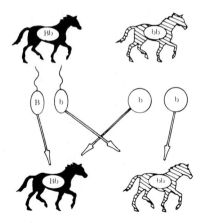

Figure 6–10. Mating of heterozygous to homozygous recessive.

Recessive × Recessive

Since recessives must be homozygous, it follows that they will breed true when mated, as shown in Figure 6–11. Chestnuts bred to chestnuts will produce all chestnuts.

If the six fundamental types of mating are understood, it is easier to comprehend more complicated genetics, in which additional genes are involved.

INDEPENDENT INHERITANCE

The variety of offspring becomes more numerous as the number of genes concerned increases. If these genes are located on separate chromosomes,

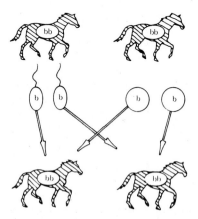

Figure 6–11. Mating of homozygous recessive to homozygous recessive.

the segregation and recombination (which occur in pairs as alleles) are completely independent. If, on the other hand, the genes are found in the same chromosome, they are said to be linked and will not segregate and recombine independently. Linked genes will be considered in Chapter 7. As an example of a dihybrid cross, dominant white spotting, *T* (where *T* equals white spotting and *t* equals solid color) can be considered in addition to the *B* gene. These two pairs of alleles are located in separate chromosomes; consequently they will segregate and recombine independently. The four possible phenotypes are shown in Figure 6–12.

If two double heterozygotes are mated, *BbTt* × *BbTt*, the phenotypic or genotypic ratio can be determined with no difficulty. For example, *Bb*

Figure 6–12.   Four phenotypes can result from a combination of two genes. The top horse is a black tobiano paint, the next is a solid black, below that is a sorrel paint, and the bottom horse is a solid sorrel.

crossed with *Bb* gives us phenotypically 3 white spotted to 1 solid color, with the following total variations in phenotype:

3 black $\begin{cases} \text{3 white spotted} = \text{9 black and white} \\ \\ \text{1 solid color} = \text{3 solid black} \end{cases}$

1 chestnut $\begin{cases} \text{3 white spotted} = \text{3 chestnut and white} \\ \\ \text{1 solid color} = \text{1 solid chestnut} \end{cases}$

The numbers before the phenotypes in the left-hand column must be multiplied by those before the phenotypes in the right-hand column in order to obtain the proper ratio.

In the dihybrid cross (a cross in which two pairs of genes are heterozygous in each of the animals used in making a mating), the offspring occur in four

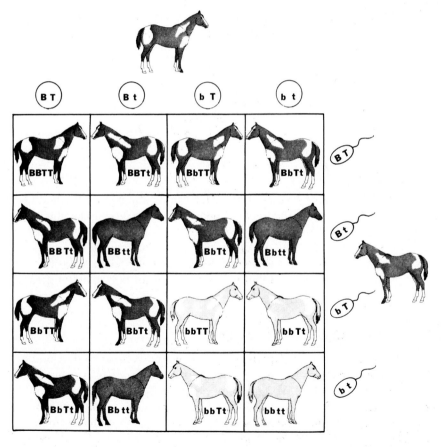

Figure 6–13. A dihybrid cross with black and sorrel spotted horses.

phenotypes: those showing both dominant genes, the largest group; those showing one dominant; and those showing both recessives, the smallest group (Figure 6–13).

Two heterozygous black tobiano horses are bred together. Each produces four different types of gametes. The largest group of progeny shows the expression of the dominant and epistatic gene. (Epistasis is discussed in the following chapter.) In this case, the largest group is composed of the black tobiano. One-fourth of the progeny show the expression of one recessive gene (solid color) and one-fourth show the expression of the other recessive (chestnut). Chances are that only one of these will be solid colored, because three-fourths of them will also carry the tobiano gene. Only one-fourth of the solid colored horses will be chestnut because chestnut is recessive to black.

The genotypic ratio when two double heterozygotes are mated can be shown if it is recalled that *Bb* crossed with *Bb* gives 1 *BB*, 2 *Bb*, and 1 *bb*; and *Tt* crossed with *Tt* gives 1 *TT*, 2 *Tt*, and 1 *tt*.

1 *BB*	1 *TT*	1 *BBTT*
	2 *Tt*	2 *BBTt*
	1 *tt*	1 *BBtt*
2 *Bb*	1 *TT*	2 *BbTT*
	2 *Tt*	4 *BbTt*
	1 *tt*	2 *Bbtt*
1 *bb*	1 *TT*	1 *bbTT*
	2 *Tt*	2 *bbTt*
	1 *tt*	1 *bbtt*

The numbers before the letters in the left-hand column must be multiplied by those before the letters in the right-hand column in order to get the proper ratio. Nine different kinds of genotypes result when mating animals that are double heterozygotes: four different classes of homozygotes, four classes that are homozygous for one pair of genes and heterozygous for the other pair of genes, and one class that is heterozygous for both pairs of genes.

PROBABILITY AND HEREDITY

Half the time, a heterozygous individual would be expected to produce the gamete carrying the recessive gene. If this heterozygote is mated to the recessive and two offspring are produced, the expectancy is that one of the offspring will be recessive. If this is used as a basic concept, it is evident that $\frac{1}{2}^7 = \frac{1}{128}$, or with seven offspring in which a dominant is tested by mating to a recessive and none of the offspring show the dominant trait, the odds of this result occurring by chance alone are less than 1 in 100. However, when a dominant animal produces a recessive offspring, it shows clearly that the dominant animal is heterozygous. It would be impossible for a homozygous dominant animal to produce any recessive offspring.

It is possible to determine the exact chances that a foal will have a particular genetic combination involving a number of genes. As an example, if a gray mare is bred to a black tobiano stallion (Figure 6–14), the chance of a particular phenotype occurring is a product of the chances of all the individual genes involved. The solid black foal will occur from this mating 1 in 32 times. The chances of another black tobiano foal is also 1 in 32, and the chances of a solid buckskin is again 1 in 32.

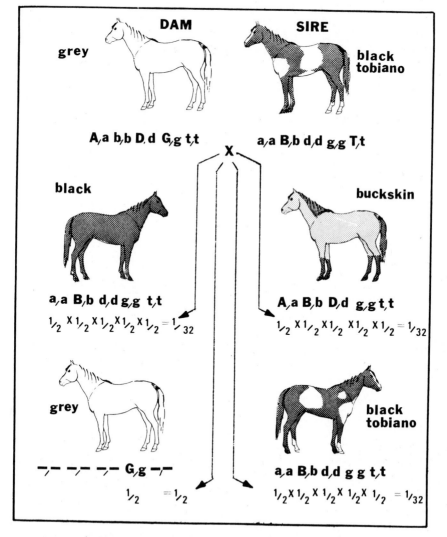

Figure 6–14. A diagram showing the method of determining the probability of various genetic combinations of a particular mating when the genotype of the parents is known.

The exact genotype must be known in order to obtain an accurate determination of the chances of any combination. Often this genotype can be determined if the phenotype of the ancestors is known. The mare in Figure 6–14 was out of a gray mare, but her sire was a very light buckskin. The mare herself was a beautiful little palomino when born. She later turned gray. The *D* gene, which determines the dun color, is in this case assumed to be heterozygous *D*. The mare is heterozygous for gray because the dominant trait was carried by only one parent. Being a palomino at birth indicated that she was heterozygous for the *A* gene (see Chapter 9).

Blood protein typing has become a very exact science (see Chapter 25). Since minute differences in the genes can be detected with modern techniques, it has become possible to describe quite a number of alleles of each gene producing a blood protein. Because genetics is predictable, the

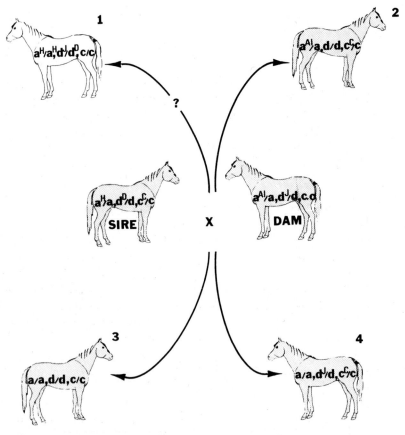

Figure 6–15. A diagram illustrating how the principles of qualitative inheritance can rule out the possibility of an offspring coming from an alleged parent. In this case three genes of blood type are used. Horse number 1 could not have come from the dam shown.

parentage of a horse can be traced by examining the alleles of the various blood protein genes it carries.

The alleles of three blood protein genes are used in Figure 6–15 for both sire and dam. A number of other genes determine blood type in the horse, but only three are used here to illustrate how they are inherited. In the illustration, offspring 2, 3, and 4 have all received one allele of each gene from each parent. Offspring number 1 could not have been out of the dam shown because he has neither allele for the A gene that would be passed to him from his dam. This determination is the basis of blood type identification.

PROGENY TESTING

The long generation time involved with horse breeding makes it necessary to be as efficient as possible in breeding for specific traits in order to make noticeable progress in the breeder's lifetime. Thus, it is important to use homozygous breeding stock whenever possible. Progeny testing, which involves keeping breeding records on planned matings so that the genotypes can be determined, is a good way of determining homozygous breeding stock.

One offspring can give a great deal of information relative to the genotypes of its parents. As an example, a black-gray and white stallion crossed with a black-gray and white mare produces a solid chestnut offspring. From this birth it can be seen that both the stallion and the mare are triple heterozygotes of the genotype $TtGgBb$.

Sometimes it is impossible to mate the dominant animal to the recessive because the recessive type may be sterile or nonviable (lethal). If this is the case, one can use as test animals any dominant that has produced the recessive, since such an animal has proven itself to be heterozygous by producing the recessive offspring. If known heterozygotes are mated to the dominant in question and any recessive offspring are produced, the dominant individual being tested by mating to known heterozygotes is definitely heterozygous. If such matings are made and 10 to 12 offspring are produced with none of them showing the recessive, one can be reasonably certain (odds of 19 to 1) that the dominant animal under test is homozygous. If one recessive is produced by such a mating, it proves that the animal under test is heterozygous. If one prefers odds of 99 to 1 that a dominant is homozygous by the test of mating him to known heterozygotes, then one needs 16 to 18 offspring, none of which show the recessive. Often one can know that a dominant mare is heterozygous because she has produced a foal showing the recessive trait.

A method which can be used for testing a stallion showing a desirable dominant trait to determine whether he is homozygous or heterozygous for this desirable dominant trait is to mate him to several of his female offspring. If a stallion is homozygous for the desirable dominant, all of his offspring produced by mating the stallion back to his daughters will show the dominant trait. The stallion that is heterozygous, on the other hand, even if mated to homozygous mares should produce female offspring half of which would be heterozygous and the other half homozygous.

It should be pointed out that the mating of a sire to his daughters as a means of testing to determine whether he carries a recessive gene is costly because of the large number of offspring necessary to give a reasonable test. However, this mating will test for any and all recessive genes. A stallion tested in this manner that did not sire any foal with undesirable traits would have given reasonable proof that he was free of undesirable recessive genes.

As an example of how to test a stallion for homozygosity for a particular trait, the black color can be used. If a black stallion is bred to a chestnut mare, it would be expected that 50 percent of the time a black foal would result when the stallion is heterozygous. A chestnut mare is always homozygous for the recessive B gene, while a black can be either homozygous, or heterozygous for the B gene with no outward difference in appearance.

When the black stallion in question is bred to a second chestnut mare and she also produces a black foal, the chance that the stallion is homozygous increases to 75 percent. When 10 black foals have been produced from 10 chestnut mares, the chance that the stallion is homozygous for black has increased to 99.9 percent (Table 6–1).

Many people want to obtain an animal possessing a combination of several traits by combining two breeds or individuals each possessing some of the desired traits. It is felt that such a combination and the interbreeding of the offspring should result in a wide array of desired traits in some of the animals. Although this concept is sound, the numbers needed to obtain

Table 6–1.   Progeny Testing to Determine Homozygosity of a Gene. The chance of homozygosity increases with increasing number of offspring. Test breedings are made to homozygous recessives.

Number of Offspring	Chance That Sire Is Homozygous, %
1	50
2	75
3	87.5
4	93.7
5	96.8
6	98.4
7	99.2
8	99.5
9	99.7
10	99.9

animals of this nature must be considered. For example, if two heterozygous animals for black, *Bb*, are mated, a minimum of four offspring is required to obtain all the kinds of genotypes and phenotypes. If two double heterozygotes, *BbTt*, were mated, a minimum of 16 offspring would be required to obtain all the kinds of genotypes and phenotypes. The mathematical equivalent is $(4)^n$ where n equals the number of pairs of genes involved. For example, a mating of heterozygotes where 10 pairs of genes are involved would require a minimum of 1,048,576 offspring to get all the possible kinds of genotypes and phenotypes.

When two heterozygotes, *Bb*, are mated, 1 *BB* to 2 *Bb* to 1 *bb* is the genotypic ratio obtained, and 3 black to 1 chestnut is the phenotypic ratio. Thus, three genotypes and two phenotypes are produced by such a mating. If two pairs of genes are involved, when heterozygotes are mated, nine genotypes and four phenotypes are obtained. These combinations can be expressed mathematically as $(3)^n$ for the number of genotypes that are produced and $(2)^n$ as the number of phenotypes produced (Table 6–2).

It is apparent that with five pairs of genes involved, a minimum of 1024 offspring would be needed to obtain all the possible kinds of genotypes and phenotypes. There would be 243 genotypes produced but only 32 different phenotypes. Since the phenotypes can be seen and not the genotypes, the homozygous genotypes desired cannot be differentiated from those that are heterozygous for the desired phenotypes.

In most matings both parents are not double heterozygotes. In addition, three or four pairs of genes may be involved instead of only two. This situation does not create a difficult problem in determining genotypes from progeny testing if one remembers the six fundamental types of matings and knows how to compute ratios. The following example illustrates how it can be done where P equals spotting, p equals solid color, g equals nongray, G

Table 6–2. The Number of Genotypes and Phenotypes Increases With the Number of Genes Involved.

No. of Pairs of Genes Involved in a Trait	No. of Offspring Required to Determine Ratios	No. of Possible Genotypes	No. of Possible Phenotypes
1	4	3	2
2	16	9	4
3	64	27	8
4	256	81	16
5	1,024	243	32
6	4,096	729	64
7	16,384	2,187	128
8	65,536	6,561	256
9	262,144	19,683	512
10	1,048,576	59,049	1,024

equals gray, *B* equals black, *b* equals chestnut, *T* equals trotter, and *t* equals pacer.

$$PPGgBbtt \times ppggbbTr$$

solid black gray trotter   chestnut and white pacer

all trotters

1 gray
- 1 black
  - 1 white spotted
  - 1 solid color
- 1 chestnut
  - 1 white spotted
  - 1 solid color

1 nongray
- 1 black
  - 1 white spotted
  - 1 solid color
- 1 chestnut
  - 1 white spotted
  - 1 solid color

When the four phenotypes for each group are combined, the result is a ratio of 1 black-gray-and-white trotter to 1 solid black-gray trotter to 1 chestnut-gray and white trotter to 1 solid chestnut gray trotter to 1 black-gray and white trotter to 1 solid black trotter to 1 chestnut-and-white trotter to 1 solid chestnut trotter.

The knowledge of how to work crosses when the genotypes of the individuals used in making the matings are known leads to the next step: determining the genetic makeup of the parents when their phenotypes and the phenotypes of their offspring are available.

A black-gray and white-spotted stallion was bred to a herd of solid chestnut mares and produced:

                15 black-gray and white spotted foals
                16 solid black-gray foals
                15 black-and-white spotted foals
                16 solid black foals

The following facts are known about the stallion and the mares by their phenotypes alone. The black-gray and white spotted stallion is *B-G-P* (P equals pinto) from appearance. The solid chestnut mares are *bbggpp* as known from their phenotype. It is immediately apparent that all the foals were black; as a result it can be assumed with confidence that the stallion was *BB* in genotype. The foals are about in a ratio of 1 gray : 1 nongray, as well as 1 white spotted : 1 solid color. Therefore, we know that the stallion had to be *GgPp*. His complete genotype can now be written as *BBGgPp*. Since the genotype of the stallion and the mares and the phenotype of their offspring are now known, the genotype of the offspring can be determined.

First, what is known about the genotypes of the offspring from their phenotypes should be recorded:

15 black-gray and white spotted	$=B\text{-}G\text{-}P$
16 solid black-gray	$=B\text{-}G\text{-}pp$
15 black-and-white spotted	$=B\text{-}gg\text{-}P$
16 solid black	$=B\text{-}gg\text{-}pp$

Since the dams of these offspring were all recessive for all three pairs of genes, it is necessary to insert only the recessive gene for the blank left for the phenotypes. The genotypes can now be written completely:

BbGgPp
BbGgpp
BbggPp
Bbggpp

BIBLIOGRAPHY

Bogart, Ralph: Improvement of Livestock. New York, The Macmillan Co., 1959.

Evans, J.W., et al.: The Horse. San Francisco, W.H. Freeman & Co., 1977.

Hutt, Frederick G.: Animal Genetics. New York, The Ronald Press, 1964.

Winchester, A.M.: Genetics: A Survey of the Principles of Heredity. Boston, Houghton Mifflin Company, 1966.

Winters, L.M.: Animal Breeding. New York, John Wiley, 1948.

7

genetic complexities

Genetics would be a much simpler science if all were black and white with no grays. This is not the case, however, and the resultant variety of horse types makes life much more interesting. Genes are not always dominant or recessive in their expression. Often they exert only a partial effect. Horses with identical genotypes for body color may sometimes display color differences because of the effect of modifying genes.

The more one learns about genetics, the more one realizes that a gene does not function as an isolated entity; its expression is made possible only because other genes lend support. Basically this support feature is why most genetic traits must be discussed on a quantitative basis rather than simply as qualitative characters.

GENE INTERACTIONS

Gene interaction occurs in several aspects. Genes may interact with their alleles (dominance interactions), with genes in other chromosomes (epistatic interactions), with the cytoplasm of the cell, or with the environment.

The allelic or dominance interaction involves three major types of dominance: complete dominance, lack of dominance (or incomplete dominance), and overdominance. It may be said that each allele at a particular locus of a heterozygous individual is attempting to affect the trait. If dominance is complete, the dominant allele completely overpowers its partner, and the result is that only the effect of the dominant allele is evident. A good example is in the expression of black in a horse that is heterozygous, *Bb*, for black. An example of lack of dominance is the dilution gene. If this gene is acting on *bb* (chestnut) animals in the homozygous condition, it removes virtually all the pigment and results in a "blue-eyed" or "glass-eyed" white, the creamello (Figure 7–1). In the heterozygous, *Dd*, condition, the dilution gene changes the chestnut horse to a dark cream color, the palomino. The recessive condition, *dd*, has no diluting effect, so the *bbdd* horse is chestnut in color.

The various types of dominance can be illustrated with the use of bar graphs if one assigns a numerical value to each genotype. Assuming that aa has a value of 2, then in the case of complete dominance Aa and AA will have a value of 6 (Figure 7–2). Lack of dominance would give values as follows: aa equals 2, Aa equals 4, and AA equals 6 (Figure 7–2b). With overdominance the heterozygote is superior to either homozygote, giving the following values: aa equals 2, Aa equals 6, and AA equals 5.

It should be pointed out that when dominance is not complete, the heterozygote may not be expressed as an intermediate between the two homozygotes; instead, its expression may more nearly approach that of one of the homozygotes.

Overdominance is best illustrated by the hybrid vigor that results when two unrelated highly inbred individuals are mated. It is probably one of the reasons for hybrid vigor and might be illustrated in the following manner: If

bay   **dd**

buckskin   **Dd**

creamello   **DD**

Figure 7–1. The dilution gene (*D*) is a good example of how genes interact because it produces different phenotypes with different genetic backgrounds. It has no effect in its recessive form as shown at the top. The many variations of the *D* gene are discussed in Chapter 10.

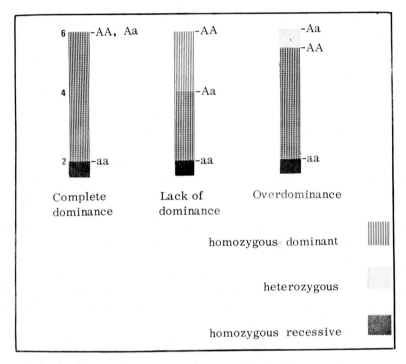

Figure 7–2.   Types of dominance explained with a bar graph.

gene *A* represents extremely rapid growth for a short period of time and *a* represents a lower rate of growth over a longer period, then neither *AA* nor *aa* would have the total gains that would be expressed by the *Aa* genotypes, in which rapid gains are made over a longer period of time. It has also been postulated that the *A* gene might function most effectively under one environment, while *a* functions most effectively under a different environment. With changes usually occurring in the environment, the homozygous animals, *AA* and *aa*, function effectively only part of the time. The heterozygous, *Aa*, individual, on the other hand, can function effectively under differing environmental conditions.

One's point of view, or values, may influence how genes are classified relative to type of dominance. For example, in the production of garden peas the average person prefers small peas to large peas. When peas are crossed, there is a tendency for the hybrids to produce large peas. Vegetable growers express this tendency as negative heterosis, or hybrid vigor. Actually, from a biological standpoint, it is positive heterosis or hybrid vigor, because the larger pea provides greater nourishment for the young pea plant, and thus allows it to start its growth with greater force.

The dilution gene must be considered in light of what is preferred by the breeder. If he is interested in the production of palomino horses, the palomino color is of great importance and would be considered that which is desired. In this case, one would classify the dilution gene as one expressing overdominance effects because the heterozygous condition, *Dd*, is desired. The breeding of palominos together will result in the production of 1 "glass-eyed" (creamello) white to 2 palominos to 1 chestnut. Thus, the selection for the palomino color and intermating of those with it is totally ineffective in the consistent production of palominos. The effective way to produce all palominos would be to mate a "glass-eyed," *DD* stallion to chestnut, *dd,* mares with flaxen mane and tail. This type of mating should give 100 per cent palominos. In this example, one produces what is desired by mating two kinds that are not desired (Figure 7–3).

One needs to understand how the types of dominance influence progress that can be made by selection. If one starts with a low frequency of the *A* gene where dominance is complete, selection will be highly effective initially, but as soon as a frequency of 0.5 of the *A* gene is reached in the population, selection will be relatively ineffective. Progress from selection after the frequency of *A* is 0.5 is made only by preventing the use of *aa* animals for breeding. A population that is homozygous for a completely dominant gene would probably not be developed by phenotypic selection alone.

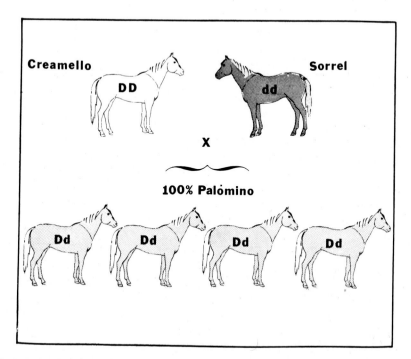

Figure 7–3.   Breeding creamellos and sorrels produces all palominos. Breeders trying to raise palominos should take note that breeding palominos to palominos is not very efficient.

To provide an example of the principle just discussed, it will be assumed that very few black horses exist, and that most horses are chestnut. If black is the preferred color and if a breeding herd were developed from the population of horses that had mostly chestnuts but a few blacks in it, the percentage of black animals in this herd could quickly be increased. However, when the point is reached at which most of the horses are black, but several chestnuts are being produced, it would be difficult to eliminate the b gene and get a population of black horses that are homozygous for black. Where dominance is lacking, phenotypic selection is totally effective until the homozygous condition for the desired trait is reached.

Only a portion of what is sought after in selection is obtained in the next generation. This portion, or fraction of 100 percent, is explained in terms of heritability. With a trait controlled by genes lacking in dominance, selection is highly effective; consequently, such traits have a high heritability. They are governed by additively controlled genes, because every addition of the A gene into the population contributes to improvement.

At very low frequencies, selection will be effective for traits controlled by completely dominant or overdominant genes. Thus, in this situation heritability would be high. When the frequency of a gene approaches 0.5, the

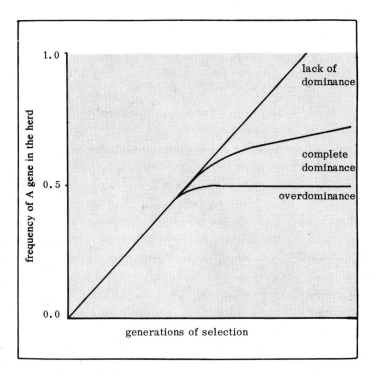

Figure 7–4. The effects of types of dominance on the progress that can be made in changing gene frequencies.

heritability decreases with complete dominance or overdominance, and it becomes zero with overdominance when the frequency of 0.5 is reached. In other words, selection would be completely ineffective. The effects of types of dominance on the progress that can be made in changing gene frequency are illustrated in Figure 7–4.

In non-allelic or epistatic gene interactions, some genes can completely mask or cover up other genes. The gene that covers up the expression of one or more genes in other chromosomes is the epistatic gene. The genes that are prevented from showing by epistatic genes are called hypostatic.

Epistatic genes may be dominant, in which case either the homozygous or the heterozygous condition of the epistatic gene will mask or cover up the expression of other genes. Epistatic genes may be recessive; if so, they must be in the homozygous condition to cover up the expression of other genes. Some epistatic genes will prevent many genes from properly expressing themselves, whereas other epistatic genes affect the genes in only one other locus. An example of an epistatic gene that prevents many genes from proper expression is dwarfism. Animals might have the inheritance for above-normal growth and outstanding conformation; but if they are homozygous for recessive dwarfism, they will simply be dwarfs, and one would never know that the desirable genes for growth and conformation are present.

DIHYBRID CROSSES

A dihybrid mating between two heterozygous bay horses is shown in Figure 7–5. Two horses of the same color produce foals with four different phenotypes. Throughout this book the term sorrel is used to define a much lighter phenotype than chestnut. In some aspects the A gene can be considered epistatic to the B gene, although evidence of the B gene remains with the black mane and tail of the bay. In reality the AA sorrel may be a slightly different phenotype than the Aa sorrel. Only one chestnut of 16 is produced with this cross. A chestnut is recessive for both the A and B genes. Only one of the black horses is homozygous for the black genes; he will be a true-breeding black. See Chapter 9 for a complete description of coat color inheritance.

Appaloosa breeders are often concerned with a practical problem concerning epistatic genes when some of their stock carry the gray (g) gene. They find that the gray blocks out the Appaloosa pattern. An explanation of how this works was put forth by Hatley; he illustrated the masking of the Appaloosa pattern by the gray (Figure 7–6). The two fillies are full sisters both of which received the Appaloosa gene W^{ap}; however, the filly behind also received a gray gene (G) which even at 6 months of age has almost entirely masked the effects of the W^{ap} gene.

Genes are often epistatic in more ways than one. An example is the Lw gene in horses described by Pulos and Hutt (they named it the W gene). It is

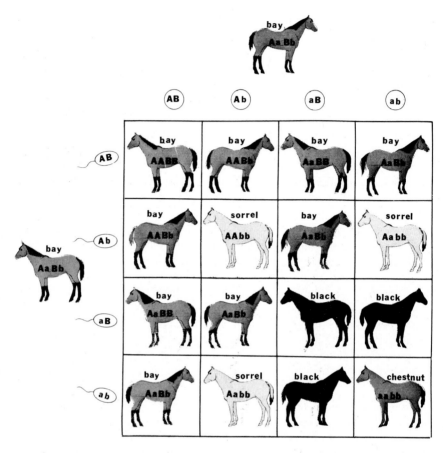

Figure 7–5.   Dihybrid cross in the horse, using the *A* and *B* color genes.

Figure 7–6.   The gray gene is epistatic to Appaloosa coloring. (Courtesy of G.B. Hatley.)

the gene which produces the popular "albino" horse. This gene produces a white horse when heterozygous but is lethal during early embryonic development when it is homozygous. Pure-breeding homozygous progeny, therefore, are not detected because the embryo dies at such an early age that absorption within the uterus is common. To the breeder it may not appear that a lethal gene exists within his herd because he sees no dead foals.

If two heterozygous albinos, both of whom carried a W^{ap} (Appaloosa) gene are mated, the phenotypes of the offspring would be as shown in Figure 7–7. The homozygous Lw zygotes would die in utero, showing that the W^{ap} allele is epistatic to the rest of the color genes. The heterozygous Lw foals are white, and are epistatic to all of the other color genes. The W^{ap} gene is epistatic to the solid color genes, while the solid color genes, A and B, are hypostatic to the other two shown.

The *typical dihybrid* ratio is 9:3:3:1 (Figure 7–5). A dihybrid cross, where only two genes are considered, may produce varying ratios of

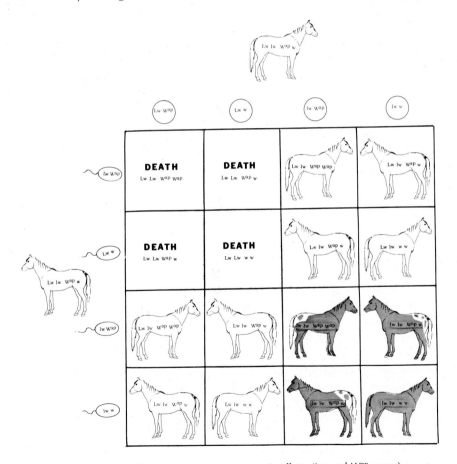

Figure 7–7.   A dihybrid cross showing epistatic effects (Lw and W^{ap} genes).

Table 7–1. Variations of Dihybrid Cross.

Genotypes	AABB	AABb	AaBB	AaBb	AAbb	Aabb	aaBB	aaBb	aabb
A & B both dominant (typical dihybrid)			9			3		3	1
A & B both intermediate	1	2	2	4	1	2	1	2	1
A intermediate, B dominant	3			6	1	2	3		1
aa epistatic to B or b (recessive epistasis)			9			3		4	
A epistatic to B or b (dominant epistasis)				12				3	1
aa epistatic to B or b bb epistatic to A or a (duplicate recessive epistasis)			9				7		
A epistatic to B or B B epistatic to A or a (duplicate dominant epistasis)				15					1

phenotypes depending upon the interaction of the two genes involved. All of the possible variations are shown in Table 7–1.

An example of a dihybrid cross in which *both genes are intermediate* is found in the breeding of palomino horses. Here the bay and brown gene, the alleles being A and A^t respectively, may affect the final coloring of the palomino (see Chapter 10). The A^t allele produces the very dark brown horse when homozygous and when under the influence of a B or Bb genotype. In the example the genetic background is bb so that a homozygous A^t would produce a dark chestnut, a heterozygous A^t (with A) would produce a lighter chestnut, and a homozygous A would produce a sorrel.

Acting in conjunction with the dilution gene, D, the phenotype ratio would be $1:2:2:4:1:2:1:2:1$ (Table 7–1). The D gene here substitutes for the B gene for purposes of correlating all of the examples. With this cross, each genotype produces a separate phenotype. In the example the various genotypes would produce phenotypes as described below:

A^tA^tDD = dark creamello or light grulla (1)
A^tA^tDd = dark palomino (2)
A^tADD = medium creamello (2)
A^tADd = medium palomino (4)
A^tA^tdd = dark chestnut (1)
A^tAdd = chestnut-sorrel (2)
$AADD$ = light creamello (1)
$AADd$ = light palomino (2)
$AAdd$ = sorrel (1)

An example of a dihybrid cross in which *one gene has an intermediate effect and the other is a simple dominant* is the B color gene working with the D gene. Here the D gene substitutes for the A gene used in Table 7–1. The various genotypes of this cross produce a phenotype ratio of 3:6:1:2:3:1, as described below:

$DDBB$ = light grulla $\left.\right\}$ 3
$DDBb$ = light grulla
$DdBB$ = medium grulla $\left.\right\}$ 6
$DdBb$ = medium grulla
$DDbb$ = creamello   1
$Ddbb$ = palomino   2
$ddBB$ = black $\left.\right\}$ 3
$ddBb$ = black
$ddbb$ = chestnut   1

An example of a recessive gene in the homozygous condition covering up the expression of genes in another locus, *recessive epistasis*, is found in paint horses. Here the O allele when homozygous produces a solid white foal (see Chapter 9). The paint horse in Figure 7–8 shows non-white areas revealing a solid-color phenotype. The o^eo^e foal (Figure 7–8) is solid white, masking all evidence of the solid-color phenotype. If two double heterozygous paints are mated, the phenotypic ratio of their offspring would be:

$O\ o^eBb \times O\ o^eBb$

3 paint
O (O or o^e)

3 black × 3   = 9 black paint

1 chestnut × 3   = 3 chestnut paint

1 white foal
o^eo^e

3 black (masked) $\left.\right\}$
                               4 white foals
1 chestnut (masked) $\left.\right\}$

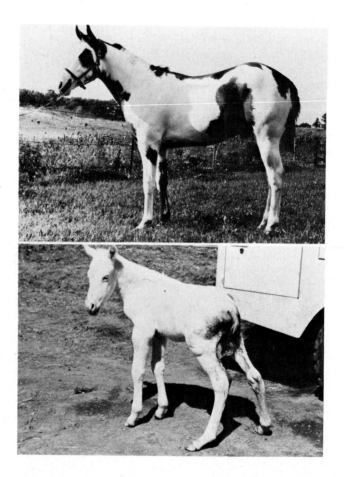

Figure 7–8.   The white overo foal, below, has a coat color which is epistatic to both the paint pattern, shown above, and the solid-color phenotype.

In this case the white phenotype ($o^e o^e$) masks the black and chestnut solid-color so that the 3 and 1 classes become all white, while the color phenotypes are evident when the all-white trait is not present—a $9:3:4$ ratio.

The gray gene working against the Appaloosa gene, explained earlier, is an example of *dominant epistasis*. In this case the two groups of the 3 and 1 are all white. In breeding Appaloosas, the mating of two double heterozygous $GgW^{ap}w$ will be expected to produce the following ratio:

$$GgW^{ap}w \times GgW^{ap}w$$

3 gray
G (G or g)

  3 blanketed
  W^{ap}

  1 non-blanketed
  ww

$\left.\right\} \times 3 = 12$ gray

1 non-gray

  3 blanketed
  W^{ap} = 3 blanketed

  1 non-blanketed
  ww = 1 colored

bay   **dd**

buckskin   **Dd**

creamello   **DD**

Figure 7–9. Gray is epistatic to the blanket pattern.

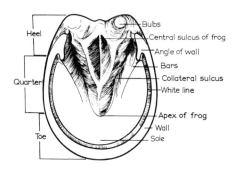

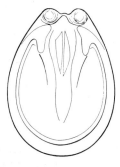

Figure 7–10.   An example of duplicate recessive epistasis. Normal foot (above), contracted heels (below). (From O.R. Adams: *Lameness in Horses,* 3rd ed. Philadelphia, Lea & Febiger, 1974.)

As soon as the gray gene expresses itself, the blanket pattern disappears (Figure 7–9), so that it shows only in the (gg) non-gray animals, because G is epistatic (covers up the expression) to blanket. The two groups of 9 and 3 are combined into 12 grays.

An example in horses of a recessive gene in the homozygous state covering up the expression of genes in another locus, *duplicate dominant epistasis,* is contracted heels (Figure 7–10). This condition is due to two pairs of recessive epistatic genes. The following symbols for these genes would seem to be appropriate:

Ch^1 = normal Ch^2 = normal
ch^1 = contracted heels ch^2 = contracted heels

Either ch^1ch^1 or ch^2ch^2 will cause contracted heels regardless of what the other gene is. If two heterozygous normal horses are mated, the following should result:

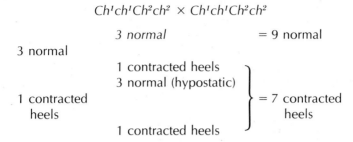

$$Ch^lch^lCh^2ch^2 \times Ch^lch^lCh^2ch^2$$

3 normal	*3 normal*	= 9 normal
1 contracted heels	1 contracted heels 3 normal (hypostatic) 1 contracted heels	= 7 contracted heels

This illustration explains why normal mated to normal can produce horses with contracted heels. It would also be possible for those with contracted heels to produce normal offspring.

An example of *duplicate dominant epistasis* in horses would be the mating of albino or lethal white and grays. Color (non-white) is produced only when both genes are recessive. The ratio of phenotypes will thus be $15:1$; only *lwlwgg* foals will have color, and this will occur 1 in 16 matings.

MULTIPLE ALLELES

The wild type of horse, Przewalski's horse, is homozygous for the A^t gene. It is bay with slight zebra markings on the legs. As long as no mutations occurred, all horses were of this type. Continuous background radiation is present, and this radiation tends to alter the genetic makeup of organisms, which results in a mutation. The gene A^+ mutated to A in horses, creating a new allele, i.e., A^+ to A (Figure 7–11). It was not until a mutation occurred that one could have known that the gene A^+ existed. The radiation background is continuous, and this locus made other changes. Another mutation resulted in A^t and another gave rise to the a gene. The result was multiple alleles. It must be clearly understood that all this action occurred at one locus in the chromosomes, as shown by the diagram in Figure 7–11. It should be kept in mind that when mutations to A^t and a occurred, they provided more possibilities for the kinds of genes that could exist in the horse population, but did not increase the number of genes at this locus for any one horse. Thus, horses now could be:

$$A^+/A^+$$
$$A^+/A^t$$
$$A^+/A$$
$$A^+/a$$
$$A^t/A^t$$
$$A^t/A$$
$$A^t/a$$
$$A/A$$
$$A/a$$
$$a/a$$

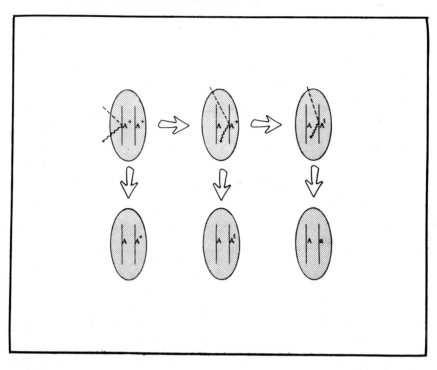

Figure 7–11.   A diagram of radiation causing a new allele. The process illustrated occurred
    over thousands of years.

The genes in a multiple allelic series are transmitted in exactly the same
manner as other alleles. There are more kinds of matings with multiple
alleles. It has been postulated that multiple alleles are present in many of the
performance traits.

The best examples of multiple alleles in horses are found in the blood
types (see Chapter 25). In these traits the genes act as co-dominants in that
each gene expresses itself. For example, in the transferrin alleles if the
animal is Tf^DTf^D, it will show one band by starch gel electrophoresis. Also, if
a horse is Tf^FTf^F, it will show one band but at a different location from the
band shown by the horse that is Tf^DTf^D. If the animal is heterozygous Tf^DTf^F,
it will have two bands showing, one at the Tf^D location and one at the Tf^F
location. A chart of frequencies of the various alleles of the blood factors for
a number of breeds is shown in Table 7–2.

Certain basic principles about mutations should be understood. Genes
may mutate, for example, from A to a and also from a to A. The return of the
mutant form to the original is referred to as the gene's reverting to the
original type. The rate of mutation from A to a is a constant, but it may differ
markedly from the rate of mutation from a to A. Mutation rate is generally
low but varies from gene to gene. It is estimated that rates in general vary

Table 7-2. Frequencies of Blood Factors in Five Breeds of Horses. Shetland Pony (SP), Thoroughbred (T), Arabian (A), Quarter Horse (QH), and Standardbred (ST). From Suzuki.

Genetic systems	Blood factors	SP	T	A	QH	ST
A	A_1	0.534	0.924	0.983	0.728	0.829
	A_2	0.869	0.924	0.983	0.896	0.993
	A′	0.583	0.052	0.094	0.362	0.564
	H	0.181	0.007	0.028	0.052	0.050
C	C	0.879	0.921	0.972	0.872	0.886
D	D	0.250	0.000	0.000	0.152	0.050
	J	0.227	0.251	0.171	0.346	—[1]
K	K	0.327	0.113	0.005	0.076	0.350
P_1	P_1	0.567	0.351	0.552	0.479	0.271
	P_2	0.621	0.497	0.569	0.586	0.257
	P′	0.094	0.170	0.028	0.090	0.057
Q	Q	0.519	0.742	0.398	0.262	0.029
	R	0.911	0.609	0.860	0.920	0.973
	S	0.354	0.633	0.744	0.523	0.720
T	T	0.698	0.892	0.819	0.843	0.633
U	U	0.534	0.248	0.354	0.490	0.429
Number tested		391	407	181	290	140

[1]Insufficient numbers were tested, but J does not occur in this breed.

from 20 to 30 mutations per million gametes. In a species such as horses, where rate of reproduction is extremely low, one should not expect to see many new mutations.

Ionizing radiations, the alpha, beta, and gamma radiation of radioactive substances and x-rays, will all induce mutations. The rate of mutation stimulation by x-rays is dependent upon the dose. The dose is dependent on intensity and time; therefore, an intense rate for a short time gives the same effect as longer radiations at lower intensities so long as the doses are equal. Other causes, such as ultraviolet rays, high temperatures, chemicals, and drugs, may bring about mutations.

Scientists are frequently questioned about using radiation to increase mutation and to obtain desirable mutations. Plant geneticists have used radiation effectively for stimulating both macro- and micro-mutations. In these organisms in which rate of reproduction is very high, the value of any individual is very low, and enormous numbers can be produced in small areas. Horse breeders should remember that when irradiation is used to create mutations so that selection can be more effective, several things may happen: (1) the production of desirable mutations may not be stimulated, and (2) radiation damage may cause sterility or reduce fertility to the point that one cannot capitalize on the mutations created.

MULTIPLE FACTORS

Two types of traits exist in horses: those showing discontinuous variation, which are called qualitative traits, and those showing continuous variation, which are called quantitative traits. Many of the traits of importance to horse breeders are qualitative and include such characteristics as color, anatomical and physiological abnormalities, lethals, and certain behavioral patterns. These traits are inherited in discrete classes, and definite phenotypic ratios are obtained when heterozygotes are intermated. The quantitatively inherited traits are also important to horse breeders. Such characteristics as milking ability of mares, speed, pulling power, and growth rates are quantitative traits. Here one finds no definite ratios appearing, but rather variation from one extreme to the other.

Although there are no clear-cut ratios with quantitatively controlled traits, when two extremes are mated to produce the first generation, and these are intermated to produce the second generation, the number of pairs of genes involved can be estimated. Rabbit ear length (a trait not highly affected by the dam, as is body size in horses) can be used as an example: If a rabbit with ears 2 inches long is crossed with one that has 12 inch ears, the F_1 (first generation) will have ears about 7 inches long. If these rabbits with 7-inch ears are intermated, the F_2 (second cross) generation is obtained. Suppose that exactly 1000 rabbits are produced with varying ear lengths. Out of these, three with 2-inch ears and five with 12-inch ears occur, giving an average of 4 rabbits at each extreme. From this figure, the number of genes

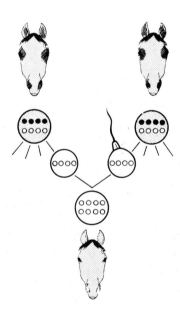

Figure 7–12.   Skin color is lost to offspring that receive only alleles for light color from parents. Dark skin originated from years of selecting for it.

involved can be estimated quickly . The minimum number of individuals needed in the F_2 to give all possible genotypes and phenotypes for varying numbers of genes involved must be kept in mind in order to perform this calculation. If 1000 is divided by 4, the result is 250. This figure is quite close to the 1 in 256 or 4 pairs of genes involved and is much closer than either 64 (3 pairs of genes) or 1024 (5 pairs of genes).

When the 7-inch F_1 rabbits are intermated, some offspring may be produced with 1-inch ears and some with 14-inch ears. This type of result has been obtained in genetic research and has been termed transgressive variation, because some of the F_2 transgresses or goes beyond the extremes of the parental types. The genetic explanation for transgressive variation, using the example in rabbit ear length, is that genes for short ears existed in the long-eared parent, and genes for the long ears in the short-eared parent. This pattern can be diagrammed as follows:

$$aabbccDD \times AABBCCdd$$

short-eared parent   long-eared parent

F_1

AaBbCcDd
(all progeny)

F_2 aabbccdd   AABBCCDD
          (ears shorter than   (ears longer than
          the short-eared   the long-eared parent)
          parent)

In the F_2 (second cross) after F_1 progeny were mated together, some rabbits were aabbccdd and had shorter ears than the short-eared parent, while some were AABBCCDD and had longer ears than the long-eared parent.

Genes involved in quantitative inheritance are called polygenes because many genes influence the trait, each with a small effect. In general, the quantitative traits do not show dominance; that is, first-cross animals when extremes are mated are intermediate. When the first-cross animals are intermated and a large number of pairs of genes are involved, probably there will never be sufficient numbers in the F_2 to have all of the genetic expression evident. With only 10 pairs of genes there would need to be more than a million offspring in the F_2 to have all the genetic expression in the F_2.

Arabian horse breeders are proud of the dark skin of their horses. It is a part of their heritage. The darker skin gives the dark appearance around the eyes and nostrils regardless of the hair color.

It has been shown in humans that skin color is a result of polygenes. It is assumed for illustration that four genes (Figure 7–12) are involved in horse skin color, but there are probably more. The variation in skin color depends upon how many of the "dark" genes are inherited from each parent. Horses with skin of medium darkness might produce offspring without colored skin (Figure 7–12).

MODIFYING GENES

Certain genes, known as the main genes, bring about the expression of a trait. Other genes, called the modifying genes, modify the expression of the main genes. Some genes modify the expression of several genes, while others modify the genes at only one locus or even one of the alleles at a locus. For example, the dilution gene, D, in horses, when it is in the heterozygous condition, causes a dilution in the basic coat colors, causing blacks to be buckskin or grulla, bays to become duns, and chestnuts to become palomino. It is an example of a modifier that affects the expression of several main genes. Dwarfism will modify the expression of genes affecting growth and conformation; therefore, this recessive gene in the homozygous condition is a general modifier.

An example of a gene that modifies only one other gene is illustrated by distal leg spots (Figure 7–13). This gene has no effect in horses that do not have white feet and legs, but causes small colored spots just above the hoof in otherwise "white-stocking" horses. The results of matings observed indicate that this specific modifier is dominant.

Horses have a dominant form of white spotting. The gene producing this effect was reported by Crew and Buchanan-Smith when they pointed out that other genes modified the amount and distribution of the white markings

Figure 7–13.   A distal leg spot.

Figure 7–14.   Variations in tobiano spotting occur because of modifying genes. The basic pattern remains the same although extremes from almost all white to almost all pigmented are found.

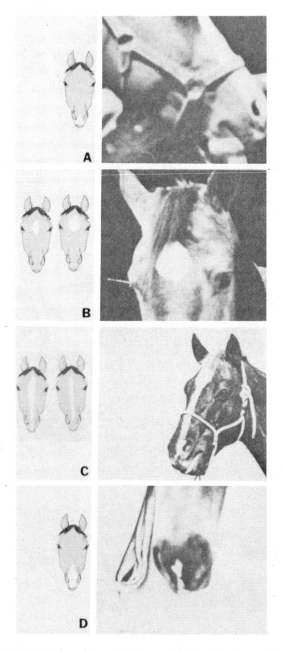

Figure 7–15.   Head markings: (a) chin spot, (b) star, (c) strip, and (d) snip.

Figure 7–16.   Variations in overo spotting occur because of modifying genes.

(Figure 7–14). Crew and Buchanan-Smith called it the piebald gene. The American Paint Horse Association now calls the gene tobiano. Blunn and Howell have described four modifying factors affecting white markings on the face of horses. They report that chin spot is recessive and causes a white spot on the chin. When animals with chin spots were mated, they produced offspring with chin spots. Animals without chin spots, particularly those with one parent having the chin spot, when mated produced some offspring with chin spots.

Other modifiers of white spotting in horses—star (a white spot in the forehead), strip (a small strip of white running downward from between the eyes), and snip (white marking from nostril to lip)—are dominant (Figure 7–15). In no case have any of these markings been found in the offspring of parents lacking these traits. Mating of horses with star, strip, or snip produced offspring with a little more than half showing the traits. Apparently two stallions used in their studies were homozygous for the three dominant modifying genes. It is of interest to note that four modifying genes affect white spotting, all of which affect facial markings.

Other modifiers that affect white spotting also influence the amount and distribution of white on other parts of the body. Some horses have spots over the body (Figure 7–16) and are called pintos, or overo, as termed by the American Paint Horse Association. In other horses the white markings are restricted largely to the legs and cause the "white stocking" effect. The American Paint Horse Association assumes that overo is recessive white spotting. In the Appaloosa there are other types of white markings, which are discussed in Chapter 9.

PLEIOTROPIC EFFECTS

Genes are not tiny entities conscious of the end result of their activities. They are simply biochemistry in action. They do not have a point of view as do humans. We look at a white horse and think in terms of a white horse gene when in reality the gene is responsible for many other characteristics of the horse's body of which we are not normally aware.

Genes do not bring about their effect on only one area of the body, but throughout the system. The classification of genes is according to effects that are of greatest interest to humans. This generalized effect throughout the body is referred to as pleiotropy. Some genes display pleiotropic effects to a greater degree than others, but all genes have generalized subtle effects on the entire body. This concept was brought out in the discussion on gene interactions.

Pleiotropic effects can be of particular interest to the horse breeder. Often a desirable genetic effect is accompanied by an undesirable pleiotropic effect. The wise breeder will learn to detect this situation and will in turn select against what he once classified as a desirable trait, or in some cases he may simply have to learn to live with the bad in order to have the good.

The most outstanding examples of pleiotropic effects are found in white or near white horses. Color inheritance is discussed in detail in Chapter 9, but it can be pointed out here that so-called white horses are caused by at least three different genes. When the *D* gene is homozygous, a light yellowish white horse is produced, which is known as a creamello, perlino, or "glass-eyed white." The second type of white, or the white horse referred to as albino, is caused by the *Lw* gene in the heterozygous condition. A *Lw* homozygous horse is never found because the *Lw* homozygous condition causes such severe pleiotropic effects in the developing embryo that it dies early in gestation. Since this is always the case, albino horses can never be true breeding. When albinos are bred together, one-fourth of the conceptions result in embryonic death, one-fourth are colored, and one-half are white heterozygotes.

The third type of white horse is caused by the recessive overo gene. Its expression is affected quite drastically by other modifying genes. This modification leads to overo paint horses with various amounts of white on the body. Those with large amounts of white bred to overo paints will often produce a pure white foal and, at this point, pleiotropic effects come into play causing the white foal syndrome.

This syndrome in white foals has not been studied in depth. It is known, however, that several abnormalities of development always lead to the death of the foal within a few days of birth. The white foal syndrome is discussed in detail in Chapter 10.

LINKAGE

Many horse breeders have wondered why some foals of a certain stallion resemble him so closely, while others miss almost completely. A bay color, for instance, sometimes seems to be inherited along with the other distinctive characteristics of a particular family.

Thoroughbred breeders have, for many years, harbored feelings that one color or another tends to be associated with speed. Within the past few years, two separate registries have been developed for buckskin horses. Their claim, although somewhat exaggerated, is that the buckskin color is unalterably associated with many of the most desirable characteristics in a horse.

Can this be explained genetically? Some genetic experts who breed mice and fruit flies find it difficult to believe that color could be genetically linked with other recognizable physical characteristics in horses, but there is no expert who knows all the answers when it comes to horse breeding. There simply have never been enough controlled genetic studies to determine even the most basic facts about the genetics of the horse. All that is known has been extrapolated from mouse and fruit fly experiments. The long-time horse breeder who has been astute in his observation may be aware of many equine genetic realities with which the experts are not acquainted.

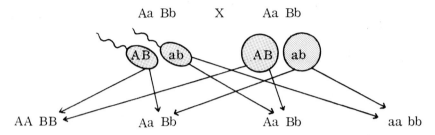

Figure 7–17.   The transmission of linked genes. Only two types of gametes are produced in a dihybrid cross.

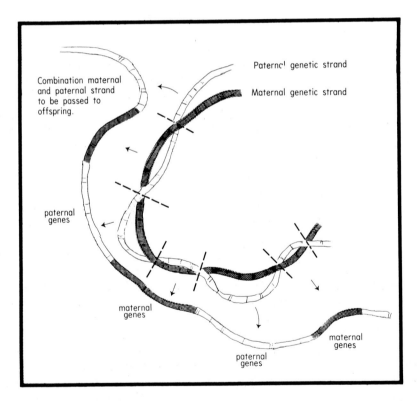

Figure 7–18.   A schematic diagram depicting the consequences of the process of crossing over during the development of gametes. The maternal and paternal genetic strands combine, segment, and reunite to form a new strand which goes into the formation of spermatozoa or ovum.

The close association of genes, and the way in which they might be transmitted as one, is illustrated in Figure 7–17. The ratio produced from a dihybrid cross is 1:2:1, exactly what it would be if only one gene were involved. In mendelian terminology, there is no segregation of the gametes.

A process termed crossing over is known to occur with the chromosomes of the sex cells in the development of both the ovum and the spermatozoa. This cellular process brings about the mixing of parts of each maternal chromosome with homologous parts of each paternal chromosome (Figure 7–18). Each of the 32 pairs of chromosomes in every sex cell undergoes this exchange of material. This crossing over results in a genetic "upset the fruit basket," so that each chromosome passed on to the offspring is a combination of genes from both the sire and the dam. The X and Y, or the sex, chromosomes are an exception as little, if any, crossing over occurs between these two.

Crossing over occurs between chromosomes because of a physical break in the genetic strand which contains the genes. Since these breaks seem to occur and then repair again in the same location within homologous chromosomes, it is often assumed that this breaking and crossing over does occur more often in certain predetermined places within the chromosome.

Even the smallest segment of one of the chromosomes pictured would contain dozens or even hundreds of genes. Thus, a tiny segment of any one chromosome could be the genetic source for an entire group of traits or characteristics.

If crossing over occurs more commonly at some locations within the chromosome than at others, it would seem natural that some traits would be inherited as a group rather than completely at random. It is feasible that a color characteristic could be one of that group.

When this type of inheritance occurs, it is said that certain genes are linked, which means that particular genes are passed from one generation to the next as a group; and as far as these genes are concerned, the "upset the fruit basket" phenomenon has not occurred.

Figure 7–19 is a hypothetical example of how linked genes would be inherited. No one has specifically identified which traits seem to be inherited as a group. The illustration uses the Arabian stallion, Abu Farwa, as an example. Many of the fans of Abu Farwa believe that his flashy chestnut color was linked with certain other traits which identify an ancestor as having "the Abu Farwa look."

For purposes of illustration, the Abu Farwa "look" has been over-simplified. Only a few characteristics are described, most of which are no doubt influenced by many genes. The dished face, the "action," the long neck, and the Abu Farwa hip are characteristics described by most breeders familiar with that champion stallion.

As shown in the illustration, the linkage type of inheritance would allow for the non-gray (or in this case chestnut) color to be passed on to offspring

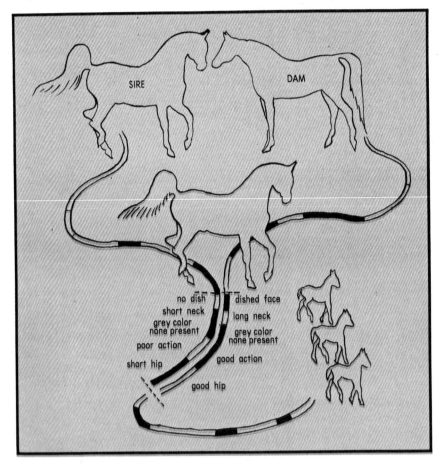

no dish / dished face
short neck / long neck
grey color / grey color
none present / none present
poor action / good action
short hip / good hip

Figure 7–19. Genetic material passes from a sire and dam in a haploid condition (only one of each gene) and combines in the offspring to create a pair of genes known as the diploid condition. The dotted lines represent possible points of crossing over. This crossing over allows the offspring to pass on genes which are a combination of both the sire and dam genes. The third-generation foals shown would receive genetic material from their sire as well as genetic material from their dam (not shown).

only if accompanied by certain other traits. A gray mare bred to Abu Farwa would pass gray on to her offspring only if her version of each of these same characteristics accompanied it. A gray foal in such a cross would have the mare's head, the mare's action, and the mare's neck.

This type of linked inheritance has never been proven scientifically as a consistent pattern or mode of inheritance. Sooner or later, with enough breeding under carefully controlled situations in mice, a linkage group is broken up owing to the process of crossing over. Breeding studies in the horse are so insufficient that it has never been possible to prove scientifically that permanent linkage groups do persist through many generations of breeding.

In some selection experiments, little progress is made for some time because animals having the proper combinations cannot be found. Suddenly one appears, and then for a period progress is made. It may be that a crossover in one of the small chromosomes allowed the proper combination. Having a good combination stay together after it has been located (no crossing over) is surely an advantage.

BIBLIOGRAPHY

Blunn, C.T., and Howell, C.E.: The inheritance of white facial markings in Arabian horses. J. Hered. 27:293–299, 1936.

Bogart, Ralph: Improvement of Livestock. New York, The Macmillan Co., 1959.

Crew, F.A.E., and Buchanan-Smith, A.D.: The genetics of the horse. Bibliographia Genetica 6:123–170, 1930.

Evans, J.W., et al.: The Horse. San Francisco, W.H. Freeman & Co., 1970.

Hatley, G.B.: Appaloosa Horses: Color Patterns, Breed Characteristics. Moscow, ID, Appaloosa Horse Club, 1962.

Pulos, W.L., and Hutt, F.B.: Lethal dominant white in horses. J. Hered. 60:59–63, 1969.

8

cellular genetics

$\mathbf{A}$ basic knowledge of cells is necessary for an understanding of genetics. During development of an early embryo, genes bring about changes in the structure and function of cells. These cellular changes then produce what we see as a difference in hair color, muscle conformation, or any other characteristic. An understanding of cells and how they function can therefore bring insights into the mechanisms of inheritance.

A cell can be considered as a tiny factory designed to do a particular job for the animal as a whole. A spermatozoon is an example of a cell whose job is to carry the male genes to the female sex cell so that an embryo can develop with characteristics from both parents. A red blood cell is another example of a cell with a specific function, carrying oxygen from the lungs to other parts of the body where it is needed. There are literally hundreds of distinct types of cells of the body, each with a specific job. Some that will be referred to later in regard to specific genetic changes are plasma cells, which produce antibodies; melanocyte cells, which produce melanin granules for hair color; and cells of the mammary gland, which produce milk for the foal.

Cells become what they are, basically because of their proteins. These molecules help to form the structural parts of the cell, the chemical structure of which has been dictated by the genes in the nucleus. Proteins also make up enzymes, which direct all chemical reactions within the cell. The variety and extent of chemical reactions within an individual cell are almost beyond comprehension. Chemical reactions break food molecules down in order to obtain energy. Other reactions occur as the various cellular products are made.

CELLS

Several types of cells in the horse's body, all of which produce specific products necessary to the life of the animal, are shown in Figure 8–1.

Notice that each cell contains a nucleus, mitochondria, a Golgi body, and other structures. Each tiny part of a cell has its individual job that contributes to the overall function of the cell. The nucleus contains the genes. Chemical messengers are sent from there to other parts of the cell, where proteins are constructed according to the directions of the messenger.

Outside of the nucleus there are some important cellular structures. Mitochondria are tiny sac-like structures where energy for the many activities of the cell is stored. The area within the cell but outside the nucleus is termed the cytoplasm. Most cells contain a Golgi body somewhere within the cytoplasm, which is involved in production of a particular cellular product. For instance, the developing spermatid (the immature form of a spermatozoon) contains a Golgi body which produces the enzyme hyaluronidase.

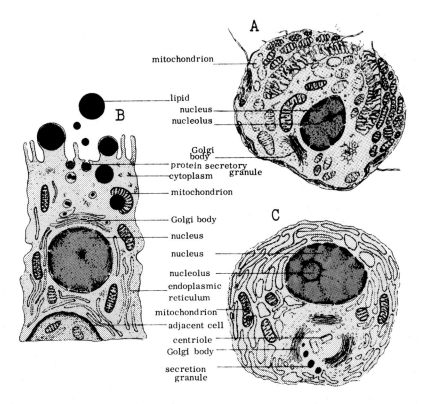

Figure 8–1.   Examples of cells and their products: (a) gastric parietal cell producing HCl, (b) mammary gland cell producing milk, and (c) plasma cell producing antibodies.

Cells do contain substances other than protein; however, these sugars, fats, and other chemicals are built as directed by enzymes, which are always protein. This fact is important. Genes exert their effect by dictating the types of protein that shall be made in the cell.

An extremely simplified way of picturing a typical cell is to compare it to a bag of enzymes to which various food substances are added and from which various cellular products leave. The transformation of food substances into a particular cellular product is brought about by enzymes. All intricate parts of the cell are important only as they assist the cell to produce its specific product or to perform its specific function.

Several structures of the cell cannot be made without a pattern, even with the help of enzymes. A template, or previous pattern, to build these structures must be passed from cell to cell. These basic units of heredity are found in chromosomes, mitochondria, and possibly some other structures.

All cells develop from previous cells. As cells reproduce, generation after generation, they change their nature. The ancestors of cells in the early embryo show less variation in structure and function than cells of the adult body. As an embryo develops, the cells form groups making up specialized types of tissue, each with its own specific function. As the cells become specialized, the vast majority of these genes are "shut off," and the cell is directed by a relatively few specific genes. As an example, a cell in the skin, known as a melanocyte (Chapter 9), produces melanin, which causes the hair color of a horse. A specific gene directs this melanin producing activity, but only in melanocytes.

All of the cells in the body of a horse have developed from the original fertilized ovum, which was a single cell, unlike any other cell of the body. As the first cells divide and the resultant cells multiply, some of them begin to change their function. Some become heart muscle fibers; others become cells of the eye. This process of change is called differentiation and does not stop until the death of the animal. Differentiation brings about the development of the embryo; it causes maturing of the foal. Differentiation brings on puberty and is even responsible for the changes that occur during old age. This process, therefore, is a blessing as well as a curse.

Two possibilities concern the future of any cell: eventual death, or cellular division to form two new cells. Generally speaking, aside from considerations of cell division, the life span of a cell is relatively short. The cells that line the small intestine live for only a few days and then must be replaced by a new generation. Mature red blood cells live only about 100 days. Other cells, such as nerve and muscle, may live as long as the horse itself. In fact, the death of these long-lived cells is usually what is involved when a horse dies of old age.

A horse is what it is because of what its cells are. All inheritance, therefore, is inheritance of some type of cellular characteristic. The more one learns about the structure and function of cells, the more one knows about genetics.

BIOCHEMICAL GENETICS

In a gross analysis, differences between animals are often determined by measurement of the skull, ears, or other anatomical parts. The number of vertebrae or even color is sometimes used as a means of differentiating breeds of horses. These gross characteristics give only a rough idea of the overall degree of difference between the two animals.

More subtle differences are measured by physiological factors. Three blood characteristics, shown in Figure 8–2, differentiate thoroughbreds, saddle horses, and draft horses. The terms "hot-blood" and "cold-blood" are used quite often to describe a particular kind of horse. These characteristics, shown in Figure 8–2, give a little more meaning to the terms. It may be that blood volume and the amount of hemoglobin are important factors making up the "hot-blood" horse.

However, the true explanation of any abnormality or any species and breed differences lies in biochemistry. Since genes direct the synthesis of a particular protein segment by means of an intricately coded messenger, if one studies the makeup and differences of proteins, the most basic differences can be determined.

All protein in the body of a horse is made up of combinations of 21 amino acids. These small molecules all have the same type of attachment on one end and another kind of attachment on the other end. The ends that fit together attach to each other linking up the amino acid molecules somewhat like boxcars in a train.

An example of how gene function works is found in one of the hormones produced by the anterior pituitary, the growth hormone, which is also known as somatotropin. This hormone circulates throughout the body and stimulates the growth of long bones of the colt. It also influences the growth

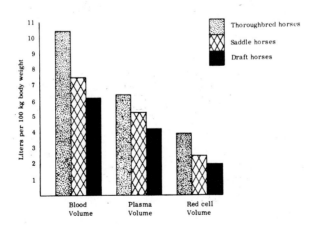

Figure 8–2.   Blood volume comparisons in horses show the differences between "hot-blood" and "cold-blood."

of the heart muscles and other soft tissue of the body and has an effect on various other chemical reactions occurring in cells.

One gene is responsible for making the protein somatotropin. This gene is DNA (desoxyribonucleic acid), with the proper code for the growth hormone. Somatotropin is not an enzyme but a protein product; and, as far as is known, the messages for the making of all proteins are in the DNA of the genes.

The long, thread-like DNA molecule seems to be divided somehow into units, each of which contains the original code for a particular type of protein. These units are the genes. Each gene therefore is responsible for production of a specific protein, which might be either an enzyme or a structural part of a future cell, or, in this case, a hormone. A typical protein is made up of several hundred subunits (known as amino acids). The code in the DNA is used to make a complementary messenger called RNA (ribonucleic acid). This messenger contains directions for placing each subunit of the protein in its proper place.

Oliver and Hartree have determined just what the subunits of equine somatotropin are. The relationship of these subunits in somatotropin of the horse and of the cow are illustrated in Figure 8–3. While the general shape

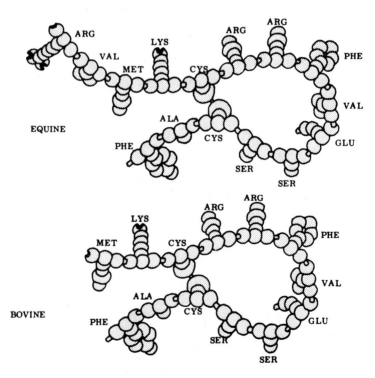

Figure 8–3.   Somatotropin of the horse and cow compared. The acronyms are abbreviations for the various amino acids.

of the molecule is similar, differences between the two species occur in the amino acid sequence. The letters on the subunits are abbreviations for various amino acids. It is evident that the gene for somatotropin in the horse has a somewhat different code from the same gene in the cow. In this case, the equine gene is a bit longer, containing directions for the production of two extra amino acids on somatotropin.

Each gene contains many individual codes (known as codons) for just one protein (or string of amino acids). This means that each gene has many sites where it might be altered. As knowledge about specific genes increases, one finds that many of them vary from one animal to another by one or two codons. A good example of this is the variety of patterns of color in bay, brown, and black horses. The bay body color is produced as a result of the A gene which alters black color on the body hairs. The recessive a gene allows the black color all over the body. The gene A^t produces the brown horse (Figure 8–4). All of these alleles of the A gene produce a similar type of protein and each has slight variations in the codons of the DNA.

Since chromosomes contain genes, it is essential that each new cell which is formed receive an exact replica of each chromosome. Before cell division, a carbon copy of each chromosome develops. At cell division, each

Figure 8–4.   Some alleles of the A gene in the horse, and their phenotypic effects.

daughter cell receives an identical set of chromosomes. This process of cell division is known as mitosis—the only time when chromosomes can be observed in somatic cells. During the time between mitotic divisions, known as an interphase, the DNA is confined to the nucleus of the cell and is not formed into chromosomes.

HEMOGLOBIN

All differences or abnormalities in horses stem from differences in proteins. The protein that makes up the hemoglobin in the red blood cell of the horse has been studied. It serves as an excellent example of how the genes bring about a particular phenotype.

Generally speaking, hemoglobin is made up of four chains of amino acids, along with four non-protein molecules called heme groups. It is the latter that picks up oxygen and transports it throughout the system. However, the protein portion also plays a role.

The immature red blood cell produces many protein chains under the direction of the structural hemoglobin genes. It has been hypothesized by Kilmartin and Clegg that there are three hemoglobin structural genes in the horse. Evidently all three genes, by means of a messenger, direct the production of their particular amino acid chain.

After the amino acid chains are formed, they bind together, four in a molecule. Two in the molecule are known as beta chains; their particular amino acid sequences are identical. Both of these are made under the direction of the same Hb-B gene; and so far as research has shown, all horse beta chains have the same amino acid sequence (Table 8–1).

On the other hand, a number of differences are found in the alpha chains. These differences have been previously attributed to various alleles in the Hb-A gene; however, Braend cites evidence that casts doubt on this idea. He separated the alpha chains from the beta chains of horse hemoglobin and by means of electrophoresis was able to show differences in the alpha chain. Electrophoresis causes proteins to migrate through a gel. The particular amino acids making up the chain and also the length of the chain determine the rate of migration. Braend detected a slow alpha chain and a fast alpha chain. He termed the fast chain A_1 and the slow chain A_2. The hemoglobin in the gel after electrophoresis is shown in Figure 8–5.

Braend found three unexpected phenotypes in relation to the alpha chain (Figure 8–5). Normally, if A_1 and A_2 were both alleles of the same gene, the three phenotypes would be A_2, A_1A_2, and A_1. Using Kilmartin and Clegg's idea that there are two separate locations on one of the chromosomes for the Hb-A gene, Braend was able to explain the three unusual phenotypes (technically each gene location of a chromosome is known as a locus).

His explanation required the hypothesis of a modifying gene which affected the action of only one of the loci for Hb-A. His idea is that the A_1, or fast-moving chain, and the A_2, or slow-moving chain, are produced under

Table 8–1. Hemoglobin.

H-val-glu-ser-gly-glu-glu-lys-ala-

asp-val-lys(asp-try-leu-ala-val)leu-ala

glu-glu-glu-val-gly(gly-glu-ala)leu-gly-

glu-thr-try-pro-tyr-val-val-leu-leu-arg

arg-phe(phe-glu-ser-phe-gly-asp-leu-

asp-gly)met(val-ala-asp-pro-gly-ser

-pro)lys-val-lys-ala-his-gly-lys-lys-val-

his-his-val-gly-glu-gly-phe-ser-his-leu

-(leu-asp-asp-leu)lys-gly-thr-phe-ala(ala-

his-leu-lys-asp-cys-his-leu-glu-ser-leu

-val-asp-pro-glu-asp-phe)arg-leu-leu-gly-

his-arg-ala-val-val-leu-ala-leu-val-asp

-phe-gly-lys-asp-phe-thr-pro-glu-leu-glu-

val-gly-ala-val-val-lys-glu-tyr-ser-ala

-ala-asp-ala-leu-ala-his-lys-tyr-his-OH

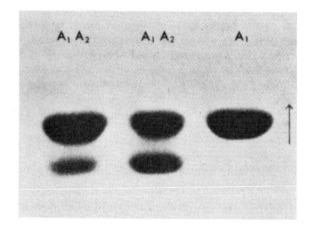

Figure 8–5.   Three hemoglobin phenotypes as shown by starch gel electrophoresis. (From Separat ur Hereditas.)

the direction of genes at these two different loci. Thus, the loci can be termed A_1 and A_2. The modifying gene evidently affects production of A_2 only. This gene has been given the symbol of Hb^m.

The amino acid differences of the alleles of hemoglobin A at both the Hb-A locus and the Hb-A_2 locus are shown in Figure 8–6. Evidently the donkey has only one locus for Hb-A. Electrophoresis shows this to be Hb-A_2; and actually only one allele is involved, Hb-A_2^a. Zebras and mules, like the horse, have both loci, according to Osterhoff.

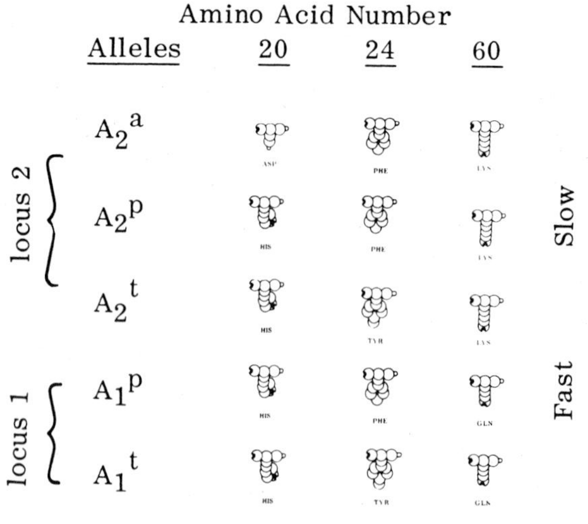

Figure 8–6.   Amino acid differences of hemoglobin-A.

The Hb^m allele evidently interferes with production of Hb-A$_2$, while the Hb^{m+} allele does not. Therefore, a horse heterozygous for these two alleles would allow some production of Hb-A$_2$. Thus, the A$_1$ phenotype of Figure 8–5 must be homozygous for Hb^{m-} because there was no production of Hb-A$_2$. Phenotype A$_1$A$_2$ would be heterozygous for Hb^{m-}, partially inhibiting production of A$_2$.

Braend checked several breeds of horses for Hb^m gene frequencies and traced inheritance of the gene in several families. It followed mendelian inheritance.

The Hb^m gene acts much like some of the regulator genes described by Jacob and Monod in bacteria. These scientists received a Nobel prize for their endeavors, which brought to light the intricate control and feedback mechanisms involved in gene action. They made the concepts of regulator and structural genes popular. In this case, $Hb-A$ is a structural gene because it contributes to the structure of hemoglobin; and Hb^m is a regulator gene because it produces an element which regulates or controls the action of another gene.

It is unlikely that these differences in hemoglobin bring about much functional difference in horses. However, they illustrate the mechanisms of gene action which probably also occur in various types of abnormalities.

MEIOSIS

During development of the germ cells a modification of cell division known as meiosis occurs. As a result of meiosis each germ cell contains only one complete set of chromosomes instead of pairs. The germ cell is then able to unite with one from the opposite sex to provide paired sets of chromosomes necessary for the development of an embryo and mature animal. In the male these cells are called spermatozoa, and in the female they are known as ova.

One of 32 pairs of chromosomes in the domestic horse cell is shown in Figure 8–7. The light figure represents a chromosome received from the grandsire and the dark one the chromosome received from the granddam, each duplicated to form two chromatids. The four almost identical chromatids form a tetrad: chromatid No. 1 from the sire; chromatid No. 2 an exact replica of No. 1 produced by the cell in preparation for division; chromatid No. 3 from the dam; and chromatid No. 4, an exact replica of No. 3, also made in preparation for cell division. Two cell divisions occur quickly so that each new cell will receive one chromatid from each pair of chromosomes. Meiosis in the male ends with four spermatozoa, each with 32 chromosomes of one tetrad each. Meiosis in the female produces ova, containing one chromatid from each pair of chromosomes. The other three chromatids of the female gamete are cast off in polar bodies and are eventually lost.

Three genes are pictured as located on this chromosome. It may be that all

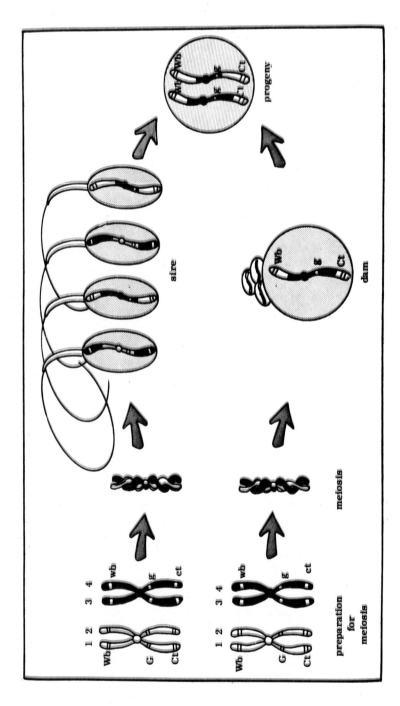

Figure 8–7.   A diagram of the mixing of genetic material during meiosis.

three do not occur on the same chromosome, but they are pictured this way to illustrate a principle. The first is the "weak back" gene. Wb represents what is known as a dominant allele; wb represents the recessive allele. Both are alleles of the same gene. Weak back is a recessive characteristic (Figure 8–8) which will not show up as long as the horse has the dominant allele Wb, which produces a straight back.

The gray gene has two alleles, represented in the Figure 8–8 diagram. Gray is a dominant characteristic; therefore, any horse that has at least one dominant G allele will become gray. The horses that have two recessive g alleles usually will not turn gray.

The ct gene causes an extremely curly hair coat on the horse. Blakeslee analyzed the pedigrees of curly haired horses and found the curly coat to be a recessive characteristic.

In the diagram the grandsire on both sides contributed the three dominant alleles, and the granddam on each side passed the three recessive alleles, making both parents heterozygous for all three genes. In simple terms, heterozygous means the carrying of different types of alleles for a specific gene. Note that chromatids 2 and 4 contain exact replicas of the genes of the chromosomes from which they are patterned. This is always true.

An exchange of chromosome material occurring during meiosis is also shown in Figure 8–7. If this shuffle of alleles did not occur, entire chromosomes would pass from generation to generation without alteration, and the possibilities for variety in genetic type would be much less. This meiotic shuffle of alleles, known as crossing over, means that no chromosome of the germ cells reflects exactly one that came from a parent; rather, it represents a mixture of both maternal and paternal characteristics.

A type of inheritance that might occur with the three genes discussed is also shown in Figure 8–7. Both the sire and the dam would be gray without curly hair or weak backs. The offspring produced from the combination shown in the illustration would not be gray nor would they have curly hair and weak backs. These characteristics might have been different in the foal had a different spermatozoon fertilized the ovum or different chromatids

Figure 8–8.   The wb gene for weak back is recessive, and must be received from both parents to be expressed in a horse.

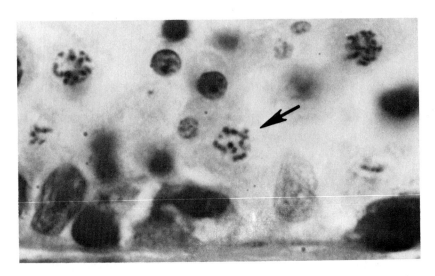

Figure 8–9.   Chromosomes of the horse during meiosis. Photographed from spermatocytes of the horse. The chromosomes (arrow) are more easily observed during cell division.

been passed to the polar bodies of the ovum. The probability of each type of offspring occurring is explained in Chapter 6.

Male sex cells from the horse during meiosis are shown in Figure 8–9. There are two divisions during meiosis, and the cell at the end of the pointer is in the first division. Both chromosomes of a pair adhere, rather like the two sides of a zipper, in a process known as synapsis. While the chromosomes are in this close relationship, there occurs a crossing over of material between the two. The actual chromosome number is not reduced until after the second division, although the associated pairs cause the cells in this stage of meiosis to appear to be half as many in number. Various stages of first division meiosis are shown in the illustration. After the second division the cells are haploid, which means that the chromosomes are not paired; there is only one of each.

CHROMOSOMES OF THE EQUINE

Each cell of the body normally contains an entire set of chromosomes. Cells must be prepared in a special manner in order to catch them in the proper stage for chromosome observation. They are treated so that the cells rupture and spill the chromosomes out on a microscope slide for viewing with a light microscope. This technique has become quite popular in the past decade and has greatly enhanced the study of karyotypes or chromosome spreads.

A technique developed to demonstrate whole chromosomes with the electron microscope has shown them to be made of tiny strands which contain the DNA (Figure 8–10). The individual strands cannot be seen with the light microscope. It will take further study with a variety of techniques before the complete ultrastructure of the chromosome is determined.

The domestic horse (*Equus caballus*) has 32 pairs of chromosomes in each cell. The number is constant throughout the species except in unusual cases, which will be discussed under chromosome anomalies. Shetlands, Percherons, and all other breeds of the domestic horse have the same number of chromosomes. A typical karyotype (set of chromosomes) is shown in Figure 8–11.

Every species of animal has its own karyotype. The karyotype is obtained and analyzed by means of photography, through a microscope, of a live cell that has been specially treated to catch it in cell division. At this time, the chromosomes are in a condensed form so that they can be stained and photographed. The cells are treated so that they break apart, spilling the chromosomes out in a haphazard fashion, as shown to the left of Figure 8–11. A photograph of the chromosomes in this position is then cut up, and the individual chromosomes are paired up and pasted on a sheet to be photographed again in neat orderly rows. Only in this way can the karyotype be visualized. In this manner the partner chromosomes can be identified rather accurately.

Both the male and female have 31 identical or homologous pairs of chromosomes known as autosomes, in contrast to the divergent sex chromosomes. Normal females have two large chromosomes, designated as X; and normal males have one X chromosome and one Y chromosome. The area where the arms come together is called the centromere (Figure 8–12). Each

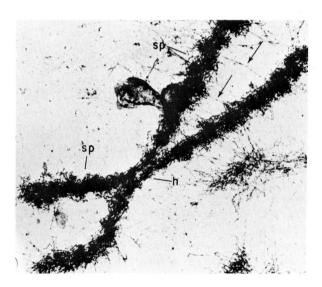

Figure 8–10.   Electron microscope view of the human chromosome. (From The Journal of Cell Biology.)

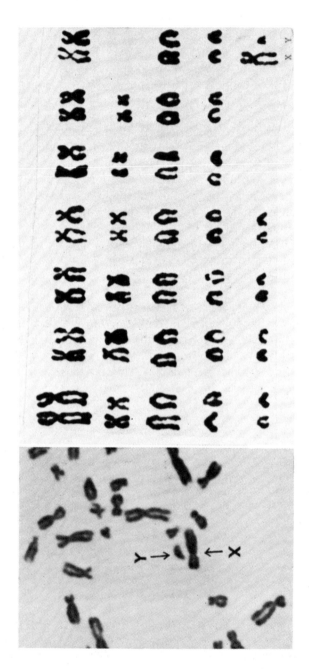

Figure 8–11. Karyotype of *Equss cabalus*. Under the microscope the chromosomes appear as shown on the left. They are photographed and cut apart and pasted into pairs to show an orderly karyotype on the right.

Figure 8–12.   Chromosomes are known to occur in many shapes and sizes. The central portion where the arms come together, the centromere, may occur at any position along the length of the arms; however, it is constant for any particular chromosome.

chromosome has a centromere which has special attraction for the centromere of its mate. In 36 autosomes and the Y, the centromere is located at the point of the V. It is acrocentric in shape. The 24 chromosomes in which the centromere is in the center are classified as metacentric in shape, and appear in the shape of an X. The designation of a chromosome as an X chromosome refers to one sex chromosome and not to the shape (Figure 8–24). A chromosome such as the X, with the centromere intermediate location, is classified as submetacentric in shape.

During meiosis the pairs of chromosomes separate, with one of each pair going to each germ cell. In the male, this means that half of the sperm cells will contain 31 autosomes and a Y chromosome. Upon fertilization the zygote (fertilized ovum) will contain 62 autosomes, and either two X chromosomes and thus become a female, or 62 autosomes, one X and one Y chromosome, and thus become a male.

A technique known as paracentesis has been developed in the human to determine the sex of a fetus before it is born. This technique takes advantage of the fact that male cells have an X and a Y chromosome, while female cells have two X chromosomes. A hypodermic needle is inserted through the abdominal cavity into the uterus. A sample of amniotic fluid is drawn out and placed in a special medium for culture in an incubator. The fluid contains some cells from the embryo, which grow and divide. After analysis, the sex of the embryo can be determined by the sex chromosomes of the cells. Females will normally have two X's in each cell and males an X and a Y. This technique could be used in the horse.

Species are frequently difficult to define. Often the characteristics of species are described by certain anatomical features, much as one breed is distinguished from another. However, differences in species usually are more numerous and extensive than differences in breeds.

Generally, species differ in chromosome number. This is true in the various equine species, which include the horses, asses, and zebras. The equine species and their chromosome number are shown in Table 8–2, and the karyotypes of several equine species are shown in Figure 8–17.

Hybrid crosses have been obtained among most of the equine species. The most popular has been the mare-and-jack cross giving the mule. In the days when work animals were a necessity, this cross gave what was considered the best characteristics of the two species.

The ass has 62 chromosomes, while the horse has 64. As explained earlier, the chromosome number halves during meiosis so that the sex cells contain 32 chromosomes in the horse and 31 in the ass. In combination, the uneven number of 63 is produced, but at least one of every pair from both species is present in the somatic, or nonreproductive cells.

Crosses among other equine species bring about a similar unbalanced condition of the gametes, or sex cells. The hybrid offspring are always healthy, but reproductively sterile. Any breeding program, therefore, which utilizes interspecies crosses, such as zebra with horses or donkeys with zebras, has no chance of developing into a new breed. Both species must be continually maintained to produce the hybrids.

Interspecies crosses have been successful in plants even when the chromosome number of the two species is not the same. With the use of colchicine or other drugs, the haploid condition of the gametes is doubled causing all of the chromosomes to become paired. The uniting of diploid gametes from different species results in healthy offspring which are fertile because each chromosome is paired. A new species is thus created with as many chromosomes as found in both parents combined.

Meiosis in the mule rarely occurs because the chromosomes are not all in pairs; and as their number is halved, the sex cells usually fail to obtain a complete set. This unbalanced condition seems to be lethal to the sex cell in the mule, or it may be that so much non-homologous DNA exists in the

Table 8–2.   Equine Species and Chromosome Numbers.

Scientific Name	Common Name	Chromosome Number
Equus przewalski	Mongolian wild horse	66
Equus caballus	Domestic horse	64
Equus hemionus	Mongolian wild ass	56
Equus kiang	Tibetan wild ass	
Equus onager	Persian wild ass	56
Equus asinus	Donkey or domestic ass	62
Equus zebra	Zebra of Cape Colony	32
Equus grevyi	Somaliland zebra	46
Equus burchelli	African zebra	44

chromosomes of the horse and the ass that lethal translocations are established (translocations are discussed under the topic of chromosome anomalies). One fact supporting this possibility is that a cross between what are probably more closely related species, the Mongolian wild horse with the domestic horse, is not lethal to the sex cells. Benirschke examined the testicles of such a hybrid and found normal spermatogenesis.

Koulischer and Frechkop originally reported this cross between *E. przewalski* and *E. caballus*. A colt was produced that thereafter bred a domestic mare which subsequently produced a one-fourth hybrid offspring.

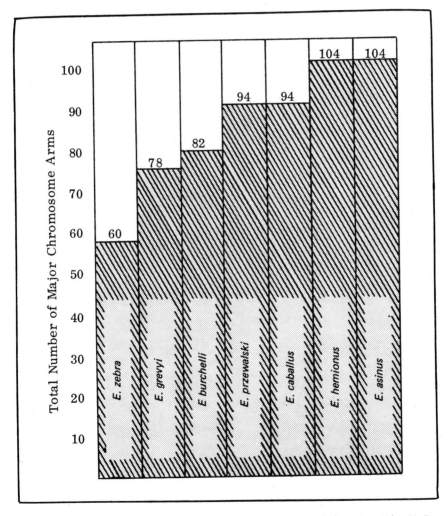

Figure 8–13. Some relationships of chromosome arms among the species. (After K. Benirschke: Sterility and fertility of interspecific mammalian hybrids. *In* Comparative Aspects of Reproductive Failure. New York, Springer-Verlag, 1966.)

The one-half hybrid had 65 chromosomes in each cell. In this case the odd number of chromosomes does not seem to produce the sterility which occurs in mules.

Benirschke, after studying various mammalian hybrids, concluded that failure of gametogenesis, the production of sex cells, may be largely due to failure of synapsis during meiosis, that is, an inability of the chromosomes to pair properly. He has found that similarity in shape and size does not necessarily guarantee fertility when crossed. It is surely the amount of homologous DNA that is the decisive factor. Benirschke has also studied the total number of major chromosome arms in each of the equine species that have been karyotyped. His findings are shown in Figure 8–13. Metacentric and submetacentric chromosomes would have two major arms, and acrocentrics would have only one major arm. Studies of other mammalian species indicate that two acrocentrics may fuse to form a metacentric or a submetacentric. Evidently this must be the major chromosomal difference between *E. przewalski* and *E. caballus*.

This process is technically known as centric fusion (Figure 8–14). It could well be that centric fusion occurred somewhere in the evolution of *E. caballus* from an *E. przewalski* ancestor, transforming acrocentrics into submetacentrics and reducing the total chromosome number.

During the process of centric fusion, one centromere is left with very short arms with only a small amount of heterochromatin attached. Since this part of the chromosome probably contains no functional genes, its eventual loss in no way hinders the reproductive capacity of the animal. Therefore, over several generations the small centromere formed as a leftover product of a previous centric fusion is eventually lost. This loss reduces the number of

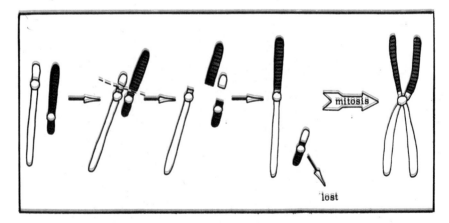

Figure 8–14.   The process of centric fusion. The diagram begins with two unrelated chromosomes, which break apart in close proximity and fuse back together in an unnatural position, resulting in exchange of parts.

Figure 8-15.   An equine hybrid family: (a) zebra, (b) mare, and (c) zebroid hybrid.

Figure 8-16.   Exmoor pony mare and Kiang hybrid.

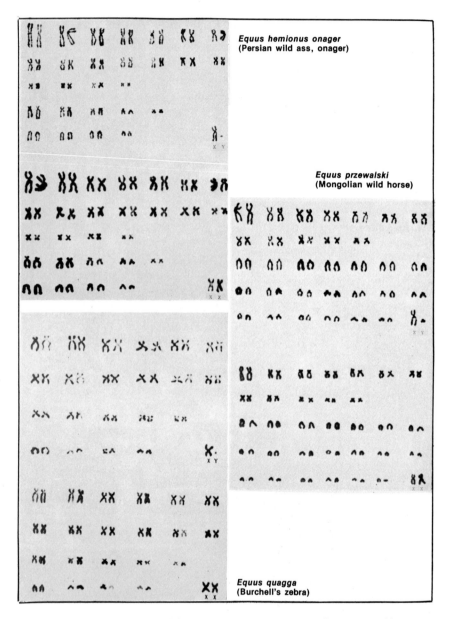

Figure 8–17. Karyotypes of three equine species. (Courtesy of Kurt Benirschke.)

centromeres in the karyotype, but the number of major arms remains the same.

In the same connection, Benirschke points out that *E. hemionus,* with 56 total chromosomes, has the same number of major arms as *E. asinus,* with 62 total chromosomes. Hybrids of this cross would perhaps be fertile. The extreme variation in that N.F. (number of major chromosome arms) among the equine species indicates that chromosomal rearrangement is more complicated than shown in Figure 8–14, and that various inversions and translocations have undoubtedly occurred.

Although there have been numerous reports of fertile mules, none have been well documented histologically. Benirschke surveyed 47 pairs of mule ovaries and found very few ova. Theoretically it should be possible for the chromosomes to segregate during meiosis in such a way that a germ cell would contain all of the chromosome material necessary for development and viability. Therefore, the possibility of a fertile mule cannot be completely ruled out.

Roberts reported the breeding of a Somaliland zebra, *E. grevyi,* to a number of domestic mares. The zebroids that were produced were all sterile (Figure 8–15). Other crosses of the same species coupled with cytogenetic studies have shown these hybrids to have 55 chromosomes. Earl Stuber of Phoenix, Arizona, once bred a zebra to Shetland ponies. He called the offspring Zeonies. An equine Kiang hybrid is shown in Figure 8–16.

The African zebra, *Equus burchelli* (which has 44 chromosomes), has been crossed with the donkey (which has 66 chromosomes); the hybrid has 53 chromosomes and is sterile. Chromosomes of three equine species are shown in Figure 8–17.

CHROMOSOME ANOMALIES

Breaks in the arms of chromosomes occur quite often. Usually these breaks "heal" quickly, regaining normal chromosome structure; however, occasionally when breaks in two chromosomes occur at about the same time, the broken pieces of the chromosomes may "heal" improperly. The broken pieces may switch places and adhere to the wrong chromosome, in a process known as translocation. A reciprocal translocation occurs when nonhomologous chromosomes mutually exchange material.

Translocations themselves are not harmful because all of the genetic material is still present. None has been lost from the cell, but it is present on a different chromosome. Translocations lead to unbalanced genetic conditions in later generations. The process which leads to an unbalanced condition is shown in Figure 8–18.

Six types of offspring may occur when a parent carries a reciprocal translocation. The parent is not affected abnormally. A hypothetical translocation between chromosomes 1 and 10 of the horse is shown in the figure. (Chromosomes are usually numbered beginning with the largest down to the

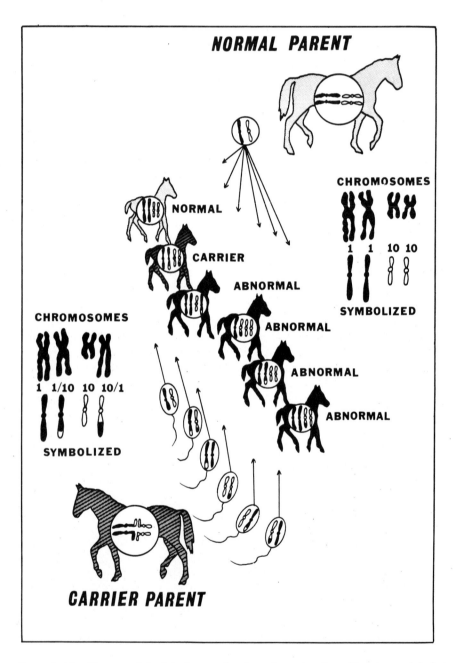

Figure 8–18.   Diagram of the inheritance of reciprocal translocations. The abnormal foals generally die at such an early stage of development that it appears that the mare has a sterility problem.

smallest.) All the other chromosomes are normal and therefore are not shown. The carrier parent has only one normal number 1 chromosome and only one normal number 10 chromosome. The other members of these pairs are both abnormal; the other number 1 chromosome carries a small portion of number 10, and the other number 10 carries a small portion of number 1. These abnormal pairs are referred to as a 1/10 translocation and a 10/1 translocation respectively.

When the chromosome number is halved at meiosis, not all of the number 1 and number 10 material may end up in a single gamete (sperm or ovum), but since six possibilities are available for the gametes, one gamete may receive both of the normal chromosomes, resulting in the development of a normal foal after conception.

Another possibility is that one gamete may receive both of the translocated chromosomes, 1/10 and 10/1, which after conception will provide the foal with the same chromosome complement as its carrier parent. This animal is then also a carrier.

The other possible gametes receive one normal and one of the translocated chromosomes, 1 + 1/10; 1 + 10/1; 10 + 1/10; and 10 + 10/1. All of these gametes when combined with a normal gamete from the opposite sex will result in an abortion before term.

Occasionally two breaks may occur in the same chromosome. If the middle portion of the chromosome, between the two breaks, becomes inverted before the breaks "heal," the resultant condition is technically known as an inversion. This situation changes the order of the genes on the chromosome and can interfere with normal synapsis during meiosis.

Payne reported a condition in a mare in which a suspected inversion of one X chromosome led to abnormalities of both X chromosomes after meiosis. The process is illustrated in Figure 8–19, which shows hypothetical

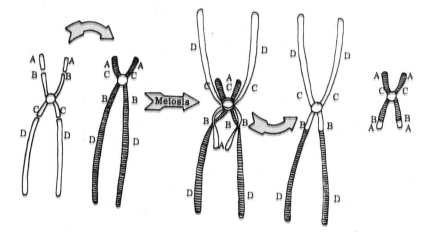

Figure 8–19.   Diagram of a possible formation of an abnormal chromosome.

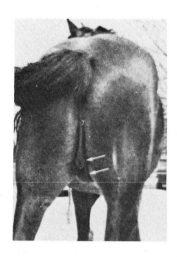

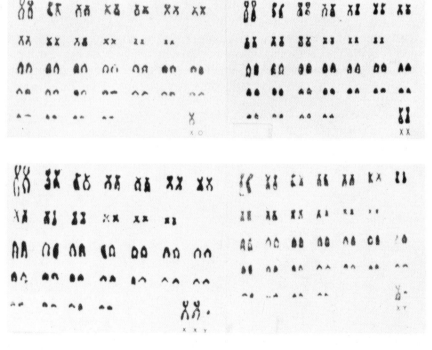

Figure 8–20.   An equine intersex. The animal has both a vulva and a testicle. The karyotypes, which show abnormal sex chromosome numbers, were obtained from the testicle of the intersex.

genes A, B, C, and D. The inversion is evident because genes B and C have inverted their position. Because the centromere is involved in the inversion, the overall arm ratio or general shape of the chromosome is changed. During meiosis the abnormal X chromosome attempts to pair up, gene for gene, with the normal X chromosome. The process of meiotic crossing over results in two abnormal X chromosomes in the primary oocyte. The small X was subsequently passed to an ovum which was evidently fertilized by an abnormal spermatozoon, one which had accidentally received no X chromosome. The offspring, a thoroughbred filly, developed into a good race mare but later proved to be infertile. Upon examination, her karyotype revealed no normal X chromosomes, only the one abnormal X.

Chromosome abnormalities seem to be more common than formerly thought. One of the most frequent problems is the failure of the X and the Y to separate during meiosis. This inability to separate is known as nondysjunction. One gamete receives no sex chromosomes, while the other receives them all.

B.K. Basrur photographed the intersex in Figure 8–20, which also shows the abnormal karyotypes of this horse. Nondysjunction during cell division has given rise to a variety of sex chromosome relationships. Karyotype analysis of various cell types of the body reveals not only normal complements but XXY complements in some cells and XO (one X and no Y) complements in others.

Sterility of both mares and stallions can be either complete or partial from genetic causes. Recessive genes can cause physiological or anatomical abnormalities which interfere with the normal processes of breeding or reproduction. These abnormalities would include a lack of development of the genitals. In other cases lethal genes can cause early embryonic death, with the only external signs being the mare coming back into heat late in the summer after she was thought to be pregnant. Abnormal chromosomes, as in the case cited, can cause abnormal development of the genitalia and hormonal processes resulting in complete sterility.

A stallion carrying an autosomal chromosome abnormality will display partial sterility, as diagrammed in Figure 8–21. Generally only about 18 percent of the mares bred to the stallion will settle normally and foal. Of the balance, some mares (about 10 to 20 percent) will not settle due to nongenetic problems unrelated to the stallion's fertility. The remainder of the mares will conceive but the embryos will carry severe chromosome imbalances which soon cause death and subsequent abortion.

In a mare, the signs of this chromosomal sterility will be a mare that is usually settled very late in the breeding season, or not at all. The sequence of events is diagrammed in Figure 8–22.

Centric fusion is illustrated in Figure 8–14. It seems to be an important process in the evolution of most species, reducing the total chromosome number without significantly reducing the number of genes. The centromere with small arms is eventually lost because it contains no vital genes. The

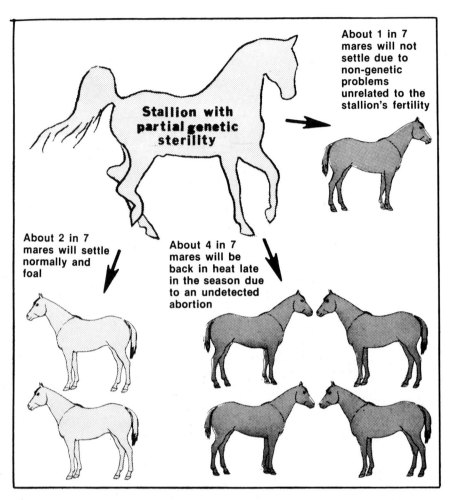

Figure 8–21.   A diagram showing the results of partial genetic sterility in a stallion. This type of problem could be a result of the stallion being a carrier of an autosomal chromosome abnormality.

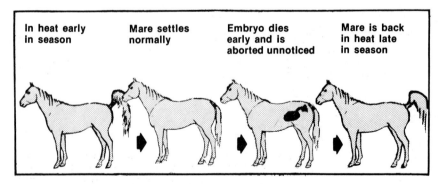

Figure 8–22.   A diagram showing a typical breeding pattern in a mare carrying an autosomal chromosome abnormality.

INHERITANCE OF CENTRIC FUSION TRANSLOCATIONS

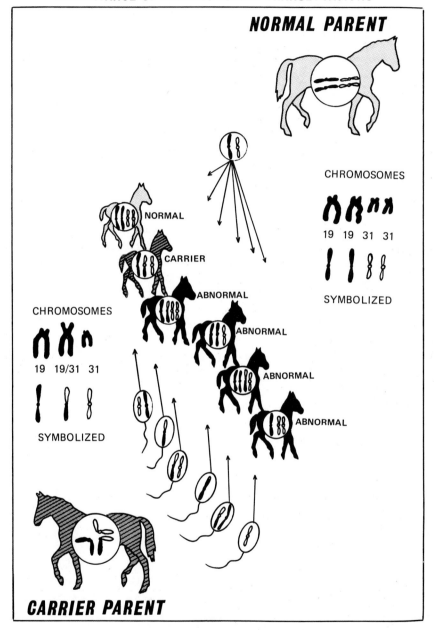

Figure 8–23.   A diagram of the inheritance of centric fusion translocations. In these cases foals may be carried to term because all of the chromosomal material is present. Tiny bits of genetic material may be missing causing abnormalities in the live foals.

chromosome material near the centromere is referred to as heterochromatin, and many indications suggest that it is genetically inert.

Once centric fusion has occurred and an animal becomes a carrier, some of the gametes will have abnormal karyotypes. This occurs as illustrated in Figure 8–23, using as an example a hypothetical fusion between number 19 and number 31. These chromosomes are both acrocentric and together they become a submetacentric chromosome. As with the reciprocal translocation, six distinct types of gametes are produced from the number 19 and the number 31 chromosome: one gamete may completely lack a number 19 chromosome and, when combined with a normal gamete of the opposite sex, will produce a zygote with only one number 19 chromosome (technically termed monosomy 19). If a gamete without a number 31 chromosome combines with a normal gamete, monosomy 31 develops. Both of these conditions would be lethal in a developing embryo and abortion would follow.

A gamete of the carrier parent might have a normal number 24 chromosome along with the 19/31 translocation, or a normal number 19 with the 31/19 translocation. Either would produce a trisomy upon fertilization, which would also be abortive. A trisomy is the presence of three like chromosomes within a karyotype.

Some of the gametes of the carrier will contain the 19/31 chromosome without the normal number 19 or number 31. Upon fertilization a carrier foal is produced with the same karyotype as the carrier parent.

The final possibility is for some of the gametes to contain a normal number 19 and a normal number 31 chromosome. Of course, foals with normal karyotypes will develop from these.

It is probable that a relatively large number of horses carry a centric fusion or another translocation. They could produce some normal or carrier foals without ever being suspected. An occasional abortion could be attributed to other causes.

INHERITANCE AND SEX

Sex is associated with inheritance in three ways: certain genes are located in the sex chromosomes and they are sex-linked; other genes are influenced in their expression by sex hormones and are called sex-influenced; other genes express themselves only in one sex, such as cryptorchidism in the stallion and milk production of the mare, and are called sex-limited.

Sex is determined by a pair of chromosomes, the sex chromosomes of the female being XX, or homogametic, while the male is XY, or heterogametic. Since homo means alike and hetero means unlike, homogametic means that the gametes produced are alike. The sex chromosomes are diagrammed in Figure 8–24. The other chromosomes are referred to as the autosomes in contrast to the sex chromosomes. Among mammals, the Y chromosome does not generally possess many genes.

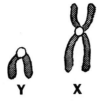

Figure 8–24.   The sex chromosomes of the horse. The Y is a small acrocentric chromosome; the X is a large metacentric chromosome.

It would be assumed that mating a stallion to a mare would give a sex ratio of 1 male to 1 female, as outlined in Figure 8–25. Federico Tesio, in his book, *Breeding the Racehorse,* lists totals of colts and fillies foaled:

Thoroughbreds foaled in England in 1928	1819 colts 1792 fillies
Thoroughbreds foaled in England in 1925	1428 colts 1389 fillies
Thoroughbreds foaled at Tesio's farm 1900–1945	271 colts 260 fillies
Showing an excess of	77 colts

This figure turns out to be slightly more than 1 percent more colts than fillies born on the overall average. This sort of ratio has also been found in human births.

Considerable evidence shows that more males are conceived than females and that prenatal and early postnatal mortality is greater in the male than in the female. It is easy to understand why there is greater early life mortality in the male, when one considers sex-linked lethal genes (explained later); however, the greater conception of males has been a puzzle for some time.

Scientists have postulated that since the Y chromosome is very small compared with the X chromosome, the Y-bearing gamete should be lighter and could conceivably move more rapidly than the X-bearing sperm. If this were the case, the Y-bearing sperm would win the race to the egg more often than the X-bearing sperm. Attempts at separating X and Y bearing sperm by centrifuging have been totally ineffective.

It has been noted by Russian and French scientists that the electric charges on the X-bearing sperm are the same as those of the egg, while the charges on the Y-bearing sperm are opposite. It is well known that the like charges repel, but opposite charges attract each other. When semen is put in

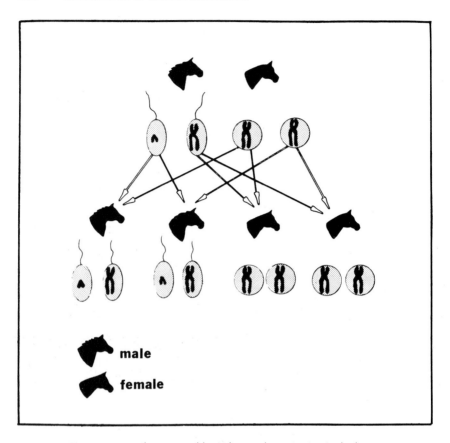

Figure 8–25.   Chromosomal basis for sex determination in the horse.

an electrolyte diluter and a current is passed through the tube containing the semen and diluter, the Y-bearing sperm tend to go to the cathode, while the X-bearing sperm tend to go to the anode. It was thought at first that this procedure would be an excellent way of controlling the sex of the offspring in farm animal production. The method, though apparently effective in separating X-bearing from Y-bearing sperm, is quite unsatisfactory because electrolytes damage sperm so badly.

Only a few cases of sex-linked genes are known in horses. Archer has reported a sex-linked hemophilia (bleeders disease) in the horse. In Figure 8–26, *H* is illustrated as normal blood clotting, with *h* being the gene for hemophilia. The offspring produced are shown for several types of matings.

This kind of mating is similar to ordinary inheritance in which a homozygous dominant is mated to a recessive and produces all dominant offspring. It differs from autosomal inheritance genetically because the female offspring produced is heterozygous, but the male is hemizygous

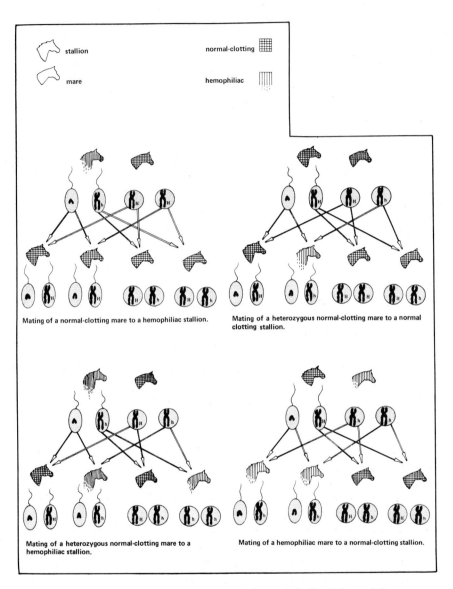

Figure 8–26.   The inheritance pattern of the sex-linked trait, hemophilia.

dominant. Hemizygous refers to the fact that the male has only one sex-linked gene or half what the female has.

When a normal-clotting carrier female is mated to a normal-clotting male, the offspring produced will have an overall phenotypic ratio of 3 normal clotting to 1 hemophiliac. Both females have normal clotting, while one male has normal clotting and the other is a hemophiliac.

If a heterozygous normal-clotting mare produces offspring by a hemophiliac stallion, the male and female offspring will show a ratio of 1 normal clotting to 1 hemophiliac. This type of mating is similar to autosomal inheritance where a heterozygote is mated to a recessive and produces 1 dominant to 1 recessive within each sex.

Suppose a hemophiliac carrier mare produces foals by a normal-clotting stallion. In this case all the male offspring will be hemophiliacs and will be like their dam, while all the female offspring will be normal clotting and resemble their sire. This is known as criss-cross inheritance and is the basis for differentiating sex-linked traits from others. If the recessive is brought into the mating through the female parent, XX or homogametic, and the dominant into the mating through the male parent, XY or heterogametic, the male offspring will all look like their mother, and the female offspring will look like their father if the genes are sex-linked. If the genes are not sex-linked, one will either obtain all showing the dominant or 1 dominant to 1 recessive within each sex, depending upon whether the dominant parent is homozygous or heterozygous.

The sex hormones can influence the expression of genes, particularly when they are in the heterozygous condition. A good example is found in baldness in man. A man that is either homozygous dominant or heterozygous for baldness will be bald and he must be homozygous recessive not to become bald. A woman that is either homozygous recessive or heterozygous for baldness will have normal hair and she must be homozygous dominant to be bald. Castration of heterozygous males stops hair recession, but administration of testosterone will reinitiate the loss of hair (baldness) after castration. This influence of testosterone on a sex-influenced trait in cattle has been well documented by Gilmor. He performed castration and testosterone injections in cattle heterozygous for the sex-influenced gene affecting blackish color, which causes the blackish to be dark if testosterone is present and light if it is not. There are many indications that white spotting on the rump of some Appaloosa horses is influenced by the male hormone so that males "color up" more readily than females.

Early in this century, Bruce Lowe theories had great influence on thoroughbred breeders. He had observed some basic differences in the way mares transmit inheritance and the way it is transmitted by stallions. Without knowing about sex chromosomes, he observed some of the subtle results of their differences. Although many of his conclusions were farfetched, he was ahead of his time in many of his observations. He is quoted by John Wall in *Breeding Thoroughbreds:*

In speaking of the breeding of great race fillies, Bruce Lowe said an unusually clever sire inherits his brilliancy from his mother. In the case of a daughter, the inheritance is apt to be the characteristics of her father. Good fillies especially are the result where running and effeminate strains are mixed with those same strains of the sire. The best blood of the stallion's sire should be nicked as well as his family member.

This topic is covered in detail in Chapter 12.

BIBLIOGRAPHY

Archer, R.K.: True haemophilia (haemophilia A) is a thoroughbred foal. Vet. Rec. 73:228–340, 1961.

Basur, P.K., Konagawa, H., and Gilman, J.P.W.: An equine intersex with unilateral gonadal agenesis. Can. J. Comp. Med. 33(4):297–306, 1969.

Benirschke, Kurt: Sterility and fertility of interspecific mammalian hybrids. In Comparative Aspects of Reproductive Failure. Edited by K. Benirschke. New York, Springer-Verlag, 1966.

Braend, M.: Genetic variation in horse hemoglobin. Hereditas 58:385–392, 1967.

Gilmore, L.O., and Feckheimer, N.W.: Gene-hormone Interactions on Hair Pigmentation in Cattle. Genetics Today. Proc. XI Int. Congress. Gen. (The Hague) 1:264, 1963.

Hughes, J.P., Benirschke, K., Kennedy, P.C., and Trommershausen-Smith, A.: Gonadal dysgenesis in the mare (a report of five cases). J. Reprod. Fertil. (Suppl.) 23:385, 1975.

Hsu, T.C., and Benirschke, K.: An Atlas of Mammalian Chromosomes. New York, Springer-Verlag, 1967.

Jacob, F., and Monod, J.: Genetic regulatory mechanisms in the synthesis of proteins. J. Mol. Biol. 3:318–356, 1961.

Kilmartin, J.V., and Clegg, J.B.: Amino-acid replacements in horse haemoglobin. Nature (London) 213:269–271, 1967.

Koulisher, L., and Frechkop, S.: Chromosome complement: a fertile hybrid between E. przewalski and E. caballus. Science 151:93–95, 1966.

Oliver, L., and Hartree, A.S.: Amino acid sequences around the cystine residues in horse growth hormone. Biochem. J. 109:19–24, 1968.

Payne, H.W., Ellsworth, K., and DeGroot, A.: Aneuploidy in an infertile mare. J. Am. Vet. Med. Assoc. 153:1293–1299, 1969.

Rauchbach, K.: Hermaphroditism in the horse. Mh. Vet. Med. 16:220–221, 1961.

Tesio, Federico: Breeding the Racehorse. London, J.A. Allen & Co., 1958.

Wall, John: Breeding Thoroughbreds. New York, Scribner Sons, 1946.

9

coat color inheritance

The original wild horses probably had subdued colors, with only an occasional mutant showing a pattern of white. If early man was anything like today, he admired the creatures which showed striking colors. After domestication, much of the selection process was probably based on coat color. Some of man's earliest concepts of inheritance grew out of observation of color transmission in horses.

Today a buyer is often willing to pay a fantastic price for a horse with color markings to his liking. Because of this, many color registries have developed: the albino, palomino, buckskin, paint, pinto, Appaloosa, and others. This adds up to a large number of horse breeders basing their entire breeding programs on coat color.

Other registries, such as the American Quarter Horse Association and the Arabian Horse Registry, make an attempt to keep certain color patterns out of their breed. They allow white coloring of the legs to extend up only a certain distance, and they refuse to register horses with body spots regardless of parentage.

Registries often use known facts of color inheritance to check the authenticity of registration applicants. As an example, the Arabian Registry will not register a bay from two chestnut parents, and a gray must have at least one gray parent to be eligible.

Although many stud book records are available to study color transmission through many generations, much work still must be done in analyzing them to determine the facts. The greatest amount of study has involved the basic coat colors, and most of the facts in this area have been established. Further study of registry records regarding the inheritance of white patterns would be helpful in gaining a more complete picture in this area.

Most of our present knowledge about white patterns has been extrapolated from known facts in other mammals. The majority of coat color genes are surprisingly similar from one species to the next. Color patterns and inheritance in an animal such as the mouse, therefore, when studied and properly compared, can produce perhaps more accurate information than more short-termed studies using the horse.

BASIC COAT COLORS

Every horse has more than a dozen genes working directly to bring about a particular coat color and pattern and probably twice that number exerting some indirect effect on coat color and pattern. In many ways these genes each affect one another, bringing about a large number of color phenotypes. Some genes affect hair color, some affect the color pattern on the body, and some affect both color and pattern.

As discussed in Chapter 6, some color genes are dominant. An example is black symbolized with a B. Other genes are recessive, such as the chestnut. Actually this gene is the recessive form of B, and is symbolized as b. Other color genes are intermediate in their effect, such as the A^t allele of the bay gene which causes the brown horse (intermediate between black and bay.)

Most of the color genes show their effects only with their so-called abnormal alleles. The D gene, for instance, in its mutant form causes a dilution of the coat color to a dun, but the normal allele of the D gene shows no effect.

The common coat colors of the horse are quite familiar to most horsemen. The names for the various color types vary from one locality to another. Following is a description of the coat color types and the proper names for each.

Black. The black color is rather basic and simple. It is surprising how many breeders confuse the brown and black coat colors. A horse should not be designated black unless the black color is complete on the muzzle and flanks. This is exclusive of white markings.

Brown. A brown horse can be nearly black. In fact many brown horses are mistakenly registered as black. A dark brown horse will have some reddish or brown hairs scattered about the head and body, particularly on the muzzle and flanks. Brown horses vary from nearly black to nearly bay. Some breeders who do not understand what a brown is, have mistakenly registered such horses as bay, simply assuming them to be dark bays.

Chestnut. The designation chestnut should be reserved for what some call "liver chestnut." The practice of grouping all light coat colors, sorrels, together and calling them "chestnut," as done by the Arabian breeders, makes for much confusion. Chestnuts are dark in color—not light. Since many Arabians have a heavy dark skin pigment, a lighter coat color will tend to look darker. There are variations of chestnut colors, from light to dark, but none of these are as light as sorrel. The darkest chestnut horses are sometimes confused with brown horses. The lightest chestnut horses can easily be mistaken for sorrel horses. The mane and tail are always a bit darker than the body, but not black. The legs of the chestnut are the same color as the body.

Grulla. Will James' Smokey, the Cow Horse, was a grulla (sometimes spelled grullo.) The body is a solid "mousy" gray color, but is not easily confused with the true gray horse. Grulla is very even in color, somewhat like "battleship gray." The mane, tail, and legs are always black. The grulla color varies from dark, which is easily confused with black or brown, to light, which sometimes passes as a normal gray. The lighter grulla horses may have a black line down the back from the mane to the tail.

Yellow Dun. This color is nearly brown but varies from a muddy yellow to nearly grulla in appearance. It is a darker color than the red dun. The mane, tail, and legs are always black.

Red Dun. This color is lighter and more red than is the yellow dun. The mane and tail are dark, but of a color similar to the chestnut. Often there is a narrow dark stripe down the back from the mane to the tail. The body color is often the same as that of the buckskin, but the mane and tail are lighter. The legs are the same shade of color as the body.

Buckskin. The buckskin name comes from the color of tanned deer hide, which approximates the body color of a buckskin horse. The mane, tail, and legs are black. The darker buckskins approach the color of a yellow dun. A very light buckskin is sometimes seen, with the body being more of a cream color and the mane, tail, and legs much lighter in color.

Blood Bay. The blood bay has a rich reddish mahogany body color with black mane, tail, and legs.

Yellow Bay. The body color of the yellow bay is a bit more dull than that of the blood bay. The mane and tail are black but sometimes the legs are lighter. These horses can be confused with the buckskin.

Dark Bay. The dark bay has a scattering of dark hairs throughout the body. The mane, tail, and legs are black.

Sorrel. The sorrel is lighter in color than the chestnut, and darker than the palomino. The mane and tail may be the same color as the body or flaxen (blond). The color varies from a copper-red to a tan. In general, the legs are the same color as the body.

Palomino. A body color varying from dark cream to a newly minted penny with a white mane and tail denotes a palomino. The leg color is the same as the body. Palominos can be classified as light, medium, and dark.

Creamello. A creamello horse is nearly white. The lighter varieties are close to pure white at certain times of the year. The darker varieties approach the palomino color. The mane and tail vary from white to a muddy yellow color.

Basically five genes appear to bring about hair color; others generally have to do with various white patterns over the body. The basic hair-color genes have been designated by the first of the alphabet: A, B, C, D, and E.

The A locus has at least three alleles, according to Castle and Singleton The recessive allele, a, has no visible effect and is the so-called normal allele. The A allele causes a black horse to be a bay and may be responsible for the difference in types of chestnuts and sorrels. The A^t allele (according to Castle) produces the brown pattern on an otherwise black horse and may even produce modifications of both chestnut and sorrel.

The B locus determines the type of melanin produced throughout the body. The dominant allele produces eumelanin, giving a black color; the recessive allele produces phaeomelanin, which causes a liver color unless it is modified by other genes, such as A.

The C locus may have only one allele in the horse, that being the dominant allele causing no abnormality in color. The recessive allele occurs in most other mammals, including man, producing the albino. Since the true albino gene always causes a pink-eyed white, it should not be confused with phenotypes having pigmented eyes.

Wright, studying the dilution of hair color in guinea pigs, described a series of alleles at the C locus years ago. A Spanish geneticist, Adriozola, picked up on this terminology in describing palomino dilution in the horse. He attributed the palomino effect as due to a C allele that he called C^{cr}. Since then other geneticists have followed his lead and continued to describe the various dilution effects in the horse as caused by a C^{cr} allele. Recently research reports with very limited numbers of horses have tried to show two separate dilution genes in the horse; C^{cr} and D.

The D locus is named after its diluting effect on normal body color. The recessive allele does not have a diluting effect. When homozygous, the D allele has an increased dilution effect. It dilutes the black horse to a grulla color. The brown horse is diluted in a similar manner. The bay becomes a medium buckskin when heterozygous and a light buckskin when homozygous. The chestnut when heterozygous for the D gene may pass as a sorrel and would be a red dun when homozygous. The sorrel with dark mane and tail becomes a red dun when heterozygous, and almost white when homozygous. The mane and tail always retain some color, however. The sorrel with flaxen mane and tail becomes a palomino when heterozygous and an almost white horse when homozygous. A very light sorrel with flaxen mane and tail often passes for a so-called albino when homozygous for the D gene. A dark chestnut with flaxen mane and tail (quite rare in most breeds except the Shetland) when homozygous for the D gene would perhaps resemble a typical heterozygous palomino produced from a sorrel. Evidently two separate recessive genes are involved in the inheritance of the light mane and tail.

The tarpan horses, a wild type in Europe, were in the past all of the same color, somewhat of a mixture between a buckskin and a grulla. Their uniformity in color over many generations indicated that they were homozygous for all color genes. Evidently, the *D* gene is homozygous as well as the *B* gene. Probably an allele of the *A* locus gives a reddish mixture to the grulla body color.

The *E* locus has been shown to be an important gene in color inheritance of the guinea pig. Little proposed a number of alleles for this locus in the dog to explain such color patterns as the mask and brindle. Castle maintained that horses also have several alleles at the *E* locus. The kind of effect that these alleles have, however, is pure speculation at this point. Castle thought that the recessive *e* allele when homozygous could produce a bay in the absence of the *A* gene. This would mean that either the *e* or *A* alleles could change a black to a bay. The *e* allele with the *A* allele would change a blood bay to a yellow bay and the *e* allele with an *aabb* genotype would change a chestnut to a sorrel. Further clarification of the action of these genes is given later in the discussion on the cellular aspects of coloration.

Considering only the aforementioned basic color genes, some rather important conclusions can be drawn by the color breeder. One of the most important is that since it is much easier to establish a recessive trait as a homozygous condition within a herd, those colors which are recessive would be the easiest and quickest to establish in a pure breeding form. The possible genotypes for the more common body colors of the horse are listed in Table 9–1. These same genes bring about the particular color of most spots in the Appaloosa and paint patterns. If a breeder desires a black horse with a white blanket containing black spots, he must concentrate on

Table 9–1. Possible genotypes for the more common color patterns of the horse.

PHENOTYPE	GENOTYPES
Black	aaB-ddee
Brown	A^t-B-ddee
Blood Bay	A-B-ddee
Yellow Bay	A-B-ddE-
Dark Bay	aaB-ddE-
Chestnut (liver)	aabddee
Chestnut (medium or light)	aabbddE-
Sorrel	A-bbddee
Sorrel	aabbddE-
Sorrel	A-bbddE-
Sorrel	aabbDdee
Grulla	aaB-D-ee
Brown Grulla	A^t-B-D---
Buckskin	A-B-D-ee
Red Dun (or Palomino)	A-bbDdE- or aabbDDee
Creamello, or a yellowish white body color	--bbDD--

breeding a herd to be homozygous for aa and BB as well as for the blanket gene. (The various Appaloosa genes are discussed later.)

MODE OF ACTION

In order to obtain a true understanding of hair coloration, it is necessary to discover what occurs on the cellular and biochemical levels. This includes the study of development and growth of hair, the development and migration of the melanocyte which produces the melanin granule, and all of the various enzymes and other biochemicals associated with these processes.

The development of the hair follicle in the fetus begins with a thickening of the skin epithelial cells. The cells which are just under the skin, called mesenchymal cells, then begin to aggregate in these thickened areas (Figure 9–1). Usually during this period of development, pigment cells or melanocytes are found just beneath the surface of the skin. As the hair cone begins to form from the epidermal cells, some of the melanocytes move into the

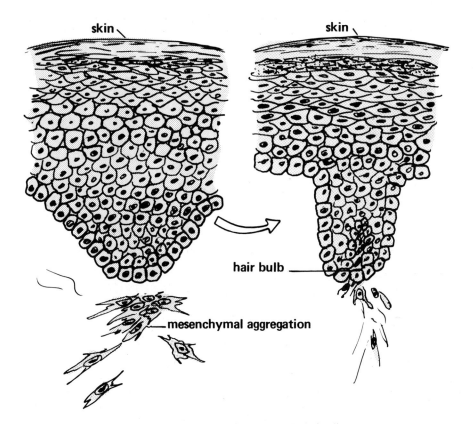

Figure 9–1.   Aggregation of mesenchymal cells.

hair bulb, where they will begin to inject melanin granules into the growing hair.

The mature hair is composed of a cortex and a medulla. The cortex is made of flattened cells composed mostly of keratin, in which pigment granules may sometimes be found. The medulla is essentially a hollow cavity with periodic cells or rows of cells across the center. Pigment granules are usually quite abundant in these cells.

It is important to remember that there are several kinds of hair on most species of mammals, including the horse. The body hairs consist of the underhairs, which are called "down-hairs," and the longer overhairs or the "guard-hairs." The morphology and the melanosome type and distribution differ in these two types of hairs. As will be discussed later, both of these factors lead to color differences. The mane and tail hairs in a horse are of a third type, giving rise to yet another color variation.

Structural abnormalities of the hair or hair bulb can interfere with the transfer of melanosomes from the melanocyte to the hair. Gruneberg has found an example of this phenomenon occurring in the mouse. The abnormal hairs occur in a rather random fashion throughout the body coat, giving a splotchy appearance.

The mature melanocyte is much like a unicellular gland which produces melanin granules (Figure 9–2). These cells originate from precursor cells called melanoblasts. They do not differentiate from the epidermis, but rather

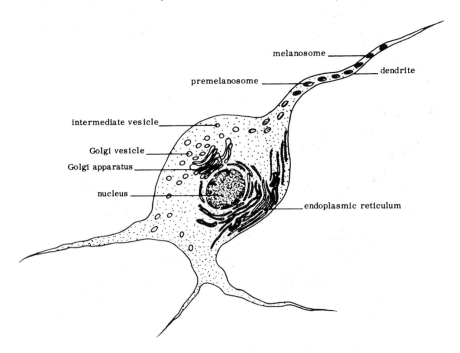

Figure 9–2.   The melanocyte.

migrate there during fetal life from the neural crest. The neural crest is an early embryonic structure which lies along each side of the neural tube and, therefore, has the same origin as nerve cells. Some spotting genes exert their effect on the cells of the neural crest, as will be discussed later. Often this type of spotting is accompanied by pleiotropic effects (side effects) of the nervous system.

Although melanoblasts originate from a few original cells at the neural crest, they may eventually, through mitotic division, become very numerous at the dermo-epidermal junction. From there they can migrate into the hair follicles or remain at the junction injecting melanin granules into the epidermal cells. As the melanocytes become extremely differentiated, they flatten out and move to the surface, becoming indistinguishable from the squamous epithelial cells. Searle points out that melanocyte density usually has no relationship to skin pigmentation; however, it may be responsible for any observable variation. Movement of the melanoblasts during development is illustrated in Figure 9–3.

Schiable gives evidence that the movement of melanoblasts is restricted to definite patterns in the mouse fetus. He shows in Figure 9–4 that the body

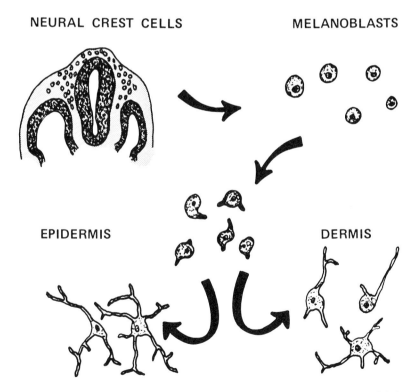

NEURAL CREST CELLS   **MELANOBLASTS**

EPIDERMIS   **DERMIS**

Figure 9–3.   Movement of melanoblasts during embryonic development. The upper left shows a cross section of a developing embryo, with the neural crest at the top.

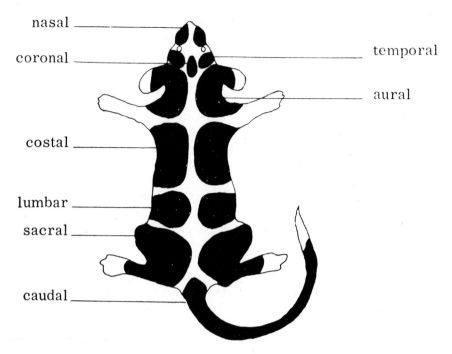

Figure 9–4.   The origin of color patterns in the mouse. (After R.H. Schiable: Developmental Genetics of Spotting Patterns in the Mouse. Ph.D. Thesis, Iowa State University, Ames, IA, 1963)

has 14 unique areas in terms of melanoblast migration. Consequently, some areas of the body may lack melanocytes and this deficiency results in a characteristic white spotting. One such process is exemplified by the belted gene. The hair follicles in both costal areas of the mouse contain no melanocytes. It is not clear whether the site of effect is in the area of the neural crest or in the epidermis. One may conclude from the evidence, however, that an entire clone of melanoblasts at the neural crest is inhibited from migration. Consequently, a particular area is white except for spots where melanoblasts may have migrated in from an adjacent area and proliferated. Good evidence of melanoblast migration is given by Billingham and Silvers.

Most Appaloosa breeders are familiar with the movement, enlargement, and even sudden disappearance of spots in that breed even after birth. Wagner gave specific examples of changing spots in a number of horses. One mare had small white migrating spots, ranging in size from very small to the size of a silver dollar. During a 3-month period at 5 years of age, a spot on each side of the croup enlarged to the size of a man's hand and then disappeared. Another Appaloosa mare lost the dark pigmented spots on the sides of her face at the age of 8, and 4 years later the spots returned. This characteristic in many Appaloosas is strong enough evidence to conclude that with some particular genetic backgrounds the melanoblasts continue to migrate throughout adulthood.

The Golgi apparatus is the cellular organelle responsible for production of all glandular products, and in the case of the melanocyte the product is the melanin granule. It begins in the endoplasmic reticulum of the Golgi apparatus, where synthesis of a protein occurs in conjunction with the ribosomes (Chapter 8). Seiji describes a process of condensation that occurs in which a phospholipid combines with the protein, producing a small granule (approximately 0.05 μ in diameter) in the Golgi vesicle. In the case of eumelanin, this small granule continues to enlarge to an ovoid shape of about 0.7 μ by 0.3 μ, which is designated as a premelanosome and contains no melanin. Albino melanocytes contain no granules developed beyond this point. Mayer postulates that this granule consists of a complex aggregate of protein subunits in the form of concentric or spiral sheets bound to a lipid component surrounded by an outer membrane. Eventually melanin is deposited at various sites on either side of the inner membrane. Melanin itself is formed from tyrosine in a manner outlined by Harris. Evidently the protein fibers do not form a well-organized premelanosome. Phaeomelanin is deposited in a rather random fashion on the tangled mass of fibers in the Golgi vesicle. Its exact chemical nature is unknown. Phenotypically the difference is expressed as a yellow-appearing hair rather than black, as with eumelanin. A summary of the process of melanin granule formation is given in Figure 9–5.

Once the melanin granules are formed, they move to the dendrites of the melanocyte and from there are transferred to another cell. This movement and transfer can be in a rapid and regular manner or, as in some genotypes, may be periodically interrupted because of an abnormally shaped dendrite or other dendritic abnormalities. In some genotypes the dendrite may be constructed so that the melanin granules are placed in abnormal locations within the recipient cell, or the granules may be abnormally clumped. Several examples of these abnormalities will be given later. There is evidence that the chemical cellular environment of the melanocyte may be able to turn the production of melanin granules on and off.

The color pattern of many mammals is known to change from season to season, or with advancing maturity. A fawn, for instance, is born with spots which disappear within a few months. Quite probably systemic hormones change the cellular environment in this case, bringing about eventual pigmentation of the white areas.

Some of the color genes in the Appaloosa may be partially sex influenced. In this respect the genes are somewhat like those of the peacock or pheasant, where the male is highly colored and the female is plain colored. Although the distinction is not as clear in the Appaloosa horse, most breeders agree that the male tends to "color up" more readily than the female.

Nomenclature	Morphology	Process
Ribosome	• 100-150 Å	RNA and protein biosynthesis.
Golgi vesicle	o 0.05μ	Phospholipid and protein 'condensation'.
Intermediate vesicle	 0.5μ	Phospholipid and protein in a 'pro-tyrosinase' structural form.
Premelanosome	 0.7-0.3μ	Final structural arrangement (End product of albino melanocyte).
Melanosome	 0.7-0.3μ	Melanin formation on phospholipid and protein structure with tyrosinase activity.
Melanin granule	 0.7-0.3μ	Finished product, melanin with no tyrosinase activity.

Figure 9–5.   Formation of the melanin granule. (After A.G. Searle: Comparative Genetics of Coat Colors in Mammals. New York, Lagos Press, Academic Press, 1968.)

It is clear that the genes affecting color and color patterns can exert their influence in a number of ways. Basically these modes of action can be divided into three groups: (1) changes in the cellular environment, (2) structural changes in cellular organelles, and (3) changes in specific enzymes.

The *A* locus causes changes in cellular environment and is known as the agouti series in most mammals. The result of yet undescribed cellular environmental changes causes melanosomes and vice versa. In some mammals this change may occur with the seasons or it may have some connection with the changes in the rapidity of hair growth.

The *B* locus causes structural changes in cellular organelles; specifically it is responsible for production of the protein framework of the melanosome. Changes at this locus bring about variations in melanosome size and structural organization which affect the phenotypic color of the animal. Genes that affect the structure of the melanocyte and those that affect the structure of the hair are known also to cause phenotypic color changes in the animal.

The *C* locus changes in specific enzymes. This albino series of alleles causes a lack of melanin production. Since melanin production is known to

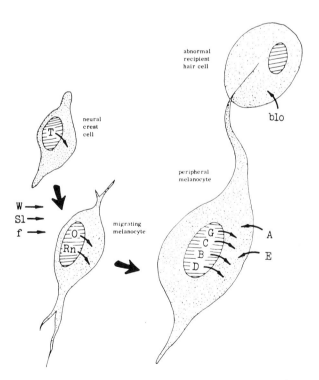

Figure 9–6.   Diagram of the coat color development process. The arrows indicate where each gene exerts its affect.

occur in a series of biochemical steps with specific enzymes necessary at each step, a lack of any one of the enzymes could block melanin production.

More color genes are described in the next section of this chapter. These various genes exert their action in at least half a dozen sites (Figure 9–6). As more information is discovered about how a certain gene exerts its effect, detailed description is made easier, and comparison with other mammals is possible. More color genes will surely be described in the future, and some now prescribed to a particular locus may later be found to be better described at another locus.

As shown in the diagram, there are currently six known sites of gene action in color inheritance: at the neural crest, through the cellular environment, in the migrating melanocyte, in the peripheral melanocyte, within the melanosome protein framework, during the melanin synthesis process, and within the hair itself.

THE COLOR LOCI

In mammalian coat color genetics, where comparisons are made from one order or family of mammals to another, the classic definition of a locus is often stretched to the breaking point. Loci that have the same mode of action in one species at about the same point in the color formation process and produce similar phenotypic results as that of another species are commonly referred to as homologous. This may or may not be the case. However, until more details are learned about the biochemistry of each step, our present concept of the various allelic series of coat color is helpful.

The A Locus. The A allele in the horse is responsible for the factor in the cellular environment of the melanocyte which causes a switch from normal melanin production to a reddish form of phaeomelanin. This switch mechanism is not active in some areas of the body nor in the mane and tail hairs. The a^t allele reduces the overall area of the body, which is changed to phaeomelanin production. Other alleles may be similar to the a^t allele, or it may be incompletely recessive, giving an intermediate effect when heterozygous. The latter seems most probable in light of the fact that most a^t color patterns fall into two groups. One might be described as a dingy or dark bay, while the other is very close to a solid black, with the only reddish hairs being on the muzzle and flanks. Some evidence exists that the A allele has a lightening effect on the phaeomelanic chestnuts and sorrels. The a allele has no effect on normal melanin production.

The B Locus. The B allele is responsible for the production of a melanosome protein framework which allows eumelanin to deposit on it. As previously stated, the b allele produces a protein which will not allow eumelanin to be deposited on it, and for some unknown reason phaeomelanin is deposited instead. Chestnut and sorrel horses are generally considered to be the result of a bb genotype with the various shades and hues of this color due to the influence of the *A* series and the *E* series, which has already been discussed.

The C Locus. There is much controversy over the existence of alleles in the horse at this locus. If there is a true albino, the *c* allele would act by interfering with melanin synthesis as it does in the mouse. Geneticists who describe equine hair color dilution as due to C^{cr} have offered no explanation on the cellular level.

The D Locus. The *D* gene in the horse is not exactly comparable to the *D* gene in the mouse. The phenotypic effects are similar but there is a lack of dominance in the horse. This has led Singleton to propose that the dilution effect in horses is caused by one of the *C* alleles. Gremmel, however, has found that in palominos, buckskins, and grullas, the melanosomes were crowded to one side of the hair cells. This crowding on one side is quite similar to the action of the *D* gene in mice which causes the melanin granules to clump. Russell found that this clumping was caused by a misshapen melanocyte. It seems quite possible that an allele at this particular locus in horses is acting in a similar manner.

The E Locus. The *E* gene exerts its effect in a manner similar to that of the *A* gene: through the cellular environment of the melanocyte. Castle felt that the *e* allele would mimic the effects of the *A* gene, thus producing the bay pattern on blacks and the sorrel color on chestnuts. The *E* allele would have no phenotypic effect. He also postulated the existence of a third allele, which he described as $e^{''}$; however, it is difficult to find evidence to support the concept of this allele. Some of the spotting colors of Appaloosas, as explained later, seem to offer evidence that the *A* and *E* loci both act by altering the chemical and cellular environment of the melanocyte, but each by a different mechanism.

The G Locus. The *G* allele produces the gray horse. Proliferation of melanoblasts, once they have migrated throughout the body, seems to be inhibited. The recessive allele has no phenotypic effect. A number of gray phenotypes are shown Figure 9–7. It may be that each of these phenotypes involves separate alleles, or perhaps the variety of grays is caused by modifying genes.

Gray in horses is dominant to nongray. Foals with the gray gene *G* are born pigmented but start showing gray with the first shedding of hair. Gray horses show more gray with age and many become completely white in old age. According to Salisbury, animals that are *GG* tend to gray more rapidly and to become white to a greater extent than *Gg* gray animals.

Horses of all types may be gray but draft horse breeders are more familiar with the black gray that occurs in the Percheron. Percheron horses are usually black and the *G* gene gives a black gray. When the gray gene is present in bay horses, it gives a red gray and in chestnut colored horses the gray gene results in chestnut gray animals. The gray gene in palominos will tend to cause them to appear lighter and eventually to appear white as the horse ages. The gray gene can mask the Appaloosa pattern by causing the

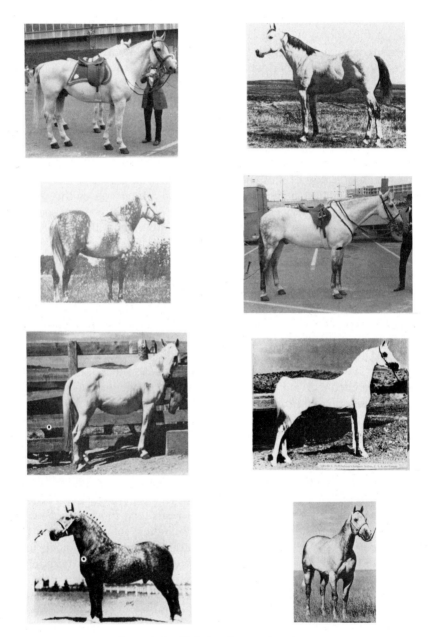

Figure 9–7.   Variations in gray phenotypes.

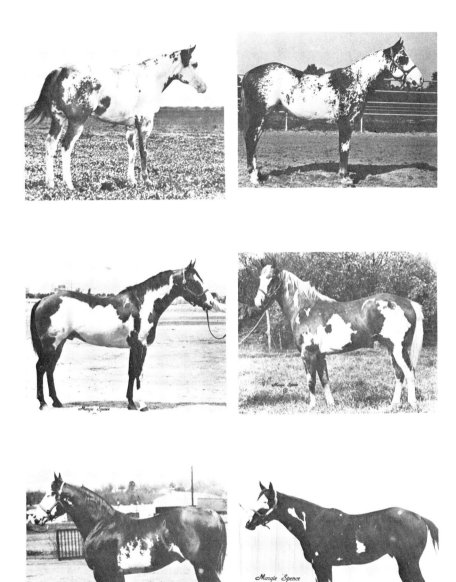

Figure 9–8.   Variations of the overo pattern.

dark spots on the white background or the dark background that has white spots to become light or white.

As pigment is removed from the hair by the gray gene, it sometimes accumulates in the skin about the anus and in the large gut of the horse. If melanomas develop, there can be strangulation of the gut or cancerous metastasis to other areas.

The O Locus. A recessive spotting gene known as o (overo) also has a variety of phenotypic expressions due to a number of modifiers (Figure 9–8). Combinations of the dominant and recessive body spotting with their various phenotypic varieties give an almost infinite number of possibilities for body spotting patterns on the horse. Overo may be an allele of the W locus discussed next. The recessive gene, *pl* (pigmented legs), which causes stocking legs is probably a modifier of the o gene. A mutant allele, o^e, causes the white foal syndrome, described in Chapter 10.

The Rn Locus. The roan locus is easily confused with other loci which bring about a sprinkling of white hairs throughout the body. Roan may occur in any color. There are black roans (often referred to as blue roans), bay roans (often referred to as strawberry roans), chestnut roans, sorrel roans, and palomino roans. The roan gene tends to make the golden color of the palomino less attractive because of the interspersion of white hairs. The Appaloosa pattern is damaged greatly by the roan gene because roaning in the dark spots on a white background or in the dark background with white spots interferes with the sharpness of the spots. Also, when roaning is an interspersion of small clumps of white hairs with the pigmented ones, it tends to give two kinds of white spots on a colored background. The roan effect would be more damaging to the leopard Appaloosa pattern (Figure 3–2) than to the blanket pattern because interspersion of white hairs in small colored spots would tend to make the spots show up to a lesser advantage.

In the Appaloosa horse, the roan characteristic seems to be closely associated with some of the spotting genes. For example, some spotting genes in the presence of certain modifiers will cause a clear and distinct spotting pattern, although these same genes passed to offspring with quite different modifiers may express themselves as a roaning. The opposite situation has also been observed many times in the Appaloosa where a "varnish" roan may pass a definite spotting pattern to its progeny. This is strong evidence that at least one Appaloosa spotting gene works by somehow interfering with the mitotic activity or proliferation of melanoblasts. Variations in the roan phenotype are shown in Figure 9–9.

Geneticists have suspected that a homozygous *Rn* may be lethal. Hintz and VanVleck have shown evidence that this is true. The mechanism of action is undetermined, but until more facts are known, it would be unwise to use roan horses in a breeding program.

The S Locus. For years the S locus has been used to designate various spotting genes. Even though tobiano and overo are thought of as being genes

Figure 9–9.   Variations in roan phenotypes.

of a separate locus, it may be helpful to retain the S locus designation and think of the genes shown in Figure 7–15 as modifiers of it.

The Sl Locus. It may be that Rn and the Sl alleles belong to the same locus. A number of phenotypes can be described as steel or varnish roans (Figure 9–10). Speculation on various alleles at this locus is given later. Evidently, the genetic mechanism involved is an interference with the mitotic activity of the melanocytes so that there are not enough available to supply melanin granules to all hairs. Surely a number of modifying genes are involved when the phenotypes range from a slight silvering to almost pure white.

Figure 9–10.   The effects of the *Sl* gene.

The *T* Locus. The *T* allele in the horse produces what is known as tobiano spotting. In other mammals a similar pattern is a piebald spotting and is designated as the *S* gene. The extent and location of the spotting are affected by many modifying genes as yet undescribed. These modifying genes are not evident unless accompanied by the dominant *T* gene.

Specific modifying genes are known for face, body, feet, and leg spotting (see Chapter 25). There is some controversy as to whether the dominant body spotting gene should be called *S* (spotting), *Pi* (piebald), or *T* (tobiano); also, there is a possibility that each of these is a separate gene causing similar phenotypic effects. Whatever the situation, a number of modifiers can cause the variations of effect (Figure 9–11).

The *W* Locus. It seems highly probable that the horse has several alleles at the *W* locus. In the mouse this series produces various types of dominant spotting with some pleiotropic effects. These tend to interact with piebald *s* and *Sl* to give black-eyed whites. Many types of Appaloosas are strikingly similar, not only in appearance, but in gene interaction and pleiotropic effects. Perhaps the most striking characteristic pattern of the Appaloosa breed is the white "blanket" over the hips and tail region. Proper designation of this gene should probably be W^{ap}. The size of the "blanket" is undoubtedly determined by modifying spotting genes.

There is strong evidence from records of Appaloosa inheritance patterns that several genes contribute to the white blanket characteristic by acting together. They would produce the all-white body color characteristic of the leopard Appaloosa. Evidence is drawn from the fact that leopards often produce blanketed colts when bred to solid-colored mares, and two blanketed Appaloosas bred together may often produce a leopard pattern. Also, some evidence exists that a homozygous W^{ap}, with proper modifiers, may be a leopard type Appaloosa (Figure 9–12).

The *F* Locus. An interesting gene in the mouse, in comparison with the Appaloosa, is the flexed tail gene *f*. It is a recessive gene producing a kinked

Figure 9–11.   Variations in the tobiano pattern.

Figure 9–12.   The Appaloosa, Plaudit's Red, is probably homozygous for the W^{ap} gene.

Figure 9–13.   The effects of the *Blo* gene.

tail. Some Appaloosas have a kinked tail, which seems to be associated with increased spotting. It is highly probable, therefore, that the *f* gene exists in some Appaloosas.

The *Blo* Locus. Blotchy (*blo*), a gene in the mouse which gives an irregular blotch spotting appearance because of abnormal hair development, may also exist in the horse. Some horses show a characteristic blotchy pattern (Figure 9–13).

BREEDING APPALOOSAS

A complete and detailed description of Appaloosa color inheritance awaits more research in melanocyte and hair follicle histology, as well as completely unbiased studies of inheritance patterns. The breeders are making a great effort to understand color inheritance in their horses, although many are laboring under false impressions that only hinder accurate analysis of the genetics involved. Since all of the answers are not yet known, more rapid progress will be made if those facts that have been established are used without bias to determine still other facts.

Appaloosa genes can be divided into two types; the most important is the loss of pigment in some areas of the body, and the other an increase in

Figure 9–14.   Minimum Appaloosa markings: (a) striped hooves, (b) white sclera, and (c) mottled skin. (Courtesy of the Appaloosa Horse Club.)

pigment in localized areas. The loss of pigment seems to take two forms: small, rather distinct white spots or small flecks of white (almost a roaning), and large areas of white, sometimes covering the entire body. The large white patches do not have a sharp, distinct edge as do the spots of most tobiano paints, but rather the edges show a roaning with small flecks of white intermingled with the colored area.

The minimum phenotypic expression required for a horse to be considered Appaloosa is a mottling of the skin, horizontally striped hooves, and an unpigmented sclera (Figure 9–14). The skin mottling shows up as small unpigmented areas of skin, where it would be otherwise pigmented, occurring usually about the genitalia. From these minimum markings, almost an infinite variety of types exists, ranging to the opposite extreme.

The Leopard

Although everyone breeding Appaloosas is experimenting to a certain extent, very little pure research has pointed the way to facts of Appaloosa color genetics. Breeding for the leopard pattern is somewhat more predictable than trying to produce the blanket with large spots. R.W. Miller studied Appaloosa Horse Club records, at one time involving nearly 10,000 Appaloosa horses. The results of his study indicated a dominant gene that produces a blanket and a separate gene which produces spots in the blanket.

Some breeders have found this relatively simple explanation to hold true for them in their breeding endeavors. Others have realized that much more must be involved. The differences in opinion are probably due to the differences in various color genes carried by their original breeding stock. At least a half dozen color genes may affect color or modify color patterns in Appaloosas. Here the term color refers to white. Although a misnomer, the word color among many Appaloosa breeders refers to white. A "highly colored" colt will have a lot of white.

Appaloosa breeders have learned that color sells. The greatest amount of contrast is the most striking and most desirable to the majority of buyers. A pure black horse has no contrast. Give him a white blanket and every eye will be drawn to him. By the same token, the pure white horse lacks contrast even though he may be an Appaloosa by blood. The best contrast is created when the non-white areas are as dark as possible. Black with a white blanket, to most, means the best contrast. Some people may prefer a blood bay or dark chestnut with a blanket to a black with a blanket. But very few will disagree that dun coloring diminishes the desirability of an Appaloosa because there is less contrast.

A common misconception among Appaloosa breeders is that a high concentration of Appaloosa blood is desirable in producing the best color. This idea has probably done more harm to the Appaloosa breed than any other single concept. The problem is that several other genes mimic the blanket gene; and when they occur in combination with it, they will so drastically increase the amount of white that the pattern loses much desirable contrast.

Another aspect of the color pattern that affects contrast is the occurrence of and size of spots within the blanket. As Miller determined in his study, a second gene determines whether spots will appear in the blanket. In the presence of this gene, a group of modifying genes determines the size of the spots. This continuous spectrum of spot size characterizes the Appaloosa breed (Figure 9–15). Very little is known about the genes that determine spot size. Because of the way that spot size is inherited and the large array of sizes, the most probable genetic explanation is that several genes combine to produce the final effect in size.

This same type of inheritance is found with skin color in humans; a number of genes combine their effects to produce a particular shade. Height in humans is also known to be determined by several genes working in combination. In these instances, exact genetic description is more difficult than breeding for it. Generally, horses with large spots bred to horses with large spots will produce large spots. It doesn't take a genius to figure this out,

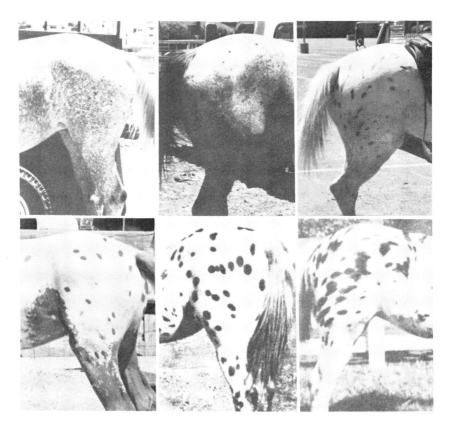

Figure 9–15.   The continuous spectrum of spot size in the Appaloosa.

but it is surprising how often breeders fail to consider this point when selecting breeding stock.

A favorite research animal for the geneticist is the laboratory mouse. These creatures have been subjected to experimental radiation time and time again in order to produce mutations that would have otherwise taken millions of years to occur. Mice with these mutations are then interbred on a mass scale so that a great deal of information is learned about the particular gene involved. Of special interest to Appaloosa breeders is one mutation that causes a coat color pattern in many ways identical to the Appaloosa pattern in horses.

This mutation causes white spotting and is called W. Breeding experiments showed this gene's expression to be influenced by so-called modifying genes. Often the colored areas were interspersed with white hairs similar to some types of Appaloosa patterns. Mice receiving a mutant W gene from both parents were all white like some leopard Appaloosas. Mice receiving a mutant W gene from only one parent had a reduced amount of white confined to the posterior portion, in some ways resembling an Appaloosa pattern. The mouse in Figure 9–16 is heterozygous for W^v but carries modifying genes which increase the blanket size.

This knowledge suggests that an appropriate designation for the Appaloosa gene in the horse is W. Because it acts a bit differently from W in the mouse, it deserves a special superscript designation, such as W^{ap}, the "ap" being an abbreviation for Appaloosa.

From the mouse studies, we can better understand why it is that many solid colored mares, when bred to leopard stallions, produce foals with very striking blankets. The stallion being homozygous (carrying a W^{ap} gene from each of his parents) shows an excessive amount of white. Since he can only contribute one gene to any pair of genes the foal receives, the foal will be heterozygous for the W^{ap} gene—a rather unstable situation in breeding for high-contrast blankets generation after generation.

Figure 9–16.   A laboratory mouse carrying the W^v gene.

The Blanket

Several other genes that occur with a rather high frequency within the Appaloosa breed work in combination with the W^{ap} gene to increase the amount of white on the body. This knowledge suggests that a breeder wanting a blanket with high contrast ought to use breeding stock with as little white as possible. It is becoming increasingly difficult, however, to find purebred Appaloosas with only a small amount of white.

Apparently, the desirability of a blanket pattern is based on the beautiful contrast created with clean white next to solid dark color; that is, any characteristic that diminishes contrast is undesirable. A lot of small colored flecks within the blanket decrease contrast and should be selected against. Also, a lot of small gray flecks within the solid color decrease contrast and should be selected against. As described previously, the silver gene, Sl, produces a varnished roan. This gene can increase the amount of white, depending on the other modifiers present (Figure 9–17). The Rn gene is common and produces a typical roan color where small white flecks occur with the solid colored areas. Roan can act as a modifier, working in concert with other genes to cause variations in the blanket pattern (Figure 9–18). Another gene known as "blotchy" and described as Blo increases the amount of white hairs on the body. The gray gene, Gr, has been known for years to be a deadly enemy of the Appaloosa pattern; however, it is still found quite often within the breed.

There may be a "blanket" gene within the breed that could be named Bl. Patterns such as frosted hips can be explained only by the presence of a Bl gene. It seems to be expressed in much variety, from slight frosting of the hips to a rather large roaning blanket. Those who are dealing with this gene in their stock find a lack of good contrast. Colored hairs are found interspersed in the white, and white hairs are found sprinkled in the solid colored area.

It is just about impossible for anyone to look at a particular Appaloosa horse and determine the exact genetic nature, or genotype, of the animal. It would be a simpler task if only one Appaloosa gene were present at a time, but this is seldom the case. The various types of W^{ap} blankets are illustrated in Figure 9–19. Since very little is known about the genetics of Appaloosa patterns, this description is pure supposition, but is offered here as a starting point for those breeders who would like to breed horses with high-contrast blankets.

If production of high-contrast blankets is the goal of a breeder, he should avoid certain types of breeding stock. Get rid of horses that have roaning in the solid colored area. Select those horses with clean, stark white blankets with no roaning within the white. While it would seem best to avoid leopard patterns if one wants blankets, the breeder would lose some excellent genetic material if he removed all leopard patterns from the breeding program. When using leopards, select those with no evidence of roaning or graying, and select those with spots as large as possible.

No one knows all the answers when it comes to color inheritance in Appaloosa horses, but an expedient means of proceeding is to define goals. "Do I want to breed for blankets with high contrast?" is a question each breeder should ask himself. If so, the whole breeding program and selection process should differ from the programs of those who want to breed for the leopard pattern.

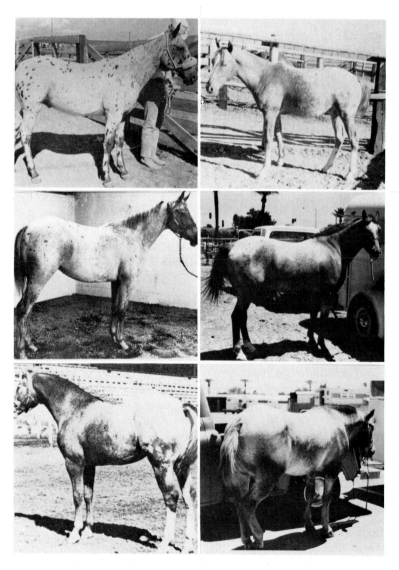

Figure 9–17.   Variations of the *Sl* phenotype in association with the *W^{ap}* gene.

To some Appaloosa breeders, the leopard pattern is more desirable. Their goal is to breed white horses with dark spots consistently. Again, good contrast means spots as dark as possible against the white background, and a clean white background without small flecks of color.

Figure 9–19 illustrates how a horse with poor modifying genes will not have a leopard pattern although homozygous for W^{ap}. This figure emphasizes the necessity of good modifiers. The type of modifiers, however, can be very important to the production of a leopard with the best contrast.

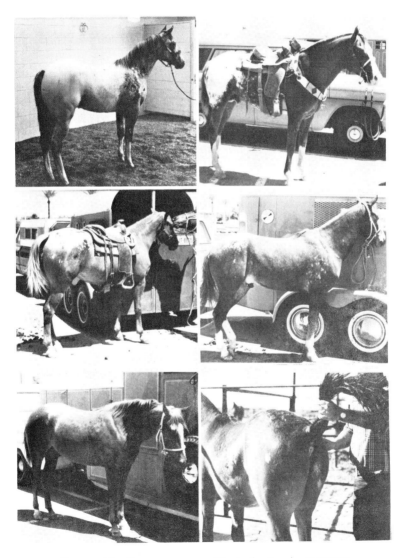

Figure 9–18.   Roan acts as a modifier to the Appaloosa pattern.

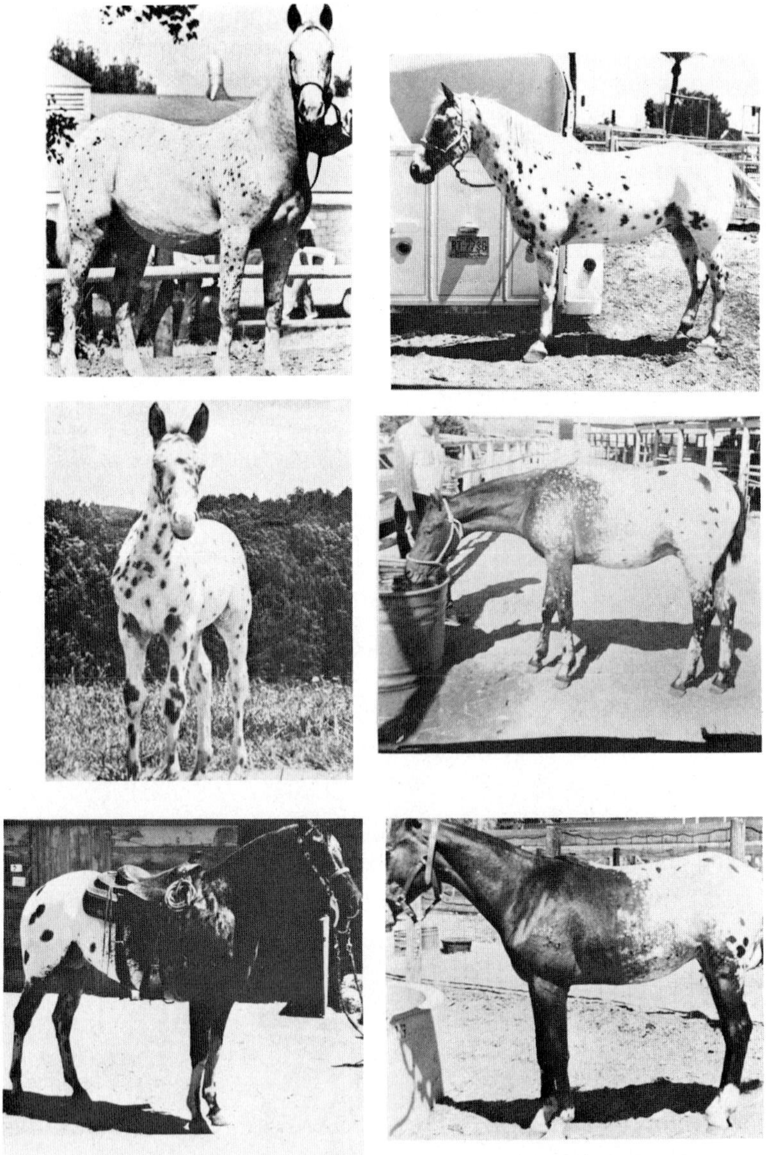

Figure 9–19.   Various Appaloosa phenotypes produced by the W^{ap} gene.

The Sl modifier should be shunned as it tends to produce "silvering" of the colored areas, thus reducing contrast. The roan modifier would also have a similarly undesirable effect.

If one can assume that ten modifying genes affect the Appaloosa pattern, and that no qualitative difference is present among them, they would tend to be inherited as a slow change from generation to generation. This mode is in contrast to a one gene type inheritance, such as a horse heterozygous for gray.

With practice it is possible to arrive at a fair estimate of how well a particular horse is supplied with modifying genes. These same modifiers increase the amount of white whenever any type of spotting gene is also present. The white area of stocking feet, for example, becomes more extensive in the presence of a larger number of these modifiers. Even the pinto horse will have more white in the presence of more of these modifiers. Also, the white marking on the face often indicates the relative number of modifiers. The bald face horse probably has more than the horse with only a strip.

Mottled skin, itself, is probably initiated by a single gene, best designated by M. The amount of mottling is determined by other modifiers. In fact there is some evidence that the M gene may not express itself, even in the homozygous state, unless accompanied by a sufficient number of modifiers. The M gene would, therefore, be a modifier that is in turn affected by other modifiers. In some respects Appaloosa breeders have erroneously regarded the M gene as the Appaloosa gene; this gene alone can never produce a leopard or a blanket. Perhaps the reason it has been such a valuable characteristic in the minds of Appaloosa breeders is that the presence of mottling, striped hooves, and white sclera indicates that a large number of modifiers are associated. These modifiers, in turn, increase the chances for good color when one of the principal Appaloosa genes is introduced.

Whichever pattern one picks as the ideal, it should be one that exists when all of the genes involved are homozygous; otherwise it will not be a true-breeding characteristic. The painting by George Phippen, which has been taken by the Appaloosa Horse Club as the standard, fits what would be a homozygous W^{ap} genotype with very few modifiers. Horses of this genotype bred to mares of the same genotype should always produce like themselves. Although this procedure sounds simple, it would be difficult to prove that a particular blanketed horse carried this genotype rather than being a genetic combination of some of the other Appaloosa genes. Progeny testing along with a close analysis of the phenotypes of his ancestry could, however, give a good indication of his genotype.

BREEDING BLACK HORSES

Since black is an appealing color for a horse, horsemen in many breeds have tried to establish a pure-breeding black herd. Such an endeavor has become popular in the Arabian breed. In the beginning, the task is relatively simple. Black horses with other desirable genetic traits can always be located. Using a black horse as a starting point has eliminated the problem of getting the a, d, and e, and other color genes in the homozygous condition. A black horse is recessive for all of these other color genes. The only problem is the B gene. Since it is dominant, a black horse may be either homozygous or heterozygous.

Once black breeding stock has been obtained, the task then becomes one of progeny testing to determine whether each animal is homozygous or heterozygous for black. As discussed in Chapter 6, progeny testing for black homozygotes is relatively easy. The most profitable place to begin is to determine the genotype of a stallion. The first year, the stallion should be bred to as many black mares as possible. As soon as one non-black foal from that stallion is produced, he is proven to be heterozygous because it takes a recessive gene from each parent to produce a non-black. As shown in Table 9–2, 16 black offspring produced from heterozygous mares will mean less than a 1 percent chance that the stallion is heterozygous; i.e., the breeder can be 99 percent sure that the stallion is homozygous for black. Fewer

Table 9–2. The chances that a particular horse carries a recessive b allele decreases the more black offspring he produces without a sorrel crop-out. The degree of certainty of a parent being homozygous for black when bred to either recessives (chestnut) or known heterozygotes increases. The number of offspring indicated here assumes that each is black. As soon as one none-black is produced, the parent is proved not to be homozygous for black.

Number of Offspring	Mating to Recessives	Mating to Known Heterozygotes
1	50.00	75.00
2	25.00	56.25
3	12.50	42.19
4	6.25	31.64
5	3.13	23.73
6	1.59	17.18
7	0.79	12.89
8	0.40	9.67
9	0.20	7.25
10	0.10	5.44
11	0.05	4.08
12	0.02	3.06
13	0.01	2.30
14	0.006	1.73
15	0.003	1.30
16	0.001	0.99
17	0.0007	0.74
18	0.0004	0.55
19	0.0002	0.41
20	0.0001	0.31
21	0.00005	0.23

offspring are needed to "prove" a stallion if recessive mares, i.e., chestnut, are used (Table 9–2). Here the breeder can be more than 99 percent sure that the stallion is homozygous with the production of seven black foals, and no non-black.

In succeeding years the mares can be progeny tested if desired. Using Table 9–2, it can be established that four black foals from chestnut stallions, with no non-black, will allow the breeder to be over 90 percent sure that the mare is homozygous. As each succeeding black foal is produced, the breeder can be more certain that the mare is homozygous. Once a non-black is produced from a known homozygous mating, the mare is proven to be heterozygous after all. The problem with progeny testing mares is that if the average life's production of a mare is only 10 foals, most of her productive life is wasted in testing with the "undesirable" chestnut.

During the first decade of breeding blacks, it will be an uphill battle trying to develop homozygous stock so that eventually all homozygous blacks are produced. During this time a breeder may find it profitable to develop a good market for his heterozygous blacks, because he will want to sell all that are known to be heterozygous.

In fact a beginning breeder of blacks may want to have some chestnut mares and a good chestnut stallion in his breeding program. The chestnut mares would help in progeny testing young stallions and the chestnut stallion would help in progeny testing young mares. Mostly blacks would be produced from chestnuts bred in this fashion, and they would all be heterozygous offspring. If a good market has been developed for the heterozygous blacks, the black breeding program could remain financially sound even during the formative years.

One of a hundred mares who have had seven black foals from chestnut stallions will produce a non-black the following year. If she has been mated to a stallion without an extensive progeny test, the breeder may suspect the stallion. The stallion, however, generally has a more extensive production record, and the breeder is more certain that he is homozygous. As an example, once he has produced 20 black foals (and no non-black) from homozygous mares, the chances of a stallion producing a non-black crop-out would then be 1 in 10,000. If the stallion has this kind of record, the mare should be considered heterozygote.

All black offspring from known heterozygotes or unproven parents should be considered heterozygotes until proven otherwise. What constitutes proof? This discretionary decision must be made by the breeder. Being 90 percent sure with mares and 99 percent sure with stallions would seem to be adequate proof to most.

BREEDING BUCKSKINS, GRULLAS, AND PALOMINOS

The D gene is a common factor in the genetics of buckskins, grullas, and palominos. This common inheritance was shown a number of years ago in my study of 10,000 American quarter horses. Recently some geneticists have tried to show that the palomino color is due to a gene other than D, and that this gene is not related to dun and grulla. It has been postulated that the white mane and tail of palominos are due to one or more separate genes, but the best evidence is that body color is the result of the same gene that produces duns and grullas.

The goal of the breeder of any of these related types must be to obtain homozygous breeding stock—only then can a pure-breeding herd be established. Grulla breeders must be as cognizant of the genetics of the black horses, as they are of grulla genetics. The principles and procedures in obtaining pure-breeding blacks were outlined previously. These same principles can be applied to the grulla color, but selecting for two genes is twice as difficult as selecting for one. The D gene effect on black acts more like a dominant gene than an intermediate one; i.e., there is little difference between homozygous and heterozygous phenotypes. In general, however, the homozygous grulla is lighter in color. Progeny testing as outlined previously will help to determine which individuals are definitely homozygous.

Breeding for the buckskin adds a third gene to be considered. The task is to obtain homozygous genotypes not only for the B gene and the D gene, but for the A gene as well. The buckskin is simply the dilution of the bay color. Because there are several types of bays, the matter is complicated. These various types reflect that several alleles are present at the A locus and that the E allele may also produce a bay. A dark bay or brown will produce a yellow dun when affected with the D gene. The dark golden buckskin is produced from a heterozygous D working on the blood bay background.

A homozygous buckskin has a washed-out appearance of not only the body coat, but also the mane and tail somewhat. This means that true-breeding buckskins will not be the most desirable color. If production of the best colored buckskins is the end product, then an effective method would be to use blood bay mares and a homozygous buckskin stallion. Some grullas, red duns, and occasional palominos would be produced unless the bay mares were homozygous for B and A. To produce the best colored buckskin every time, both the mares and stallion should be homozygous, the stallion being homozygous for A, B, and D. It would take the same amount of effort and a similar procedure to "prove" the genotype in buckskin breeding stock. The action of each gene under consideration is recognizable from each of the others in the offspring. Thus, as each foal is born, it can be checked for B, A, and D. For example, in progeny testing the first stallion, he is bred to as many heterozygous bay mares as possible. If the first seven foals all show evidence of B, A, and D, the breeder can be 90 percent sure that the stallion is homozygous for three of these genes. Evidence of B would mean a black, brown, bay, or buckskin. Evidence of A would mean a bay or buckskin. Evidence of D would mean a buckskin, dun, grulla, or palomino.

It has been generally thought that the production of palominos followed the same principles as the production of buckskins as discussed above, but using stock recessive for *b*. Figure 7–3 illustrates how 100 percent palominos can be produced by using a creamello stallion and sorrel mares if the breeding stock all carry the proper homozygous genetic makeup for production of the white mane and tail. If not, some red duns will crop out.

There is another method of producing homozygous palominos with the desired shade of body color based on the fact that a very dark chestnut requires a great amount of dilution to reach the palomino color. The chestnut, homozygous for *a* and *b*, and heterozygous for *D*, will be a sorrel in appearance. In this instance, the phenotypic effect of *A* and *D* on the chestnut is quite similar. The same horse, if homozygous for the *D* gene, will be diluted sufficiently to result in the palomino color. The *A* gene sorrel, however, when homozygous for the *D* gene will be creamello. This fact has not been well known because few dark chestnuts carry the genetic background for production of the white mane and tail. Still, it is possible to find such horses. These facts suggest that a diligent breeder who understands the genetics involved can establish a pure-breeding herd of palomino horses of the desirable shade.

BIBLIOGRAPHY

Adalsteinsson, S.: Colour Inheritance in Icelandic Ponies. Proc. Int. Symp. on Genetics and Horse-Breeding. Dublin, Ireland, 1976.

Billingham, R.E., and Silvers, W.K.: Further studies on the phenomenon of pigment spread in guinea pigs' skin. Ann. N.Y. Acad. Sci. *100*:348–363, 1963.

Blun, C.T., and Howell, C.E.: The inheritance of white facial markings in Arabian horses. J. Hered. *27*:293, 1936.

Castle, W.E.: The ABC's of color inheritance in horses. Genetics *33*:22–35, 1948.

Castle, W.E., and Singleton, W.R.: Genetics of the "brown" horse. J. Hered. *51*:127, 1960.

Evans, J.W., et al.: The Horse. San Francisco, W.H. Freeman & Co., 1977.

Fitzpatrick, T.B., Brunet, P., and Kukita, A.: The nature of hair pigment. *In* The Biology of Hair Growth. New York, Academic Press, 1958.

Geurts, R.: Hair Colour in the Horse. London, J.A. Allen & Co., 1977.

Gremmel, Fred: Coat colors in horses. J. Hered. *30*:437–445, 1939.

Greneberg, H.: More about the tabby mouse and about the Lyon hypothesis. J. Embryol. Exp. Morphol. *16*:569–590, 1966.

Harris, H.: Human Biochemical Genetics. Cambridge, MA, Cambridge University Press, 1959.

Hatley, George: Crosses that will kill your color. Appaloosa News, Feb., 1962.

Jones, William E.: Some Phenotypic Effects of the *D* Gene in the American Quarter Horse. Masters Thesis, Colorado State College, Greeley, CO, 1965.

Little, C.C.: The Inheritance of Coat Color in Dogs. New York, Comstock Publishing Assoc., 1957.

Lusis, J.A.: Striping patterns in domestic horses. Genetica *23*:31, 1942.

Mayer, T.C.: The development of piebald spotting in mice. Dev. Biol. *11*:319–334, 1965.

Miller, R.W.: Appaloosa Coat Color Inheritance. Masters Thesis, Montana State University, 1965.

Pulos, W.L., and Hutt, F.B.: Lethal dominant white in horses. J. Hered. *60*:59, 1969.

Russell, E.S.: A quantitative histological study of the pigment found in the coat color of the house mouse II. Genetics *33*:228–236, 1949.

Salisbury, G.W.: The inheritance of equine coat color: the basic colors and patterns. J. Hered. *32*:235–240, 1941.

Salisbury, G.W., and 3ritton, J.W.: The inheritance of equine coat II: the dilutes, with special reference to the palomino. J. Hered. *32*:255, 1941.

Searle, A.G.: Comparative Genetics of Coat Colour in Mammals. New York, Lagos Press, Academic Press, 1968.

Seiji, M., et al.: Subcellular illumination of melanin biosynthesis. Ann. N.Y. Acad. Sci. *100*:497–533, 1963.

Shiable, R.H.: Developmental Genetics of Spotting Patterns in the Mouse. Ph.D. Thesis, Iowa State University, Ames, IA, 1963.

Singleton, W.R., and Bond, Q.C.: An allele necessary for dilute coat color in horses. J. Hered. *57*:75, 1966.

Smith, A.T.: Inheritance of chin spot markings in horses. J. Hered. *63*:100, 1972.

Smith, Jean F. DeMounthe: Horse Markings and Coloration. San Diego, CA, A.S. Barnes & Co., 1977.

Wright, S.: Estimates of amounts of melanin in the hair of diverse genotypes of the guinea pig from transformation of empirical grades. Genetics *34*:245–271, 1949.

10

genetic lethals

All equine characteristics have a genetic aspect including conditions known as disease. Natural infectious disease resistance varies because of genetic factors. Some horses have almost complete resistance to all infectious organisms while others have little or no resistance. The genetic component in some disorders is easy to understand; in others it is subtle or scarcely discernible. Some disorders are apparent at foaling; others may appear during various stages of the life span. The horse breeder generally underestimates the number of horses that will have limited potential as a result of some genetically related defect.

The development of a horse throughout life is determined by a complex interaction between genetic and environmental factors, as explained in Chapter 5. Disease is no exception. Only the more rare diseases such as combined immunodeficiency have simple genetic causes. The more common diseases and disorders are triggered by environmental causes to varying degrees (Figure 10–1). On one extreme are disorders determined entirely by the individual's genetic constitution, such as the white foal syndrome. On the other extreme are traumatic injuries which are entirely the result of the environment. In between these two extremes lie other disorders, with varying amounts of environmental and genetic causes.

INHERITED ABNORMALITIES

Abnormalities of the horse are often classified as either lethal or non-lethal. Technically a lethal abnormality causes death; however, since all animals die sooner or later, a problem of definition arises. A horse at 20

ENVIRONMENT

trauma

tuberculosis

colic

melanoma

anhidrosis

cataract

wobbles

white foal syndrome

HEREDITY

Figure 10–1.   A diagram indicating the spectrum of diseases resulting from varying combinations of heredity and environment. Environment is more heavily weighted than heredity
in descending order; heredity is more influential in the reverse order.

years of age might die because of an inherited susceptibility to a disease. Difficulties would be encountered if this condition were classified as a lethal.

This discussion, therefore, will deal with both lethals and inherited weaknesses. Three types of lethals will be considered: (1) true lethals, those which are expressed prior to or shortly after birth (such as dominant white in horses, which causes death of the homozygous $LwLw$ foal in early gestation); (2) delayed lethals, those which are expressed later in life; and (3) partial lethals, those in which the animal inherits the character which becomes lethal only under certain circumstances (such as ruptured blood vessels in racehorses from the severe stress of racing). In some ways, inherited weaknesses are partial lethals and include reproductive disorders, poor disease resistance, locomotion disorders, and sight disorders.

Abnormalities may also be classified as dominant, recessive, or of variable expressivity. True lethals that are dominant are never perpetuated because the death of the fetus automatically eliminates them. True lethals may be incompletely dominant, as exemplified by the lethal white condition or the Lw genes, in which the heterozygous horse expresses various abnormal characteristics, such as the lack of pigment. Dominant delayed lethals could be perpetuated from generation to generation if the lethality does not develop prior to the time of reproduction. The majority of these lethals are found to be recessive. Partial lethals are found to be either dominant or recessive, because some of the animals with the abnormality will not encounter circumstances leading to death and will, therefore, reproduce and perpetuate the gene. A horse with weak blood vessels, for instance, that is never put into the stress of a race might live to an old age and sire many foals.

Hadorn has pointed out that generally the term "lethal factor" is more accurate than the term "lethal gene," because a large number of genes which cause lethality work within an environment of other genes; that is, various modifying genes affect the penetrance or expressivity of the lethal gene. Thus, lethality occurs only with the proper genetic background. The expressivity of a lethal gene may also be influenced by environmental factors, such as amount of exercise or the diet of the dam during gestation.

A condition occurring in immature horses, known as epiphysitis, may be inherited in a similar manner. The condition is brought on by environmental stress, generally trauma of some sort. It appears as a swelling on the medial side of some long bones, at the level of the epiphyseal cartilage. The cannon bone is often affected. The epiphyseal cartilage is crushed, and swelling and inflammation result. What would be normal activity for one horse with no harmful results might result in epiphysitis in another. Therefore, certain factors predispose to the development of epiphysitis. Rapidly growing fat colts are more prone to develop the trouble, as are those that have base narrow conformation.

Lethal genes may also be classified as autosomal or sex-linked. If the gene for the lethal trait is in the X chromosome, it is sex-linked (Chapter 9). It would be impossible for a stallion to have a true lethal in his X chromosome, for by definition it would have caused death earlier. The mare, on the other hand, can be *Ll* or heterozygous for a true lethal in her two X chromosomes and be perfectly normal if it is recessive. She will transmit the lethal genes to half of her sons, who will die, and to half her daughters, who will be carriers but normal. True lethals that are sex-linked tend to reduce the survival of colt but not filly offspring.

It is common to describe many foal abnormalities as congenital, a catch-all term often used to avoid description of the fundamental cause. According to *Dorland's Illustrated Medical Dictionary,* congenital simply means existing at or before birth. This fact is self-evident when abnormalities are observed shortly after the foal is born. Often the term is used to indicate an abnormality that develops because of a hostile uterine environment, from poor nutrition, toxins, poisons, infections, or the like. While uterine environment may be a major cause of congenital abnormalities, all too often the genetics of the foal itself is overlooked as a cause.

Radiation, as well as toxins and poisons, can cause chromosome changes within the cells of the developing embryo. These changes usually cause abnormal development of the embryo. Often the chromosome changes are so severe that they kill the affected cell of the embryo. If this occurs in extremely vital cells, an embryonic abnormality will result, even though the chromosome abnormality is lost.

Teratogenesis is the development of a congenital body defect. Arey classified causes of teratogenesis into four categories: physical agents such as irradiation, chemical agents, infectious agents, and immunity agents. Recent studies have shown increased incidence of chromosome aberrations associated with teratogenesis in all of these categories but the last.

The preceding agents can also cause gene mutations in individual cells, which would play an important role in teratogenesis only from one generation to the next. A one-gene mutation might have few consequences on the somatic cellular level because many other normal cells are available to carry out the function of the abnormal one. On the other hand, a one-gene mutation in a germ cell will duplicate its abnormality in each cell of the developing embryo, setting up errors of metabolism which can, in turn, lead to gross anatomical abnormalities.

Genetic factors transmitted as maternal inheritance may be responsible for some congenital abnormalities. (Maternal inheritance will be covered in Chapter 12.) Many cases of crooked legs and weak foals could be due to maternal inheritance.

It is well known that the abortion rate in horses is higher than in most other livestock. While the frailty of the placental attachment is probably a major factor, it may well be that chromosomal aberrations are also a significant cause. Much more study is needed, however, to prove this point. In humans, on which much more study has been done, it is reported that 20 percent of all spontaneous abortions are associated with chromosomal aberrations.

Since no two animals are exactly alike, the definition of abnormal is somewhat relative. Of course, the extremes are easy to determine, such as hydrocephalus, albinism, or dwarfism. Most differences between horses, however, would be better described simply as polymorphisms. This term indicates that there are many different shapes or structures of a particular animal or its anatomical parts.

TRUE LETHALS

The true lethals cause death of the foal before or shortly after birth. The appearance of true lethals in a herd can be very discouraging to a breeder. Some cause early abortion, such as the white lethal of the albino horses, in which case the owner is aware only of a reduced colt crop. Other lethals cause abortion later in embryonic development; an example is schistosoma reflexum or a lack of amniotic development. It is quite difficult to detect and describe lethals that occur before term, as generally the embryo or fetus is lost or absorbed and no examination can be made. The majority of true lethals, therefore, that have been described occur at birth or shortly thereafter. A list of the described true lethals of the horse is presented in Table 10–1, and some of the more common conditions are described next.

Anal Atresia. The lack of an anus is an abnormality known as anal atresia. It is caused by abnormal embryonic development, probably initiated by some genetic mechanism. Its rare and sporadic occurrence indicates that it may be a simple dominant. Surgical correction is usually unsuccessful.

Atresia Coli. Occasionally a foal is born with a deformity where the large intestine consists of two sections with no connection between them; this condition is known as atresia coli. The contents of the intestinal tract are unable to pass to the exterior.

Evidence suggests that more than one factor causes this abnormality. McGee has observed it in thoroughbreds, and others have observed it in white foals. The latter is described in more detail under the "white foal syndrome." In the white foals there seem to be other complicating factors besides the atresia coli.

Even without other complicating factors, the condition is always fatal. McGee has attempted surgical correction on numerous occasions, but the foal died in every case. Recently veterinarians at Kansas State University have saved foals with atresia coli by surgery.

E-lymphocyte Deficiency. Lymphocytes are cells of the body which fight infection. They bring on the immune response to various microbial invaders. There are two lines of development of lymphocytes in the foal (Figure 10–2). One line develops in the thymus, and the resultant cells are known as T-lymphocytes. Each lymphocyte type has its own method of combating disease.

The B-lymphocytes produce immunoglobulins, described as IgM, IgG, IgG(T), and IgA. McGuire has described an inherited deficiency of B-lymphocytes in foals. It results in an inadequate immune response that begins to develop as the antibodies obtained by the foal from the colostrum are depleted. Foals with this genetic problem die within a few months.

Combined Immunodeficiency. The disorder, combined immunodeficiency (C.I.D.), has been found only in Arabian foals to date. It is a complete deficiency of both B-lymphocytes and T-lymphocytes. Death occurs before 4 or 5 months.

The foal is protected for the first month or so of life and cannot be distinguished from the normal foal. A C.I.D.-affected foal is shown in Figure 10–3. Note the rough coat and snotty nose. It is soon to die.

Somewhere in the past, a single gene mutation occurred in an Arabian. This recessive mutant gene was passed on down the genetic line, generation after generation, and somewhere it wound up in a very popular stallion to which many breeders have bred back incessantly. This practice has produced a high percentage of carriers within the Arabian breed, and a phenomenal death rate (from C.I.D.) of 2.5 percent of all Arabian foals born.

Two carriers bred together will produce a C.I.D.-affected foal one out of four times, on the average. Fifty percent of the time this cross will perpetuate itself with another carrier. Only one out of four foals will revert back to the normal genetic situation, losing all traces of the abnormality in its progeny

Table 10–1. True Lethals in the Horse

Defect	Description	Incidence	Reported By
1. Abrachia	Absence of front legs.	Rare	Mauderer (1938)
2. Absence of eye orbits	Skull deformity with no eye orbits developing.	Rare	Prawochenski (1936)
3. Atresia coli	Blockage of colon due to incomplete development	Occasional (thoroughbreds)	McGee (1958)
4. Atresia recti	Absence of anus. Anus ends in a blind pouch.	Occasional	Fuchslocher (1971)
4. B-lymphocyte deficiency	Deficiency of a specific lymphocyte resulting in an inadequate immune response.	Occasional	McGuire (1976)
6. C.I.D.	Combined immunodeficiency. A deficiency of both B and T lymphocytes resulting in inadequate immune response.	2.5% (Arabian breed)	McChesney (1973) McGuire (1973) Poppie (1977) Thompson (1975)
7. Convulsive syndrome	A nervous disorder with convulsions and/or blindness. Often described as a "barker," "dummy," or "wanderer."	Occasional (mostly English thoroughbred)	Mahaffey (1957) Cosgrove (1955)
8. Contracted foal	Severe twisting of joints and spine.	Approx. 50 cases reported	Finocchio (1973) Rooney (1966)

9. Cyclopia	Fused eyes at center of head and only one brain hemisphere.	One case	Wilkins (1974)
10. Epitheliogenesis imperfecta	Epithelial defects usually just above the carpus or on the tongue.	Rare	Berthelsen (1935) Crowell (1976)
11. Hydrocephalus	Increased cerebrospinal fluid in ventricles of brain, with or without cranial enlargement.	Occasional	Behrens (1954) Ramsey (E.M. '63)*
12. IgM deficiency	Deficiency of a specific immunoglobulin resulting in an inadequate immune response.	5 cases reported	Perryman (1977)
13. Isoerythrolysis	Serum antibodies of the mare destroy the red blood cells of the foal.	Common	Doll (E.M. '63)*
14. Schistosoma reflexum	Herniated abdominal wall and skin with lack of amniotic fluid.	Rare	Weber (1947) Irwin (1975)
15. Spinal degeneration	Degeneration of spinal nerve roots with motor weakness.	Rare (Shetland ponies)	Innes (1962)
16. Stiff forelegs	Joints below the knee flexed and stiff.	One-half of offspring from an Anglo-Arab	Prawochenski (1936)
17. White foal syndrome	White foal with pink eyes and insufficient development of colon and abdominal nerves.	Common (Paint horses)	Jones (1973)
18. White lethal	Heterozygous horses are white in color. Homozygous embryos are aborted early.	Common (Albino registry)	Pulos (1969)

*From Rooney, J.R.: Disease of the bone. *In* Equine Medicine and Surgery. Santa Barbara, American Veterinary Publications, 1963.

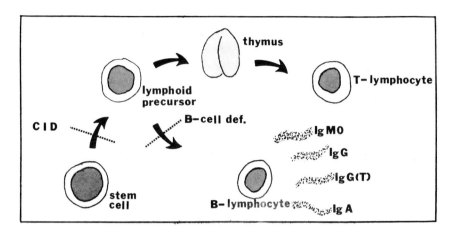

Figure 10–2.   Lymphocytes of two varieties develop in the foal; thus, genetic deficiencies can occur at more than one point in the development.

Figure 10–3.   An affected C.I.D. foal which died within a few weeks. (Courtesy of Dr. Mark Ratzlaff, Washington State University.)

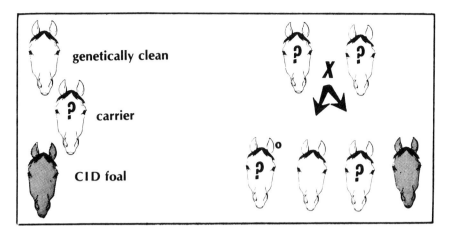

Figure 10–4.   A diagram showing transmission of C.I.D.

(Figure 10–4). At present there is no way to detect carrier animals, except by progeny testing, i.e., breeding to a known carrier.

Hydrocephalus. Congenital hydrocephalus is an abnormal condition of the foal wherein an increased amount of cerebrospinal fluid within the ventricles of the brain causes a severe enlargement of the cranial cavity. This condition is uncommon and generally causes dystocia in the mare.

Congenital hydrocephalus is usually caused by abnormal development of parts of the nervous system. It has been impossible to follow any sort of inheritance pattern as the condition is 100 percent fatal. There have been no reported incidences of more than one case occurring within any one herd, and no indication that the incidence increases with the amount of inbreeding. This evidence would, therefore, suggest that hydrocephalus is caused by a single dominant gene. The condition is known to develop in varying degrees of severity, which indicates that modifying genes are involved. In mild cases the foal may stand at first, but the condition usually becomes progressively worse.

Immunoglobulin M Deficiency. Genetic deficiency of immunoglobulin M (IgM) has been known to occur in humans, where it has been determined to be sex-linked. Five cases in Arabian and quarter horse foals were reported in 1977 by Perrymen and McGuire. Not enough information was available to determine the mode of inheritance or, in fact, whether it was inherited. It points up the importance, however, of an accurate diagnosis between the various immune deficient conditions in foals.

Schistosoma Reflexum. Weber has reported a case of what he termed schistosoma reflexum. Abnormal development of the foal occurred (Figure 10–5), with a complete lack of amniotic development. The fetus was aborted at 6 months.

Figure 10–5. Schistosoma reflexum in a foal.

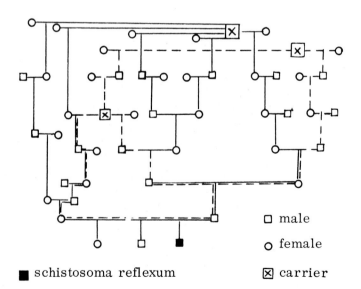

□ male

○ female

■ schistosoma reflexum   ☒ carrier

Figure 10–6.   A pedigree showing inheritance of schistosoma reflexum.

The previous year's foal from the same mating had aborted at 10 months but no information was available as to the cause. Because of this occurrence, and lack of data, Weber decided to construct the chart of the pedigree, shown in Figure 10–6. He showed that this condition can occur with excessive inbreeding; recessive lethals eventually become homozygous causing a bizarre abnormality.

Stiff Forelegs. Prawochanski described a lethal condition which he called stiff forelegs. It seems to be inherited as a dominant, in that nearly one-half of the offspring of an Anglo-Arab stallion, Menzin, were foaled with the abnormality. The joints below the knee were flexed and stiff (Figure 10–7); as a result the foals could not stand and nurse.

White Foal Syndrome. Recessive white pinto spotting evidently is due to a number of alleles each of which may give a slightly different pattern of spotting. One, for instance, may cause the roaning at the edges of the color, while another may produce a rather sharp change from color to white. One notorious allele of the overo color pattern produces the lethal white foal when homozygous. It is completely recessive in every respect so far as is known, and carriers are impossible to detect in all cases. All other overo alleles are epistatic to the o^l allele. Carrier horses would be either o/o^l or O/o^l. In the first case the animal would be spotted and, in the second, would be solid colored but both would carry the lethal white allele. These animals would be perfectly healthy themselves. A foal receiving an o^l from each parent would have the white foal syndrome.

These foals are born healthy and vigorous (Figure 10–8). They are all white with blue eyes. Invariably they have a constriction of a section of the

Figure 10–7. Stiff foreleg in a foal. (From *The Journal of Heredity*.)

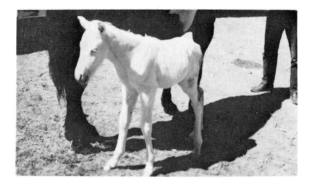

Figure 10–8.   The white foal syndrome.

large intestine. Generally this constriction is a foot or more in length and so extensive that no intestinal material can pass through. Death always results within 3 days. A number of attempts at surgical correction have been made, all unsuccessful. Evidently there are abnormalities of the innervation to the bladder, colon, and possibly other internal organs. I have definitely observed a lack of myenteric and submucosal plexuses in white foals.

Overo x overo crosses can result in foals which do not have the white foal syndrome. Some overo horses are entirely free of the o^e allele, carrying an o/o genotype, which is the desired genotype for overo breeding stock. It is

Figure 10–9.   A so-called "albino horse" which carries the white lethal gene.

extremely discouraging to overo breeders that so many lethal white foals are born.

White Lethal. Pulos and Hutt first described the white lethal gene in horses although many breeders had recognized its existence for years. The White Horse Registry (formerly the Albino) is set up to register and promote the white horse, most of which carry this gene in the heterozygous condition (Figure 10–9). When homozygous, the early embryo dies in utero.

Pulos and Hutt chose to use *W* as the symbol for this gene; however, for several reasons, *Lw* would be a better designation. It has been the general practice in genetics to use identical symbols to denote genes of the various mammalian species whenever similarities exist. The genes *A, B, C, D,* and *E* all originated in description from studies in the mouse and other lab animals. One gene of the mouse has been designated *W*; however, the only similarity between this and the white lethal condition is white coloring. The homozygous condition in the mouse is not always lethal and some areas of pigmentation occasionally occur. In contrast, lethal white in the horse never shows body pigmentation. Some of the Appaloosa phenotypes more closely resemble that of the *W* mouse. It seems logical, therefore, to use the symbol *Lw* for the lethal white condition in the horse. No known homologous gene exists in the mouse.

A peculiar type of inheritance results from crosses between two white heterozygotes. One fourth of the progeny are homozygous and these die early in development without the knowledge of the breeder. This loss leaves a ratio of 2 white foals to 1 solid colored foal. The conception rate is reported to be extremely low under these circumstances.

DELAYED LETHALS

Certain abnormalities of the horse could be classified as delayed lethals (Table 10–2). These do not generally cause death until sometime later in the life of the horse. Generally recessive traits, they crop up only rarely because of the intensive selection practiced by horse breeders for centuries the world over. Dominant traits are easily eliminated through selection since even the heterozygotes show the characteristic.

Some of these lethals, such as cleft palate, spinal ataxia, and blindness, are more apt to cause early death than others. In some respects they are not much different from a true lethal such as stiff forelegs. Classification in this area is somewhat arbitrary. Other delayed lethals such as melanosis may never bring about death, or at least not until old age. This trait, then, might be similar to some partial lethals, such as laminitis, a lack of disease resistance, or chronic pulmonary emphysema. Generally, however, the distinction is made by whether or not death is precipitated by some environmental factor. Our classification of lethals is often quite arbitrary. A number of the delayed lethals are described next.

Table 10–2. Delayed Lethals in the Horse.

Defect	Description	Incidence	Reported by
1. Accessory lung	Tumor-like mass in thorax.	Rare	Smith (1974)
2. Ameloblastic odontoma	Tumor in maxillary region.	Rare	Lingard (1970)
3. Cataracts	Opacity of lens at birth causing blindness.	Occasional	Weber (1947) Gelatt (1974)
4. Cerebellar hypoplasia	Lack of complete development of cerebellum resulting in eventual lack of coordination.	Common (Usually Arabs)	Baird (1974) Bjorck (1973) Cook (1971) Dungworth (1966)
5. Cleft palate	Fissures of soft and hard palates.	Occasional	Batstone (1966) Kendrick (1950)
6. Cortical depression syndrome	A metabolic disorder with incoordination.	Rare (Shetland ponies)	McGrath (1963)
7. Eye developmental defects	Several varieties of defects described, such as absence of iris and cornea.	Occasional	Eriksson (1955) Garner (1969)

8. Heart developmental defects	Many irregularities such as interventricular septal defect, patent ductus arteriosus, persistent foramen ovale, coronary artery misplacement, misplacement of pulmonary artery, etc.	Common	Bartels (1969) Knauer (1973) Rooney (1964) Vitums (1970) Glazier (1974)
9. Hemophilia-A	Failure of the blood to clot properly.	Rare	Archer (1961) Sanger (1964)
10. Idiopathic epilepsy	Recurrent episodes of sudden loss of consciousness.	Rare	Neal (E.M. '63)*
11. Kidney agenesis	Absence of kidney.	Rare	Hofliger (1971)
12. Melanoma	Posterior spinal tumors developing in gray horses.	Occasional	Jones (1973)
13. Oldenburg ataxia	Degeneration of areas of cerebellum causing incoordination.	Occasional (Oldenburg breed)	Koch (1950)
14. Parotid duct atresia	Parotid duct ends in a blind sac.	Rare	Fowler (1965)
15. Polydactyly	Presence of more than one digit on a leg.	Rare	Evans (1965)
16. Spinal dysraphism	Degeneration of the spinal cord.	Rare	Cho (1977)
17. Wobbles	Spinal ataxia with degeneration of cord.	Occasional	Beasley (1973) Dimock (1950)

*From Rooney, J.R.: Disease of the bone. In Equine Medicine and Surgery. Santa Barbara, American Veterinary Publications,1963.

Cerebellar Hypoplasia. A genetic abnormality of foals known as cerebellar hypoplasia has been increasing rather rapidly over the years. All reported cases have been in the Arabian breed. The condition often involves a severe underdevelopment of the cerebellum of the brain, which results in failure of the cerebellum to maintain muscle synergism, leading to signs of ataxia, which is manifested as overreaching and paddling, and head tremors. The head tremor becomes exaggerated when the animal is excited. Symptoms of the condition may be present at birth but may not appear for 3 months or more. As the foals mature, they develop a tendency to go up and over backward when startled.

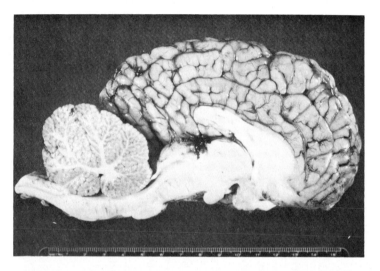

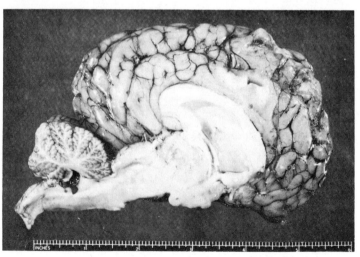

Figure 10–10.   Cerebellar hypoplasia in a young Arabian. The normal is above. (Courtesy of Dr. A.W. Dade.)

The condition is evidently due to a single recessive gene. An affected foal must, therefore, have received a mutant gene from each parent. Some of the more popular Arabian lines may be carrying the gene. The incidence has been rather high in some herds. For example, 6 percent of one year's crop of 67 foals and 8 percent in another crop of 36 foals have been reported. Recent research indicates that a virus may somehow be involved in many cases.

As shown in Figure 10–10, the cerebellum from an affected horse is significantly smaller than that from a normal animal. This difference in size is also reflected in cell number, resulting in reduced efficiency of this portion of the brain. In some cases a virus destroys the brain cells.

Cortical Depression Syndrome. An inherited condition occurring in Shetland pony foals has been reported by McGrath. The affected foals are extremely drowsy and show signs of incoordination, often accompanied by quivering. The symptoms are believed to be caused by some type of metabolic disturbance, which may be genetic.

Heart Defects. Various types of heart defects have been observed in newborn foals. No one has yet described an exact pattern of inheritance for these lethals. Such abnormalities as interventricular opening (Figure 10–11), an opening between the pulmonary artery and the aorta, abnormal openings

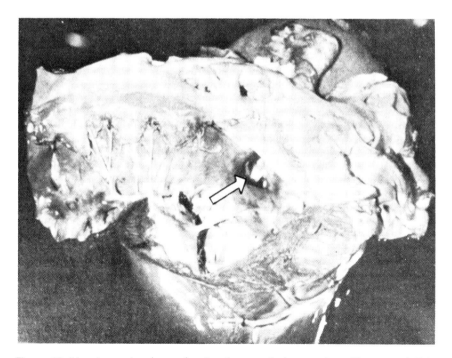

Figure 10–11. An equine heart showing interventricular opening. (Courtesy of D.K. Detweiler.)

between an atrium and a ventricle, and abnormal valves have all been described.

These defects always seem to hamper the function of the horse by cutting down the amount of oxygen reaching the tissue. Thus, the animal tires easily. Many of the heart defects are not immediately lethal, but death generally occurs within a few years.

Hemophilia. Hemophilia is a failure of the blood to clot properly. Since there are a number of factors involved in the process of blood clotting, there are various types of hemophilia. One of the most well known is hemophilia A, caused by a deficiency of an important blood globulin known as AHG. The production of this globulin is directed by a gene on the X chromosome, making it a sex-linked trait.

Archer described the first reported case in the horse, which occurred in a thoroughbred foal. Later Sanger and his co-workers reported a case in a standardbred foal (see Figure 10–13). They found evidence that 3 of 7 other male foals from the same dam died of this bleeding disease. Since these foals were sired by various stallions, it is evident that the mare carried the h gene on one of her X chromosomes. The family pedigree of the standardbred foal is shown in Figure 10–12.

The symptoms of the disease are generally observed upon foaling. Jaundice is present, and localized swellings under the skin occur in almost any area of the body. These swellings are hematocysts, or blood-filled cavities within the tissue. They result from small vessels rupturing under the

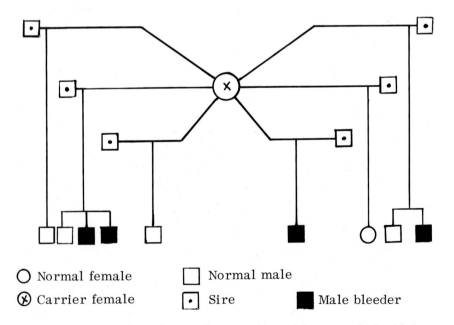

O Normal female   □ Normal male

⊗ Carrier female   ☐• Sire   ■ Male bleeder

Figure 10–12.   A pedigree showing inheritance of hemophilia in a standardbred foal.

skin; and without proper clotting, bleeding continues until a large blood cyst is formed. Eventually internal bleeding develops, in which case death quickly follows. The foal shown in Figure 10–13 displays the typical hematocysts; he died at 6 months of age.

Melanoma. A melanoma is a tumor of the melanocytes which form the pigment of the body. It is extremely difficult to pinpoint the exact cause of any cancer; however, a predisposing factor of the melanoma is the gray (*G*) gene. The majority of horses that develop melanomas are gray in color.

Melanomas usually develop in old horses as a dark-colored growth in the region of the anus. Generally a number of growths develop and spread to other areas of the body. Metastasis to the spleen, lungs, and lymphatics can soon lead to death.

Polydactylism. A rare occurrence in horses is the development of a second toe on one foot (Figure 10–14). In other animals this same abnormality usually leads to several extra toes, resulting in the name polydactylism. The exact genetic mechanism has not been determined, nor is it known whether the characteristic is dominant or recessive.

Spinal Degeneration. Shetland pony breeders have occasionally observed an abnormal condition of newborn foals, with symptoms of staggering and eventual paralysis leading to death. Innes and Saunders carefully studied a

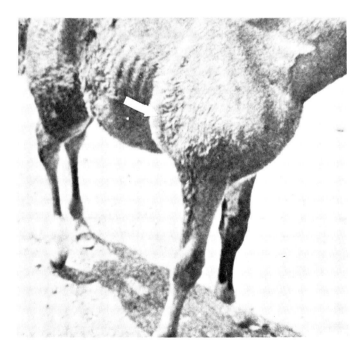

Figure 10–13.   A foal with hemophilia. Notice the large swelling from subcutaneous bleeding. (Courtesy of V.L. Sanger.)

 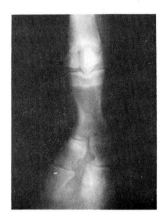

Figure 10–14.   Equine polydactylism. (Courtesy of Dr. Oscar Swanstrom.)

case and gave it the comple description of "pseudo-amyloid" multiradicular degeneration of spinal nerve roots. They reported it to be much like a rare condition in children, which is inherited as a simple dominant.

Affected foals are weak, have a wobbling stance, and show a disinclination to move; the appetite is good, however. Autopsy reveals a degeneration of the spinal cord.

Wobbles. An abnormal development in the shape of the vertebrae of some horses leads to the "wobbles syndrome." The condition is found in young animals, and does not seem to be limited to a particular breed. It may be associated with the sex of the animal as 75 percent of those affected are males.

The abnormality is an asymmetrical formation of the wings and articular processes of the cervical vertebrae. The condition may not become evident until after a fall or severe flexion of the neck, in which case the abnormal vertebrae impinge upon the spinal cord, causing severe damage to it. The result is posterior incoordination and a weaving gait in the rear legs.

PARTIAL LETHALS

The term, partial lethal, is used here in a very broad context. In most cases the trait may never go so far as to contribute to death of the animal; rather it may have an indirect effect in reducing the number of offspring from animals who display the trait. This result is probably more true in the wild state and under conditions where poor horse breeding procedures are being practiced.

Parrot mouth, for instance, may contribute to a shorter life span. Horses with shorter life spans have fewer progeny, on the average. Over a long period of time this kind of trait would cause elimination of those horses who carried it. In that respect it is classified as a partial lethal.

A list of many of the known partial lethals is shown in Table 10–3. Many of these have been known to horsemen as simply poor conformation characteristics. However, the fact that they occur among horses in such great abundance is indicative that most horse breeders have failed to recognize the subtle danger in them. The more successful breeders are selecting against these partial lethals, some of which are described in detail in the following paragraphs.

Cataract. A cataract is a loss of transparency of the lens of the eye or its capsule. Occasionally, congenital cataracts occur in the horse. Weber reported a series of five cases, all of which could be traced back to one sire. Its pedigree is shown in Figure 10–15. He attributed the condition to a single recessive gene.

Fibular Enlargement. Variations in the size and development of the fibular bone in horses are commonly reported. Speed described an extreme enlargement of the fibula, causing lameness in Shetland ponies. He reports that the condition recurs in certain families, and that it may miss a generation or two and then reappear. He feels that it may be related to dwarfism.

Hereditary Multiple Exostosis. A hereditary condition involving multiple bony growths in the horse resembles hereditary multiple exostosis in man.

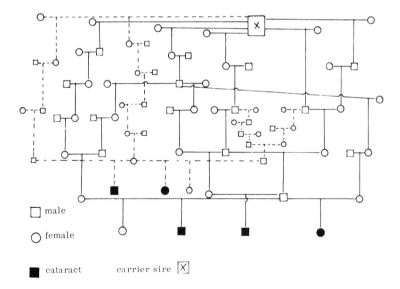

Figure 10–15. A pedigree showing inheritance of cataract in a horse.

Table 10–3.   Partial Lethals in the Horse

Defect	Description	Incidence	Reported By
1. Atlanto-occipital fusion	Fusion of the atlanto-occipital joint in the vertebral column.	Rare	Leipold (1974)
2. Contracted foot	Unilateral contraction of heels.	Common	Adams (1974)
3. Cryptorchidism	Retained testicle.	Common	Cox (1975)
4. Dentigenous cyst	Tumor-like cyst containing tooth remnants.	Rare	Markus (1974)
5. Ectropion	Eversion of the eyelid.	Occasional	Bryans (E.M. '63)
6. Entropion	Inversion of the eyelid.	Occasional	Bryans (E.M. '63)
7. Fibular enlargement	Variation in fibular bone size, sometimes causing lameness.	Rare (Shetland ponies)	Speed (1958)
8. Hip dysplasia	Shallow acetabulum of hip joint with lameness.	Occasional	Jogi (1962) Manning (1963)
9. Lordosis	Swayback development of fetus.	Rare	Rooney (1969)
10. Microcornea	Abnormally small cornea of the eye.	Occasional	Bryans (E.M. '63)
11. Multiple exostosis	Bony enlargements on the ribs and sometimes of the pelvic bones.	Occasional	Morgan (1962) Shupe (1970) Hanselka (1974) Gardner (1975)

	Description	Frequency	Reference
12. Nuclear cataracts	Opacity of the central portion of the lens of the eye. Sometimes Y-shaped.	Rare	Bryans (E.M. '63)
13. Optic nerve hypoplasia	Underdevelopment of optic nerve causing partial to complete blindness.	Rare	Bryans (E.M. '63)
14. Parrot mouth	Upper jaw overgrowth.	Common	Jones (1963)
15. Patellar agenesis	Failure of development of patella bone in the stifle.	Rare	Kostryra (1963)
16. Patellar luxation	Underdevelopment of the lateral ridge of the femoral trochlea, causing a lateral positioning of the patella.	Rare	La Faunce (1971) Rooney (1971)
17. Periodontitis	Malplacement of molar teeth predisposing to inflammation around roots.	Rare	Gobel (1960)
18. Pervious urachus	Failure of urachus to close after birth allowing dripping of urine. Usually spontaneous recovery.	Common	Jones (1973)
19. Pharyngeal defect	Abnormal flap over pharyngeal opening to guttural pouch causing tympany of pouch.	Rare	Mason (1972)
20. Retinal detachment	Detachment of retinal layers present at birth, causing blindness depending on severity.	Rare	Koch (1952)
21. Scrotal hernia	Protrusion of gut through inguinal canal into scrotum at birth or later.	Common	Johnson (E.M. '63)
12. Umbilical hernia	Protrusion of gut at umbilicus under skin.	Common	Jones (1973)

The ribs and limb bones are the principal sites of growth. Morgan and co-workers reported several cases among related horses. None of the bony growths interfered with joint movement and there was no associated soreness or inflammation. Gardner studied the condition and found it caused by a single, dominant, autosomal gene. He defined the site where lesions occur (Figure 10–16).

Hip Dysplasia. The head of the femur (uppermost bone of the rear leg) fits into a depression in one of the hip bones, creating a ball-and-socket joint. Such a joint is not very strong if the depression (technically termed the acetabulum) is quite shallow. The resultant condition is known as hip dysplasia.

Jogi reported on a standardbred colt in which the condition resulted in lameness of the hind limb. Manning reported a case in a Shetland pony with symptoms of lameness and an altered gait. Figure 10–17 is a radiograph of the hip joint showing dysplasia. The acetabulum, or cup-shaped cavity that holds the femoral head, is extremely shallow.

Ocular Defects. R.H. Smythe has studied vision peculiarities in the horse. He points out that the horse possesses a limited degree of forward binocular vision, and that his eyes do not focus as do man's; instead he lifts his head up and down to allow the light to fall upon the most effective portion of the retina. Horses with deep orbits and wide-set eyes, both of which are inherited characteristics, possess less binocular vision than horses that are narrow between the eyes and have shallow orbits. The latter type of anatomy tends to provide better sight for jumpers and hunters.

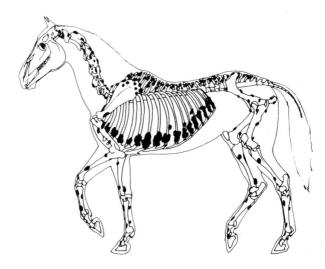

Figure 10–16.   A diagram of the horse skeleton showing locations where bony tumors most frequently occur with hereditary multiple exostosis. (Courtesy of Dr. Gardner.)

Priester made a study of congenital ocular defects found in veterinary schools in horses, cattle, cats, and dogs. Although a much higher incidence of inherited problems occurred in certain breeds of dogs, he found quite a list of ocular defects in the horse (Table 10–4).

Parrot Mouth. An abnormality of the teeth, known as parrot mouth, occurs quite often in foals. Figure 10–18 shows how the upper incisors extend much further forward than the lower. The abnormality seems to be inherited as a dominant, although the pattern is not typical.

Good horsemen are continually on the lookout for this condition, culling all foals which appear with the abnormality. Under normal conditions a parrot mouth colt would be unthrifty; however, the heavy feeding of hay and grain eliminates the handicap of an overshot upper jaw, and foals thrive quite well. It would be foolish, however, to breed the condition into a herd.

Umbilical Hernia. Umbilical hernia (Figure 10–19) is extremely common in horses. Since it usually disappears within the first few months of the foal's life, it is not considered to be a serious abnormality.

Occasionally the hernia will persist, endangering the life of the young horse. The hernia is a protrusion of the abdominal cavity through the

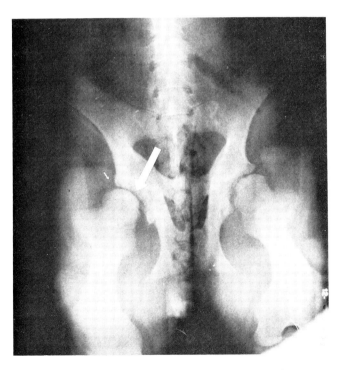

Figure 10–17. Hip dysplasia in a Shetland pony. The radiogram shows a very shallow acetabulum. (Courtesy of J.P. Manning.)

Table 10–4. A List of Congenital Ocular Defects Found in Horses in a Survey by Priester

Compound and Unspecified Eye Defects	8
Eye defect, not otherwise specified	2
Ectasia syndrome	0
Coloboma, eyeball	1
Anophthalmos-microphthalmos	5
Hydrophthalmos, infantile glaucoma	0
Cornea and Sclera	2
Defect of cornea, not otherwise specified	0
Lack of cornea	0
Opacity of cornea	1
Dermoid cyst of cornea	1
Dislocation of cornea	0
Hernia of cornea	0
Pigmentation of cornea	0
Keratosis of cornea	0
Defect of sclera	0
Uvea	2
Defect of iris, not otherwise specified	0
Coloboma of iris	1
Polycoria	0
Heterochromia	1
Coloboma of choroid	0
Retina and Optic Papilla	2
Defect of retina, not otherwise specified	0
Partial absence of retina	0
Coloboma of retina	0
Detachment of retina	0
Lack of pigment of retina	1
Optic papilla, not otherwise specified	1
Pigmentation of optic papilla	0
Lens	13
Defect of lens, not otherwise specified	0
Coloboma of lens	0
Hernia of lens	0
Ectopia of lens	1
Cataract	12
Ocular Neuromuscular Defects	0
Pupillary mechanism	0
Neuromuscular defect, not otherwise specified	0
Nystagmus	0
Other Congenital Eye Defects	3
Amaurosis	3
Amblyopia	0
Persistent hyaloid artery	0
Persistent pupillary membrane	0

Table 10–4. A List of Congenital Ocular Defects Found in Horses in a Survey by Priester (Continued)

Eyelids, Conjunctiva, and Lacrimal System	4
Defect of eyelid, not otherwise specified	0
Lack of eyelid	1
Duplication of eyelid	0
Entropion	1
Ectropion	0
Defect of membrana nictitans	0
Defect of conjunctiva, not otherwise specified	0
Adhesions of conjunctiva	0
Defect of lacrimal system	2
Total	34

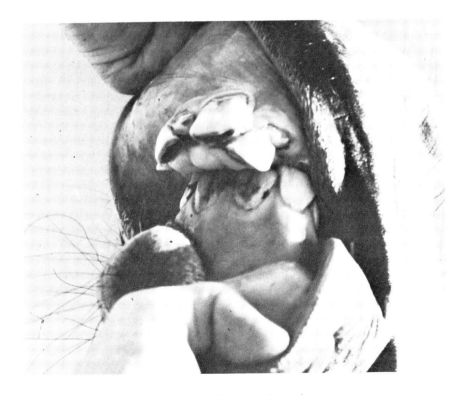

Figure 10–18.   Parrot mouth in a horse.

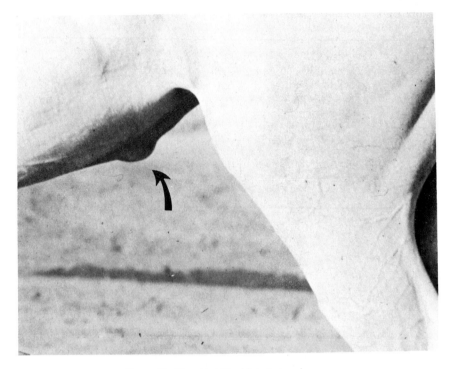

Figure 10–19.   Umbilical hernia in a horse.

musculature. Often a part of the intestine lies just under the skin and is in danger of strangulation. Should this occur, that portion of the gut will die, opening the contents of the intestinal tract to the abdominal cavity, thus leading to death from infection. Surgical repair of the umbilical hernia can be successful.

INHERITED WEAKNESSES

Many of the inherited weaknesses (Table 10–5) do not seem to fit the term lethal. These conditions are not much more than a weakness that under certain conditions develops into blemishes or unsoundnesses.

The most outstanding examples are found in the legs and feet. Weakness in the ligaments of the stifle joint can lead to dorsal patellar fixation. Poor conformation of the hock joint can lead to capped hock, or various types of spavins. Crooked legs can lead to the development of knee injuries. Splints, although caused by injuries, would not develop if it were not for the small bones which lie alongside the cannon bone. These bones are remnants of an earlier evolutionary stage, and are not important in the horse of today.

Other inherited unsoundnesses are such factors as a predisposition to heaves or chronic pulmonary emphysema. Koch has explored the genetic

Table 10–5. Inherited Weaknesses in the Horse.

1. Anhidrosis
2. Bursting of nasal blood vessels
3. Chronic pulmonary emphysema
4. Contraction of digital flexor tendons
5. Curb
6. Deviation of the metacarpophalangeal joint
7. Enchondrosis
8. Flat-feet
9. Fusion of vertebrae
10. Hypothyroidism associated with bone lesions
11. Knock knees
12. Knuckling at the fetlock
13. Lack of disease resistance
14. Laminitis
15. Lordosis
16. Myopia
17. Navicular disease
18. Osteoarthritis
19. Osteoporosis
20. Ring bone
21. Sesamoid bone irregularities
22. Side bones
23. Spavin
24. Summer sores
25. Supernumerary teeth
26. Testicular hypoplasia
27. Weak flexor tendons

implications of the condition and has produced evidence that it may be recessive, although several genes are probably involved.

Cook has done some research relating to the inheritance of laryngeal paralysis in the horse. This condition causes the "roaring" often heard in race-horses.

Many factors are associated with the ability of an animal to resist diseases. One of the most important is production of antibodies against foreign invaders. The plasma cells of the body produce the antibodies under normal circumstances. Abnormalities in the production of plasma cells would hinder the process of antibody formation. This occurs occasionally, and results in reduced resistance. Many such cases may be inherited.

FIGHTING BIRTH DEFECTS

Much has been written about fighting birth defects in people because birth defects are a leading killer of babies. It is astonishing to read of the high percentage of babies born with birth defects. Even more surprising is the realization that the percentage of birth defects in horses is also very high.

Years ago breeders of Oldenburg horses in Germany were perfecting their particular kind of horse by inbreeding with old Friesian blood. Occasionally one of their foals developed a peculiar incoordination that progressed to death. They came to recognize it as a genetic problem which developed from linebreeding to a particular family.

In the early 1950s breeders of English thoroughbreds occasionally were disheartened with the birth of a freakish foal with a convulsive syndrome. They called them "barkers," "dummies," or "wanderers." Some of the better bred foals had to be put down because of the defect. They were linebreeding to Derby winners, but somewhere along the way one of these winners carried a fatal genetic mutation. Today we recognize there are substantial environmental factors in the condition.

About the same time the breeding of Shetland ponies in the United States had become fashionable. Many breeders were practicing linebreeding to some of the popular champions, when a lethal spinal defect began to crop out in some of the foals. It was very discouraging and a big financial loss to some of the breeders.

A decade ago breeding of overo paints was developing into quite a fad. Breeders thought of them as the sports car of the horse world. The price was high because of a scarcity of the eye-catching paint horse. No one realized at the time that the scarcity was caused by a lethal gene carried by many overo horses which produced a homozygous condition called the white foal syndrome. Even though these paint horses are still gaining in popularity today, breeding them is a very difficult and discouraging task because of this genetic defect in the breed.

In breeding white horses, or so-called albinos, to each other, two out of three offspring inherited the parent's white and one-third were solid colored. This peculiar ratio baffled geneticists until 1969 when it was discovered that one-fourth of the foals conceived were dying in utero. It was then that the breeders realized why they had such a low conception rate in their herds.

Arabian horse breeders have for many years been the supreme masters of inbreeding. In fact, they have often proclaimed that the Arabian horse can stand more inbreeding than any other breed. However, Arabians are not completely immune to the consequences of uneducated linebreeding. It has happened that one of the very popular bloodlines carried a deadly lethal mutant gene. This mutant lay dormant, as any recessive gene will, until inbreeding began to concentrate the chances of the gene in any one individual. Soon many horses in the Arabian breed were carrying the lethal gene now described as C.I.D. (combined immunodeficiency).

At first this condition was treated by Arabian breeders as a minor problem that largely affected the other guy. Today, however, Arabian breeders, faced with the realization that this gene has increased to the point where at least one-fourth of their animals carry it, are worried and are supporting research to solve what has become a major problem within the breed.

A look at the cattle industry reveals many similar examples of birth defects, so one cannot point a finger at a certain breed or species. Inbreeding is a necessary tool for developing and perfecting a breed. Genetic defects will always be a troublesome side-effect. Rather than despairing, or hiding the existence of these defects, the intelligent breeder will ask, "How can I fight them?"

The secret is to eliminate the carriers. To do so, a breeder must know the mode of inheritance for a particular defect. The dam of an affected foal is not necessarily a carrier. A few years back a Belgian stallion was discovered to have an absence of the iris. All of his foals carried the same defect, determining that the trait was dominant. The mares producing the foals were not responsible for the problem, so the solution was to avoid breeding to the carrier stallion.

The C.I.D. defect is known to be recessive and it is a simple matter to cull the mares and stallions who produce C.I.D. foals. The problem is that these animals are very valuable, and breeders keep on with them hoping to obtain that one-in-four that is genetically free of the defect. This practice only increases the number of carriers, and will never be an acceptable procedure until carriers can be identified before they are bred.

A genetic defect known as wobbles is of variable expressivity, as discussed earlier. This means that even though a foal may carry all of the genetic factors necessary to produce the condition, the defect is altered by certain environmental conditions, or influences from other genetic systems in the body. The defect is seen in all breeds; and although it is sporadic in occurrence, it does seem to follow a genetic pattern. There is a sudden onset, generally with ataxia of the rear legs. Some horses show only a weakness of the hind legs. Occasionally wobbles develops in the front legs. It is found most often in younger horses, beginning sometimes as early as 3 months. The genes involved seem to be recessive, but associated with a nutritional deficiency, possibly copper. For some reason the occurrence is more common in heavily muscled horses.

Since not all genetic birth defects are simple dominant and recessive traits, at times it can be difficult to identify carriers. A defect causing blindness in foals was reported in Germany in 1952. The pattern of inheritance was studied considerably and it was determined to be polygenic based on the fact that several variations in severity of retinal detachment were found among affected foals. Polygenic inheritance works like skin color in humans (see Chapter 7). A number of genes contribute to the trait. An individual can be carrying one or more of these contributing genes, resulting in a wide variety of types. Under these circumstances, it becomes difficult to know where to draw the line when culling carriers. Should a good mare with one gene from the polygenic trait be culled or bred to a clean stallion, thus producing a good foal?

The type of inheritance pattern diagrammed in Figure 10–20 is polygenic. The darker horses represent those with the most extreme form of a bad genetic trait. When sufficient numbers are analyzed, several degrees of

Figure 10–20.   A diagram of polygenic inheritance.

severity occur in the offspring. Some are only slightly affected, others are as severe as the affected parent, and others are intermediate.

Inheritance of a dominant gene is covered in Chapter 6. One-half of the offspring are going to be affected. These that do not show the trait are genetically clean.

The inheritance of recessive traits is rather insidious. Neither parent shows signs of carrying the genetic defect. It just appears suddenly in 1 out of 4 offspring of parents that are both heterozygous for the trait. This point is important to remember. A recessive defect which crops out suddenly like this one always comes through both parents. Neither the mare nor the stallion is totally at fault.

Fighting birth defects in horses can be a much simpler task than fighting them in humans. The horse breeder can, and must, be ruthless in culling carriers. It is not necessary to kill the carriers, just to keep them out of the breeding program.

It is better for a breeder to do a good job of selection of breeding stock than to shun inbreeding as a technique. The better horses are produced by inbreeding, but so are the genetic defects. The most successful breeder will learn how to fight birth defects and continue a program of linebreeding.

BIBLIOGRAPHY

Archer, R.K.: True haemophilia (haemophilia A) in a thoroughbred foal. Vet. Rec. 73:338–340, 1961.

Arvey, L.B.: Developmental Anatomy. Philadelphia, W.B. Saunders Co., 1968.

Aurich, R.: A contribution to the inheritance of umbilical hernia in the horse. Berl. Munch. Tieraertzl. Wochenschr. 72:420, 1959.

Bain, A.M.: Foetal losses during pregnancy in the thoroughbred mare: a record of 2,562 pregnancies. N.Z. Vet J. 17:155, 1969.

Bekschner, H.G.: Horses' Diseases. Sydney, Angus and Robertson, 1969.

Bielanski, W.: The inheritance of shortening of the lower jaw in the horse. Przegl. Hodowl. 14:24, 1946.

Bornstein, S.: The genetic sex of two intersexual horses and some notes on the karyotypes of normal horses. Acta. Vet. Scand. 8:291, 1967.

Bouters, R., Vandeplassche, M., and deMoor, A.: An intersex horse with 64XX/65XXY mosaicism. Equine Vet J. 4:150, 1972.

Britton, John W.: Birth defects in foals. Thoroughbred 34(3):288–386, 1962.

Butz, H., and Meyer, H.: Epitheliogenesis imperfecta neonatorum equi. Dtsch. Tieraerztl. Wochenschr. 64:555, 1957.

Cook, R.: Temperament gene in the thoroughbred. J. Hered. 36:82, 1945.

Dimock, W.W.: "Wobbles"—an hereditary disease in horses. J. Hered. 41:319, 1950.

Dorland, S.: Illustrated Medical Dictionary, 24th ed. Philadelphia and London, W.B. Saunders Co., 1968.

Dungswith, D.I., and Fowler, M.E.: Cerebellar hypoplasia and degeneration in a foal. Cornell Vet. 56:17, 1966.

Dunn, H.O., Vaughan, J.T., and McEntree, K.: Bilaterally cryptorchid stallion with female karyotype. Cornell Vet. 64:265, 1974.

Eriksson, K.: Hereditary aniridia with secondary cataract in horses. Nord. Vet. Med. 7:773, 1955.

Fischer, H., and Helbig, K.: A contribution to the question of the inheritance of patella dislocation in the horse. Tierzucht 5:105, 1951.

Flechsig, J.: Hereditary cryptorchism in a depot stallion. Tierzucht 4:208, 1950.

Foley, C.W., Lasley, J.F., and Osweiler, G.D.: Abnormalities of Comparison Animals: Analysis of Heritability. Ames., IA, Iowa State University Press, 1979.

Gilman, J.P.W.: Congenital hydrocephalus in domestic animals. Cornell Vet. 46:482, 1956.

Hadorn, E.: Developmental Genetics and Lethal Factors. New York, John Wiley & Sons, 1961.

Hamori, D.: Inheritance of the tendency to hernia in horses. Allatorv. Lapok 63:136, 1940.

Hosoda, T.: On the heritability of susceptibility to wind-sucking in horses. Jpn. J. Zootech. Sci. 21:25, 1950.

Hutchins, D.R., Lepherd, E.E., and Crook, I.G.: A case of equine hemophilia. Aust. Vet. J. 43:83, 1967.

Innes, J.R.M., and Saunders, L.Z.: Comparative Neuropathology (Illustrated). New York, Academic Press, 1962.

Jogi, P., and Norberg, I.: Malformation of the hip joint in a standardbred horse. Vet. Rec. 74(14):421–422, 1962.

Koch, P.: The heritability of chronic pulmonary emphysema in the horse. Dtsch. Tieraerztl. Wochenschr. 64:485–486, 1957.

Koch, P., and Fischer, H.: Hereditary ataxia in Oldenburg foals. Tieraerztl. Umsch. 5:317, 1950.

Manning, J.P.: Equine hip dysplasia osteoarthritis. Mod. Vet. Pract. 44(5):44–45, 1963.

Mauderer, H.: Hereditary defects in the horse. Dtsch. Tieraerztl. Wochenschr. 46:469, 1938.

McFeely, R.A.: Chromosome abnormalities in early embryos in the pig. J. Reprod. Fertil. 13:579–581, 1967.

McGee, W.R.: Veterinary Notebook. Lexington, KY, The Blood Horse, 1958.

McGrath, J.T.: Nervous Disorders of the Horse. Proceedings 1962 AAEP Convention, 1963, pp. 157–164.

McGuire, T.C., et al.: Combined immunodeficiency: a fatal genetic disease in Arabian foals. J. Am. Vet Med. Assoc. 164:70, 1974.

Morgan, J.P., Carlson, W.D., and Adams, O.R.: Hereditary multiple exostosis in the horse. J. Am. Vet. Med. Assoc. 140:1320–1322, 1962.

Prawochenski, R.: A new lethal factor in the horse. J. Hered. 27:411–414, 1936.

Pulos, W.L., and Hutt, F.B.: Lethal dominant white in horses. J. Hered. 60:59–63, 1969.

Rooney, J.R.: Disease of the bone. In Equine Medicine and Surgery. Edited by J.F.Bone. Santa Barbara, American Veterinary Publications, 1963.

Rooney, F.R., and Prickett, M.E.: Congenital lordosis of the horse. Cornell Vet. 57:417, 1967.

Rooney, F.R., Raker, C.W., and Harmany, K.J.: Congenital lateral luxation of the patella in the horse. Cornell Vet. 61:670, 1971.

Sangor, V.L., Mairs, R.E., and Trapp, A.L.: Hemophilia in a foal. J. Am. Vet. Med. Assoc. 144:259, 1946.

Scekin, V.A.: The inheritance of stringhalt in the horse. Konevodstru 2:20, 1973.

Schlaak, F.: Investigations on the inheritance of umbilical hernias in a horse breeding region. Z. Tieraerztl. Zuchtbiol. 52:198, 1942.

Smythe, R.H.: Veterinary Ophthalmology, 2nd ed. London, Bailliere, Tindall and Cox, Ltd., 1958.

Speed, J.G.: A cause of malformation of the limb of Shetland ponies with a note on its phylogenic significance. Br. Vet. J. *114*:18–22, 1958.

Sponseller, M.L.: Equine Cerebellar Hypoplasia and Degeneration. Proc. 12th Annual AAEP Meeting, 1967.

Tuff, P.: The inheritance of inguinal hernia in domestic animals. Norsk. Vet. *57*:332, 1948.

Weber, W.: Congenital cataract. A recessive mutation in the horse. Schweiz. Arch. Tierheilkd. *89*:397–405, 1947.

Weber, W.: Schistosoma reflexum in the horse with a contribution on its origin. Schweiz. Arch. Tierheilkd. *89*:255–267, 1947.

Wheat, J.D., and Kennedy, P.C.: Cerebellar hypoplasia and its sequelae in a horse. J. Am. Vet. Med. Assoc. *131*:241, 1953.

part III

the breeding program

11

inbreeding

The subject of inbreeding brings many things to mind. Some will consider it an aspect to avoid in a breeding program; others will view it as a mysterious technique understood by only the greatest of breeders. The actual definition of inbreeding is the breeding of animals more closely related than the average of the breed, or population.

Inbreeding in excess can be bad. Inbreeding without ruthless selection will be deleterious, as explained in Chapter 10. On the other hand, without inbreeding it is impossible to develop homozygosity, which brings predictability to a breeding program.

Before embarking on any type of inbreeding program, one should consider those factors that will determine its success. The three basic factors that determine the success or failure of inbreeding programs are: (1) the merit of the breeder's horse foundation material, (2) the effectiveness of the selection program, and (3) the rate of inbreeding. The merit of the horse foundation material reflects how well the horses perform, how free they are from inherited lethals and defects, and how fertile the broodmares in the herd are. If the herd is relatively free of inherited defects of all kinds, and is highly fertile and performing at a high level, then a breeder can expect success using his own stallions in a breeding program. A master at selection and one who is ruthless enough to keep only those horses of outstanding merit for breeding will probably do well with inbreeding. The flaw is that people who work with horses are often swayed in their judgment by a horse who has a particularly good personality even though it may be extremely weak in other respects. Although disposition and temperament should not be slighted, all of the attributes needed by a horse must be considered rather than only one or two. If one inbreeds rapidly (sire to daughter, son to dam, or

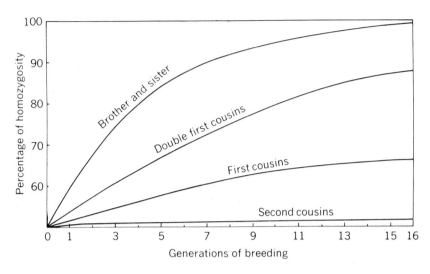

Figure 11–1.   A graph showing the effects of various types of inbreeding on homozygosity.

full brother to sister) over a period of time, the genes tend to occur in the homozygous condition at a rapid rate.

A method of computing the amount of inbreeding is presented later in this chapter. The exact amount of inbreeding can never be accurately determined because there is never enough information about the breeding of all ancestors. Many breeds of horses carry a much higher amount of inbreeding than suspected. Figure 11–1 shows how various types of inbreeding increase the amount of homozygosity in a herd. This graph assumes a beginning homozygosity of 50 percent, which may not be unrealistic in most herds. Seven generations of brother-sister matings are required before homozygosity rises above 90 percent. The constant breeding of first cousins increases the percent of homozygosity for only about 15 generations, at which time it levels off and, thereafter, is only about 18 percent higher than that of an average population.

Desirable and undesirable gene combinations will be produced in the homozygous state. Unfortunately, some undesirable recessive genes when in the homozygous condition can mask the expression of a large number of desirable genes, and result in considerable numbers of undesirable horses. With a slower rate of inbreeding, selection can be more effective, and the less desirable material can be culled gradually without a large financial loss. Also, one can avoid the extreme of fixation that eliminates the opportunity to select.

Studies with most animals have shown that mild inbreeding with rigid selection generally results in initial improvement, but with inbreeding at 25 to 50 percent (62.5 to 75 percent homozygosity) some loss in vigor usually occurs.

PREPOTENCY VERSUS HETEROSIS

A good sound prepotent horse compared to a nondescript crossbred is much like a diamond among chunks of coal; for although diamonds and coal are made up of identical carbon atoms, the diamond has a rare perfection of molecular structure which accounts for its great value. So it is with the prepotent horse: his genes are of a rare quality because of their splendid molecular structure.

The term prepotency actually means the strong ability of a particular breeding animal to stamp its likeness on its offspring. Traits that do not contribute to the well-being of an animal are severely selected against by both nature and man. As a result, the vast majority of genes which become homozygous from inbreeding are desirable genes. In short, inbreeding and careful selection must go hand in hand in order to produce a valuable prepotent animal.

The terms homozygous and heterozygous (explained in Part II) are used to describe whether the two alleles of a specific locus are identical. A homozygous pair of alleles is like identical twins, while a heterozygous gene pair is more like brother and sister. In the same way that brothers bear a family resemblance and yet have minor differences in structure, so alleles of a pair may have differences in structure. In fact, a particular allele may have a whole family of brothers with slight variation in effect on the animal body.

Inbreeding causes a higher percentage of the genes to be homozygous, and this situation brings about greater prepotency. In fact, inbreeding reduces the diversity of the alleles of all genes at random although 100 percent homozygosity is never reached, and both beneficial and detrimental effects can result from extreme inbreeding.

Inbreeding (the most common form of which is linebreeding) can result in some detrimental effects, as well as genes being expressed in a new foal even though they had been masked in a long line of ancestors. If selection against these individuals is practiced generation after generation along with heavy inbreeding, the resultant line of horses will have developed a high degree of homozygosity and will have eliminated many undesirable alleles from their gene pool.

Every serious horse breeder, regardless of the type he is seeking to develop, is striving for prepotency. More progress can be made in any breeding program that has developed prepotency.

As homozygosity increases, even with strict selection procedures, some genes will cause a decrease in vigor which is so slight that the breeder cannot select against them. As homozygosity continues to increase, each new generation becomes a bit more fragile. Size seems to decrease, along with adaptability to environmental changes. Year after year of inbreeding increases these changes.

Then crossing out to a new line, especially crossing to another breed, will effect a dramatic increase in vigor. This new strength is known as hybrid vigor, or heterosis. It comes as a result of a high percentage of the genes suddenly becoming heterozygous. Offspring of the cross seem to grow faster, function better, and reproduce themselves more readily. The crossing of two breeds results in a great increase in heterozygous genes. Since most of the deleterious genes are recessive, an individual with good vigor is a result. Heterosis diminishes quickly in succeeding generations, however, as the crossbreds are mated to individuals closely related to either parent.

The successful breeder must be able to walk the tightrope between heterosis and prepotency. Both are important in developing a truly outstanding herd of horses, even though heterosis and prepotency are antagonistic to each other. Maintaining the proper balance between the two has often been referred to as the science of horse breeding.

LINEBREEDING

The principal method of increasing homozygosity in horse breeding is the practice of linebreeding, which is thoughtfully approached by breeders who want to preserve certain desirable characteristics found in a particular animal. Generally this is a stallion with an outstanding performance record, or he may have been a halter champion. Less frequently, the object of a linebreeding program is a mare.

Linebreeding is an attempt to get as much blood of a particular stallion or mare into the animals of the herd as possible while at the same time keeping inbreeding as low as possible. The theory of linebreeding is excellent, but if one decides that a particular horse is outstanding and concentrates the blood of that horse in his herd, the merit of such a program is no better than the judgment as to whether the horse is truly outstanding. If a breeder, through selection based on performance, decides on linebreeding to certain animals, the program is sound because the animals to which he is linebreeding have proven that they transmit outstanding performance.

Linebreeding has been used effectively at the King Ranch in their quarter horse breeding program which is based on the superb qualities of "Old Sorrel" (Figure 11–2).

The Morgan horse breed was established by linebreeding to Justin Morgan. Early quarter horse breeding found the well-established breeders linebreeding to such horses as Peter McCue, Syke's Rondo, Yellow Jacket, and Traveler. As they did so, a definite quarter horse type developed. These horses were then linebred by other breeders to horses of a new generation such as Leo, King, and Poco Bueno.

Rhoads, in a 1961 article in *The Quarter Horse Journal,* analyzed the quarter horse breed and concluded that too much nondescript and thoroughbred blood had been introduced too quickly. He felt that more inbreeding was needed in order to gain a greater degree of homozygosity. Later, Denhardt challenged Rhoads' conclusions, pointing out that although the percentages in the stud books indicated a preponderance of nondescript

Figure 11–2.   The King Ranch has linebred for years to Old Sorrel, shown in an old photo.

and thoroughbred blood, the breeders more often tended to use sires from one of the established families, thus increasing the homozygosity for the breed as a whole. Both of these articles should be of extreme value to anyone studying the genetics of the horse.

Many years ago the Bedouins on the Arabian desert were linebreeding. They did not use that term, but their practices had the same effect as modern linebreeding to certain stallions. The Arabs linebred to famous mares, and as a result developed what has since been called strains. They were interested only in the bottom line of a pedigree. To them a stallion was important because of his dam. They developed very well-known strains such as Seglawi, Abeyan, and Jilfan.

Breeders have recognized the importance of linebreeding because they can observe the prepotency that results in future generations. Breeding a champion to another completely unrelated champion may produce a winner, but he will seldom be a breeding horse. He will lack prepotency. His offspring will miss the mark that he made in almost every instance.

Inbreeding tends to accentuate faults as well as outstanding traits, but not to the same extent in all horses. Some offspring will inherit most of the best traits, and few of the faults. Intelligent selection along the way improves the line.

Often breeders will linebreed to members of a particular family, not confining their aim to one individual. This approach provides several sources of the specific genetic material desired in a variety of combinations. As a result the offspring tend to have more variety at first, and then by selection more vigorous linebred stock can be developed.

An example of linebreeding to a family is found in the modern Arabian breed. For many years Dan Gainey linebred his horses to two famous sons of Skowronek. The pedigree of Gai Danizon reflects many years of linebreeding (Figure 11–3). Both *Raseyn and *Raffles were sons of Skowronek, one of the most famous Arabian horses of all time. Ferzon was the beginning herd sire for Dan Gainey. He was linebred to Ferseyn, a son of *Raseyn. Gainey used several *Raffles-bred mares with Ferzon, which increased the linebreeding to Skowronek. The result of one of these crosses was Gai Parada, a horse that won the 1977 National Arabian Championship in the United States.

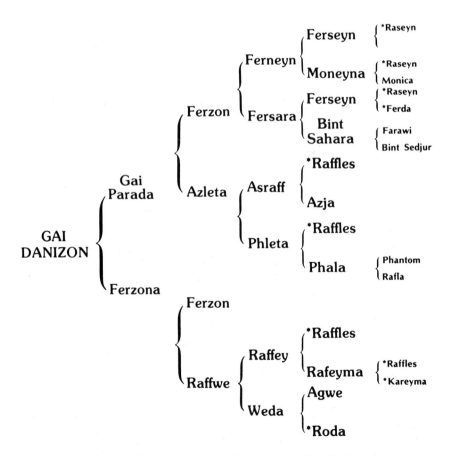

Figure 11–3.   The pedigree of the Gainey-bred stallion, Gai Danizon.

Figure 11–4.   Dan Gainey with his foundation stallion Ferzon.

In a sense, Gainey linebred to his foundation mares over the years. A jeweler by profession, Gainey (Figure 11–4) used his artistic insight and deep appreciation of quality to develop a type of horse that is now well known in Arabian horse circles the world over. The Gainey horse is the result of linebreeding.

THE BALANCED PEDIGREE

Techniques such as outcrossing, dosage, and breeding for nicks will be discussed later since each has merit. However, when a breeder is in the process of increasing homozygosity, he is inbreeding in as safe a manner as possible. One technique is to keep the common ancestor, or ancestors to whom one is breeding, at as distant a place in the pedigree as possible, and yet have as little "outside blood" as possible.

This description may seem a contradiction in terms. Yet the goal is to have as great a chance at the good traits of a famous horse as possible, while reducing the chance at the poor traits. The result will be a balanced pedigree, a good example of which is the pedigree of Gai Danizon (Figure 11–3).

There is a difference between heavy inbreeding and a balanced pedigree. A balanced pedigree is one that may contain much inbreeding but at the same time contains linebreeding to more than one horse, especially linebreeding through more than one horse to a single desirable foundation horse. The pedigree of Gai Danizon is well balanced. It shows linebreeding to Ferzon as well as to *Raffles. The whole Gainey type was developed from this kind of breeding. Mr. Gainey started with a *Raffles daughter, Gajala, who, bred to Ferzon, produced his four best foundation mares. The fact that Ferzon is a linebred *Raseyn horse and that *Raseyn and *Raffles were both sons of Skowronek bring the pedigree into beautiful balance.

Indeed, such balanced linebreeding ensures prepotency. Many generations of thoughtful breeding are displayed in the pedigree of Gai Danizon, the whole representing not only the work of Dan Gainey, but also that of Frank McCoy, Roger Selby, and H.H. Reese; and behind them, the Crabbet and Antoniny studs.

OUTCROSSING

As mentioned earlier, outcrossing tends to increase heterosis although it is not as pronounced as that produced by crossbreeding. Outcrossing is the breeding of two unrelated lines within a breed. The amount of inbreeding decreases rather than increases, so it is the opposite of linebreeding.

Charles Leicester, in his book *Bloodstock Breeding*, gives valuable information on the merits of outcrossing versus linebreeding. He classified as outbred those horses that have no name appearing twice in the first four generations of their pedigrees. In addition, he classified inbred horses into four categories: (1) those inbred at the fourth generation, (2) those inbred with the same name occurring at the third and at the fourth generation of their pedigree, (3) those inbred at the third generation, and (4) those more closely inbred.

With this classification Leicester investigated the percentages of these types within the general thoroughbred population and among various types of winners (Table 11–1). The general population contained 56.8 percent outbreds; yet each of the five categories of winners analyzed had a lower percentage of outbreds. His survey shows that although winners are more highly inbred than the general population, the parents of winners were more highly inbred than the winners. This study exemplifies the relationship between prepotency and heterosis, as explained earlier. The breeder is always faced with the paradox of breeding for the best speed through outcrossing and at the same time breeding for the best breeding stock by inbreeding.

Vuillier, inventor of the Dosage System, studied 654 pedigrees of the best thoroughbreds between 1748 and 1901. He found that 54.3 percent of these horses were inbred, by Leicester's definition.

The inbred thoroughbred, Wisdom (1873), had a poor racing record of his own—no wins, and he finished seventh of the 10 starters in the Derby of his year. He was eventually sold for 50 guineas. However, he proved to be a very successful sire, getting the Derby winner, Sir Hugo; the winner, Surefoot; the Oaks winner, La Sagasse; the Ascot Gold Cup winner, Love

Table 11–1.   Charles Leicester's Survey of Outbred and Inbred Thoroughbreds.

	% of outbreds	% inbred at the 4th generation	% inbred with the same name in the 3rd and 4th generation	% inbred at the 3rd generation	% closer inbred
		A	B	C	D
(1) Cross section of 1,000 horses taken at random	56.8	25.6	14.8	2.6	0.2
(2) 100 recent Classic winners but excluding Derby winners	52.0	22.0	18.0	7.0	1.0
(3) 56 Derby winners (1900–1955)	42.9	28.6	21.4	7.1	.0
(4) 21 horses with the greatest parental influence in the production of (3) above	24.0	33.3	33.0	9.5	.0
(5) 100 leading sires of winners taken at 5 yearly intervals (1910–1954)	44.0	31.0	22.0	3.0	.0
(6) 74 separate leading sires of winners taken at 10 yearly intervals (1910–1954)	47.3	28.4	20.2	4.0	.0

Wisely; and other excellent horses. Many of his daughters became valuable broodmares.

Outcrossing does help in producing winners, but the chances of these winners becoming great breeders are diminished. Table 11–2 shows a list of important thoroughbred winners who were outbred; also shown is the inbred progenitor of each.

COMPUTING INBREEDING

The difference between homozygous and heterozygous is diagrammed in Figure 11–5, which shows how breeding can be more predictable and what is the basis of prepotency. An animal with genes as represented on the upper left of the figure will pass only one type of gene to its offspring. An animal

Table 11–2. Important Outbred Thoroughbreds and Their Inbred Progenitors.

OUTBREDS

Name and Year of Birth	Remarks	Nearest Inbred Progenitor
A. *Hampton* (1872)	Champion sire of his time and founder of the *Hampton* sire line which includes the *Gainsborough's Son-in-Law's* and *Hyperion's.*	His dam *Lady Langden,* who was inbred to *Otis.*
B. *Chaucer* (1900)	A great sire of broodmares.	His sire *St. Simon*
C. *Polymelus* (1902)	Five times Champion Sire.	His sire *Cyllene*
D. *Swynford* (1907)	St. Leger winner and Champion Sire.	His sire's maternal grandsire *St. Simon*
E. *Sir Gallahad III* (1920)	French 2000 Guineas winner and Champion Sire in U.S.A.	His dam's sire *Spearmint*
F. *Bull Dog* (1927)	Full brother to *Sir Gallahad III* and also Champion Sire in U.S.A.	His dam's sire *Spearmint*
G. *Brantôme* (1931)	A great French racehorse and sire.	His sire *Blandford*
H. *Bahram* (1932)	Unbeaten Triple Crown winner.	His sire *Blandford*
I. *Fair Trial* (1932)	A great sire of sprinters, Champion Sire 1950.	His sire *Fairway*
J. *Mahmoud* (1933)	Derby winner and a great sire in both Europe and U.S.A.	His sire *Blenheim*
K. *Precipitation* (1933)	Ascot Gold Cup winner and an important stallion.	His sire *Hurry On*
L. *Vatellor* (1933)	A great French stallion who sired two English Derby winners.	His sire *Vatout*
M. *Djebel* (1937)	Won both the English and French 2000 Guineas and later an exceptional stallion in France.	His sire *Tourbillon*
N. *Nasrullah* (1940)	Champion Sire in both U.K. and U.S.A.	His sire *Nearco*
O. *Dante* (1942)	Derby winner. Died in 1956 after a good start at stud.	His sire *Nearco*
P. *Citation* (1945)	The largest stake winner in the world up to 1955, his winnings amounting to 1,085,760 dollars. This record was surpassed by Nashua in 1956 with a total of 1,288,565 dollars.	His dam's sire *Hyperion*

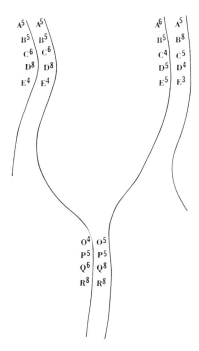

Figure 11–5. A diagram illustrating the difference between homozygous and heterozygous. The lines represent chromosomes and the letters represent genes. Each gene may have a number of varieties, known as alleles, which are represented by the numbers in superscript. The homozygous condition is represented in the upper left, where the gene pairs have matching alleles. The heterozygous condition is represented by the upper right, where none of the genes are matched with identical alleles. The normal situation is represented at the bottom, where some of the gene pairs have matching alleles.

with genes as presented on the upper right will pass first one allele and then another in an unpredictable manner.

The general assumption is that about half the genes in an animal's makeup are homozygous. This percentage of homozygous genes is more likely to occur in purebred horses. Most of the deleterious genes are recessive and will not produce a bad effect unless homozygous. As the diagram shows, both "good" and "bad" genes can become homozygous, and herein lies the danger in inbreeding. An example of a "bad" gene is the dreaded C.I.D. condition, which causes foals to be born without the ability to produce antibodies. Many other "bad" genes are not quite so drastic in their effect.

A breeder must known what to seek and what to avoid when he is inbreeding. The goal of inbreeding is to concentrate desirable genetic material and preserve it for future generations. Breeders whose only purpose is to produce horses who can perform may find that outcrossing brings better results. If healthiness and vigor are an end product for a breeding program, then outcrossing may be the best technique for the breeder. However, the breeder who is seeking to preserve a group of characteristics found concentrated in a certain bloodline must focus his attention on inbreeding, and select against the "bad" genes that crop out.

It is common to find some breeders bragging about the high percentage of inbreeding in their horses, particularly when they are linebred to a famous individual. The reason they boast is that they know the higher the percentage of inbreeding, the greater the prepotency. In truth, selection can increase the amount of homozygosity or prepotency for a certain group of characteristics since the determination of which allele will be passed on at fertilization for any particular gene is entirely at random. Elimination of the undesirable alleles as they occur will naturally increase the percentage of the desirable alleles in the breeding herd.

It is helpful to be able to determine the amount of inbreeding in any particular animal. Years ago, a mathematical formula was developed to determine the coefficient of inbreeding. This coefficient is not a percentage, but can be easily converted to a percentage. The formula is rather simple considering the complexity of inbreeding that could occur in a five-generation pedigree. The formula, $F_x = S(\frac{1}{2})^{n+n'+1}(1+F_A)$, where the coefficient symbol is F_x, calculates the amount of inbreeding from 0.0 to 1.0. Because of the basic premise that 50 percent of the genes are already homozygous, the accurate method for figuring the percentage of homozygosity is to divide the coefficient by 2, add 0.50, and multiply by 100. This calculation narrows the total possible range from 50 percent to 100 percent. A coefficient of 0.50 would represent a 75 percent homozygosity.

From another perspective, it is common to talk of a percentage of inbreeding—which would be different from a percentage of homozygosity. Since inbreeding is defined as breeding animals more closely related than the average, the 50 percent homozygosity is considered a starting point, or zero inbreeding. Thus, a coefficient of 0.50 represents 50 percent inbreeding. Using this type of nomenclature, the coefficient, ranging from 0.0 to 1.0, can be directly converted to percentage by multiplying by 100.

Calculation of the coefficient can be simplified by a systematic approach, as described by W.W. Green. He suggests the use of a work sheet with headings as follows:

Common Ancestor	On Sire's side-n	On Dam's side-n'	+1 Raise ($\frac{1}{2}$) to this power	( )	F_A	$(1+F_A)$	Multiply ( )×( )

In figuring n and n', the generations are not counted in the customary manner. What is normally considered the first generation (sire and dam) is counted as the 0 generation because they are actually the genetic components of the animal analyzed. Since the grandparents are one generation

Table 11–3. The Method of Computing the Amount of Inbreeding for Stud Prospect.

	0	1	2	3
State Prospect	Big Daddy	Good Stud	Old Stud	Foundation Stud / Foundation Mare
			Old Mare	Foundation Stud / Foundation Mare
		Good Mare	Old Stud	Foundation Stud / Foundation Mare
			Old Mare	Foundation Stud / Foundation Mare
	Big Momma	Good Stud	Old Stud	Foundation Stud / Foundation Mare
			Old Mare	Foundation Stud / Foundation Mare
		Good Mare	Old Stud	Foundation Stud / Foundation Mare
			Old Mare	Foundation Stud / Foundation Mare

Common Ancestor	On Sire's side n		On Dam's side n'	+	1 =	Raise (½) to this power	( )	$F_A(1+F_A)$	multiply ( ) × ( )
Good Stud	1	+	1	+	1 =	3	(.125)	.25(1.25)	.15625
Good Mare	1	+	1	+	1 =	3	(1.25)	.25(1.25)	.15625
Old Stud	2	+	2	+	1 =	5	(.03125)	0 (1)	.03125
Old Mare	2	+	2	+	1 =	5	(.03125)	0 (1)	.03125
Old Stud	2	+	2	+	1 =	5	(.03125)	0 (1)	.03125
Old Mare	2	+	2	+	1 =	5	(.03125)	0 (1)	.03125
F. Stud	3	+	3	+	1 =	7	(.0078125)	0 (1)	.0078125
F. Mare	3	+	3	+	1 =	7	(.0078125)	0 (1)	.0078125
F. Stud	3	+	3	+	1 =	7	(.0078125)	0 (1)	.0078125
F. Mare	3	+	3	+	1 =	7	(.0078125)	0 (1)	.0078125
F. Stud	3	+	3	+	1 =	7	(.0078125)	0 (1)	.0078125
F. Mare	3	+	3	+	1 =	7	(.0078125)	0 (1)	.0078125
F. Stud	3	+	3	+	1 =	7	(.0078125)	0 (1)	.0078125
F. Mare	3	+	3	+	1 =	7	(.0078125)	0 (1)	.0078125

$$F_{\text{Stud Pros.}} = .5000000$$

$$100 \times .50 = 50\% \text{ inbreeding}$$

$$\left(\frac{.50}{2} + .50\right) \times 100 = 75\% \text{ homozygosity}$$

$$(\tfrac{1}{2})^1 \ = \ .5$$
$$(\tfrac{1}{2})^2 \ = \ .25$$
$$(\tfrac{1}{2})^3 \ = \ .125$$
$$(\tfrac{1}{2})^4 \ = \ .0625$$
$$(\tfrac{1}{2})^5 \ = \ .03125$$
$$(\tfrac{1}{2})^6 \ = \ .015625$$
$$(\tfrac{1}{2})^7 \ = \ .0078125$$
$$(\tfrac{1}{2})^8 \ = \ .00390625$$
$$(\tfrac{1}{2})^9 \ = \ .001953125$$
$$(\tfrac{1}{2})^{10} \ = \ .0009765625$$
$$(\tfrac{1}{2})^{11} \ = \ .00048828125$$

Figure 11–6.   Coefficient values for the Power of ½.

removed from the sire and dam, they are counted as generation 1. From there to the right in the pedigree, the generations are numbered successively. F_A is the inbreeding coefficient of the common ancestor, which must be figured first. If it cannot be determined, zero should be used.

Table 11–3 is a hypothetical pedigree accompanied by the simplified method of calculating the percentage of inbreeding. Note that in the calculations, an animal is listed more than once if it is repeated more than once in the pedigree. The repeating is only related from the top half to the bottom half of the pedigree. The fact that Old Stud, for example, is repeated in the top half of the pedigree is calculated in the F_A of Good Stud and Good Mare.

One might wonder about the value of using more than a five-generation pedigree. The answer lies in the decreasing values of the powers of (½) (Figure 11–6). Even at $(\tfrac{1}{2})^9$, only 0.0019 is added to the coefficient. Unless the F_A is known, and is rather high, for a distant ancestor, the change in the final coefficient will not be significant.

It is important to remember that the coefficient of inbreeding is not a magical system. It is at best only an estimate. The estimate is based on an average of all genes. Where selection for certain desired genes has occurred over many generations, the homozygosity may be much higher than the coefficient indicates.

GENE FREQUENCIES

There may be a need for calculating the frequency of a gene in a population. The Hardy-Weinberg law states that in a population mating at random, the population is in equilibrium and in the relationship of $p^2 + 2pq + q^2$ with $p + q = 1$. If a lethal occurs in a herd four times in 100 foals born, and if this is expressed as frequency, then the frequency of the lethal is

0.04. If the square root of 0.04 is extracted, it equals 0.20, which is the frequency of the lethal gene in the population.

A serious breeder should know the amount of inbreeding of each of his offspring in order to know how rapidly homozygosity is developing. He needs to know how much each animal is inbred so that he may keep the best animals within each level of inbreeding. For instance, when it is time to cull the foal crop, if the bottom half or bottom two-thirds are culled regardless of the amount of inbreeding, the less highly inbred individuals will be at a disadvantage. If on the other hand, some of the better of the less inbred colts are saved for breeding, by the time a high percentage of inbreeding is reached in their offspring, a much higher quality is possible. The frequency of the dominant (normal) allele to the lethal is 0.8, because $p + 0.20 = 1$ or $p = 1 - 0.2$, which is 0.8. If this is expanded in the equation ($p^2 + 2pq + q^2 = 1$), then $(0.8)^2 + 2(0.8 \times 0.2) + (0.2)^2 = 2(0.8 \times 0.2) + (0.2)^2 = 0.64 = 0.32 = 0.04 = 1$. Thus, this equation can be used to calculate the frequency of a deleterious gene in a population. Although inbreeding will not affect the frequency of an allele in a population, selection will affect it. In fact, this result is the goal of selection: to increase the frequency of the desirable genes at the expense of the less desirable genes.

C.I.D., as discussed earlier, prevents the foal from developing an immune system to fight off infection within the body. The exact nature of the problem is detailed in Chapter 9.

Veterinary research workers have made preliminary reports showing that 0.025 percent of all Arabian foals born are affected with C.I.D.; i.e., 0.025 percent of Arabian foals born are homozygous for the recessive gene.

By use of the Hardy-Weinberg principle, one can determine the percentage of carriers, or heterozygotes, in the Arabian horse population. Let A equal the frequency of the dominant allele (*Def*) and B equal the frequency of the recessive allele (*def*). Since each horse has two alleles for each locus of his genotype, the distribution of the two alleles in the population is as follows:

$$(a + b)_2 \text{ or } a_2 + 2ab + b_2$$

Researchers have determined the value of b_2 to be 0.025, which is the percentage of foals that are affected. Thus,

$$b = b_2 \text{ or } 0.025 \text{ or } 0.158$$

Since the alleles of the gene in the population are either a or b, then

$$a = 1 - b \text{ or } 1 - 0.158 \text{ or } 0.842$$

Thus, 0.158 percent of the gene pool in the Arabian horse population consist of the *def* allele and 0.842 percent are of the *Def* allele.

With these results, one can then determine the frequency of the heterozygous (*Def def*) horses in the population as follows:

$$2ab = 2 \times 0.842 \times 0.158 = 0.266$$

Figure 11–7.   A group of C.I.D. carrier mares used for research at Washington State University.

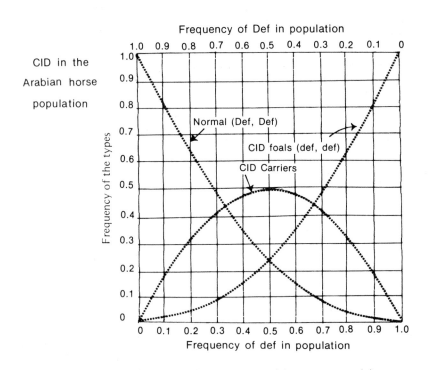

Figure 11–8.   A graph showing how the percentage of homozygous and heterozygous individuals vary as the allele frequency varies in a population. C.I.D. is used as the example.

This means that 26.6 percent of the population are heterozygous for the gene or, in other words, are carriers. Figure 11–7 shows a group of C.I.D. carrier mares used for research.

Figure 11–8 graphs the relationships of these alleles in the population as they change according to the Hardy-Weinberg principle. Note that the number of horses dying of the defect becomes the same as the number of carriers when the frequency of *def* reaches about 67 percent.

BIBLIOGRAPHY

Cruden, Dorothy: The computation of inbreeding coefficients. J. Hered. *40*:248, 1949.

Denhardt, R.M.: Some second thoughts on Dr. Rhoads' "Genetic Trends." Quarter Horse J., July 1961, p. 22.

Dusek, J.: Inbreeding in thoroughbreds in relation to performance. Zivoc. Vyroba *11*:21, 1966.

Fletcher, J.L.: A study of the first fifty years of Tennessee walking horse breeding. J. Hered. *37*:369–373, 1946.

Fomin, A.B.: Heterosis obtained by the purebreeding of Orlov trotters. Genetika (Mosk.) *11*:131–133, 1966.

Green, W.W: The Coefficient of Inbreeding—Its Calculation. Arabian Horse Owners Foundation. Barnesville, MD, 1967.

Krasnikov, A.S., Parfenov, V.A., and Silova, A.V.: The effectiveness of inbreeding in the breeding of purebred and halfbred Kabarda horses. Ivy. Timiryayev. Sel: Khoy Akad. *3*:166–180, 1966.

Leicester, Charles: Bloodstock Breeding. London, J.A. Allen & Co., 1957.

Ocsag, L.: The inheritance of speed in crosses between American trotters and Hungarian halfbreds. Agrartud. Egy. Meyogaydastud. Karanak Kozl. 83–90, 1964.

Pearson, P.B., and Lush, J.L.: A linebeeding program for horse breeding. J. Hered. *24*:185–191, 1933.

Rhoads, A.O.: The American quarter horse. Quarter Horse J., March 1961, p. 33.

Steele, D.G.: Are pedigrees important? Blood Horse *38*:574, 1942.

Steele, D.G.: Modern horse breeding behind the Bamboo Curtain. Quarter Horse J., Dec. 1962, p. 174.

12

breeding theories and systems

All successful horse breeders are using a system of breeding. Those who haphazardly jump from one breeding system to another can never hope to make progress toward their goals. Some successful breeders seem to have a "secret" system; in fact, some have been accused of taking that "secret" to their graves. A careful study of the systems of the more successful breeders, however, reveals patterns that can be readily categorized.

There are only two general categories, or systems, of breeding for the purebred horse breeder: (1) continuous outbreeding, which involves stallions that are unrelated to the mares of the herd, or (2) some form of inbreeding.

Those who are successfully producing nonregistered horses also seem to have two categories of breeding systems: (1) grading up and (2) crossbreeding. Grading up is the use of good purebred stallions of the same breed generation after generation. Crossbreeding is either the crossing of families or lines within a breed or the crossing of breeds.

GRADING UP

In the breeding of nonregistered horses, concern is with either grading up by using outstanding stallions of the same breed generation after generation or crossbreeding to bring together certain combinations which are desirable for the type of horses that are being bred.

The grading up of horses is generally a sound practice because one can substitute genes for outstanding performance for the genes in the foundation stock. As illustrated in Figure 12–1, with six generations of grading up practically all of the genes of the outstanding purebred have been substituted for the genes with which the program started. It must be emphasized that for this method to be successful, one must use stallions of outstanding merit in each generation. Too frequently breeders with rather nondescript mares will breed to stallions of low merit in order to avoid a high stallion fee. It is not necessary to breed to a stallion of exceptional merit initially, but using a stallion of low merit at any time should be avoided.

Many of the coat color breeders are using the technique of grading up with much success. A paint horse of quarter horse type, produced in this manner, is shown in Figure 12–2. Paint horse breeders are taking advantage of the many generations of breeding available in the quarter horse by continually crossing back to purebred quarter horses and selecting only among those with the proper color pattern. This type of breeding is probably the only realistic answer for breeders of Appaloosas, palominos, paints, and other breeds. As will be pointed out in Chapter 13, it is difficult to make progress in selection when a large number of traits are being selected. Therefore, the color breeder can take advantage of the years of breeding and selection in another breed (if a breed can be found of the type desired). Within six generations, a breeder can incorporate over 98 percent of the genes of that breed into his own stock, and at the same time maintain the desired color.

A popular registry of today is the part Arabian. These horses are produced by grading up and can carry 87.5 percent Arabian blood (or more). A

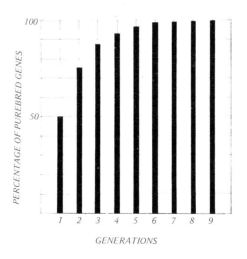

Figure 12–1.   A graph showing the effects of grading up.

Figure 12-2.   A paint horse of quarter horse type.

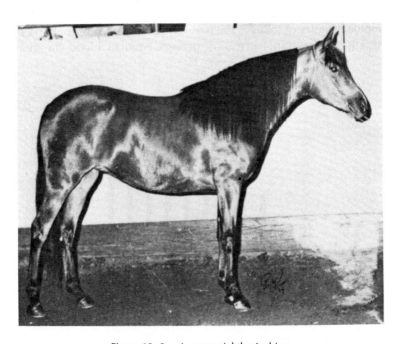

Figure 12-3.   A seven-eighths Arabian.

seven-eighths Arabian, shown in Figure 12–3, is difficult to tell from a purebred Arabian.

Two hundred years ago, some Arabian horse breeders were using a system of grading up in Poland. It was extremely difficult to obtain purebred horses from the desert during those times, but a few good stallions had been imported. This situation was ideal for the breeding technique of grading up. A number of farms were carrying on breeding programs many hundreds of years old. Mares that they considered quite valuable were crossed with the imported Arabian stallions.

Although some Polish breeders felt it was imperative to develop a breeding program based on both purebred mares and stallions, those using the grading up system soon had horses impossible to distinguish from the purebreds. Today, "purists" want to point fingers at some Polish-Arabian horses and call them tainted. However, many generations have passed since the days of Polish grading up, and these horses are now "purebred."

The continued breeding of horses with a grading up system brings the horses involved one step closer to purity with each generation. As illustrated in Table 12–1, impure genes are halved with each generation until, with the thirteenth, the breeding herd is even purer than Ivory soap.

There is no attempt made here to debate the desirability of "almost pure" horses versus the absolutely pure; to each his own. From a practical point of view, 200 years of grading up produces a horse indistinguishable from a purebred. No breed has been "pure" forever.

Proper selection within each new generation can produce increased effectiveness of the system. If the original mares have a few genes that might

Table 12–1. Percentage of the Genes of the Purebred, by Generations of Use of Purebred Stallions in a Grading-up Program.

Generation	Percent of Purebred Genes
1	50
2	75
3	87.5
4	93.75
5	96.88
6	98.49
7	99.25
8	99.63
9	99.82
10	99.91
11	99.96
12	99.98
13	99.99

be especially desirable, these genes can be selectively maintained through-out years of breeding so that eventually one can attain an almost pure strain of horse, but one with some unique characteristics. On the other hand, if the purebred characteristics are all that is desired, selection over the years can eliminate every visible characteristic of the original mares.

Mixing the blood of two breeds of horses might be compared in some ways to mixing skim milk and cream. Let us say that the purebred Arabian cross is the cream and with each new generation the old mixture is halved and an equal amount of cream is added. To make the illustration more analogous, let us say that the top half of the old mixture is used each time to represent the effects of good selection. Under these conditions it would not take long to eliminate almost every trace of skim milk.

The key to successful grading up is the use of a purebred stallion on each generation. The crossing of two half-bred horses is not grading up.

It is almost impossible to visualize what can happen in 12 or 13 generations of horseflesh. This period is more than a lifetime for most, and few horse breeding programs have maintained such a prolonged system.

CROSSBREEDING

The other non-purebred system of breeding, available for both horse and mule producers, is crossbreeding. Crossbreeding is used by horse breeders to make genetic combinations which further the purpose for which the horses or mules are being produced. The crossing of existing breeds has been used to put certain colors or color patterns into a breed. Crossbreeding has also been used to give hybrid vigor, as with the Arabian horse used to produce horses with great endurance and stamina.

The greatest use of crossbreeding was in the production of work stock in

Figure 12–4. A mammoth jack.

which the stallion was bred to jennies to produce hinnies or the jack was bred to mares to produce mules. The great majority of the crosses were jack to mares to produce mules. Large draft mares of the Percheron, Clydesdale, or Belgian breeds were bred to mammoth jacks (Figure 12–4) to produce draft mules. Smaller and more active mares were bred to jacks to produce farm mules and mules used in mines.

The mule is remarkably vigorous as a work animal and has many desirable characteristics. When work stock was used in highway and railroad construction, it was possible to work mules all day and turn them loose in a corral at night where they had access to hay, grain, and water — with no problems. The mule would not develop digestive difficulties from improper eating under such a program whereas horses managed in this manner would be plagued with illness. Mules are not so nervous and flighty as horses; if they get tangled in a barbed wire fence, they will not cut themselves severely attempting to get free. They will wait patiently for their master to come to their rescue. Mules do have more of a stubborn nature, so they have never been used extensively as riding animals. The horse has always been the popular riding animal.

THE BRUCE LOWE FIGURE SYSTEM

The first *Thoroughbred Stud Book* covered the years 1791 to 1814 and listed about a hundred foundation mares, i.e., mares with unknown dams. About half of these foundation mares had no survivors by the end of the 19th century.

An Australian by the name of Bruce Lowe became obsessed with the mare's contribution to the best bloodstock. During the late 1800s he spent much time tracing the pedigrees of all of the winners of the Derby, Oaks, and St. Leger. He made charts of the tail-female line of each, back to the foundation mares. From his research he classified the more prominent mares. At his death, Lowe left a monumental, but uncompleted work, *Breeding Race Horses by the Figure System*. The work was completed by Allison in 1895.

A general explanation of the system was given by Arthur Portman in *The London Horse and Hound:*

> The descendants in tail-female of Tregonwell's Natural Barb Mare were found to have most often won the three classic races of the Nineteenth Century. Next in the order of winners came the descendants of Burton's Barb Mare, then those of the dam of the Two Tru Bleues and so forth. These families were numbered No. 1, No. 2, No. 3, and so forth to No. 43. Eventually 50 families were described.

The family number of any horse indicated which of the original mares the horse was descended from on the maternal side, never on the paternal side, and so the number of a horse is always the same as the number of its dam. Thus Ladas was by Hampton, a No. 10 horse, but Ladas belongs to the No. 1 family mare, descendant in the female line of Tregonwell's natural barb mare. The first five families were called by Lowe the "running" families. Another grouping is that of the "sire" families, in which the stallion element is supposed to be powerful. Lowe originally classified his running families from 1 to 5. Further consideration led him to classify family Nos. 3, 8, 11, 12, and 14 as sire families.

Many of the thoroughbred breeders of the early 20th century used the Figure System and believed in it wholeheartedly. As a result, the winners of that period traced to the low numbered mares, or the "running" families, because those mare families had been bred to the best horses. They ignored the normal procedures of good breeding and selection, such as selecting by performance records. In defense of Bruce Lowe, he had never intended to generate such pure "pedigree nuts." The system was to have been only one of several helps in the breeding of good racehorses. In the United States, the use of the system tended to weed out some good American mares since they could not be found in the original *English Stud Book*.

Without a knowledge of modern genetics, Bruce Lowe surmised that incestuous breeding to the best strains would concentrate a larger vital force in an individual than would be the case otherwise, although it may not be possible to utilize the force until a future generation.

After analyzing pedigrees, Lowe proposed an explanation for the success or failure of most of the great racehorses of his time. As an example, he felt that Gladiateur, although a very successful racehorse, failed as a sire because of a lack of "sire blood" in the bloodlines in his pedigree. Lowe insisted that a great producing stallion must have sire figures in one of the first three top removes. He pointed out that Gladiateur's pedigree is riddled with No. 5 family blood, which is not a sire line.

Lowe had struck upon an important underlying principle discovered many years before by the Arabs: the mare is a more important consideration in inheritance than is the stallion. For many decades fanatical followers of the Figure System developed and followed very strict rules and procedures, based on pedigree analysis. They let these rules dictate which mares to breed to which stallions. In recent times the Lowe Figure System has fallen into disrepute.

Genetic discoveries of recent years offer support to the idea that the mare contributes more to the inheritance of the offspring than the stallion does. Estimates are that the mare contributes roughly 10 percent more. This estimate comes from cytogenetic discoveries which have to do with differences in the male and female germ cells, the sex chromosomes, and the fact that the uterine environment affects the embryo.

Fertilization of an ovum is diagrammed in Figure 12–5. The relative size difference is approximately accurate. Very little other than chromosomal material from the male goes into the makeup of the embryo, while a

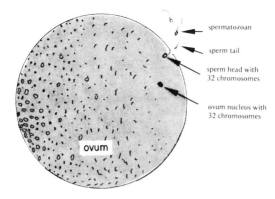

spermatozoan

sperm tail

sperm head with
32 chromosomes

ovum nucleus with
32 chromosomes

ovum

Figure 12–5.   Size comparison of ovum and spermatozoa at fertilization.

multitude of ingredients from the female provides the starting point for embryonic development. The lesser importance of the male contribution is exemplified in some artificial situations in certain species (including the rabbit) in which an ovum or egg can be made to develop into a viable embryo after stimulating the female chromosomes to double. In this situation the embryo develops completely from the female.

Great differences can be observed in offspring from mares bred to zebra stallions and offspring from zebra dams bred to horses. In each case the progeny resembles the dam much more closely than the sire. The same observation can be made with crosses of donkeys and horses. Even crosses between widely contrasting horse breeds show marked differences in the offspring, depending upon which breed is the sire and which is the dam.

There are at least two major reasons for these differences in inheritance. The first is the disproportionate amount of genetic material contributed by male and female. The second is the influence which the uterus has on the developing embryo. Just as "we are what we eat," so is the embryo strongly influenced by the type and amount of nutrients supplied to it during development.

Probably the most important cause of genetic differences between male and female is the phenomenon of the X chromosome. Figure 12–6 shows how sex is determined by the sex chromosomes, the X and the Y. In the horse, the X chromosome is much larger than the Y chromosome, and it represents approximately 5 percent of the total chromosomal material. As the Arabs would have put it had they known the facts, "The X chromosome belongs to the strain;" or as Bruce Lowe would have said had he known, "The X chromosome is the vehicle for transmission of family traits."

During development of both male and female germ cells, the chromosomal material from the previous sire and dam are mixed, resulting in the transmission of a mixture of genetic material from both sides of the

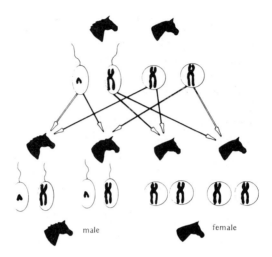

Figure 12–6.   The method of sex inheritance in horses as determined by the sex chromosomes.

family tree. This mixture does not occur, however, with the X chromosome. Here is 5 percent of the inheritance that comes directly from the female line.

It is true that fillies receive an X chromosome from their sires. However, it has been passed intact from the sire's dam, with no mixture of material from the sire's sire. The X chromosome is truly female all the way. As Bruce Lowe reasoned, although he knew nothing of chromosomes, the sire's family number is important and it represents the inheritance passed by means of the X chromosome. With both sire and dam out of the same family, a filly may

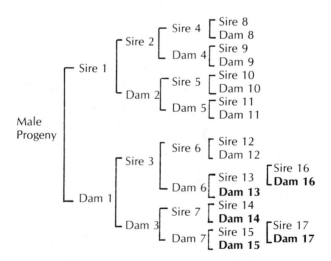

Figure 12–7.   The five family lines (in bold type) of male progeny.

be purely of one family. If this family is extremely desirable, then it becomes important to study pedigrees before breeding.

The problem is that no true families may exist at present. As far as the X chromosome goes, there is no way to determine which family or foundation mare the X chromosome came from originally. For example, in Figure 12–7, the male progeny shown received an X chromosome from Dam 1. She, however, had two X chromosomes, either of which she might have passed on. This could be the original X chromosome, passed through her sire, from Dam 16 or it could be the one from Dam 13, also on her sire's side. The X from Dam 3 might have come from her sire and thus Dam 14. The X from Dam 7 could have come from Dam 17 or Dam 15. Thus, in the pedigree shown, the male progeny could belong to any of 5 different families, considering only five generations. Figure 12–8, a female progeny, illustrates how a mare can be a mixture of any two of nine families, considering only five generations.

Both the Arabs and Bruce Lowe show a fallacy in their thinking when they trace a strain or family strictly in tail-female descent. If the bulk of strain or family inheritance is due to the X chromosome, then, as shown in Figure 12–8, any two of nine lines could actually make up the heritage of a mare. If significant genetic inheritance is passed through the ovum and is sustained by means of uterine nutrients, then the tail-female line takes on special significance. (Tail female is the bottom mare at each generation of the pedigree.)

Those who enjoy pedigree studies might find great reward in searching the pedigrees of famous mares, looking for inbreeding in the proper lines. If each of the dams printed in bold type in Figure 12–8 were the same dam, then the pedigree would be inbred for family. Even such inbreeding would not

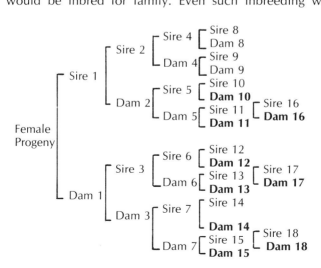

Figure 12–8. The nine family lines (in bold type) of female progeny.

guarantee that the mare possessing such a pedigree carries identical X chromosomes. No guarantee is possible because the dam represented in bold type received an X chromosome from both her sire and dam and it would not be likely that the foundation mare was also so highly inbred that she carried identical X chromosomes. In short, there is no way to trace a strain or family with total accuracy.

This statement does not imply that the idea of strains or families has no merit. Since the dam is responsible for more than half the inheritance, inbreeding for strain or family is valuable if viewed in percentages. After all, inheritance is by chance; in some cases certain traits are lost along the way and in other cases certain traits can be carried for generations. The breeder who places all his emphasis on one or two mares is extremely lucky if he produces what he is seeking. The breeder with 100 mares has a better chance for inbreeding to produce what he desires. Selection for certain family traits among a large group of similarly inbred mares will, over the long haul, produce some individuals that truly represent a family.

Families as Bruce Lowe described them have no meaning because he traced back too many generations on the tail-female line in describing a mare's makeup.

ARABIAN STRAIN BREEDING

The standard in horse breeding was set long ago by the Arabs. Their horses were without equal. In A.D. 60, the historian Pliny wrote: "Everyman of the Arabs was a warrior, and nothing could touch them because of the fleetness of their horses." Five hundred years later, Lufeyl el Kheyl, famous for his verse on horses, wrote of Arabians, "Swift runners, charging, rushing, skimming ahead, slender yet strong."

Perhaps the secret of the Arabs' success at horse breeding lay in their fanatical admiration for a good mare. They used stallions from good mares, and all records were kept with reference to the mare lines. One of the most repeated Arab legends concerns the alleged foundation of the Arabian breed with five superlative mares owned by Salamon. These mares, known as the Al Khamseh, had been immortalized for hundreds of years, and their fame may be the reason why the ancient Arabs placed such heavy emphasis on the mare in their breeding program.

By following the lineage of the mare, the Arabs developed what came to be known as strains. Early 19th century travelers to Arabia learned of the many strains of Asil (meaning pure) horses. An English Consul General in Basra and Baghdad in 1810, Colquhoun was the first European to attempt to list the many strains of Arabian horses. Rosetti, who spent 40 years in Syria, Northern Arabia, and Egypt in the early 19th century, wrote:

> The strains belong to one breed or genus (Arabian) but they differ in points of conformation and individual characteristics and

peculiarities of the shape, which they transmit, but even outstanding differences in one mare do not make her the fountainhead of a new strain, though her foals may be named after her as a substrain. Usually after three or four generations these substrains disappear and only the strain from which they originally descend retains its name and carries on.

Some of the more popular strains were the Kuhaylan, the Ubayyan, the Saqlawi, the Hamdani, the Hadban, the Munique, and the Jilfan.

Most modern Arabian horse breeders scoff at the value of strain breeding today even though most Arabian stud books around the world have kept record of the strains of their horses until relatively recent times. A few breeders have now become interested in trying to "recreate" the strains as near to their makeup when they were on the desert as possible. They feel that the breed has lost valuable seed stock by not preserving "strain bred horses."

The secret to recreating the Arabian strains will be in strain breeding only to certain lines of the pedigree—the lines which carry a tail-female X chromosome. As shown in Figure 12–8, by the fourth generation only 9 lines out of the 16 carry a tail-female X chromosome. As long as these lines remain of the same strain, there is ample room for outcrossing to fresh lines.

It should be possible after a half dozen or so generations to have Arabian horses relatively pure-in-the-strain in the proper lines. What breeding has gone in this direction so far indicates that the strains show a definite type, even after two or three generations pure-in-the-strain.

STAMINA INDEX

An important principle has been illustrated in the development of quarter horse racing capacity. For many years breeders favored short-distance runners. They bred their quarter mares to thoroughbreds who were short runners themselves; i.e., their best distance was from 3 to 6 furlongs. This type of selection pressure over the years has resulted in the development of a quarter horse who can outrun any thoroughbred at one-fourth of a mile. Many quarter horses also race competitively with thoroughbreds at distances of up to a half mile or more.

This breeding result only confirms what racehorse breeders have known for years — that stamina, or winning speed, at a specific distance is highly heritable. Peter Burrell, a world-renowned authority on thoroughbred breeding, once said:

> Races are run at all distances from five furlongs to two-and-a half miles, requiring different types of horses; the sprinter being as different from the stayer as a dairy cow from a beef cow. So high is the standard required today at each distance that the individual horse has become specialized for that particular distance, and we

must regard the horse who is top class at all distances as something of a freak. . . It is not difficult to breed sprinters, but it is an undisputed fact that the more speed one breeds into a horse the less distance he stays, and with the stayers—also a type that breeds true—one can breed them to stay further, but they then begin to lack the turn of speed so essential to the final dash to the winning post.

The Stamina Index is, in reality, the average winning distance of a horse. It can be predicted rather closely in a given foal by a study of the average winning distance of its parents. Breeders today may be overlooking the potential of the Stamina Index as a tool for improving the racing ability of their stock. Often they use it only as a helpful indication of where the horse may excel on the track. As an example, if a colt's average winning distance works out to be 8 furlongs, the trainer feels that it might be advantageous to hold off on breaking him at 2 years, and may wait until the racing distances lengthen at 3 or 4 years of age.

It is not necessarily correct to assume that stamina is a result of inheriting a certain number of factors for distance and that four factors from the sire and eight factors from the dam result in six factors for the offspring. Since the concept is probably not that simple, the conclusions are also false. The inheritance of stamina may work in a similar manner but some of the factors may be antagonistic to each other.

These genetic stamina factors must, in the final analysis, result in a measurable anatomical form, physiological difference, a unique running style, a specific temperament, or all of these. As an example, muscle structure of the gaskin relates to the prowess of the quarter horse. Such characteristics as large air passages, large lung capacity, and increased hemoglobin content of the red blood cells lend themselves to increased stamina. There are also genetic variations in muscle fiber types, which influence speed.

It may be more accurate to think of stamina inheritance as the passing on of genes from a group of traits which contribute to winning at a certain speed, just as has happened with the quarter horse. It may be that each speed has its own set of traits. It would be a mistake to carry this line of thinking to absurdity by breaking the distances up too closely. However, Tom Ainslie, the well-known handicapper, wrote: "Every horse's best distance can be measured with a yardstick." The point is that the breeding of a 5-furlong horse and a 15-furlong mare will not necessarily result in the best traits for winning at 10 furlongs. Over the years this has appeared to be generally true; however, refinement of this concept may result in even more predictability in breeding.

It would seem that if a specific group of traits are needed for winning consistently at 8 furlongs, then it would make more sense to breed two 8-furlong horses than to breed a 4-furlong horse and a 12-furlong mare. The first procedure would have a better chance of producing exactly the traits needed, especially if the sire's and dam's parents also were winners at 8 furlongs.

Most thoroughbred breeders are not particularly concerned about what distance their horses can run, as long as they win. Their stable contains horses capable at a variety of distances. Thus, very few individual horses inherit a specialized set of traits at a specific distance. If more breeding were done with a specific distance in mind, more equine specialists would be developed.

The fact that the better horses can win at about any distance does not necessarily contradict the preceding idea. It may mean only that thoroughbred horses are all rather heterogeneous as to stamina, or distance ability; and speed is a relative thing in relation to the speed of other thoroughbreds.

If the Stamina Index truly represents staying power, or distance capability, then the inheritance of that power or capability will follow established genetic principles. The traits of a sire have a stronger influence on progeny than those of a grandsire. The old-time breeder, Sir Francis Galton, described this principle years ago: on an average both parents between them contribute one-half to the offspring's makeup, the four grandparents subscribe a quarter between them, the great-grandparents, one-eighth, and so on indefinitely. Therefore, in examining a pedigree for stamina, each succeeding generation has a diminishing effect.

Stamina is only one factor to consider in breeding. It cannot make up the entire breeding program. However, selection toward consistent distance ability in sire and dam should make some difference over several generations.

PARENTAGE AGE EQUALITY

Horse breeders sometimes grab at straws for secrets of successful matings. One of these attempts was mating mares with stallions of the same age. It was reasoned that with the parents having the same "vitality" to transmit to the foal, the results would be better. Proponents of this theory would point at some horses that did poorly at the track when the age difference of the parents was great.

To put the matter to rest, Sir Charles Leicester prepared a chart showing the age differences in sire and dam of all the winners of the Derby. In addition, he calculated the age differences of sire and dam of all the starters in the Derby. As shown in Figure 12–9, the difference between winners and starters is not significant.

In his book, *Bloodstock Breeding,* Leicester ended the controversy:

> Although there are some minor variations, taking it fine and large
> the percentage of Derby winners in the various parental age groups

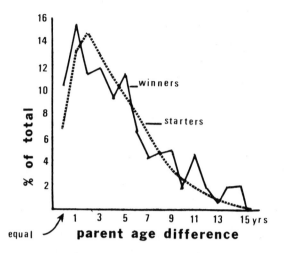

Figure 12–9.   Parent age differences compared between winners of the Derby and starters.

correspond with the percentage of mares mated within these groups. The discrepancies would probably be eliminated or reversed if another cross section of Thoroughbreds were taken. I think we may conclude that the apparent good results of the plan are an illusion and merely the natural result of the frequency of matings.

DOSAGE SYSTEM

In the 1920s, Colonel Vuillier described what is now known as the dosage system, in the French publication, Les Croisements Rationnels. He had analyzed the pedigrees of many of the better thoroughbreds of the time and found a relationship in the balance of blood of certain progenitors in them. Like prescription medicine, there was a dose of this and a dose of that.

Colonel Vuillier computed that 15 stallions and one mare appeared in pedigrees of first-class horses with approximately the same frequency (Figure 12–10). The secret of the system as a tool in breeding selection was to choose mates for horses that were strong in their weak lines. As Bruce Lowe did with his system, Vuillier proposed the system only as an adjunct to other systems, such as the stamina index and conformation. The idea possibly began as a balancing system for the Bruce Lowe system, in that Bruce Lowe considered only tail-female whereas Vuillier considered the sire line (with the exception, for some reason, of one mare, Pocahontas, foaled in 1837).

Vuillier recognized the genetic principle proposed by Galton: the effect of an ancestor on a progenitor diminishes in relationship to how far back it appears in the pedigree. Using a 12-generation pedigree, Vuillier calculated the strength of a particular horse according to the generation in which it appeared. Each horse in the twelfth generation was one strain, in the

(H)	Pantaloon (1824)	140
(T)	Touchstone (1831)	351
(O&B)	Birdcatcher (1833)	288
(H)	Pocahontas (1837)	313
(T)	Newminster (1848)	295
(T)	Hermit (1864)	235
(K)	Galopin (1872)	405
(B)	Bend Or (1877)	210
(K)	Votaire (1826)	186
(H)	Gladiator (1833)	95
(H)	Bay Middleton (1833)	127
(M)	Melbourne (1834)	184
(B)	Stockwell (1849)	340
(T)	Hampton (1872)	260
(O)	Isonomy (1875)	280
(K)	St. Simon (1881)	420

Figure 12–10. Vuillier's 16 important horses are shown with dates of foaling in parentheses after, and the amount of dosage found on the average in the pedigrees of high class horses. The letters in parentheses in front of the names refer to foundation sires to which these horses trace.

eleventh two strains, and so forth, as shown in Table 12–2. Each time the name of one of the select 16 horses appeared in the pedigree, its dosage was calculated by strains, depending on the generation in which it appeared. The dosage was increased as the name appeared more than once. By this method of calculation, Vuillier determined the "standard" dosage, as shown in Figure 12–10, for each of the 16 horses.

Table 12–2. Vuillier's Diminishing Values
According To Generation

Generation	Value of name (in strains)
1	2,048 strains
2	1,024 strains
3	512 strains
4	256 strains
5	128 strains
6	64 strains
7	32 strains
8	16 strains
9	8 strains
10	4 strains
11	2 strains
12	1 strain

Vuillier proposed that the value of a name in a pedigree should depend upon how close up it was found. His standard dosage for each of his 16 horses was figured in terms of "strains."

Many breeders struggling to find a formula for success have analyzed the Vuillier dosage system carefully. The basic principle involved seems to make sense: certain horses can contribute certain essential traits to a progenitor, and a group of important ancestors, each contributing their own essential traits, could produce the "ideal" horse. The questionable part of the system is whether Vuillier's dosage calculation can determine the influence of a progenitor in a pedigree.

In critiquing the system, Sir Charles Leicester wrote in *Bloodstock Breeding:*

> The whole system is sometimes adversely criticized from the fact that (Vuillier's) method of calculation is mathematically erroneous. If each of the 4,096 ancestors of the twelfth generation count 1/4,096 the total influence of this generation comes to 1/4,096 × 1, i.e., the complete build up of the horse, and so on all through the pedigree. Thus each generation is self-sufficient, leaving no room for the influence of any other generation.

Realizing this miscalculation of Vuillier, more modern breeders have clung to the principle that the sire line is important, but they have arrived at more logical ways to determine the influence of the sire line. Some have traced the lines back from Vuillier's 16 horses to the original sires. Since some of the 16 are related, or belong to the same sire line, the total number of lines to consider is reduced. Figure 12–10 shows the sire line of each of the 16 Vuillier horses. The letters in parentheses denote foundation sires as follows: Herod (H), Touchstone (T), Oxford (O), King Fergus (K), Birdcatcher (B), Mathem (M). Realizing the value of other lines, some breeders have added Waverly (W) and others.

Some breeders, attempting to better the dosage system, have converted the dosage to percentages of what Vuillier proposed. Others have taken strictly the fifth generation of each pedigree and figured the sire line of each of these ancestors, then calculated how often each line is found. This method then determines the dosage in a consistent, easy-to-figure manner.

Modern proponents of the dosage system use it in conjunction with some form of the Bruce Lowe system because they recognize that the sire line does not tell the whole story. Use of the two systems together brings the entire pedigree into focus, so the thinking goes.

The use of pedigrees to determine breeding value, whether it be sire lines or families, has an inherent fallacy in the basic principle on which such theory is predicated. The fallacy is in assuming that inheritance of racing ability operates according to the "blood theory" of old. This theory, proposed 2000 years ago, assumed that the blood carried substances which were determiners of the many traits of the progeny. The idea was that this mixture was more fluid than specific entities such as marbles or beads.

This blood inheritance concept made it easy to think in terms of percentages of blood from a certain parent. Actually Galton's law relating to the diminishing influence of succeeding generations is based on the blood theory. It supposes that each progenitor in the fifth generation contributes equally to the distant progeny. The blood inheritance theory has not stood the test of time, and it is now known that traits are inherited as a result of an offspring receiving one or more genes for that trait. Generally speaking, there is very little variety in the gene types themselves, the difference in individuals coming about because of the variety of combinations of genes. As described in Chapter 6 and illustrated in Table 6–2, as many as 1024 phenotypes can result from only 10 gene pairs.

During the past 300 years, with intensive selection occurring in development of the thoroughbred, genes for the important traits have become increasingly homozygous. As this occurs, some of the alleles (Chapter 6) are completely eliminated from a specific line, whether it be a sire line or family. As an elementary example, suppose Herod, a popular thoroughbred pro-

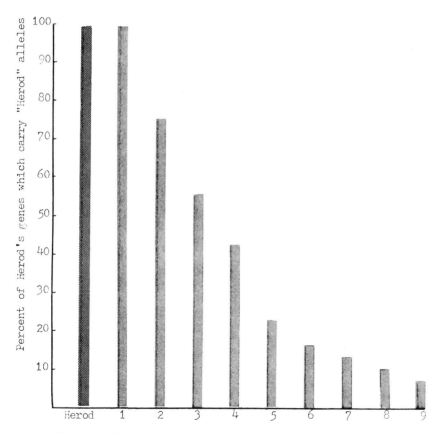

Generations of breeding with nondescript mares, homozygous for alleles other than "Herod" alleles.

Figure 12–11. The loss of "Herod" alleles in succeeding generations of breeding with nondescript mares homozygous for alleles other than "Herod."

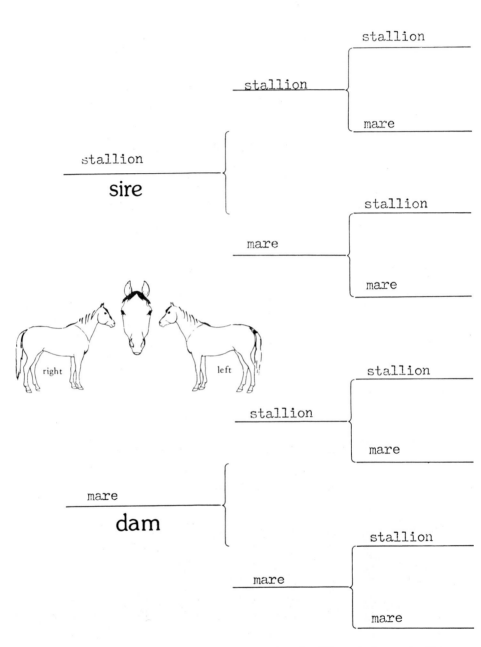

Figure 12–12. A pedigree which has dosage figured in the fifth generation, and which represents 100% Herod breeding.

stallion	(H) Comanche	
	mare	(H) Le Sancy
mare	(H) Halma	
	mare	(H) Hindoo
stallion	(H) Chouberski	
	mare	(H) Hamburg
mare	(H) Bruleur	
	mare	(H) Dunboyne
stallion	(H) Rey Hindoo	
	mare	(H) Sir Modred
mare	(H) Kingman	
	mare	(H) Hanover
stallion	(H) Bruleur	
	mare	(H) Chouberski
mare	(H) Joe Madden	
	mare	(H) Kingman
stallion	(H) Yankee	
	mare	(H) Vagrant
mare	(H) Bruleur	
	mare	(H) Shirley
stallion	(H) Halma	
	mare	(H) Cambyse
mare	(H) Half Time	
	mare	(H) Hamburg
stallion	(H) Hindoo	
	mare	(H) Montague
mare	(H) Joe Madden	
	mare	(H) Duke of Mag.
stallion	(H) Burgomaster	
	mare	(H) Shirley
mare	(H) Rey Hindoo	
	mare	(H) Halma

genitor, was homozygous for all the important genes he carried. Further, suppose that each mare he was bred to was homozygous for a different allele of each of these important genes. Actually this assumption is made when pedigree analysts calculate the importance of Herod on the top as 100 percent and mare X on the bottom as 0 percent. As illustrated in Figure 12–11, the "Herod" alleles remain in the first generation in such a cross but they are heterozygous. Then crossing back to another mare X, one-fourth of the "Herod" alleles are lost. Each succeeding generation in which a cross back to a mare X is made, one-fourth more of the remaining "Herod" alleles are lost, until after 10 generations of such breeding, a colt of the Herod line would carry less than 6 percent of Herod's genes; i.e., over 92 percent of his genes would be those from "nondescript" mare lines.

Such is the fallacy of the basic assumptions of complete faith in sire lines. When no credit at all is given to the mare, it is not many generations before a famous ancestor has no more than "paper potency." At this point a breeder who believes strongly in dosage may argue that the mares all have sires, and that the influence of these sires is also figured into the system. This may sound like a good argument, but very few dosage proponents can show a system where the sires of the dams all the way back in the pedigree are calculated.

As an example, take the system that uses the fifth generation of a pedigree. According to the dosage theory, a pedigree (Figure 12–12) would be 100 percent Herod. With such a system a particular stallion is labeled "Herod" although he is many generations from Herod. This system assumes that the mares all along the subsequent generations have contributed nothing good or bad. If these (H) horses are an average of 10 generations from Herod, they will carry only 5.6 percent of Herod's genes. Add this to the influence of the unmentioned mares behind the fifth generation and the pedigree becomes more like 2.8 percent Herod. The dosage breeder may argue further that with all the Herod blood in thoroughbred pedigrees, some of the mares along the line are going to help out with Herod blood. Although true, since the importance of the whole dosage system is based on determining percentages of "dosage," of what value is the system when it cannot show true genetic relationships? The point is that if the true genetic relationship of the pedigree in Figure 12–12 is only 2.8 percent Herod, it is highly probable that some other line (good or bad) might have a greater genetic relationship to the pedigree than Herod. This theory depends upon the composite breeding of the many mares in the distant parts of the pedigree.

It all points up the fact that it is really impossible to assess the genetic value of distant relatives. If the sire and dam, and their immediate parents are not desirable animals, no ancestral dosage is going to help. The genetic makeup of a sire line is ever-changing. After 50 years of breeding there will be no resemblance whatsoever to the original line. The sire line does not have the X chromosome going for it as does the tail-female line. The male chromosome (the Y) carries so few genes that their significance is negligible.

NICKS

The art of horse breeding centers around the discovery and use of nicks. Producing the best horses consistently is generally a result of having found a good nick. Essentially a nick is a breeding where the strong points and weak points of sire and dam complement one another to produce offspring better than either parent.

Like many breeding theories, results may be an illusion with nicks, but some are certainly well documented. Thoroughbred breeders are aware of the famous English nick of Phalaris crossed with Chaucer mares. Counting horses who won at least £ 2000 or more, offspring of this cross won a grand total of £141,391. Phalaris and the prolific Chaucer mares (daughters) do produce winners, of course, when crossed with other types. However, all of the other offspring of Chaucer mares won only a total of £145,179, slightly more than the amount the one sire produced with these mares. The great sire Phalaris was bred to other types of mares and produced many winners. But all of his other offspring (not out of Chaucer mares) won less than offspring from the nick of Phalaris × Chaucer mares.

Other thoroughbred nicks have been noted. An early nick was St. Simon and Touchstone line mares. Some more recently have been Pharis and Asterus mares, Round Table and Nashrulla mares, Fair Play and Rock Sand mares, Hail to Reason and Djeddah mares, and Broomstick and Peter Pan mares.

The success of nicks can be explained genetically (Fig. 12–13). At least the theory can be explained, although specific reasons for the success of each cross are not known.

In this illustration five hypothetical genes are considered. These genes have been chosen as an example only, and there is no implication that single

SIRE   *DAM*
LL, NN, HH, RR, mm, tt, bb × *ll, nn, hh, rr, MM, TT, BB*

OFFSPRING

Ll, Nn, Hh, Rr, Mm, Tt, Bb

L = Large lung capacity
l = Smaller lung capacity
N = Large nose and throat air passages
n = Smaller nose and throat air passages
H = Large heart
h = Smaller heart
R = A great amount of hemoglobin in red cells
r = Smaller amount of hemoglobin in red cells

M = Large muscular body
m = Smaller body
T = Strong tendons
t = Weaker tendons
B = Strong bones
b = Weaker bones

Figure 12–13. An illustration of how a genetic nick can combine strong points to produce offspring better than either parent.

genes for these traits even exist. Suppose that the stallion involved in a good nick was homozygous for four of the genes necessary for speed and stamina: large lung capacity, large nose and throat air passages, large heart, and a great amount of hemoglobin in red cells. However, the stallion carried recessive genes of another type for the size and strength of the body. In this situation the stallion has more availability of oxygen than he needs when running. His only limiting factor is his body size and structure. He could be rather fast in his own right, but not outstanding.

Suppose that the mare in the good nick is homozygous for three genes necessary for a large strong body, but she is recessive for inferior genes for oxygen supply. She would probably not have good stamina, but might be an excellent sprinter.

When bred together, as shown in the illustration, the foals receive a desirable allele from each parent, some from the stallion and some from the mare. Suppose, for purposes of illustration, that these genes are all dominant. Then the foals of this cross will have the excellent oxygen supply capacity of the sire combined with the excellent body size and strength from the mare to utilize all of the oxygen available. The result is, on the average, high-class offspring from mediocre parents.

Many other sets of genetic mechanisms could have been used as an example that a trait cannot be fully utilized because of the lack of a complementing trait—the genetic basis of the nick technique.

DEVELOPING THE BREEDING PROGRAM

Most breeders know that some weaknesses exist either in their entire herd or in some of the individuals of the herd. One will generally try to find a stallion who is particularly strong in the trait or traits that need strengthening in the herd. One must also be sure that another weakness is not brought into the herd. The stallion used to strengthen the weaknesses in the herd must not have undesirable traits. If a stallion with one undesirable trait is used to try to correct another undesirable trait already in the herd, two undesirable traits may result instead of only one. It would be better to use a stallion with no undesirable traits, even though he is not outstanding in the trait that needs correcting in the herd. Thus, it is imperative to use stallions that are not weak in any trait which contributes to the value of the animal in terms of breeding objectives. To emphasize only one trait for correcting a weakness while another weakness is brought into the herd simply means that continual attempts must be made to correct weaknesses, and even so, after 25 to 30 years of breeding little or no improvement will be evident.

To be effective in a breeding program, it is necessary to be completely objective and ruthless. A breeding program is no place for sympathy toward an animal or for personal pets. Every attempt should be made to see that the environment is the same for all of the animals. A foal who is particularly appealing, given special care and training, may develop into a desirable animal; however, many foals that are less appealing in early life might also develop into desirable animals if given special care and training. A desirable animal that has had special care and training may transmit no better than a less desirable animal that has not had this treatment. Selecting a horse for breeding that has had special care and training may be selecting only for special care and training. Certainly these environmentally produced differences in horses are not inherited and will not be transmitted. In fact, it is often wise to select animals who were developed under the type of environment in which they are expected to perform. If stock horses are being developed for handling cattle in rough, rugged country, selection under such conditions is more desirable than under conditions that are not severe (Figure 12–14). Animals who have inherited weaknesses will tend to break down under the rugged environment; as a consequence, they will not be used for breeding. Such animals might never show these inherited weaknesses under a less rugged environment.

For most horsebreeding programs, the ideal environment has plenty of forage of good quality covering an area that compels the horses to take considerable exercise over fairly rough land in order to obtain the forage. Where the land area is highly productive and level, the animals may have to

Figure 12–14.   Stock horses should be selected from horses raised in rough, rugged country, an environment in which they are expected to perform.

be forced to exercise a great deal. Forced exercise will tend to keep the animals from becoming overly fat and will give greater strength to the feet and legs so that they can better carry any excess weight they may acquire. Too frequently horse breeders tend to overfeed with concentrates and to exercise their animals sporadically. Strenuous exercise given to an animal who has not exercised for a considerable period of time may be harmful. Regular exercise, even if it is quite strenuous, will not harm an animal who is genetically sound. However, it may bring out inherited weaknesses possessed by some horses.

One should provide an environment that will be useful in distinguishing the animals which are genetically most capable for the purpose they are to fulfill. Animals who are being bred for endurance in traveling should be made to travel long distances every day to determine whether they can do so without developing unsoundnesses (Figure 12–15). This practice is useful as a selection tool in deciding whether to keep the horse for breeding. Likewise, animals who are being bred for jumping ability should be made to jump as early in life as they can be trained so that those lacking the ability to jump or those who go unsound from jumping can be culled from the breeding program before they leave any offspring. Of course, a horse must

Figure 12–15.   In breeding endurance horses, only stock which remains sound under use should be used.

be mature and soundly conditioned before serious jumping is done. Horses who are bred for high speed at short distance should be forced to run at their highest possible speed. If they are made to run fast often, it will be possible to detect those animals who lack the ability and also those who become unsound from the strain of rapid running. Again, a horse must be properly conditioned before high speed is forced. It is always best to know the capabilities of a horse before it is selected for breeding. Usually, much variation is present among the horses, and one should keep only those that meet the needs of the type of horse being bred.

BIBLIOGRAPHY

Ainslie, Tom: Ainslie's Complete Guide to Thoroughbred Racing. New York, Trident Press, 1968.

Bormann, P.: A comparison between handicap weight and timing as measures of selection in thoroughbred breeding. Zuchtungskunde 38:301, 1960.

Finochio, Louis: Thoroughbred Nicking Patterns. Published by author, Alhambra, CA, 1978.

Hollingsworth, K.: Conclusion: The greater a colt's earnings on the track, the greater his chance of success as a stallion. The Blood-Horse, Nov. 4, 1967, p. 3267.

Leicester, Charles: Bloodstock Breeding. London, J.A. Allen & Co., 1957.

Lesh, Donald: A Treatise on Thoroughbred Selection. London, J.A. Allen & Co., 1978.

Napier, Miles: Blood Will Tell: Orthodox Breeding Theories Examined. London, J.A. Allen & Co., 1977.

Napier, Miles: Thoroughbred Pedigrees Simplified. London, J.A. Allen & Co., 1973.

Tesio, Federico: Breeding the Racehorse. London, J.A. Allen & Co., 1958.

Varoloa, Franco: Typology of the Racehorse. London, J.A. Allen & Co., 1974.

Viljcinskii, A.D.: Criss-cross breeding as an effective method of increasing the working quality of horses. Konevodstvo 25(1):11–14, 1955.

13

selection methods

Few horse breeders have set goals for their programs with a view that is much beyond next year's foal crop. Those who do are trying to develop a "type" that they know will not come into reality until many generations have passed. Those who have been successful over the years seem to have had a consistent selection system, the core of which is a good method of selecting, each year, the horses which will be retained for breeding. Selection is extremely important if the program is to continue toward its goal with each generation.

The goal itself can be an important consideration in whether a program will be successful. The characteristics of the specific "type" that a breeder is trying to develop can be vital. More than one trait is being sought in the "ideal" horse, e.g., size, specific muscling type, temperament, and correct conformation. Even those who only care about whether the horse wins on the racetrack need to divide that one trait into several specifics in order to make more progress toward their goal.

TANDEM SELECTION

One of the simplest selection systems is the tandem system. The procedure is to select entirely for trait 1 until it is established at a satisfactory level, then select entirely for trait 2 until it is established at a satisfactory level, after which selection for trait 3 occurs.

This system is useful for color breeders. The establishment of a color breed is compatible with the gaining of one desirable trait at a time. In this case color has top priority. At the beginning selection is entirely for color. Black, palomino, or Appaloosa would all be the same situation. Coat color that involves more than one gene can be obtained step by step. (Details are found in Chapter 9.) Once the color has been properly established in the herd, the breeders can go on to another trait. Size, as an example, may be the next step in the selection system. Within a few years when the desired height is obtained in the breeding stock, concentration can be shifted to another trait such as soundness and straightness of legs. To be successful, this system must begin with a very wide base and a large enough number of original horses to provide the variety for later selection of other traits.

It is not often that a single breeder can count on his herd alone as a complete basis for selection. Tandem selection occurs automatically within a color breed—after a fashion. The Appaloosa breed, as a whole, provides a population from which a breeder may select for height, or for some other trait, so he can made rapid progress as he takes advantage of selection from among the thousands of Appaloosas available.

As the breeder progresses, he must be sure that the horses brought into his herd have all of the traits previously established in the herd. Otherwise the tendency may be to work in circles.

This system of selection may prove to be unsatisfactory if a large number of traits are important to the breeder—or if the few traits chosen have a negative genetic association. This problem is discussed later in the chapter.

THE INDEX SYSTEM

The best selection technique for a breeder who has many items on the selection list is known as the Index Method. It involves the developing of an index or point system for each trait considered important in the selection process. Those traits with a higher priority can be assigned a higher total number of points, thus balancing the more important traits carefully against those deserving less consideration. This system of selection is basically what occurs in the mind of astute horse breeders, even though they may not put the idea on paper.

The Index Method of selection entails the yearly development of a chart similar to the one shown in Table 13–1. Here, ten fillies are graded for selection. If the breeder is selecting only 20% for replacement, then the two highest scoring fillies will be saved. Since each breeder will have his own list of traits, each will vary as to the possible value of each trait. This choice of traits is where the perceptive breeder will excel. He will understand the importance of each trait relative to the others. Action or type may be more important to one breeder than another. One breeder may put more relative emphasis on general quality than height, and so on.

The more traits used in selection, the less progress can be made in any one year. The exact relationship of progress to number of traits is graphed in Figure 13–1. Selection for four traits is half as efficient as for one. Selection for as many as 16 traits will reduce selection efficiency by one-half again. If the average diligent breeder is concerned about a dozen or more traits, it is

Table 13–1. A Grading Chart of the Type Used in the Index Selection System

Trait	Total Possible Value	Filly 1	Filly 2	Filly 3	Filly 4	Filly 5	Filly 6	Filly 7	Filly 8	Filly 9	Filly 10
Height	100	80	100	75	80	100	0	100	50	100	100
Weight	30	25	30	25	30	30	10	30	15	30	30
Tail set	200	100	50	175	100	150	200	100	100	0	200
Head	200	100	50	200	100	150	200	0	100	50	200
Straightness of legs	250	250	200	200	200	250	200	250	125	250	0
Disposition	155	100	50	140	100	0	150	130	75	150	150
Action	200	100	25	200	100	50	200	150	100	200	200
General conformation	100	50	25	90	70	80	50	50	50	50	100
Value of pedigree	80	10	60	80	80	80	80	50	40	80	80
Speed	75	50	40	50	50	75	25	25	40	75	60
Color	10	10	5	8	5	10	5	10	5	10	10
General quality	100	50	45	90	60	50	80	50	50	75	80
Total	1500	925	680	1333	975	1025	1200	945	750	1070	1210

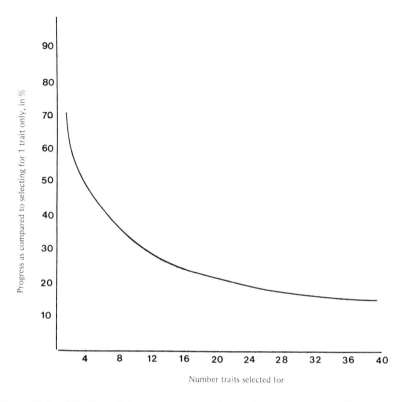

Figure 13–1. Selection efficiency decreases as the number of traits selected for increases.

going to take him about four times longer to reach his goal than a breeder selecting for only one trait.

With a well-thought-out index system, it might be most advantageous to select breeding stock on the basis of those that reach a certain index level rather than selecting a set percentage of the crop each year. Some years will produce more good replacements than other years, and the index system, if carefully executed, should accurately identify the better individuals.

Grading should be consistent from year to year, and it is helpful to describe in the beginning the exact method of grading with all the variables of each trait identified in terms of how they affect the total value of that trait. For instance, if straightness of legs is determined to carry a weight of 250 (16.6 percent of the total), then the 250 should be broken down into a number of factors, each of which carries specific values—perfection being worth 250. How much would be taken off for cow-hocked conformation? How much for pigeon toes, and so forth?

Careful consideration should also be given to the overall percentage of each trait in relation to the total. Has too much or too little weight been given a particular trait? This determination can often be made by carefully grading a number of animals with the proposed index. Some conclusions can also be drawn from the chart as to the desirability of the weight given these various traits. Suppose that one decides to retain all fillies with a total score of 1000 or more. Thus, filly 5 with zero disposition would be retained, raising the question of whether enough weight has been given to disposition. Filly 10 has the second highest overall score, and yet is tremendously defective as far as straightness of legs is concerned. Should this be so?

The answer is not engraved somewhere on golden plates and established as irrevocable fact; rather it is for each breeder to decide what type of horse he wants as a final product. It may be that color will carry more weight in the minds of some, while tail set carries less weight; still others would choose completely different traits.

An important consideration in assigning a certain value to a trait is its heritability—a ratio of how much a trait is produced by the environment in relation to genetic factors. Weight of an animal is a good trait to illustrate the concept of heritability. Weight is strongly influenced by environmental factors, such as amount of food available and weather. But even under ideal conditions, each horse has a certain genetic potential for weight at a particular age. Body weight has been estimated to be about 25 percent heritable. This figure is a rough estimate, as are all heritability figures. True heritability is almost impossible to determine accurately.

Coat color is a trait more highly heritable than body weight, and the shape of the head and body conformation are quite highly heritable. In short, traits which are less heritable should be awarded less value in proportion to their importance when establishing an index.

One more caution: Over the years the standards of judging halter classes at shows, or choosing thoroughbred yearlings at sales, have evolved from one type to another. Breeders engaged in long-term programs should not adjust their goals to reflect the changing fads of the show, or sale ring. Following trends results in a program that leads in circles, with the breeder never reaching his prescribed goal.

MINIMUM CULLING LEVEL SELECTION

Rather than altering the criteria, and the relative weight of each, as the herd approaches a high level of desirability, the breeder would be better advised to switch to an entirely different system of selection. One known as the minimum culling level is effective when the number of traits selected for is small and when only a small percentage of the crop is kept. This system is particularly useful in stallion selection from a very high-quality herd. Table 13–2 is a hypothetical example of one that would be developed in determining selection of worthy stallions under these conditions. The main disadvantage of adhering to this system is in not being able to utilize the outstanding performance in two or three traits at the expense of inferiority in

Table 13-2. Hypothetical Data of Four Potential Stallions. Those horses not reaching the minimum in any one trait are culled.

Trait	Minimum	Stallion 1	Stallion 2	Stallion 3	Stallion 4
Height	14.3	14.2	14.3	15.0	14.3
Weight	1000	1000	950	1100	1000
Disposition	7	10	8	10	7
Conformation	8	9	10	7	8

one trait, so this system is more useful with a high-quality herd, as not much variation will be present in the best individuals.

The minimum culling level method is extremely useful for all-or-none traits, such as defects and reduced fertility in horses where the mare either does or does not have a foal. Traits that show continuous variation, such as speed and body size, could be selected by setting minimum levels below which all horses are culled. An advantage to this sytem is that a breeder can cull at the time a trait is measured so that all animals do not have to be kept until measurements are made on all traits. For example, one could set a wither height of 15 hands and a weight of 1000 pounds for a 2-year-old as a minimum. Those horses not meeting these minimum levels would be culled, and the others would be trained and tested for speed for the next culling.

Table 13-2 shows that stallion 1 would be culled because he fell short 1 inch in height. If the system were followed consistently, his excellent disposition and conformation would be for naught. Likewise, stallion 2 would be culled in spite of excellent conformation because of his light weight. Stallion 3, while excellent in size and disposition, would not be used for breeding because he falls a bit short on conformation. Stallion 4 would be the only one qualifying for the breeding program, and he is not as good in several points as some of the other animals culled.

The dilemma of this system is that no provisions exist for a priority rating of the traits. This system does not allow a breeder to obtain outstanding performance in two or three traits at the expense of inferiority in one trait. This problem is a handicap to the system because some characteristics are harder to come by than others and ought to be rated higher on the priority list.

The relatively simple process of breeding a high-priced stallion to high-priced mares and then selling high-priced foals does not necessarily constitute a sound breeding program. At any rate, it does not generally command deep respect from fellow breeders. A program that begins with stock chosen on the basis of specific traits, rather than price, and continues on the basis of a predetermined system, rather than the dictates of fickle buyers, will prove to be much more profound.

NEGATIVE GENETIC ASSOCIATION

The selection list of criteria is developed around several general factors such as appearance, pedigree, performance ability, and progeny testing. As the years pass, pedigree should take on a lesser role in selection since it can be controlled and planned by the breeder in advance, so no genetic variation should occur. Appearance will generally provide the most possibilities for selection criteria. Appearance can show soundness, with individual selection items such as straightness of legs, strength of tendons, and amount of bone. Conformation criteria, such as height, length of body, legs, hips, and head, should be rather specific. The appearance of the horse also helps in establishing selection criteria relating to style and disposition. Details about selecting for conformation will be found in Chapter 15.

The breeder must be aware of other considerations about selection. Some traits have a negative genetic association; i.e., they are somewhat antagonistic to each other. Selection for trait 1 will automatically cause a decline in trait 2 and vice versa; selection for trait 2 will automatically cause a decline in trait 1. An example of this antagonistic action would be the selection for a straight or upright shoulder on the one hand and a long smooth stride on the other hand. These characteristics certainly do not go together. Instead, the sloping shoulder contributes to a longer stride. Figure 13–2 shows this so-called "teeter-totter of selection."

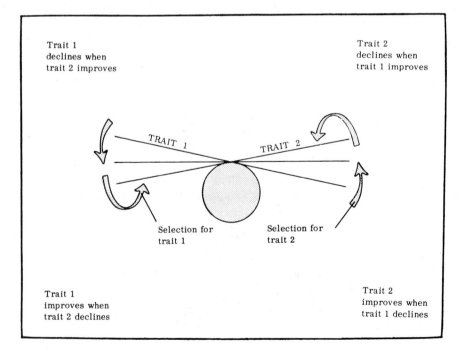

Figure 13–2.   The teeter-totter of selection.

RATE OF PROGRESS

The only way to improve the breeding herd is to select the best part of each new generation and sell the poorest part. In traits that have no dominance or even some dominance, when highly heterozygous parents are intermated, a normal curve is usually produced (Figure 13–3). The largest portion is intermediate in merit and the smaller portions are inferior or superior in merit.

If selection is made from among the upper 60 percent of the mares and the upper 10 percent of the stallions, this selection is essentially from the upper third of the general population. When these selected animals reproduce, they will also produce a population which shows a frequency curve similar to the original, but the average of the population will be slightly superior to the average of the original population.

Progress from one generation to the next is made by means of proper selection (Figure 13–4). Suppose a breeder begins with a herd of 100 mares. He breeds these mares until he has another replacement herd of 100 fillies, which become the first generation. Through selection their average merit has increased. They, in turn, produce a second generation which has an average higher merit, and so the breeder progresses toward his goal.

Selection is defined as differential reproduction of what is desired; consequently, it means more than simply keeping an animal for breeding. Even though attempting selection, one's efforts are sometimes futile because of chance. For example, one might keep a highly desirable prospect for a stallion and that horse may receive an injury so severe that he does not survive. Effectively, this animal was culled from the breeding program by

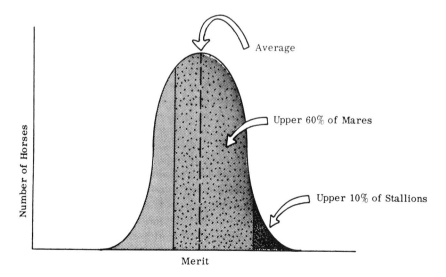

Figure 13–3.   Frequency distribution of phenotypic merit in horses.

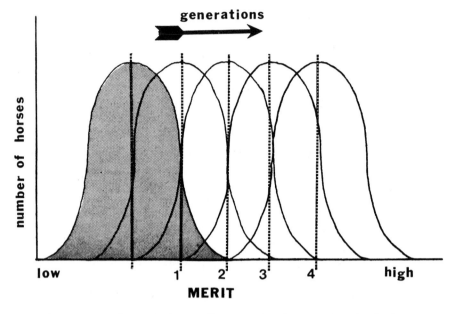

Figure 13–4. The merit of succeeding generations improves through selection.

chance. A breeder may select several stallions for use in his program, but breed to only one because he has the best personality, in effect selecting against the other stallions. Effective selection implies using the horses in the breeding program once they have been selected. Some factors may influence the progress made in our selection. These factors are of sufficient importance to be considered before discussing the selection program in horse breeding.

Selection is not always equally effective from one breeding program to another. At times, rapid progress can be made toward the breeder's goals, while sometimes little progress is made even over many years of breeding.

Number of Genes Affecting a Trait

The number of pairs of genes affecting a trait has a large effect on selection progress. For example, if only one pair of genes is involved, one should make rapid progress in achieving the desired goal, after which no need exists for further selection. If a large number of genes affect a character, progress can be made over a long period of time. As has been pointed out, if 10 pairs of genes affect a trait, all in the heterozygous state, more than a million offspring would be needed to obtain all possible phenotypes and genotypes. In horse breeding where herd sizes of 100 or less are used, all of the possible phenotypes with 10 pairs of genes involved would not be present.

Progress will be slow in improving desired traits except for those controlled by only one or two pairs of genes. It will be possible to bring about improvement only by consistent selection over many generations. Since the lifetime of an individual is short in relationship to the time needed to bring about improvement in horses, it is highly desirable to establish ideals for a breed of horses so that a pattern of selection can be practiced over several human generations. If possible, herds should be kept intact and the program of selection continued even when the owner dies.

Number of Traits Considered in Selection

If one is concerned with the improvement of only one trait, all his efforts can be concentrated on that trait and considerable progress can be made. If another important trait is considered when selecting, one cannot put all his efforts into obtaining either of the two traits. Selection pressure is lower for each of the two traits than it would be for either of them if only one were under consideration. The result is that with two traits instead of only one, progress will be reduced for each trait (Figure 13–1). The progress that can be made for traits, as more are added to the list, is $1\sqrt{n}$ where n = number of traits.

The practical consideration in determining what traits are important is to keep the number of traits at a minimum while not overlooking an important trait. Thus, any trait that does not contribute to the value of the animal in terms of the purpose for which it is being bred should not be considered for selection. On the other hand, no trait that does contribute to the value of the animal in terms of the purposes for which it is bred should be ignored, because such a trait may decline by chance alone or by negative association with one of the other traits for which selection is being used.

This criterion certainly puts the horse breeder in a position of having to make important decisions relative to the value of traits in his selection program.

Goal of Breeder

A breeder with conflicting and shifting goals will make little or no progress over the years (covered in detail in Chapter 4). Show horse breeders will have the most trouble in this respect since show standards seem to change from year to year. The popular style a decade ago is obsolete today. Even racehorse breeders have the same problem, though perhaps over a longer period of time. A thoroughbred breeder who tends toward sprinters for a few years and then leans toward stayers is not being consistent. His progress toward the production of better horses will be significantly hampered.

This consistency is one of the best reasons for a selection system. It works somewhat like a computer. Once the program is set, there is no wavering. Horses are selected by the formula, and personal feelings can be left out of it, once the original formula is selected.

Rate of Production

If selection is to be effective, horse breeders must use methods which will result in a high level of fertility. No farm animal has a lower level of fertility than the horse. Some of the low fertility is due to improper management and feeding practices. The mare has a long heat period of about 6 days. If she is bred once early in heat, the sperm may be dead by the time she ovulates. If too long a time elapses before breeding and the mare has ovulated some time before she is bred, the egg may have passed the stage at which fertilization is possible. By the use of artificial insemination to conserve the sexual strength of the stallion, a mare could be inseminated every other day during heat. This use of insemination should result in greater chances for pregnancy than breeding only once during heat. If only one breeding or insemination during heat is used, the third or fourth day of heat would be preferable to either earlier or later times in the heat period.

Many mares are kept in a fat condition, are not given much exercise, and are fed imbalanced diets. Regular exercise, the use of good grass pastures, and the restriction of heavy grain feeding should contribute to higher fertility.

The possibility that some cases of lower fertility are due to genetic causes must also be considered. Some mares breed and settle, but the foal is lost because of inherited lethal genes. Some mares are infertile because of inherited abnormalities of the reproductive tract. The use of stallions out of mares exhibiting low fertility is not recommended because these stallions may propagate more breeding problems. A young stallion out of a mare who has been a regular breeder and has had normal, live foals at each foaling is a good prospect for increasing fertility in the herd.

An example of the role of fertility in the selection of all-important traits follows. Breeder A has a herd in which 40 percent foals are produced, while breeder B, with the same number of breeding animals, obtains an 80 percent foal crop. It is evident that breeder B has twice as many young horses from which to select his replacement animals as breeder A.

Disease

Diseases are important in a selection program in three ways. Diseases may lower fertility or reduce the foal crop so that fewer animals are available for selection. Diseases may kill some of the horses and thereby reduce the number from which to select or may kill some of those previously considered as outstanding and now being used for breeding. Perhaps the most important effect of disease is that it may mask otherwise good inheritance (Figure 13–5). All animals in a highly diseased or parasitized herd will do poorly and show poor performance. Livestock producers often raise the logical question, "Why not breed animals for resistance to disease?" Certain important factors must be considered before an attempt is

Figure 13–5.   It is difficult to judge the genotype, and potential, of a diseased or undernourished horse.

made to breed disease-resistant strains of horses. In the first place, disease
resistance is not usually general; that is, some horses may be resistant to one
disease but very susceptible to others. In the second place, the microorganisms causing diseases in horses have the capacity to mutate or change
their inheritance. This mutant form can result in a strain of disease-causing
organisms that are different from the strain from which they mutated. It
follows that resistance for the original strain which has been developed by
breeding would not give resistance to the mutant form. Thus, everything that
had been done would be of no avail when the mutation occurred. The third
factor to be considered is that in horses the rate of reproduction is low and
the generation interval is long. It appears that breeding for disease resistance
is not indicated for horses, although it has been quite useful in certain plant
species.

The alternatives to breeding for disease resistance are sanitation, disease
prevention, and vaccination. Vaccinations prevent disease by artificial
immunity. The most important way of coping with disease is to prevent its
appearance on the farm by sanitation and the introduction of breeding stock
from clean farms. Animals who are to be brought onto the farm should be
isolated for a period of time sufficient to let them develop any disease to

which they may have been exposed. If these precautions are taken and if mostly replacements from the breeder's own herd are used, the chances of introducing disease are low. The promiscuous introduction of animals from any herd before isolation is an almost certain way of bringing a disease problem to the farm.

Phenotype as an Indicator of Genotype

Horse breeders will generally do phenotypic selection, that is, selection based on the performance records or appearance of the animals. The degree to which the phenotype predicts the genotype of the animal determines the effectiveness of selection. For several reasons, the phenotype may not predict the genotype accurately. Environmental influences may alter the phenotypes of the animals so markedly that the phenotype may not be of much value in predicting the genotypes. An example is straightness of legs in foals. Some crookedness is definitely genetic, but often crookedness can result from a deficiency or imbalance of minerals, which is environmental.

Since it is only genotypic selection which can change gene frequency of a population, environmental influences which have marked effects on the phenotypes will reduce the effectiveness of selection. In such a situation only environmental differences are selected rather than genetic differences. When this situation occurs, the heritability of the trait is very low and selection is quite ineffective. In traits in which the environment does not have such marked effects, the phenotype may be a better indicator of the genotype. Then heritability is high and selection is effective in substituting more desirable for less desirable genes in the population (increasing the frequency of the desirable genes in the population).

The accuracy of assessing the true genetic value of breeding stock can be improved by studying data from offspring. The final accuracy, however, depends upon how heritable the trait in question is. This relationship is shown in Table 13–3. If a stallion has thin hair and scaly skin, and a breeder wants to know whether this horse will pass the condition to his offspring, the accuracy of his prediction will be zero if the condition is due to a vitamin deficiency. In that case, the heritability is zero. If, on the other hand, a breeder wants to predict how the black color will be inherited, he can be 100 percent accurate since black coloring is 100 percent heritable. Most traits are less than 100 percent heritable, as explained later in this chapter.

Taking wither height as an example of a heritable trait, one can gain an understanding of how a study of ancestors' records can be helpful. Some estimates put wither height heritability at about 50 percent. Table 13–3 shows that, in this case, measuring the height of a stallion and predicting what size offspring will result based on that figure alone will give 71 percent accuracy. This accuracy can be increased to 76 percent with use of data from the two parents as well as the stallion under consideration.

Pedigree fanatics should note that considering wither height of the four

Table 13–3. The Accuracy of Predicting Genetic Value From Own and Ancestor Records.

Records Used	Heritability		
	.1	.25	.50
Own	32%	50%	71%
Own + 1 parent or progeny	35	53	73
Only 1 parent or progeny	16	25	35
Own + 2 parents	38	57	76
Only 2 parents	23	35	50
Own + 1 grandparent	8	12	18
Own + 4 grandparents	35	53	73
Only 4 grandparents	16	25	35

Modified from Evans et al.: The Horse. San Francisco, W.H. Freeman & Co., 1977.

grandparents alone, and ignoring the stallion in question, reduces the accuracy of genetic predictions to 35 percent. The same holds true for any trait considered, including racing ability.

Even when the environment does not markedly affect a trait, the phenotype may not be an accurate indication of the genotype because epistatic genes may be preventing other genes from expressing themselves. The result of undesirable recessive genes acting epistatically to many

Table 13–4. Decreasing Efficiency of Selection With Decreasing Frequency of a Recessive Allele in a Population

Generation Number	Percentage of a Recessive Allele in the Population
1	50.0
2	33.3
3	25.0
4	20.0
5	16.7
10	9.1
20	4.8
30	3.2
40	2.4
50	2.0
100	1.0
200	0.5
1000	0.1

desirable genes will be eliminated by culling animals who are undesirable because of these epistatic genes.

An example in racing horses would be when a high-class animal with great potential in most respects happens to have a genetic weakness of the tendons. This potentially great runner will pull up lame before he ever reaches the big time. In this case the outstanding genetic traits can never be expressed because of the one undesirable genetic trait.

Another reason why the phenotype is often not an accurate indicator of the genotypes is found in the figures on the number of genotypes and phenotypes for crosses in which the parents are heterozygous for 10 pairs of genes. In such a cross there would be 59,049 genotypes but only 1024 phenotypes (Table 6–2), so several genotypes would give the same phenotypic expression.

Many undesirable genes are recessive and are expressed only in an occasional crop-out. Because of this, selection efficiency decreases as the frequency of that gene in the herd (or population) decreases. This relationship is shown in Table 13–4.

Linkage Relations

If the genes affecting two traits are located in the same chromosome, they tend to stay together. Crossing over will soon establish the association of the genes in equilibrium if crossing over is common. In some instances the genes are so close together that few or no crossovers occur; this situation is called close linkage (see Chapter 7). When close linkage occurs, the combination of genes is not brought into equilibrium, and the result is that certain gene combinations tend to segregate as a unit. If a desirable gene of one locus is closely linked to an undesirable gene at another locus, it is difficult or impossible to develop homozygous animals for the combination of the desirables.

Many studies with farm animals reveal correlations of certain traits with other traits. This pattern can be the result of linkage relations, or it may be due to physiological relationships. For example, horses that tend to stay fat under almost any management system may be low in fertility, probably because of an endocrine disturbance which causes both excessive fatness and low fertility.

Genetic Association of Traits

Correlations of traits are of two general kinds, those caused by environmental influences and those caused by inherited differences. If the environment affects two traits in the same way, i.e., if a better environment increases performance of both traits, the two traits will be positively correlated so that performance in one trait is positively associated with performance in the other trait. This type of association tends to decrease selection effectiveness.

If two traits are associated because the more desirable genes cause both traits to be at a higher level of performance and the less desirable genes cause both traits to be at a lower level of performance, the traits are also positively associated. This type of association will help in selection, because selecting on the basis of either trait will bring about improvement in both traits. On the other hand, if certain genes cause one trait to be at a high level, and another trait to be at a low level of performance, while other genes cause the reciprocal effects, the traits are negatively associated in that any animal who is strong in one trait will be weak in the other. This type of negative association interferes drastically with selection (Figure 13–2).

Inbreeding

Inbreeding tends to create homozygosity by changing the heterozygous condition to the homozygous condition in the population. Inbreeding in itself will have no effect on gene frequency. If homozygosity is obtained at a locus, selection will become ineffective for this particular gene. Homozygosity reduces phenotypic and genetic variety, and thus slows down the selection process.

As an example of how increased inbreeding can decrease the rate of progress from selection, breeding for a dominant trait, such as black coat color, can be used. In the beginning the herd will be heterozygous and the first generation will have a 50 percent frequency of the undesirable recessive allele. In the next generation, after eliminating all non-black, the herd will carry a frequency of 33.3 percent for the recessive allele. Each generation of selection continues to reduce the frequency of the undesirable allele, but as shown in Table 13–4, the rate of reduction diminishes drastically with each succeeding generation. At first the frequency of the undesirable allele is halved in two generations. After 50 generations, it takes 50 more generations to halve the frequency of the undesirable allele once more.

Foundation Animals

Selection will not create new genes nor will it cause genes to change or mutate. All that selection does is change the frequency of genes in a population. For example, if A and a are in the population each at a frequency of 0.5, it might be possible to increase the frequency of A to 0.99 while, at the same time, the frequency of a is reduced to 0.01 in the population. If all animals in the population were of the genotype AA, then the frequency of A is 1.0 and of a is 0.0.

If one uses a foundation herd of horses who are all pure for a gene that is not desirable, selection in this herd will never have an opportunity to alter the genetics of the herd for this particular gene.

To ensure a genetic variation in a population, it is necessary to start with a population that has a wide genetic base. Attaining this type of population could be achieved by crossing breeds and then selecting to bring about improvement in performance. Crossing of breeds may not be desired by the horse breeder, and he may want to improve his herd within a breed. He could start with a wider genetic base if he mated unrelated animals from about four or five different bloodlines within a breed. He would then gain more genetic variation in his herd and should offer more possibilities for bringing about improvement from the subsequent selection than if he did his selection within one bloodline.

Although the Bedouins of the Arabian Desert practiced heavy inbreeding to produce their individual strains, or families, those who later purchased these desert horses interbred the strains. This wider genetic base has produced a certain amount of heterosis.

The Arabian breed today, even in a worldwide context, traces to a relatively few foundation animals. However, the genetic base is wide because of the genetic variety in these original animals.

Level of Performance

If a herd is performing at a low level, it is not difficult to locate stallions of good merit to improve it. When a herd is built up to a high level of performance, it becomes difficult to locate stallions who will contribute to further improvement. Thus, as perfection is approached, improvement of the herd becomes more and more difficult. When a breeder develops a nearly perfect herd, his selection becomes one of preventing a decline in the herd more than improving it. Thus, a breeder of an outstanding herd is concerned principally with eliminating undesirable recessive genes that are in the heterozygous condition. These genes may often fail to express themselves until two heterozygous individuals are mated and produce a homozygous recessive offspring.

Size of Herd

Horse breeders with very small herds have difficult problems in improving through selection. In a small herd, chance plays a major role. If breeding is done within the herd, there will be a tendency to inbreed rapidly, because it is uneconomical to keep four to six stallions in a herd just to breed to 15 to 25 mares. With only 10 to 15 mares, a breeder would tend to use only one stallion. The use of only one stallion would present two hazards; very close inbreeding would occur if the stallion were selected from within this herd, and the accuracy of the decision of this stallion having merit is crucial. Using only one stallion, a breeder cannot afford a mistake, because all the foals in the herd will reflect this mistake. When only one stallion is used in a herd, it is imperative that this stallion have no genetic weakness and that he be truly outstanding. The only problem is that the stallion's quality is never known for certain. It might be reasonable to assume that a stallion carries no deleterious genes if he has been progeny tested by mating to 35 of his daughters and no undesirable outcrops have occurred. At present, few, if any, stallions have been given this progeny test.

The small breeder does have a chance of increasing his rate of improvement by considering the entire breed as his field of selection. He can take mares to outside stallions who have the merit desired. Most breeds today, such as thoroughbred, quarter horse, standardbred, Appaloosa, Arabian, and others, have a large number of outstanding horses. The thoughtful breeder can take advantage of the progress already made in other breeding programs.

Selection Intensity

The rate of genetic progress is affected by the intensity of selection. Dale Van Vleek, in *The Horse,* writes:

> The factor for intensity of selection is not quite proportional to the fraction culled. The intensity factor increases at an increasing rate as the fraction selected declines. These factors (Table 13–5) are a relative measure of how much the selected group will exceed the average of the population from which selected animals are taken. For the intensity factors to apply, selection must be based entirely on predicted genetic value. That is, selection of a random 10 out of the top 50 of 100 horses is not 10 percent selection but is equivalent to selection of the best 50 percent, the same as selecting the best 50 of 100, since selection is random out of the best 50.

Table 13–5.   A Chart of Selection Intensity Factors.

Select Top Percentage	Selection Intensity Factor	
100	.00	No culling
90	.20	Usual level for selection of mares.
85	.27	
75	.42	
70	.50	
60	.64	
50	.80	
40	.97	Range for selecting dams of
30	1.16	stallions.
10	1.40	
5	1.75	
4	2.06	Possible range for selecting
3	2.27	stallions.
2	2.42	
1	2.67	

Modified from Evans et al: The Horse. San Francisco, W.H. Freeman & Co., 1977.

THE SELECTION PROGRAM

The Computer Horse Breeders Association of Sherman, Texas, measures various parts of the body of a horse, feeds the information into a computer, and receives a rating for the horse as to how it stacks up against a "so-called" ideal conformation. It is all very exciting, but computer mating still leaves a lot to be desired. The computer program has not been designed that will analyze all the many characteristics of individual mares, weigh them against the characteristics of several stallions, and accurately predict the best mating. Such a program, if available, would have no way of allowing for each breeder's idea of "the perfect horse."

The Computer Horse Breeders Association has designed its program around the ideal of speed on the racetrack. The group is quite accurate at evaluating potential speed with conformation measurements. However, only if a breeder has a 100 percent mare would he necessarily want to breed to a 100 percent stallion. Such is probably never the case. Rather, the mare is a bit short here, a bit long there, and her temperament generally lacks something. The task for the breeder is then to find a stallion a bit strong in her weak points and a bit too short where she is a bit too long.

The average horse breeder feels as qualified to choose a stallion for his mares as anyone else, especially since only the breeder knows his goals and objectives. From one point of view this is true. If "like begets like," then choosing a stallion is simply selecting the stallion who most closely fits the idea of the ideal.

But the adage, "like begets like," is much too general a statement to be supported by scientific fact. Like tends to beget like, but a hundred variables cause a stallion to produce better than himself in some cases and worse than himself in others.

The experienced horse breeder knows that successful nicks in breeding do occur so that a horse with certain faults and strong points can be bred to a mare with complementing strong points and faults, resulting in an offspring better than either parent (see Chapter 12).

Often an outstanding individual mare or stallion will carry a recessive trait for some weakness that may show up in the offspring. It can be a conformational defect such as a weak back or a crooked leg, or it may be a dreaded genetic disease such as C.I.D. This is one of the negative considerations in breeding mares to stallions owned by others. These recessive weaknesses are seldom revealed to the breeder with stud fee in hand. A breeder with enough mares can purchase his own stallion and will soon find out first hand about any recessive weaknesses.

A breeder who has been in the business long enough will get to know what his mares can produce. He will discover their recessive weaknesses. Since recessive traits are expressed in the offspring only if both sire and dam contribute a gene for the characteristic, determining a mare's recessive weaknesses is a big help in knowing what hidden weaknesses to look out for in the stallion. One of the values in breeding an older mare is that by studying her offspring carefully, a more intelligent mating can be made.

Inbreeding tends to strengthen prepotency. A study of a stallion's pedigree will give an idea of how much prepotency one can expect. If the sire and the dam of the prospective stallion were closely related, a greater degree of prepotency can be expected than if sire and dam were of more-or-less unrelated lines. A stallion who shows superior type and conformation and also a certain amount of inbreeding warrants careful consideration. One of the most common types of inbreeding practices is the breeding of half-brother and half-sister (see the hypothetical pedigree in Figure 13–6).

This Prospective Stallion would be 50 percent the blood of Mr. Perfect. Chances are that he will pass on more of Mr. Perfect's good traits than would a son of Mr. Perfect (even though both are 50 percent the blood of the foundation animal) because of the selection that generally takes place to obtain a good stallion. Son of P. was probably not gelded because he showed more good qualities of Mr. Perfect than bad. Other sons of Mr. P. were gelded because they received more bad genes than good. Prospective Stallion as shown in the figure has a good chance of receiving Mr. Perfect genes from both his sire and dam. Therefore, many of his Mr. Perfect traits will be carried in a homozygous condition. If Prospective Stallion has been selected carefully from among many offspring from the same type of linebreeding, a high percentage of those homozygous genes which he carries will be for the good traits rather than the poor traits originally carried by Mr. Perfect. The fact is that Mr. Perfect was not really perfect. Let's assume that out of four genes only genes *A* and *B* are really desirable. Had the proper selection been made among many colts, Prospective Stallion could have received a genetic makeup as shown in Figure 13–7.

In this example, inbreeding has produced an animal capable of passing on the good traits of Mr. Perfect and fewer of his bad traits. Because the genes are homozygous in the inbred individual, the offspring are going to be much more alike, making breeding much more predictable. The selection of a proper stallion for a mare must be considered in light of what is to be done with the offspring. Is the product of the mating to be primarily for breeding stock, to be sold for the highest dollar, or to be used for some type of performance, such as racing?

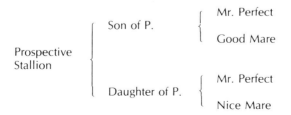

Figure 13–6. Hypothetical pedigree showing a common type of inbreeding.

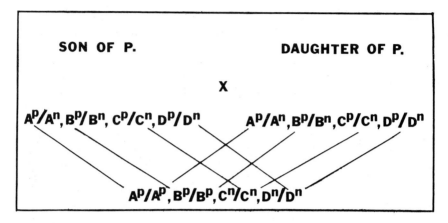

Figure 13–7.   An example of how inbreeding can produce desirable homozygosity.

If the primary goal is to raise breeding stock, then it may be important to balance the pedigree with proper linebreeding, in order to increase prepotency in the offspring. Only top notch mares of excellent pedigree should be considered for this purpose

Often high-quality breeding stock brings the highest dollar, but not always. Many buyers are oriented toward performance. Can the horse win? Some buyers cannot really think for themselves when it comes to conformation and type; therefore, they pay a lot of money for a popular bloodline. If a breeder is aiming for this segment of the market, then good looks and a popular sire are determining factors in choosing a stallion. With the right nick, even mediocre mares can produce high-selling foals.

If performance of the offspring is the only consideration, then the amount of inbreeding makes no difference. The proper nick resulting in a well-balanced individual is the important consideration. It may be helpful to prepare a chart specifically noting the various points of both mares to be bred and potential stallions. The Breeding Evaluation Chart (Figure 13–8) contains most of the vital information for an intelligent decision about stallion selection.

The real joy of horse breeding comes in the spring as the foals begin to drop. The expectation of what this year's foals are going to be like always seems to be more exciting than the realization of how last year's foals turned out. Much time, effort, and money have been expended at this point to produce the foal who is about to arrive. Will it be a good one?

At this time of year, every breeder is formulating plans about how to match his mare with the proper stallion, the uppermost thought being to breed a good horse. But what is a good horse? A show ring winner? The best producer of show ring winners? The horse with the best legs? The most athletic horse? At one time "good" meant any characteristic that improved a

Breeding Evaluation Chart

HORSE _____   REG. NO. _____   SEX _____

Characteristic	good	fair	poor	parent faults sire	dam	Nature of Defect
Head proportions						
Eye size & location						
Ear size & shape						
Teeth apposition						
Throatlatch						
Neck length & set						
Shoulder slope						
Withers						
Back						
Loin						
Hip length						
Hindquarters						
Underline						
Upper arm length						
Upper leg length						
Lower leg length						
Muscling						
Tail set						
Heart girth size						
Rib cage size						
Straight front legs						
Knee						
Fetlocks						
Hock						
Slope of pasterns						
Foot						
Straight rear legs						
Stifle						
Cannon						
Flank						
General quality						
Breed type						
Balance and Symmetry						
Action						
Temperament						

Possible recessive defect:

_____

_____

_____

_____

_____

_____

_____

_____

_____

Figure 13–8.   A breeding evaluation chart.

horse's chance for survival in the wild. Now the horse breeder has changed that.

Longevity has always been Mother Nature's yardstick. While wit, speed, and strength have contributed to longevity, these characteristics have had to be tempered with hundreds of other traits such as good teeth, disease resistance, and keen eyesight. Sound conformation too was of paramount importance as unsoundness reduced longevity.

Today the horse breeder develops his own yardstick, but to be successful in breeding a "good" horse, the breeder must have a clear definition in mind as to what constitutes "good." The thoroughbred people use speed as their yardstick while Appaloosa breeders look to color. The point to remember is that the final product can be judged good or bad only in relation to the original goal of the breeder. A draft horse cannot be judged poor because it is not fleet-footed anymore than a racehorse can be judged poor because it cannot pull a heavy load. Draft horse breeders can be justifiably proud of producing their "type."

The truly great horse breeder attempts to produce not only a good physical horse but also a good genetic horse. To him a horse is "good" only if it has the proper physical attributes along with an ability to perpetuate its likeness. It is a common practice among shallow thinking breeders to judge a horse "good" if he is a consistent show ring winner. True, our present judging system, wherein both halter and performance characteristics are evaluated, tends to eliminate the poorer individuals. Nevertheless, many attributes important to any breed receive no consideration at all in the show ring.

If one were to list the four most important criteria for determining how "good" a horse is, they might be (1) conformation, (2) prepotency, (3) disposition, and (4) show or performance record. Many breeders would agree that show record is in the proper order of importance here. Those that argue would say that an extensive show record is good proof of proper conformation and disposition, but they should realize too that good professional training, with all the tricks of the trade, can cause an irregularity in conformation or a bad disposition to be completely overlooked. Moreover, no attempt has been made to judge prepotency in the show ring.

Conformation

Many proponents of "form predicts function," such as Dr. Beeman, Dr. Rooney, and Professor Goode, have convincingly shown that every physical characteristic from the shape and position of the eye to the straightness of the back will be instrumental in determining how a horse moves and acts. They have even shown that these physical traits will affect the disposition. For example, a horse with pig-eyes will tend to be a bit nervous with a handler working behind him. From this point of view, conformation should probably be placed as number one on any list of what goes to make a horse "good." Without good conformation there is no place to start with a horse, and no other traits are really important.

An honest, conscientious breeder has a much better opportunity to judge the conformation of his horses than any show ring judge. After all, the breeder has watched the horse since birth. He has seen his brothers and sisters develop over the years, and he knows the difference between an awkward stage in a particular horse and a permanent lack of symmetry. He has struggled with the proper trimming of the hooves and he has watched the horse for many hours at play and in training. He knows which of his horses have the best conformation and which have the worst.

Therefore, unless he simply is not honest with himself, he has no reason to take the opinion of a show ring judge over his own. A well-seasoned breeder should decide for himself which horses have conformation good enough to qualify them for a breeding program. Maybe this self-assurance is one of the secrets as to why some breeders become successful and others do not.

The novice should be aware, however, that it is easy to become blind to the faults of his horses if he has only a few and they have become pets. In this case the show ring judge will put things into their proper perspective.

Prepotency

Prepotency can be predicted to a certain extent by a study of the pedigree. It can be roughly evaluated with a look at a good-sized foal crop. It can be proven over the years by the progeny record.

Some breeders select on the basis of pedigree. They know the traits of certain key ancestors and the families they produced. This type of selection has some merit, but it would be foolhardy to base one's entire selection system on this criterion (Figure 13-9). A young stallion from a broodmare that has consistently produced outstanding offspring should be given serious

Figure 13-9.   Mares from good families produce good foals.

consideration. Generally speaking good mares from good families produce good foals.

In a large herd in which some undesirable offspring are cropping out, it may be discovered that these undesirable offspring are all produced by one line of breeding. The intelligent thing to do in such a case is to cull or eliminate this entire segment of the herd. Such culling has been done effectively in herds of other farm animals. It is important to keep in mind that if undesirable offspring are cropping out, both the sire and the dam of such offspring carry the gene for the undesirable trait. Since they do carry the gene, some of their normal offspring will also carry the gene for the undesirable trait. The elimination of the entire segment in which the undesirable trait has been produced may eliminate the undesirable gene from the herd.

While the pedigree can be analyzed at any time, the progeny record becomes evident only after a long period of time. This record is a horse's greatest claim to fame as he grows older and his beautiful conformation begins to fade. Here again the merits of any particular offspring are generally judged by their show record. Personal observation by an experienced breeder of a number of progeny from a particular horse can easily establish just how prepotent the parent has been.

Disposition

A horse with perfect conformation and tremendous prepotency is still not worth having if he has a bad disposition. Any trainer will admit that disposition severely affects the performance of a horse. What is more, some strains seem to be genetically prone to a poor disposition. A horse cannot be "good" with a bad disposition. Even in racing, a disposition makes a difference when it comes to winning. The characteristic described as "heart" is a quality of disposition.

Show Record

Every breeder would do well to study the stallion to which he plans to breed his mare. He should begin with a careful look at his conformation, remembering that an excellent show record is not proof of good conformation. Then various points of conformation in the prospective stallion should be compared to those same points in the mare, for the stallion and mare may complement each other in conformation. The pedigree of the prospective stallion is the next vital consideration. As explained previously, it can predict whether the breeder will get what he sees in the stallion. Finally, the stallion's pedigree should be one which complements that of the mare so that the future foal's pedigree will be well balanced (see Chapter 11).

Every stallion to which a breeder might want to linebreed has ancestors. Chances are that the good traits found in the stallion have come from a bit of linebreeding back to another stallion, or mare, in the pedigree. The secret to balancing linebreeding is to bring in blood from some of the better ancestors who have contributed to make the stallion great. This practice allows the effects of selection by other breeders, which has concentrated the better traits, to improve the new progeny while keeping inbreeding at a lower level.

BIBLIOGRAPHY

Baumohl, A., and Estes, J.A.: Racing class and sire success. The Blood-Horse 80:48, 1960.

Evans, J.W., et al.: The Horse. San Francisco, W.H. Freeman & Co., 1977.

Foye, D.B., Dickey, H.C., and Sniffen, C.J.: Heritability of racing performance and a selection index for breeding potential in the thoroughbred horse. J. Anim. Sci. 35:1141, 1972.

Gillespie, R.H.: Performance rates. The Thoroughbred Record, April 17, 1971, p. 961.

Hamari, D.: The inheritance of performance in thoroughbreds. Acta Agron. 12:19, 1963.

Hamari, D., and Aala'sz, G.: The effect of selection on development of speed in horses. Z. Teisy. Aucht. Biol. 73:47–59, 1959.

Hazel, L.N.: The genetic basis for constructing selection indexes. Genetics 28:476, 1943.

Hazel, L.N., and Lush, J.L.: The efficiency of three methods of selection. J. Hered. 33:393–399, 1942.

Kieffer, Nat: Inheritance of racing ability in the thoroughbred. Thoroughbred Record, July 7, 1973, p. 50.

Legault, C.: Can We Select Systematically for Jumping Ability? Proc. Int. Symp. Genetics and Horse-Breeding, 1975.

Lesh, Donald: A Treatise on Thoroughbred Selection. London, J.A. Allen & Co., 1978.

More O'Ferral, G.J., and Cunningham, E.P.: Heritability of racing performance in thoroughbred horses. Livestock Prod. Sci. 1:87, 1974.

Ocsage, J.: Which methods are suitable for evaluating the inheritance of speed potential in stud stallions of racing breeds? Acta Agron. 12:181, 1963.

Sasimowski, E.: Difficulties of Applying Genetic Methods in Horse Breeding and Possibilities of Overcoming Them. Proc. Int. Symp. on Genetics & Horse-Breeding, Dublin, Ireland, 1975.

Tesio, Federico: Breeding the Racehorse. London, J.A. Allen & Co., 1958.

Varo, M.: On the relationship between characters for selection in horses. Ann. Agr. Fenn. 4:38, 1965.

14

aids in selection

Breeders who select on the basis of one or two traits need few aids in their selection decisions. However, those who have an extensive list of traits desired, and are extremely concerned about performance, will need to use all the help they can get. Racehorse breeders, as an example, have such intense competition that selection becomes a very exact science if a breeder is to succeed.

Selection aids fall into four general categories. The appearance of the animal is often the only selection criterion. If the animal doesn't "measure up," it is culled.

The list of traits developed for the index system of selection is important; but since they are based almost entirely on what a horse appears to be at maturity, the whole picture cannot be seen.

Experienced horse breeders study pedigrees for clues on how their young horses should develop. Their knowledge of the performance of each animal in the pedigree is used as an indication of how the new generation will do.

Often the best performers do not become the greatest producers. This fact has puzzled horse breeders for centuries, so breeders frequently want to know what the sire and dam of a particular prospect have produced. The production record of parents can be an aid in selection. It has been said that appearance shows what an animal looks like, pedigree reveals how it should develop, performance records tell what the ability of the animal is, and the progeny test tells how it will breed or transmit.

APPEARANCE

The experienced horseman can see defects in conformation and abnormalities of the eyes, mouth, teeth, feet, and legs. He may be able to detect respiratory abnormalities such as heaves and roaring. However, he cannot

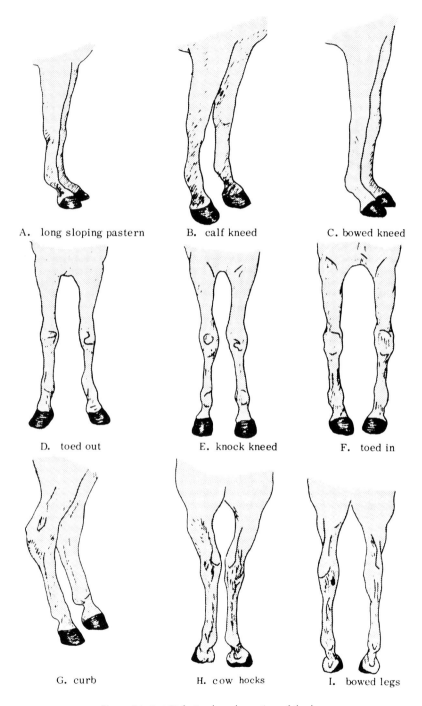

A. long sloping pastern   B. calf kneed   C. bowed kneed

D. toed out   E. knock kneed   F. toed in

G. curb   H. cow hocks   I. bowed legs

Figure 14–1. Defects of conformation of the legs.

readily tell whether a normal horse carries genes for the production of such abnormalities or defects.

Examination of horses for defects of conformation requires knowledge of normal and deviant conformation. For the correct and incorrect position of the feet and legs, see Figure 14–1, which shows the front legs from the front and side views, and the hind legs from the rear and the side.

Since all the weaknesses of conformation are probably under genetic control, a stallion used for breeding must be free from any of them. Although the exact modes of inheritance of defects in leg conformation are not known, it has been shown that the frequency of the weaknesses is much higher when one or both parents have any of these weaknesses than when both parents are of correct conformation of feet and legs.

Many lamenesses have a genetic predisposition. A hock with faulty conformation, for example, is more likely to develop a spavin than one of sound conformation. These defects are aggravated by putting excessive strain on the feet and legs of the horse, but it is unlikely that excessive strain would result in such defects when a horse has feet and legs of desirable conformation.

Unfortunately, several of the defects can be hidden from inexperienced horsemen. For example, a horse will become lame with sidebone and ringbone inflammation. By severing the nerve coming from the foot, one can prevent the horse from feeling pain, and it will not show lameness. Inexperienced horsemen might not be aware of any defect. One should always examine the tissue above the hoof and at the heel to see whether it is soft, shows bony enlargement, and has proper conformation, when considering a purchase. Contracted heels will often lead to the development of sidebones; since the condition is under genetic control, it should be selected against.

It is good to watch the horse while it is moving to see whether its action is smooth and whether stumbling occurs. Horses can be viewed from in front, from behind, and from the side as they move in the walk and the trot or pace. If a horse toes in at the front feet, it will have a tendency to swing the front feet out in a circle, a motion called paddling. If it toes out, it will tend to throw the front feet inward in a circle, a motion known as winging. A horse may tend to strike the supporting leg with the striding foot, a motion causing interference. Some horses tend to overstride with the hind feet and catch the heel of the front foot with the toe of the hind foot, a pattern known as forging. Horses with short, straight pasterns tend to walk in a stilted way and will be uncomfortable for riding. Such horses are more likely to stumble than a horse with normal pasterns.

All of these undesirable patterns of movement are to be avoided in stallions because abnormal movement usually reflects defects in conformation which are under genetic control.

Defects of the eyes are many and varied (see Chapter 10). Defects of respiration include broken wind or heaves, in which the horse has a double action in exhaling; and roaring, in which the horse makes a noise with respiration. Since both of these conditions are under genetic control, animals with these weaknesses should not be used for breeding. It is always wise to move a horse rather vigorously during examination so that defects in respiration will be exhibited.

Every serious breeder should carefully study Chapter 15, which details the various aspects of conformation. A knowledge of good conformation allows a horseman to predict how well an animal can perform, where it will excel, and what its weaknesses will be.

PEDIGREE

An accurate pedigree is essential to a breeder who is serious about his selection program. As explained in Chapter 11, a breeder needs to know the amount of inbreeding in his horses. Whether performance or the production of breeding stock is the goal, the breeder must keep a tab on the amount of inbreeding in each animal. The only sure check is through a pedigree.

Horsemen who are interested strictly in performance, even though they may not be breeders, need to make a careful examination of the pedigree when evaluating prospective purchases. Excessive inbreeding will not usually be indicative of outstanding performance ability. Breeders interested in production of breeding animals that will consistently produce good performers should look for a pattern of good linebreeding in the pedigree. They may want to avoid the appearance of any one ancestor too often in the pedigree of their horses. Whatever the breeders' concerns about performance, he can answer them by examining the horse's pedigree.

The Thoroughbred Record Sire Book now provides a Pedigree Cross-Reference Index each year. This pedigree index is helpful for planning inbreeding and outcrossing. It would be valuable if the same sort of information could be obtained about other breeds. As an example, the 1977 Index lists almost 10,000 stallions found in the pedigrees of 703 sires in the 1977 Stallion Register, published by the *Thoroughbred Record*. The explanation in the index reads:

> How can this (index) be used? Say a breeder has an *Amerigo mare who he wished to breed to a stallion with a cross of *Ambiorix. He also wants to avoid inbreeding. Looking under *Ambiorix, he finds the names of 30 stallions, from Ambassador's Image through Zen, who have a cross of *Ambiorix within their first three generations. But looking under *Amerigo, he finds that one of those 30, Set N'Go, also has a cross of *Amerigo, and according to his theories should be avoided for that particular mare.

* Ambiorix—Ambassador's Image, Amberbee-3, Amber Morn-4, Ambetella-3, Bancinto-3, Big Spruce, Bold Commander-3, Bold Dun-Cee, Bold Effort-3, Bold Roll, Bold Tactics-3, Cashon Delivery, Dawn Flight, Destroyer, Drum Fire-3, Free Swing-3, First Impression, George Navonod, Good Behaving, Impressive-3, In A Trance, Island Leader-3, Lord Gaylord-3, Majestic Light, Royal Chocolate-3, Set n' Go, Spirit Rock, Twice Worthy-3, Turf Hero-4, Zen.

The influence of an ancestor diminishes according to its position in the pedigree.

Sire and Dam	50% each
Second Generation	25% each
Third Generation	12.5% each
Fourth Generation	6.2% each
Fifth Generation	3.1% each
Sixth Generation	1.5% each
Seventh Generation	.7% each
Eighth Generation	.3% each
Ninth Generation	.1% each
Tenth Generation	.08% each

In figuring percentage of influence of ancestors, each generation must be considered as 100 percent. Therefore, the influence of a tail-male grandsire, as an example, would be 25 percent, but this influence must be expressed through the sire (the son of the grandsire). Frequently breeders draw the false conclusion that the son is a 50 percent replica of the sire. In reality, the son receives one allele of each gene from his sire. Only those sires that are completely homozygous will transmit the same quality each time.

PERFORMANCE RECORDS

If records have been kept on horses and then incorporated in the pedigree, the pedigree is a good indication of what the genetic constitution of the animal is and how it should transmit. Pedigrees without performance records are usually of little value in terms of indicating the genetic value of an animal. A pedigree without records of performance does reflect the inbreeding that was involved in the production of the individual. A pedigree containing a performance record is shown in Figure 14–2. When pedigrees do contain records, several factors need consideration when these are used for evaluation. First, a pedigree is man-made and can be considered important only if the breeders who kept the pedigree records were honest and accurate. Most horse breeders are honest, and it is unfortunate that the practices of the few who are not have led to mistrust of some pedigree records. Since every record is subject to the chance errors of the breeders who keep them, about the only solution is to assume that a record kept by a reputable breeder is accurate.

Second, records of low performance in a pedigree are as important as records of outstanding performance. By chance alone, a foal with one outstanding great-great grandsire and one great grandsire with very low performance is as likely to exhibit inheritance for low as for high performance. The name of an outstanding animal in the pedigree of a horse should not be overestimated; in fact, a prospective buyer should check the

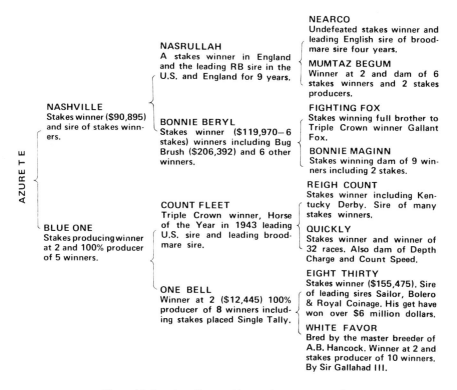

NASRULLAH
A stakes winner in England and the leading RB sire in the U.S. and England for 9 years.

NEARCO
Undefeated stakes winner and leading English sire of brood-mare sire four years.

MUMTAZ BEGUM
Winner at 2 and dam of 6 stakes winners and 2 stakes producers.

NASHVILLE
Stakes winner ($90,895) and sire of stakes winn-ers.

BONNIE BERYL
Stakes winner ($119,970–6 stakes) winners including Bug Brush ($206,392) and 6 other winners.

FIGHTING FOX
Stakes winning full brother to Triple Crown winner Gallant Fox.

BONNIE MAGINN
Stakes winning dam of 9 win-ners including 2 stakes.

AZURETE

BLUE ONE
Stakes producing winner at 2 and 100% producer of 5 winners.

COUNT FLEET
Triple Crown winner, Horse of the Year in 1943 leading U.S. sire and leading brood-mare sire.

REIGH COUNT
Stakes winner including Ken-tucky Derby. Sire of many stakes winners.

QUICKLY
Stakes winner and winner of 32 races. Also dam of Depth Charge and Count Speed.

ONE BELL
Winner at 2 ($12,445) 100% producer of 8 winners includ-ing stakes placed Single Tally.

EIGHT THIRTY
Stakes winner ($155,475). Sire of leading sires Sailor, Bolero & Royal Coinage. His get have won over $6 million dollars.

WHITE FAVOR
Bred by the master breeder of A.B. Hancock. Winner at 2 and stakes producer of 10 winners. By Sir Gallahad III.

Figure 14–2.   A pedigree with a performance record.

performance of all the others in the pedigree. An animal can have one outstanding performer in its pedigree and also have few genes in its inheritance from this important animal.

Third, a pedigree with performance records must show records on all the animals. Unless one knows otherwise, he must assume that a missing record is absent to make the horse look better than the real record would show. It is sometimes the practice to omit a low-performance record as a means of not detracting from the pedigree. Some animals have not established perfor-mance records. Whether the horses did not show promise and were not tested for performance or whether there was a good reason for not testing the animal for performance is usually not known. One should probably consider that the animal did not show promise and was not tested unless the record shows otherwise.

Fourth, performance of the sire and dam is much more important than performance of animals more removed in a pedigree. Unless an outstanding ancestor appears several times in the pedigree, it can be given only minor consideration when it is five or six generations removed from the animal being evaluated.

Performance records can take several forms in the horse. Records should be kept on whatever aspects are of importance to the breeder. Probably the most common are records such as racing speed, money earned, or contests won. Records of winning at halter would also be considered as performance records. Breeders striving for a particular color would want to keep a record of the color development and changes throughout life. This feature is especially important in the Appaloosa, where the best record would be a series of photographs showing any changes. Records should be kept on disposition, cow sense, jumping ability, and all other heritable characteristics.

These records become increasingly valuable as they accumulate generation after generation. One of the best services that a breed registry can perform is to act as a central holding file for production records of its breeders. The Jockey Club has performed this function for years with thoroughbred racing records. Most registries keep a record of color and recently some have also begun to keep performance records. Registries which promote contests to prove the abilities of horses are performing a great service for their breed.

One of the major ways the American Quarter Horse Association has contributed to the improvement of its breed has been the establishment of various performance and conformation contests and a setting up of a point system for the winners. The Register of Merit (ROM), A.Q. H.A. Champion, Supreme Champion, and the A, AA, and AAA racing ratings are tremendously helpful to the breeders as production records.

Performance records are important as the basis of improvement through selection when the heritability of the traits is high. Progress can be made more rapidly by individual or mass selection than by any other method when the heritability of the traits is 0.5 or more. Duzek estimated heritability of speed at 0.45 and wither height at 0.63. If selection is being used for increases in these traits, the records would be important because about 45 percent of whatever advantage was selected for in speed should be obtained. Perhaps the best way to improve racehorses would be to put every animal under racing tests as young animals. Since mares seem to be less influenced by environmental disturbances, training and racing fillies to evaluate their running abilities prior to selection and breeding should contribute to genetic improvement.

Traits that are intermediate in heritability (20 to 30 percent) will respond to selection, but giving some attention to records on close relatives of the animal under consideration as well as the animal's own record will result in greater improvement than selecting only on the basis of the record of the animal (Table 13–3).

Traits that are lowly heritable (5 to 15 percent) will not respond greatly to selection unless the selection differential is the difference between the performance of the selected animals and the level of performance of the population from which they were selected. There must be great variability in a trait before large selection differentials can be obtained. In most of the traits of importance in horses, there is considerable variation, so that modest selection differentials can be realized. However, since a large percentage of the mares must be kept for replacements and since the variability in traits is not enormous, only modest, not large, selection differentials are realized. Thus, for most traits of low heritability, improvement in these traits will be exceedingly slow by selection. Sometimes the heritabilities are low because of wide environmental variations to which such traits respond. Standardization of the environment among the animals of the population could increase the heritability, which would make selection based on performance records more effective.

Some traits show a response to crossbreeding (heterotic effect), and such traits might be increased more rapidly by a crossbreeding program than by selection. Hamari and Aala'sz state that speed in racehorses was improved greatly from 1806. The mile was run in 34½ seconds less after 50 years of selection, and 29¼ seconds less after another 50 years of selection. They say that speed during the next 50 years showed much less improvement and conclude that a genetic maximum for speed had been attained. Ocsag reported that American trotters crossed with Hungarian halfbreds covered 1000 meters in 24 to 28 seconds less than the purebreds. These studies suggest that if selection is no longer effective in improving a trait, crossbreeding may be necessary for increasing performance based on such a trait.

Performance has been indexed as a means of comparing all horses on the basis of one figure. The average earning index is often used in racehorses (Figure 14–3). This index is computed by first calculating the average amount of earnings by the breed each year, which is the total dollars earned divided by the total number of starters in all races by the breed for a given year. A horse under consideration is compared with the average of his breed. If his earnings are equal to the breed average, he is given a rating of 1.0. If his earning were four times that of the average of the breed, he would have 4.0 for his rating. His earnings in comparison with average earnings of the breed are averaged for the years that he has raced. Let us assume that he has ratings of 3.0, 4.0, 5.0, and 4.0 for 4 years. His average earning index would be 4.0, which would be good. Laughlin studied the racing index of 1000 thoroughbreds, developed correction factors for sex, age, weight carried, distance run, and condition of turf and was able to compare accurately the race results of horses. He found that the adjusted figure for the sires, dams, and other relatives of a foal were useful in predicting its racing ability. He developed a sire breeding value index by taking one-third of the average racing capacity of the sire and dam plus one-third of the racing capacity figure of the sire himself plus one-third of the average racing capacity of his offspring. Thus, Laughlin considered the pedigree (sire and dam figures), the production record (sire's record), and the progeny test (offspring record) as being of equal importance.

Some horses are used for harness racing as either trotters or pacers. No

Age	Starts	1st	2nd	3rd	Earnings
		JIM FRENCH			
2(SW)	12	5	1	3	$ 74,410
		I'M FOR MAMA			
2	7	3	0	2	$ 10,067
3(SW)	20	8	4	5	76,398
Totals	27	11	4	7	$ 86,465
		TAKEN ABACK			
2	4	0	1	1	$ 1,650
3	18	5	4	2	44,150
4(SW)	18	5	5	2	150,019
Totals	40	10	10	5	$195,819
		IPSE			
2(SW)	7	3	0	1	$ 18,200
		TALL TALL			
2(SW)	7	1	0	3	$ 18,030
		JOEY BOB			
2(SW)	9	3	2	1	$ 26,022
		LULIES GLORY			
2(SW)	13	3	4	4	$ 27,490

Figure 14–3.   A thoroughbred earning record.

index for trotting and pacing speed has been developed, but it appears that speed in both cases is rather highly heritable and that selection should bring about improvement. Some horses are used for jumping, and jumping ability is highly heritable. However, there appears to be no relationship between jumping ability and speed.

Considerable research has been done on the pulling ability of horses. The dynamometer was developed for measuring pulling ability. Two methods of evaluating the ability to pull have been used: load pulled in relation to body weight and ability to pull when increments of weight are added as the horses are pulling. The latter method measures pulling endurance, whereas the former measures maximum output. Kownackl says that horses can pull from 60 to 76 percent of body weight. He found large variations between and within individuals at different times. Disposition played a part in repeatability of pulling capacity because it was found that hotbloods varied more than coldbloods. It is estimated that eight tests would be needed in coldbloods to evaluate the pulling power of a horse, and four times this number of tests would be needed to evaluate hotbloods.

Varo developed a draft horse index composed of four major categories: (1) pulling power in relation to body weight, (2) score for soundness of feet and legs, (3) foreleg breadth, and (4) chest breadth. The value of his last two categories might be questioned if pulling is the major use for which horses are bred. Kibler and Brody found that oxygen consumption in horses is proportional to body weight. They also found that oxygen pulse is proportional to body weight but varies with training and work capacity. They concluded that oxygen pulse per kilogram body weight is a good index of work capacity. If oxygen pulse per kilogram is high and if it increases with work, it denotes high work capacity. If oxygen pulse per kilogram is low and it does not respond by a rapid increase with work, this denotes low work capacity.

PRODUCTION RECORDS

The thoroughbred industry in the United States, headed by the American Jockey Club, has made good improvement in the breed during the past 100 years. The Jockey Club has shown itself to be an innovator in the keeping of production records.

Thoroughbred breeders have no need to do their mating "by guess and by golly." In cooperation with *The Thoroughbred Record,* the racing statistics accumulated by the Jockey Club each year are made available to breeders. They are the most comprehensive summary of stallion records available anywhere. This tremendous aid to thoroughbred breeders can show breeders of other types of horses how important production records can be.

In the annual *Thoroughbred Record Sire Book,* every sire of a winner in North America is listed as the year passes. Each winner is identified by name and age and with pertinent information about his winning offspring (Figure 14–4). In addition the *Thoroughbred Record Annual Sire Book* lists the complete production record for each stallion (Figure 14–5). The highest class of race won by each winner is indicated by C (Claiming), O (Optional), M (Maiden), A (Allowance), H (Handicap), and S (Stakes). The claiming price; average distance run; number of starts; times won, placed, or showed; and total amount won by all offspring are provided in the record. In 1977, this record covered 68,826 races with 61,958 horses involved. The total purse money was $335,710,923, which gave an average earning per runner of $5,418.

By way of summary, *The Thoroughbred Record Sire Book* each year lists the top ranking 100 sires (in rank order) for the following categories: average earnings per start, average earnings per runner, percent of winners, number of winners, and number of races won. Broodmare sires are ranked in the same way. In addition, the 100 leading sires of 2-year-olds are ranked in the

EXCLUSIVE NATIVE

Chestnut Horse, 1965, 16 Hands   For index to pedigrees, see beginning of yellow section

HIS RECORD

ON THE TRACK

Age	Starts	1st	2nd	3rd	Won
2	7	3	2	2	$ 96,363
3	6	1	2	1	72,650
	13	4	4	3	$169,013

At 2, WON Sanford S. (5½ fur., 1:03⅗, ⅛ sec. off TR, by 4 lengths, defeating Vitriolic, Forward Pass, etc.), an allowance race at Monmouth (5 fur., :59, defeating Mara Lark, King of Ridan, etc.), a maiden race at Aqueduct (first start, 5 fur., by 3½ lengths, defeating Hand to Hand, Nez Perce, etc.); 2ND Saratoga Special S. (6 fur., carrying 122 lbs., by a nose to Vitriolic 114, defeating Pappa Steve, Forward Pass, Wise Exchange, etc.), Arlington-Washington Futurity (2nd div., 7 fur., to Vitriolic, defeating Court Recess, Royal Trace, Nash Co K., Master Bold, etc.); 3RD Futurity S. (6½ fur., to Captain's Gig, Vitriolic), Hopeful S. (6½ fur., to What a Pleasure, Royal Trace, defeating Forward Pass, Subpet, etc.).

At 3, WON Arlington Classic (mile, defeating Iron Ruler, Good Investment, Nodouble, Verbatim, etc.); 2ND Swaps H. (7 fur., by a neck, to Good Investment, defeating Loud Singer, Master Bold, Kaskaskia, etc.).

IN THE STUD — EXCLUSIVE NATIVE, 1965

		BY CROPS (ALL SEASONS)					SEASONS (ALL CROPS)		
		WNRS	$ CROP	AV PER				WNRS	
TR	FLS	RNRS	(SWS)	EARNINGS	RNR		TR	(WINS)	EARNINGS
*70	21	16	15(2)	898,573	56,161		*72	6(18)	149,244
*71	21	17	16(4)	748,866	44,051		*73	24(48)	587,059
*72	26	22	20(2)	554,978	25,226		*74	33(75)	449,855
*73	39	29	26(5)	1,756,918	60,583		*75	45(92)	557,446
*74	29	24(5)	1,109,470	41,091		*76	47(123)	1,468,845
*75	28	23	20(3)	3,183,678	138,421		*77	51(112)	1,263,965
*76	28	21	15(3)	870,764	41,465		*78	41(106)	1,969,867
*77	44	19	8(2)	195,539	10,292		*79	43(104)	2,872,505
	236	174	144(26)	9,318,786	53,556			(678)	9,318,786

Note: Exclusive Native was North America's leading sire in 1978 and 1979. He has sired 27 stakes winners through 1979, including Puerto Rican stakes winner Prodigo, not reflected in the table above.

						*Sickle
Male line (E)			Unbreakable			*Blue Glass
		Polynesian		Black Polly		*Polymelian
Native Dancer						Black Queen
		Geisha	Discovery			Display
						Ariadne
			Miyako			John P. Grier
						La Chica
		Case Ace	*Teddy			Ajax
						Rondeau
Raise You			Sweetheart			Ultimus
						*Humanity
	Lady Glory	American Flag			Man o' War
						*Lady Comfey
		Beloved			Whisk Broom II
						Bill and Coo
		Equipoise	Pennant			Peter Pan
						*Royal Rose
Shut Out			Swinging			Broomstick
						*Balancoire II
		Goose Egg	*Chicle			Spearmint
						Lady Hamburg II
			Oval			Fair Play
						Olympia
		Pilate	Friar Rock			*Rock Sand
						*Fairy Gold
Good Example			*Herodias			The Tetrarch
						Honora
		Parade Girl	Display			Fair Play
						*Cicuta
			Panoply			Peter Pan
Family No. 10						Inaugural

(Side labels: Raise a Native, Chestnut, 1961 / Exclusive, Chestnut, 1953)

MALE LINE

HIS SIRE, RAISE A NATIVE, undefeated stakes winner of 4 races, $45,955, Champion 2-year-old, Great American–NTR, Juvenile S.–ETR. Among leading sires. Sire of 52 stakes winners.

EXCLUSIVE NATIVE HAS SIRED:

AFFIRMED (dam by Crafty Admiral): 22 wins in 29 starts, 2 to 4, $2,393,818, Champion at 2, 3, and 4, Horse of the Year twice, Triple Crown, Hollywood Gold Cup-G1, etc.

LIFE'S HOPE (Tim Tam): 13 wins, 3 to 6, 1979, over $681,000, Amory L. Haskell H.-G1, Jersey Derby-G1, etc.

VALDEZ (Graustark): 7 wins in 14 starts at 3, 1979, over $443,600, Swaps S.-G1, Silver Screen-G2, Rutgers H.-G3; etc.

OUR NATIVE (Crafty Admiral): 14 wins, 2 to 4, $426,969, Monmouth Inv. H.-G1, Flamingo S.-G1, Ohio Derby-G2, etc. Sire.

NATIVE COURIER (Native Bend): 8 wins at 3 and 4, 1979, over $279,300, Brighton Beach-G3, Seneca H.-G3; etc.

ERWIN BOY (Warfare): 10 wins, 3 to 5, $199,876, Tidal-G2, Bowling Green-G2, Edgemere H.-G3; etc.

NATIVE LOVIN (*Court Martial): 6 wins in 8 starts at 2 and 3, $196,115, Land of Enchantment Futurity, Riley Allison Futurity.

SISTERHOOD (Court Harwell): 8 wins at 3 and 4, 1979, over $180,700, Gamely-G2, Boiling Springs H.–NTR; etc.

ELOQUENT (*Prince Taj): 4 wins in 8 starts at 3, 1979, over $167,800, Railbird-G3, Pasadena, La Centinella, Starlet S.; etc.

BE A NATIVE (*Cavan): 10 wins, 2 to 4, $151,742, Hawthorne-G3, Marlboro Nursery, Arch Ward, To Market S.; etc.

QUI NATIVE (Francis S.): 8 wins, 3 to 5, 1979, over $142,700, Lamplighter-G3, Appleton H.; 2nd Round Table H.-G3, etc.

ROOT CAUSE (Sailor): 9 wins, 2 to 5, to 1979, over $134,200, Primer S.; 2nd Juvenile S.; 3rd Youthful S.

SOCIETY HILL (On-and-On): 8 wins, 3 to 5, 1979, over $102,400, J. Rine Strohecker Memorial H.; 2nd Donald P. Ross H.

GENUINE RISK (*Gallant Man): Undefeated in 4 starts at 2, 1979, over $100,200, Demoiselle-G2, Tempted S.

Other stakes winners: KID CALVERT (Citation), SMASHING NATIVE (Hill Gail), LA BELLE COQUETTE (Fleet Fleet), NATIVE GOAL (Amarullah), etc.

FEMALE LINE

EXCLUSIVE. 4 wins at 3, 4, $14,075. 15 foals, 13 strs.—12 wnrs.—

EXCLUSIVE NATIVE. Stakes winners, above.

EXCLUSIVE NASHUA (c. by Nashua). 16 wins, 2 to 6, $132,029, Armed H.; 3rd Toboggan twice, Paumonok, etc. Sire.

IRVKUP (g. by *Alcibiades II). 11 wins, 2 to 9, $115,358, Jerome H., Flash S.; 2nd Roamer, Benjamin Franklin H.; etc.

MELLOW MARSH (f. by *Seaneen). 9 wins, 2 to 5, $83,049, Monrovia H.; 2nd Railbird S., Campanile H.; etc. Dam of L'NATURAL (5 wins at 3 and 4, $66,239, All American H.-EAR, 5 fur., turf, :56; 3rd Lakes and Flowers H.), YALE COED (4 wins at 2, $45,560, Starlet S.; 2nd Susan S.; etc.).

EXCLUSIVE DANCER (f. by Native Dancer). 6 wins at 2 and 3, $69,299, Prioress S., Miss Florida H.; 2nd First Lady H. Dam of GENERAL ASSEMBLY (7 wins at 2 and 3, $463,245, Travers-G1-NTR, 1¼ mi., 2:00, Hopeful-G1, Gotham-G2, Vosburgh-G2, Saratoga Special S.-G2; 2nd Kentucky Derby-G1, etc.), Ten Cents a Dance (4 wins at 2 and 3, $85,077; 2nd Firenze H.-G2, etc.).

Exclusive Ribot (c. by *Ribot). 6 wins, 2 to 4, $43,974; 3rd Golden Beach H.

Solitary Hail (c. by Hail to Reason). Winner at 2 in England and 4 in U. S., $29,517; 2nd Laurent Perrier Champaigne S.-G2.

Grand Alliance (c. by *Vaguely Noble). Winner at 2, over $20,700; 2nd Santa Catalina S.; 3rd Sutter S.

Reading Room (c. by *Ribot). Winner at 2 and 3 in France and England; 2nd Champagne S.-G2. Sire.

Other winners: Libro D Oro (f. by Francis S.), Buck Private (c. by Buckpasser), sire; Nashua Dancer (c. by Nashua).

Charvak (f. by *Alcibiades II). Unraced. Dam of MAJESTIC KAHALA (11 wins, 2 to 4, $124,037, Nettie S.-G3, etc.).

Good Example. 2 wins at 3, $11,600; 2nd Ral Parr S. 9 foals, 7 to race, all winners, including—

Headmaster. 10 wins, $66,746; 3rd Charles S. Howard S., etc.

Red Letter Day. 3 wins; 3rd Rosedale S. Dam of GALA PERFORMANCE (sire), KEEP THE PROMISE, SOMETHING SUPER, RING AROUND, GALA OCCASION, Scarlet Letter dam of COLD COMFORT, INDIAN LOVE CALL).

Fee $5,000 — Live Foal

Property of a Syndicate

STANDING AT

WINDFIELDS FARM LIMITED

Phone: (416) 725-1195   P. O. Box 67   Oshawa, Ontario, Canada L1H 7K8

Figure 14–4. A sire listing, from *The Thoroughbred Record's* "Stallion Register."

ETONIAN					B. 1954
Owen Tudor-*Windsor Whisper, by Windsor Slipper.					
Foals of 1968–12					
Doggone Cute c	1C8	8	2	0	0	2,025
Foals of 1969–18					
Quarnos g	3C8	7	2	1	0	7,095
Foals of 1970–13					
British Red c	3C8	16	1	3	0	3,938
Foals of 1971–22					
Alexeton c	H16	15	7	0	2	16,124
Bimbonian c	2C6	13	3	1	1	5,588
Double Hey c	A7	8	4	0	1	3,415
Eton Song f	13C6	21	3	4	3	22,387
Fourth Decade c	3C6	2	2	0	0	3,960
Waterdell f	2C6	16	2	1	3	4,554
Foals of 1972–7					
Broke n Hungry c	16C8	19	2	4	4	24,344
Clarence Henry g	10C8	20	6	4	2	29,284
TONI C.c	**S8**	**8**	**2**	**0**	**1**	**11,705**
Foals of 1973–8					
Etonian Kay f	2CB	15	1	2	5	3,718
Exonerator c	3C9	17	3	0	3	7,320
Faretonian f	25C8	19	4	2	4	27,445
Wind in the Sails f	H8	19	1	3	2	13,544
Foals of 1974–7					
Balla Boots f	5C6	14	1	3	1	5,260
Bobs Etonian c	A8	15	2	4	1	6,131
Eaton Totsy c	9C6	4	1	1	0	3,440
Good for Nothing f	10C9	23	2	2	2	12,461
Holy Helen f	M5	13	1	0	2	2,763
Shivering f	M6	6	1	1	1	3,808
Foals of 1975–8					
Correlate f	18C6	1	1	0	0	4,800
Gruffness f	10C6	1	1	0	0	2,600
2YO 2 rnrs, 2 wnrs		7	2	1	1	9,935
Non-wnrs, 3 & up, 8		40		0	1	1,570
TOTAL, 33 rnrs, 24 wnrs		345	55	37	40	231,814
Avg dist 7.62	Per rnr $7,025				Per strt $672

Figure 14–5.   Production record of Etonian, listed in the *Thoroughbred Record Sire Book*.

Average Earnings Per Start

	Starts	Earnings	Earnings Per Start
Exclusive Native	53	$390,376	$7,366
In Reality	59	432,596	7,338
Rainy Lake	51	325,584	6,384
Buckpasser	50	236,565	4,731
Dr. Fager	85	366,275	4,309
Don B.	96	393,391	4,098
Tumiga	63	206,486	3,278
Roberto	73	237,998	3,260
Bold Bidder	81	254,602	3,143
Gaelic Dancer	52	161,284	3,102
King's Bishop	67	197,103	2,942
Nodouble	77	214,459	2,785
Naskra	62	168,738	2,722
Never Bend	51	135,827	2,663
Proudest Roman	74	189,217	2,557
Blade	116	284,292	2,451
Vitriolic	85	198,165	2,331
Cornish Prince	68	156,920	2,308
Moonsplash	68	149,184	2,194
Explodent	77	166,464	2,162

Average Earnings Per Runner

	Rnrs.	Earnings	Earnings Per Rnr.
Raise A Native	7	$325,710	$46,530
In Reality	14	432,596	30,900
Exclusive Native	14	390,376	27,884
Rainy Lake	12	325,485	27,132
Reb's Policy	6	161,399	26,900
Pretense	8	168,920	21,115
Master Hand	6	118,690	19,782
Don B.	20	393,391	19,670
*Tobin Bronze	8	156,897	19,612
Dr. Fager	19	366,275	19,278
Buckpasser	13	236,565	18,197
Northern Bay	5	86,270	17,254
Tumiga	12	206,486	17,207
Roberto	14	237,998	17,000
Naskra	10	168,738	16,874
Tequillo	6	101,044	16,841
Bold Bidder	16	254,602	15,913
National	5	78,305	15,661
I'm For More	11	169,026	15,366
Rambunctious	8	121,369	15,171

Percent of Winners

	Rnrs.	Wnrs.	Earnings	Pct. Wnrs.
Reb's Policy	6	6	$161,399	100.0
Mister Dust-All	9	9	63,318	100.0
Sir Wiggle	5	5	62,031	100.0
Coco La Terreur	6	6	32,681	100.0
Judge	6	6	20,530	100.0
Icy Song	5	5	19,911	100.0
Illustrious	8	7	55,196	87.5
Native Rhythm	7	6	85,438	85.7
Lucky Mike	13	11	114,562	84.6
Queen's Knight	6	5	33,371	83.3
Goal Line Stand	6	5	19,854	83.3
Blue Serenade	6	5	15,943	83.3
I'm For More	11	9	169,026	81.8
Never Bend	11	9	135,827	81.8
Northern Bay	5	4	86,270	80.0
National	5	4	78,305	80.0
First Landing	5	4	44,078	80.0
Whistling Kettle	5	4	35,418	80.0
*The Knack II	10	8	21,292	80.0
Equilibrium	5	4	20,135	80.0

Gross Earnings

	Rnrs.	Starts	Races Won	Earnings
In Reality	14	59	16	$432,596
Don B.	20	96	23	393,391
Exclusive Native	14	53	14	390,376
Dr. Fager	19	85	18	366,275
Raise A Native	7	32	8	325,710
Rainy Lake	12	51	13	325,584
Blade	22	116	24	284,292
Kaskaskia	3	37	11	282,911
Bold Bidder	16	81	17	254,602
Quadrangle	19	138	24	244,334
Roberto	14	73	13	237,998
Buckpasser	13	50	10	236,565
No Double	16	77	13	214,459
Marshua's Dancer	21	110	27	212,574
Tumiga	12	63	14	206,486
Vitriolic	15	85	11	198,165
King's Bishop	16	67	19	197,103
Big Burn	17	126	28	189,831
Proudest Roman	13	74	17	189,217
Dewan	17	114	11	172,802

No. of Winners

	Rnrs.	Wnrs.	Races Won	Earnings
Distinctive	27	15	27	$149,720
Blade	22	14	24	284,292
Quadrangle	19	14	24	244,334
Marshua's Dancer	21	14	27	212,574
Irish Ruler	25	13	17	87,915
Verbatim	19	12	18	96,109
Icecapade	23	12	15	92,689
Don B.	20	11	23	393,391
Dr. Fager	19	11	18	366,275
King's Bishop	16	11	19	197,103
Big Burn	17	11	28	189,831
Tentam	15	11	20	152,911
Lucky Mike	13	11	13	114,562
Flying Lark	19	11	20	77,871
*Figonero	20	11	20	75,252
Twist The Axe	16	11	16	69,248
Well Mannered	16	11	13	69,203
Iron Warrior	16	11	13	63,656
Victoria Park	18	11	15	62,407
Spring Double	14	10	16	148,279

No. of Races Won

	Rnrs.	Wnrs.	Races Won	Earnings
Big Burn	17	11	28	$189,831
Marshua's Dancer	21	14	27	212,574
Distinctive	27	15	27	149,720
Blade	22	14	24	284,292
Quadrangle	19	14	24	244,334
Don B.	20	11	23	393,391
I'm For More	11	9	22	169,026
Winged T	16	8	22	120,115
Tentam	15	11	20	152,911
Flying Lark	19	11	20	77,871
*Figonero	20	11	20	75,252
King's Bishop	16	11	19	197,103
Duck Dance	18	9	19	127,622
Dr. Fager	19	11	18	366,275
Explodent	13	8	18	166,464
Verbatim	19	12	18	96,109
Northern Answer	15	10	18	93,104
Bold Bidder	16	8	17	254,602
Proudest Roman	13	7	17	189,217
Night Invader	20	10	17	95,455

Figure 14–6.   Top ranking sires by production records for 1977. (Data from *The Thoroughbred Record Sire Book*. Lexington, KY, Thoroughbred Record, 1977.)

Stallion	No. Sold	Amt.	Avg.
SECRETARIAT	**17**	**$3,849,000**	**$226,412**
Colts	11	2,769,000	251,727
Fillies	6	1,080,000	180,000
NORTHERN DANCER	**21**	**3,122,000**	**148,667**
Colts	8	1,905,000	238,125
Fillies	13	1,217,000	93,615
ROUND TABLE	**7**	**900,500**	**128,643**
Colts	6	745,500	124,250
Filly	1	155,000	155,000
GRAUSTARK	**16**	**1,588,000**	**99,250**
Colts	10	1,029,000	102,900
Fillies	6	559,000	93,167
HOIST THE FLAG	**8**	**735,000**	**91,875**
Colts	3	259,000	86,333
Fillies	5	476,000	95,200
***VAGUELY NOBLE**	**20**	**1,833,700**	**91,685**
Colts	11	1,000,500	90,955
Fillies	9	833,200	92,578
SIR IVOR	**24**	**2,180,000**	**90,833**
Colts	11	631,000	57,364
Fillies	13	1,549,000	119,154
DAMASCUS	**12**	**1,077,500**	**89,792**
Colts	11	1,050,000	95,455
Filly	1	27,500	27,500
BOLD BIDDER	**24**	**2,106,000**	**87,750**
Colts	8	671,000	83,875
Fillies	16	1,435,000	89,688
NIJINSKY II	**16**	**1,390,000**	**86,875**
Colts	11	895,000	81,364
Fillies	5	495,000	99,000
IN REALITY	**5**	**424,000**	**84,800**
Colts	5	424,000	84,800

Figure 14–7.   The top thoroughbred sires in 1977 for sale prices of yearlings. (Data from *The Thoroughbred Record Sire Book*. Lexington, KY, Thoroughbred Record, 1977.)

same six categories. The top 10 of each category for 1977 are shown in Figure 14–6.

Another important production record summary published by *The Thoroughbred Record* is a listing of the top 100 sires for the prices of their yearling sales. The top 10 for 1977 are shown in Figure 14–7.

The availability of such records is probably creating trends in breeding toward those horses in the record book. It is true that some excellent horses are not listed in the record because of their limited use as a sire; but on the whole, sire records of North American thoroughbreds are a tremendous asset to breeders.

Speed—Dam—daughter regression		.45	Dusek
son—dam pairs		.25	Dusek
Wither height, dam-daughter		.63	Dusek
Heart girth		.12	Dusek
Cannon bone circumference		.28	Dusek
Body weight		.27	Dusek
Genetic correlation of heart girth & body weight		.47	Dusek
Wither height	—variance ratio	.29	Ericksson
	—covariance	.41	Ericksson
Heart girth	—variance	.43	Ericksson
	—covariance	.43	Ericksson
Cannon bone	—variance	.48	Ericksson
	—covariance	.53	Ericksson
Pulling power—resistance increased 5% body wt. each 50 M		.26	Johnannsen & Rendel
Pulling power		.14	Johnannsen & Rendel
Walking speed		.41	Johnannsen & Rendel
Trotting speed		.43	Johnannsen & Rendel
Points for movement		.41	Johnannsen & Rendel
Points for temperament		.23	Johnannsen & Rendel
Wither height		.26	Johnannsen & Rendel
Heart girth		.32	Johnannsen & Rendel
Hip width		.34	Johnannsen & Rendel
Body weight		.25–.30	Varo
Pulling power	in 4 yr. old	.14	Varo
	in 5 yr. old	.24	Varo

Figure 14–8. Heritability estimates in the horse.

HERITABILITY ESTIMATES

Heritability is best defined in terms of what is obtained from what is selected for. This realized heritability is the most valuable figure obtainable, because it reveals with considerable accuracy how much improvement can be made by a given amount of selection pressure (selection differential). It is not often that one has the type of information needed for calculating heritability from response to selection.

Heritability can be estimated because related animals should resemble one another to a greater extent than nonrelated animals. Thus, full sibs or half-sibs should be more similar than unrelated animals. One can estimate heritability by half-sib analysis. Heritability estimates for a number of traits in the horse are shown in Figure 14–8.

There are two kinds of heritability estimates. Heritability in the broad sense involves calculating the ratio of all genetic variance to the total phenotypic variance:

$$h^2 = \frac{O^2G}{O^2G + O^2E} \text{ or } \frac{O^2G}{O^2P}$$

where h^2 equals heritability, O^2 equals variance, G equals genetic variance, E equals environment, and P equals phenotypic variance. This type of heritability estimate tells nothing about the amount of progress that might be expected from mass selection.

Heritability in the narrow sense is the ratio of the additive genetic variance to the total variance. The total variance can be considered as variance due to the additive effects, dominance deviations, gene interactions, or epistatic deviations and environmental effects:

$$h^2 = \frac{O^2A}{O^2A + O^2D + O^2I + O^2E}$$

Breeding stock
a mile in 130 sec.

Offspring average
a mile in 140 sec.

Figure 14–9.   Knowing the heritability of a trait, such as speed, is an aid in setting selection priorities.

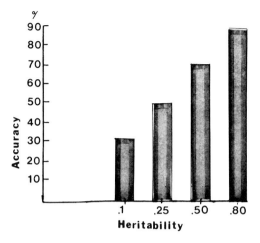

Figure 14–10.   The accuracy of predicting genetic value is equal to the square root of the heritability. (Data from Evans, J.W., et al.: *The Horse.* San Francisco, W.H. Freeman & Co., 1977.)

where h^2 equals heritability in the narrow sense, O^2 equals variance, A equals additive factors, D equals dominance, I equals interactions, and E equals environmental effects. Heritability in the narrow sense gives an indication of the amount of progress that could be expected from mass selection.

The concept and explanation of heritability are important in understanding gene interactions. Although heritability can be defined or explained in different ways, at present it will be considered the portion of what is sought that is obtained in selection. For example, one can use a herd of horses that can run a mile on an average of 150 seconds, with variations among individuals from 120 to 180 seconds. If from this herd a group of breeding stock were selected that averaged the mile in 130 seconds, a reduction of 20 seconds has been reached in selection (Figure 14–9). When these selected animals are mated, and if their offspring average 140 seconds, then 10 seconds' reduction in time will have been obtained; i.e., the heritability of speed of running the mile is 0.5, or 50 percent.

The heritability of a trait determines how accurately a breeder can assess the genetic value of a sire or dam. This accuracy is equal to the square root of the heritability (Figure 14–10). If racing ability is 0.50 heritable, then a breeder's accuracy in determining the best sire and dam from racing records will be 71 percent.

BIBLIOGRAPHY

Brzeski, E.: The evaluation of a stallion on the basis of some behavioral characteristics of his progeny during work and grooming. Acta Agr. Silvest. Ser. Zootech. (Krakow) 6(2):17–53, 1966.

Cook, R.: Temperament gene in the thoroughbred. J. Hered. 36:82, 1945.

Dusek, J.: Some biological factors and factors of performance in the study of heredity in horse breeding. Sci. Agric. Bohemoslov 3:199, 1971.

Dusek, J.: The heritability of some characters in the horse. Zivocisna Vyroba 10:449–456, 1965.

Estes, B.W.: A study of the relationship beween temperament of thoroughbred mares and performance of offspring. J. Genet. Psychol. 81:273, 1952.

Evans, J.W., et al.: The Horse. San Francisco, W.H. Freeman & Co., 1977.

Franganillo, A.R., and Pozo-Lora, R.: Heritability and repeatability of gestation length in Spanish and Arabian mares. Arch. Zootecnia 9:132–171, 1960.

Hamari, D., and Aala'sz, G.: The effect of selection on development of speed in horses. Z. Teisy. Aucht. Biol. 73:47–59, 1959.

Hartwig, W., and Reichardt, U.: The heritability of fertility in horse breeding. Zuchtungskunde 30:205–213, 1958.

Haussler, H.: The inheritance of the ability to jump in thoroughbreds in Germany. Vet. Med. Dissertation Tieraerztl. Wochenschr. (Hanover), 1935.

Kibler, H.H., and Brody, S.: Growth and Development LVIL. An Index of Muscular-Work Capacity. Mo. Agri. Exp. Sta. Res. Bull. 367, 1943.

Kownackl, M.: Changes in Ability of Horses to Show Maximum Draft Force. Rocyn. Nauk. Roln. Ser. B. 851313-322, 1965.

Laughlin, H.H.: Racing capacity in the thoroughbred horse. Sci. Monthly 38:210–222, 310–321, 1934.

Ocsag, L.: The inheritance of speed crosses between American trotters and Hungarian halfbreds. Agrartud. Eyg. Meyogaydastud. Karanak Kozl. (Godollo) 83–90, 1964.

Pirri, J., and Steele, D.G.: The heritability of racing capacity. The Blood-Horse 63:976, 1952.

Thoroughbred Record Sire Book. The Thoroughbred Record, 1980.

Varo, M.: Some coefficients of heritability in horses. Ann. Agr. Fenn. 4:223, 1965.

Watanabe, Y.: Timing as a measure of selection in thoroughbred breeding. Jpn. J. Zootech. Sci. 40:271, 1969.

15

selection by conformation

The successful horse breeder must be a shrewd judge of horse conformation. He must believe wholeheartedly in the principle that "conformation predicts performance." He has learned, perhaps by sad experience, which conformation traits lead to unsoundness and what type of conformation gives poor performance in his horses.

Disposition and "heart" often have profound effects upon a horse's winning ways. And some horsemen claim, even today, that color can affect speed and disposition. More often than not, however, conformation is the reason why one horse wins over another.

TO BREED AN ATHLETE

The description of a horse as "an athlete" is not just a trite phrase. It relates to the ease with which the horse performs; he has balance, proper symmetry, and good equilibrium. These characteristics are vital for good conformation in the horse.

Consider the horse's center of gravity—the point within the body where the greatest force of gravity is exerted. All bodies, whether living or inanimate, are influenced by the force of gravity, which draws them toward the earth or impedes forward motion. Since this force is exerted on every cell of the body, small parallel forces are acting in the same direction. The center of gravity is, therefore, the center of these parallel forces and not necessarily the center of the body from front to back or top to bottom (Figure 15–1).

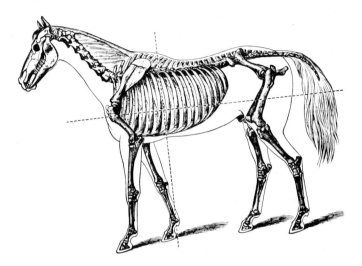

Figure 15–1.   The center of gravity in the horse.

Typical athletic movement of the horse is up and down and in a forward direction. Since the overall conformation of a horse is designed around the center of gravity, this point should move as little as possible to gain maximum forward speed. Disturbances in this center, by way of defective conformation, cause this heavy point to move further and more often than would otherwise be necessary. This off-center balance results in greater expenditure of energy, less ease of movement, and reduced athletic ability.

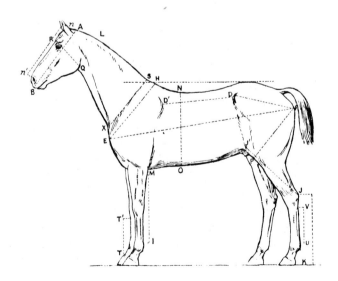

Figure 15–2.   Proportions of an athlete.

The center of gravity is changed by such conformation problems as an excessively long or short neck, exceptionally long or short legs, extreme abdominal mass, the length of the hip, and the size of the head. The various parts of the horse's body contribute to the balance. A body is said to be in equilibrium when the several forces acting on it balance one another.

Because of the location of the center of gravity it is possible to determine proper proportions for a horse with maximum athletic ability. Using the head as a unit of measure, the proportions are as shown in Figure 15–2, where it can be seen that the length of the head equals the distance:

1. From the back to the abdomen (N–O).
2. From the top of the withers to the point of the arc (H–E).
3. From the superior fold of the stifle to the point of the hock (J'–J).
4. From the point of the hock to the ground (J–K).
5. From the upper edge of the scapula bone to the point of the hip (D'–D).
6. From the lowest portion of the chest to the fetlock (M–I).

Another observation is that two and one-half times the head equals the distance:

1. From the withers to the ground.
2. From the top of the croup to the ground.
3. From the front of the chest to the point of the buttock (E–F).

The length of the hip (D–F) in a properly balanced horse equals the distance:

1. From the point of the buttock to the stifle (F–P).
2. From the lower attachment of the neck to the chest to the upper attachment of the neck to the withers (S–X).
3. From the lower attachment of the neck to the chest to the angle of the lower jaw when the head is held parallel to the shoulder (X–Q).
4. From the poll to the nostril (N–N').

A measure of one-half the head is a good guide for proper proportions, equalling the distance:

1. From the most prominent point of the angle of the lower jaw to the anterior profile of the forehead above the eye (R–Q).
2. From the lower part of the knee to the coronet (T'–T).
3. From the throat to the upper border of the neck behind the poll (Q–L).
4. From the base of the hock to the fetlock (V–U).

The horse in motion is constantly displacing the center of gravity in one direction or another. This disturbance of equilibrium is immediately restored by the formation of a new base of support. The part upon which a body rests is termed the base of support, and determines whether equilibrium will be stable or unstable.

Stable equilibrium occurs when a body, slightly displaced or pushed aside, will return to its original position. If, on the other hand, it tends to move farther away from its original position, or to topple over, its equilibrium is unstable. Proper conformational balance helps to stabilize the equilibrium of the horse and makes him a better athlete.

The anatomy of the horse in general has evolved such that the front legs are built to carry 60 percent of the body weight. Athletic ability is enhanced by this situation. When the height at the withers and the croup is not equal, the distribution of weight is shifted from front to back or from back to front legs. This sets up unnatural strains and the athlete must work harder to function. The problem will be greater or less, depending on the particular work performed. The pack horse or the endurance horse will show the strain of this defect much sooner than will the racehorse.

Body length is an important consideration and the subject of much controversy. The total anatomy of the horse is built to function at its best when the body length is about two and one-half times the length of the head. Body length is the total length of the shoulder, back, and hip, each of which may vary. Of these, a long back is generally the cause of problems.

A long shoulder or long hip can more than compensate for a short back; when this situation exists, the animal has superior symmetry and even greater athletic ability than a long-backed horse with the proper two and

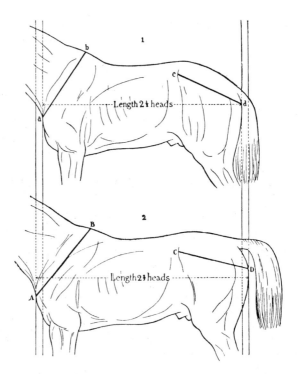

Figure 15–3.   Head length as a measure of body length.

one-half head lengths of the body (Figure 15–3). In this figure proper body length in horse 1 is the result of a short hip and a short, steep-sloping shoulder compensating for a long back. Horse 2 has much better proportions even though the total body length is two and two-thirds the head length.

The height at the withers should be approximately the same as the body length for a good athlete. Horses with excessive height over length will have a construction that is much too narrow. They will have deficient muscle, and be light of bone, making them slow in their paces. They will tend to brush and forge and will become easily fatigued. A horse with height much less than body length has a distinct lack of liberty, range, and grace in movement. The shorter legs mean a shorter pace.

Narrowness is a serious defect of conformation, both in front and behind. The breast should be wide, the shoulders muscular, and the ribs well sprung for power and endurance. The hips should be broad and full of muscle. Width in the stifle area of each rear leg is especially desirable in speed horses, as propelling power resides there.

Considering the overall symmetry of an athlete, too much muscle in front tends to encumber movement and impose undue weight and wear on the legs. Heavier muscle, however, is desirable when pulling power is sought. Muscle in relation to total conformation can be good or bad. The strength of any group of muscles is in direct relation to the mass or volume of that muscle. Still other factors of anatomy affect muscle performance. The longer the muscle, the greater will be its range of contraction, and the greater will be the distance the bones will be moved. The longer muscle means increased speed since a muscle mass, short or long, will contract at the same time-rate, given maximum stimulation. There are various types of muscle fibers within the muscle mass. Slow twitch fibers are useful for muscle endurance; fast twitch fibers are for speed. A muscle biopsy will determine the ratio between the types and help in predicting performance ability.

Muscle mass or size beyond the requirements necessary to move the weight of the horse will tend to retard speed, but will add to strength and staying power. Endurance, however, may be more related to blood supply of muscles, and muscle type, than to muscle volume.

Consider the draft horse versus the racehorse. Long muscles and long legs give speed. Heavy muscles and legs short in relation to body mass give pulling strength.

The muscles concerned with body movement are generally attached to bones. These bones connected by their joints act as a series of levers which give greater strength and speed of movement. A lever can be thought of as a rigid bar used to move bodies by means of a power or weight.

A lever has three parts: the fulcrum, the power, and the weight. When a bone is operating as a lever, one end is more or less fixed and moves on the corresponding surface of the bone with which it is articulated. This joint serves as a fulcrum. The other end of the bone is left free to move in obedience to the power and the weight.

The effectiveness of a lever in increasing force or speed depends on the relation of power to weight. On this basis, there are three classes of levers (Figure 15–4). In levers of the first class the fulcrum is placed between the power and the weight (as illustrated by the scales). In levers of the second class, the fulcrum is situated at one end, as illustrated by the wheelbarrow, and the weight occupies a position between it and the power. In levers of the third class, the weight is at one end, the fulcrum at the other, and the power in the center, as in the articulated bone previously described.

The arms of a lever are distinguished respectively as the "power arm" and the "weight arm." The former is represented by the distance between the point at which the power acts and the fulcrum, and the latter by the distance between the fulcrum and the line through which the weight acts.

The increase in power or speed is determined by the class of lever. First-class levers are those of speed and, in horses, are generally involved in extension. Second-class levers are more effective at increasing power. Third-class levers increase both speed and power, but with less effectiveness than the other classes of levers.

Examples of these levers in the horse are shown in Figure 15–5. A lever of the first class is provided in the extension of the cannon bone on the hock—when the foot is off the ground. Here the muscle representing the power acts upon the point of the hock, the fulcrum is the hock joint, and the

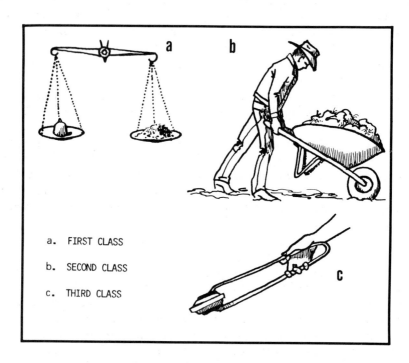

a. FIRST CLASS

b. SECOND CLASS

c. THIRD CLASS

Figure 15–4.   Classes of levers.

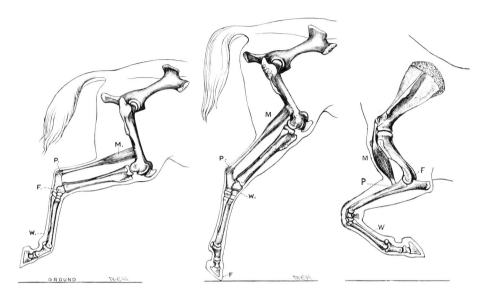

Figure 15–5.   The bones of the equine leg work as levers. M = muscle, P = power, F = fulcrum, W = weight.

parts below are the weight. When the foot is on the ground, a lever of the second class is provided. Now the point of the hock is the part on which the power acts, the ground is the fulcrum, and the weight is at the hock joint. The forearm of the horse serves as an example of a third-class lever. The fulcrum in this case is at the elbow joint, the power is at the biceps muscle, and the weight is the lower limb.

It is amazing what evolution has done with horse anatomy over the course of thousands of years. The principles of physics have been applied expertly bringing about symmetry and balance, which makes for a real athlete. Too often man, blind to this symmetry and balance, breeds horses without regard to what makes conformation suitable for maximum function. He selects, ignorantly, for such traits as color, or he tends to overemphasize one part of the horse. As a result clumsy, inept creatures come into existence. And we as breeders wonder why! To be successful a horse breeder must carefully consider what makes an athlete.

PROPER CONFORMATION

The varieties of conformation in every part of the equine anatomy indicate that many individual genes are involved. Selection pressure must be continually maintained to prevent poor conformation from asserting itself, especially in horse breeds where there is no extreme test of athletic ability.

Few activities test the physical limits of a horse as does a heavy racing schedule. Weaknesses of conformation are soon discovered at the track, and the successful breeders use only winners as breeding stock. Breeders of show horses, pleasure horses, the color breeds, and even modern draft horses, must use their powers of observation to differentiate poorly built animals from those with the best anatomy. Years of neglect in this area of selection can do great damage to any breeding program.

The Head and Neck

Although, as often stated, "You don't ride the head," it is an important part of the selection process. Given two horses with all traits equal except the head, the good-headed horse would be far superior to the poor-headed horse. The head is more than a thing of beauty or ugliness; it is a reflection of other traits—good or bad. It is an outward expression of inward temperament in all of its varied moods. The head cradles the all-important brain; it is the seat of the eyes and the nostrils through which the horse has his principal contact with the world. Also, the head contains the mouth, which is vital to the well-being of every horse. If the numerous functions with which the head is associated are to perform efficiently, ample space and proper framework need to be provided. Slight variations in the head may not seem important, but they may signal vast differences in attitude or ability.

The equine head varies in the size and shape of its many parts. More variation exists among breeds, but within a herd, and even among sibs, there occur subtle differences that are important. In some animals the head presents a clean-cut finely chiseled outline. In others the angles and lines of the bony framework are rounded off and more or less obscured, with the head, as a whole, displaying a heavy plain appearance. The former, known as a "lean" head, is characteristic of Arabians, thoroughbreds, and other horses of the "hotblood" type. The "lean" head has thin skin which closely adheres to underlying parts. The bony outline of the head is sharply defined. The muscles, vessels, and nerves stand out in bold relief, giving the head that clean, sharp expression. Just the opposite is true of the plain or "fleshy" head. Here the skin is thick, coarse, and attached more closely to parts beneath with a great amount of connective tissue and fat. The bony lines and prominences are obscured. The head is more rounded off with that "fleshy" look. This type of head is characteristic of the "coldblood" and is associated with a dull temperament. Both types are shown in Figure 15–6.

Harmony of proportion is an important consideration, whether in a light or heavy breed. An excessively large head would be more objectionable in a light breed than in a heavy breed. A workhorse is expected to put maximum force into the collar and a large head can help in this respect. A large head on a racehorse or hunter would be unsightly and would also displace the center of gravity forward. This problem would predispose the horse to stumbling and place excessive weight on the already overburdened front limbs.

Figure 15–6.   The lean head (above); the fleshy head (below).

Figure 15–7.   Broad face with lop ears (left); a narrow face (right).

When the horse's head is viewed from the front, the most noticeable feature is the forehead (Figure 15–7), which should be deep and broad since it forms the brain cavity. Also of importance are two other structures housed in the bones below the brain: the frontal sinuses, which are a part of the respiratory system. A small forehead, therefore, encroaches upon the brain capacity and the respiratory system. A larger sinus does not increase respiratory efficiency, but it may lead to fewer problems with drainage and other conditions. From the forehead down the nose, the width should be ample. Here the nasal passages can restrict the amount of air intake and hamper maximum respiration. From just above the angle of the mouth, the nostril will begin to widen , with the most desirable nostril being wide and spacious. Thin and pliable skin over the nostrils and fine and mobile lips denote quality. The more lethargic horse will have undue thickness and fleshiness of the nose and lips. The upper and lower teeth should match squarely to provide most efficient grazing.

Great variation can be observed in the profile of the head. The face may be straight, dished, or arched (Figure 15–8). This aspect of the head may be purely esthetic in value, but it does denote the breeding and ancestry of the animal. The hot-blooded horse tends to have a dished or straight face, while

Figure 15–8.   Four types of faces: (A) undulating, (B) straight, (C) Roman nose, and (D) arched.

the coldbloods are often arched. It is commonly believed that a horse with a Roman nose is inclined to be ill-tempered. This temperament may also be true of a horse with eyes set too far to the side. The size and positioning of the eyes cause variations in vision which affect the action and temperament of a horse.

There seems to be some correlation between intelligence and the placement of the eyes among the various mammalian species, those with more frontal vision tending to be more intelligent. This correlation seems to hold true in the horse. R.H. Smythe, in *The Mind of the Horse,* writes:

> In the horse the retina is irregularly concave, being nearer the cornea at some parts than at others, giving rise to what is known as a "ramped retina". . . the ciliary muscle which operates the shape of the lens in most other animals is poorly developed, and the horse depends upon this ramped retina for its ability to focus its eyes.

He maintains that free and easy movement of the head and neck makes for the best vision in the horse. These facts emphasize the importance of eye placement and the type of attachment of the head and neck in terms of good vision.

The ears can add or detract from the beauty of the head. Although the principal function of the ears is to enhance hearing, they serve to indicate the emotions and temperament of the animal. The width of the poll, the set of the ears, and the ear size and form can all determine hearing effectiveness. When set well apart on a fairly broad crest and carried with a gentle inclination forward, they set off a well-formed head to advantage. Again, ear size may be more a clue to ancestry than to function. In the quick, intelligent horse the ears are ever erect and respond to every sound, moving vigorously from one position to another in rapid succession. On the other hand, ear movements are slow and the carriage is heavy and drooping in the dull and lethargic horse.

Even when observing the best-formed head, the breeder should be concerned about how it is attached to the neck. The undesirable extremes are either too closely attached or too loosely coupled (Figure 15–9). A head should be attached onto the neck and not into it. How the head is attached will affect the carriage and movement of the head in every direction. The head movement should be free and have ample range of action.

The close-coupled head tends to curtail the range and movement of the head. In performance, the close-coupled head is generally heavy in hand, and stiff and ungainly in its side movements. The loosely-coupled head is most often seen associated with the gangly horse that is light in the middle and lacking in general muscular development.

Carriage of the head is influenced by the shape and form of the neck. As an example, the "ewe-neck" forces the head into a somewhat horizontal position which results in less bridle control and less inclination on the part of the horse to tuck his head properly. Such a head carriage causes backward

Figure 15–9.   A loosely coupled neck on the left, and a closely coupled head on the right.

Figure 15–10.   A swan neck on the left, a ewe neck on the right, and a straight neck, center.

displacement of the bit, which tends to ride improperly on the angle of the mouth or on the first molar teeth.

In the other extreme, a severely arched neck forces the head into a rather oblique angle downward, or even backward; sometimes the chin even touches the chest. It is difficult for the horse to maneuver in this position or to see where he is going. This extreme is not objectionable in the draft horse, but riding horses of this type are inferior.

The best carriage of the head for an athlete is at an angle of about 45 degrees. This position gives the best field of vision, and the most freedom of movement. The throat will be well open, allowing the greatest room for air passage. A head carried at the proper angle will also assist proper movement of the shoulder. Variety in neck conformation has come about because of the different uses of the horse. The form and shape of the neck, therefore, are specific for the type and use of the horse (Figure 15–10). Neck and head carriage also vary from high to low, with the extremes being objectionable for most use. A well-carried neck gives an expression of courage. The elevated neck carriage enhances the use of the mastoido-humeralis muscle

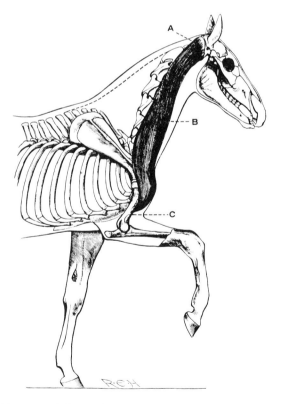

Figure 15–11.   The mastoido-humeralis muscle: (A) attachment to head, (B) body of the muscle, and (C) attachment to the humerus.

(Figure 15–11), which assists front leg movement, and helps give the great knee action that is admired by breeders of hackneys, saddlebreds, Tennessee walkers, and the like.

A head held in a lower position puts the eyes down to see what is immediately at hand. If the neck is not too short, and held almost horizontally, the horse is in a position to act quickly, since sudden movements from side to side are possible. The stock horse carries his head low.

The racehorse has a long neck, since the longer mastoido-humeralis muscle brings about a longer stride. The athletic horse has a long neck. Neck length gives flexibility and enables a quick response. It regulates the center of gravity while the body passes from one attitude to another.

The Withers

The withers show much variation according to age, sex, and type. The withers are not well developed in foals, and are generally more prominent in males. As with the head, the withers are somewhat a mark of quality, with hotbloods showing considerable elevation and refinement and coldbloods displaying more fleshy, low, and thick withers. Development of the many horse breeds has brought about a variety of wither types between the two extremes.

The height of the withers is determined by the length of the dorsal spines of the vertebrae in this area (Figure 15–12). Also the elastic ligament (the ligamentum nuche) which gives support to the head and neck is attached in this area, as are muscles which assist in movement of the neck. The freedom and quickness of neck movement is enhanced with elevated withers.

Good depth and angle of shoulder are generally found when withers are high and sloping, especially when the withers have ample length from anterior to posterior and continue well into the back. Coarse withers lead to a lack of liberty and range of shoulder movement. The action of the forelimb is restricted. Coarse withers make it difficult for a proper fit of the saddle, and sores often result.

The Back

A shorter back is generally a stronger back, but in a tall, large-boned horse, a longer back can give more length of muscles resulting in a longer stride for racing. An excessively short back has a tendency to inhibit body suppleness and elasticity much in the way a short neck does. A short back will diminish the length, and thus the capacity, of the thorax, but a deep chest and well-arched ribs will compensate somewhat (Figure 15–13).

An excessively long back is an evidence of weakness, and is frequently associated with legginess, lightness of muscle, and want of stamina. A long back tends to become depressed and hollow from the weight of a rider. The

Figure 15–12.   High withers (above), and low withers (below).

Figure 15–13.   A short back (above), and a long back (below).

long back can be compensated for by good width and muscle and by short, strong loins.

A straight back is considered to contribute to the greatest amount of power with the most perfect freedom and scope of action. The horse with an upward curve of the back, described as a roach back, tends to have weaker and underdeveloped spinal muscles, which hinder the range and speed of movement. The roach back was more common in the days when horses were often forced into heavy and prolonged work at an early age.

The hollow back is a downward curvature in which the bones of the spine are pulled down either from an inherent weakness of ligaments and opposing muscles or from a weakening by advanced age. The swaybacked horse lacks the power and pace of the more perfectly constructed horse.

Sometimes a high-withered horse will appear to be somewhat swaybacked, when in reality the spine is very straight. This deceptive appearance occurs most often when high withers are combined with a high croup.

Direction of the Back and Loins

The loins are just above the flanks and in front of the croup and haunches. Since the lumbar vertebrae have no support laterally from the ribs, it is best for this area to be as short as possible, with most of the back length being in the thorax. A short, strong-muscled loin enhances the impetus that passes from the posterior legs to the front end because it lessens the mobility and elasticity of the region.

The Croup and Hip

The croup is generally considered to be the upper portion of the hip. The long hip gives rise to longer muscles, which are necessary for speed. Hip width is a measure of power, since width in this area is principally due to increased muscle volume. Heavy muscle on the hip and croup will lead to a shorter stride.

The angle or slope of the croup shows great variety among horses. Two extremes are shown in Figure 15–14. The Arabian is well-known for its high tail set, or level croup. The muscle lengths and shapes afforded by this structure may contribute to the endurance for which the Arabian is noted. Horses of barb origin have a very low tail set. This characteristic is carried forth today principally in the quarter horse, where it contributes to great power in the early strides of a race. The thoroughbred is generally intermediate in the angle of the croup, and has staying power intermediate between the Arabian and the quarter horse.

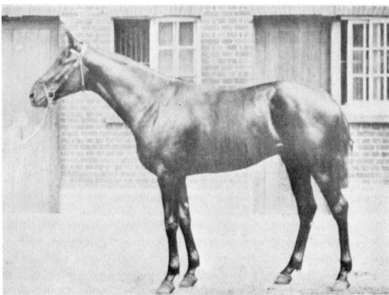

Figure 15–14.   A straight croup (above), and a sloping croup (below).

The Breast

The major difference between a narrow breast and a broad breast (Figure 15–15) is muscle size. Surprisingly little difference occurs in interior chest width or capacity. The heavily muscled breast was developed in the draft breeds as an aid to pulling power. Too much muscle in this area can inhibit speed and contribute to a shorter stride.

The Chest

The proportions of the chest, in both depth and width, are extremely important to endurance. The volume of the lungs is regulated by the space allotted. The heart size, while not entirely dependent upon chest size, is also important to maximum athletic capacity. The "spring" of the ribs, or the curvature laterally, gives increased volume to the thorax, or chest cavity. Chest proportions should always be an important consideration in selection.

The Shoulder

As with the hip, a long shoulder provides a greater base for the long muscles, which contribute so much to speed. The angle of the shoulder is

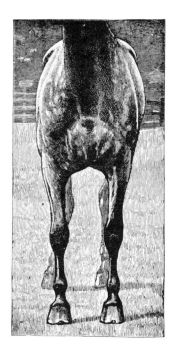

Figure 15–15.   A narrow chest (left), and a wide chest (right).

Figure 15–16.   A straight shoulder (above), and a sloping shoulder (below).

also an important factor. A sloping shoulder gives liberty and elasticity to the gait.

The straight or upright shoulder (Figure 15–16) is shorter than the sloping shoulder. It contributes to a short choppy gait and affords less resistance to the concussion of running and jumping.

The Foreleg

A long upper arm allows greater length of muscle and contributes to greater speed. The length of stride becomes greater with a longer limb. A horse with maximum knee action, on the other hand, must be a bit shorter in the forearm. The muscles of the forearm should be well developed throughout their length. When they are not, a horse is described as having weak forelegs because a lack of development of these muscles generally indicates a weakness of tendons on down the leg (Figure 15–17). The leg should be in a perfectly upright position at rest so as not to alter the distribution of the body weight.

The knee is a large and complex joint made up of a number of bones united by many connecting ligaments. It should be as free as possible from excessive connective tissue and fat, which indicate coarseness and interfere with its primary function. A large knee provides larger articular surfaces and

Figure 15–17.   A good forearm and cannon (left), and a weak forearm and cannon (right).

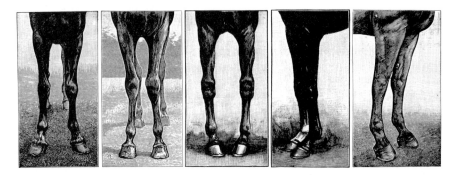

Figure 15–18.   Poor leg conformation: from left to right, toes out, knock-kneed, pigeon-toed, bowed knees, calf knees.

has a greater capacity for movement. The leg should not form an angle at the knee. A forward angle is described as bowed knees; a backward angle is described as calf knees (Figure 15–18).

Twisting of the leg at the knee is a common defect of conformation which leads to many unnatural stresses on joints and ligaments, and affects movement of the limb, causing awkwardness. This defect can be in the form of toes turned out, knock-knees, or pigeon-toes (Figure 15–18).

The cannon just below the knee has less need of length for speed than the upper arm. An extremely short cannon, however, will shorten the length of stride, simply because the legs are shorter. As compared to the forearm, the cannon should be shorter. The large tendons and ligaments of the leg are

Figure 15–19.   Good pasterns (left), long sloping pasterns (center), and short sloping pasterns (right).

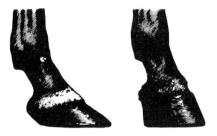

Figure 15–20.   Flat foot (left), and an upright foot (right).

located behind the cannon bone. Like the knee, the cannon area should have little fat or connective tissue, since either tends to interfere with optimal movement. Lack of fat and connective tissue results in a flat appearance of the cannon from side to side. A coarse animal will have a more rounded cannon.

The fetlock should be broad both from front to back and from side to side. This greater articular surface at the joint gives more strength.

The length and angle of the pastern determine the smoothness of gait in the same way as the shoulder does. An excessively long pastern brings more force to bear on the region, often overtaxing the strength of the ligaments and sesamoid bones. A very short pastern has less ability to absorb concussion. Various types of pasterns are shown in Figure 15–19. A relatively long and oblique pastern disperses more easily the violent force of locomotion. An upright pastern offers a hard, choppy gait.

Large feet are cumbersome and require more energy to lift. Some breeders, therefore, select for small feet, although many structures in the foot are vital to a proper base and the ability to absorb concussion. An extremely small foot for the overall size of the animal leads to unsoundness. A flat foot (Figure 15–20) is generally a weak foot, and it tends to be larger than necessary. The upright foot forces too much weight on the toe, and much of the elastic reaction of the tendons and ligaments behind is lost to the limb.

The Hind Limb

While the forelimbs are attached to the body by muscles alone, the hind limbs have a bony attachment at the hips. The thigh and angle of the femur are important to length of stride. With the thigh inclined too far forward, its range of forward movement is curtailed. When inclined too much to the rear, maximum stride is impossible.

The hock is found in a variety of shapes, and is formed by a group of specialized bones. The hock should be large, shapely, and well directed, but often it is undersized and forms too much of an angle. Like the knee, a lack of fat and connective tissue in the hock denotes quality. Too much angulation from front to back is known as curb and can lead to spavin under

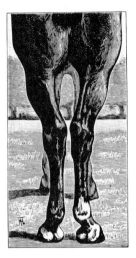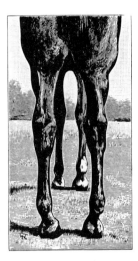

Figure 15–21.   Cow hocks (left), and bow legs (right).

heavy use. The condition in which the points of the hock are turned in is called cow hocks; a turning out is called bowlegs (Figure 15–21).

CONFORMATION PREDICTS SERVICEABILITY

One of the most disheartening problems to a veterinarian is his inability to improve serviceability in a horse. Most performance weaknesses result from poor conformation, and a horse is generally born with the problem. A horse must be built for the job he is asked to do. Sometimes he must roar around a racetrack at his top speed; he may be asked to travel 100 miles in double time; in other situations the rider may want him to turn on a dime, move immediately from a standstill to a dead gallop, and make spectacular sliding stops from a run. Without proper conformation the using horse may be unsafe to ride or eventually may become lame, sick, or injured.

A blemish is a defect which does not cause an unsound condition. It is an unsightly scar or swelling which will not go away. As long as it does not interfere with the ability of the horse to function at his best, it is only a blemish. As an example, a splint is only a blemish as long as it is small enough and too far away from the knee joint to interfere with movement of the joint or tendons of the leg.

There is a difference between working serviceability and breeding serviceability. A mare may be serviceable for the performance asked of her and yet unable to conceive or carry a foal to term due to a conformational or physiological defect. The slope of the vaginal opening may be so horizontal that feces from the anus above continually contaminate the area. This condition, which results from poor conformation, can cause a severe breeding problem. A mare may fail to produce the proper amount of progesterone during pregnancy causing degeneration of the uterine attachments and subsequent abortion. This is a common physiological breeding problem.

Many working weaknesses are shown in Figure 15–22. Each of these defects interferes in some way with full and complete movement of a joint or tendon, or the defect may cause pain with a certain movement resulting in an unconscious inhibition of action.

Riding gear and harnesses are made to fit the average horse. Those horses that deviate very much from the norm may begin to develop sores which become problems. A weak back or "mutton" withers may lead to saddle sores or even fistula of the withers. Conformation of the shoulders is especially important in draft horses where the collar of the harness can cause real damage, resulting in fistula of the withers or sweeny. Poll evil (which occurs at the top of the neck where it joins the head) and fistula of the withers often become deep, infected sores which fail to heal because of extensive dead tissue deep inside.

The bone at the point of the hip is the anterior spinous process of the ilium. This bone is the site of injury when two horses crowd through a narrow door at the same time. When this part of the ilium breaks, the condition is known as a knocked-down hip. Such an injury is more likely to

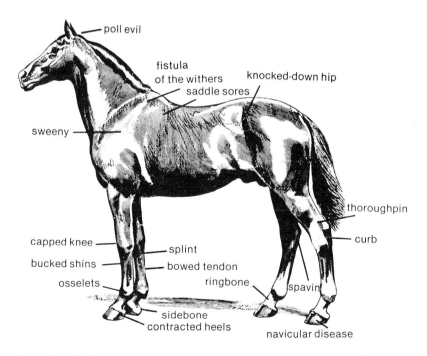

Figure 15–22.   Some common unsoundnesses resulting from defective conformation.

occur when this area has insufficient muscle padding. Improper muscle size and development in the hip can also lead to a bursitis in this area — often known as whorlbone lameness.

Conformation of the front legs is extremely important. They bear the majority of the weight, and need to be straight and properly angulated. A common conformational fault is crooked legs from the knees down. This defect puts excessive stress at the knee joint resulting in injury which often develops into an enlarged knee. Beginning as a bursitis, it can develop into a calcified knot which hampers knee action.

Bucked shins are a direct result of injury or strain when a young horse is pushed too hard in training. Since some horses are more likely to develop the condition than others, it may be that a weak attachment or excessively thin periosteum (bone covering) predisposes a horse to the problem. Proper conditioning will strengthen the periosteum and its attachment and reduce the likelihood of bucked shins. A weakness in this area cannot be observed externally.

A predisposition to splints may come from the same weakness. Splints arise from excessive irritation of the periosteum of the young horse as a result of injury or hard use. Here, also, crooked legs can put extra stress on the area, increasing the chances for a splint. The splint bones are not firmly attached to the cannon bone until about 6 years of age. Splints seldom occur after that. Conformation which adversely affects the way of going can cause splints. Interference, for example, when the horse strikes the inside of one leg with the inside of the opposite hoof, can be a cause. A horse who is base wide, or splayfooted, may show interference. Overreaching, which occurs when a rear hoof strikes the foreleg in an extended gait, may also initiate splints.

Bowed tendons occur from excessive strain. The flexor tendons at the rear of the cannon bone are injured. Swelling and hemorrhage interfere with the natural movement of these tendons in their sheath. This condition causes pain and inhibits proper movement. Long weak pasterns contribute to the problem. Horses who are too heavy for their tendon structure will more often develop bowed tendons when put into heavy training, than will lighter horses. Here again, proper conditioning will strengthen tendons and reduce the danger of bows.

Horsemen who describe the leg of a good horse as one with a flat bone are not really talking about the bone but about the size of the tendon in relation to the bone. The leg just below the knee is made up of the cannon bone and the flexor tendons from front to back. The larger the tendons in relation to the bone, the flatter this part of the leg appears.

Crooked legs, especially in the area of the fetlock, will cause such strain under heavy use that unsoundness in the form of osselets, ringbone, and sidebones may develop. Osselets are caused by inflammation of the tissue covering the bone in front of the fetlock. Ringbone occurs on the phalanges below the location of osselets and from the same causes. Sidebones result from an ossification of the lateral cartilages on the sides of the foot. All of these conditions can result from a simple injury and may be entirely unrelated to conformation; however, the crooked-legged horse is susceptible to them.

Navicular disease is an affliction of the coffin bone within the foot, or the joint behind the bone. It may begin as an inflammation of the navicular bursa between the bone and the deep flexor tendon. The majority of cases of navicular disease can be traced to faulty conformation. The common practice of breeding for small feet leads to trouble with a heavily muscled horse; the strain becomes too much for the coffin bone. Short pasterns with upright conformation will often lead to navicular disease, as will contracted heels under heavy work. This condition is often inherited. The heels of the foot are much too close together and the size of the frog is greatly reduced.

In the rear leg, sickle hocks and cow hocks often lead to curb, an enlargement of the plantar ligament at the back of the hock. Crookedness in this area will often make curb a chronic condition that interferes with good performance.

Cow hocks may also lead to a spavin, which is a disease of the hock joint. In the severe form the bones of the joint enlarge and may grow together in what is called a bone spavin. This condition causes immobility and can reduce serviceability.

A thoroughpin is a swelling above and behind the hock joint between the Achilles tendon and the bone. Horses with straight weak hocks are predisposed to this condition.

If a horse is to be used to his full potential, as in endurance riding, racing, or other performance events, he must possess sound conformation. He must be well-balanced with good withers, a strong back, a long muscled hip, and, above all, straight legs. Conformation predicts serviceability.

BIBLIOGRAPHY

Adams, O.R.: Lameness in Horses, 3rd ed. Philadelphia, Lea & Febiger, 1966.

Albert, W.W.: Suggestions for Buying and Judging Horses. Extension Circular 1057, University of Illinois, 1972.

Dusek, J.: Heritability of conformation and gait in the horse. Z. Tierzucht. Zuchtbiol. 87:14, 1970.

Hayes, M.H.: Points of the Horse. London, Hurst and Blackett, Ltd., 1904.

Hildebrand, M.: Motions of the running cheetah and horse. J. Mammals 40:481, 1959.

Kays, D.J.: The Horse (rev. ed.) Cranbury, NJ, A.S. Barnes, 1969.

Mellin, Jeanne: The Morgan Horse Handbook. Brattleboro, VT, The Stephen Greene Press, 1973.

Rooney, James R.: The Lame Horse: Causes, Symptoms and Treatment. Cranbury, NJ, A.S. Barnes, 1974.

Slade, L.M., and Canfield, R.V.: Conformation and racing potential. The Intermountain Quarter Horse, Dec., 1980, p. 76.

Smythe, R.H.: The Horse, Structure and Movement. London, J.A. Allen & Co., 1972.

Smythe, R.H.: The Mind of the Horse. Brattleboro, VT, The Stephen Greene Press, 1965.

Wakeman, D.L.: Selecting and Judging Light Horses. Light Horse Production in Florida. Bulletin, Florida Depart. Agric. 188:31, 1965.

Willis, Larryann C.: The Horse Breeding Farm. Cranbury, NJ, A.S. Barnes, 1973.

part IV

breeding management

16

equine reproduction

During centuries of evolution, the horse acquired a unique type of reproduction suited to its wild state and fleet-footedness. The foal is able to get about and travel with its mother within a few hours of birth. This characteristic and others associated with reproduction in the horse are not particularly helpful under domestication.

It would better suit the purposes of the horse breeder if reproduction were as sure as in the cow. However, gestation in the horse is long and precarious, and although the foal is strong and quite highly developed when born, chances of abortion along the way are quite high. This situation can be improved by carefully selecting both the male and female for various factors involved in reproduction.

THE FEMALE REPRODUCTIVE SYSTEM

The ovaries of the mare rest in the abdominal cavity on either side of the backbone posterior to the kidneys and may vary in size from 1 to 4 inches. A diagram of an ovary with a developing follicle is shown in Figure 16–1. The follicle is a fluid-filled cavity which contains the tiny ovum. When the follicle ruptures, the ovum usually passes into the uterine tube. The cells lining the follicle then begin to grow and change in nature, eventually forming a large, yellow-colored growth called a corpus luteum.

The fluid of the follicle contains estrogen hormones, while the corpus luteum produces the hormone progesterone. The interior of the ovary contains thousands of ova awaiting development within primary follicles. It has been thought, in the past, that normally only one follicle develops during

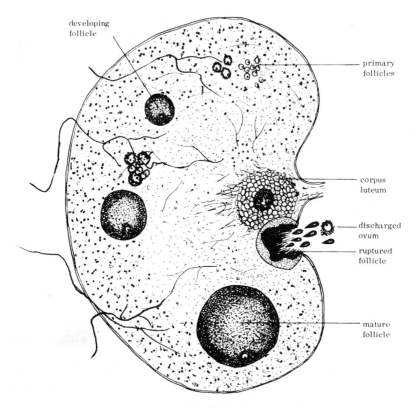

developing
follicle

primary
follicles

corpus
luteum

discharged
ovum

ruptured
follicle

mature
follicle

Figure 16–1.   Follicular development in the ovary.

each heat period. F.T. Day describes twin follicles as developing all too often. Some statistics suggest a frequency of almost 5 percent. Dr. S.J. Roberts suggests that 15 to 30 percent of equine estrous periods are characterized by twin or double ovulations. More than three-fourths of these dual ovulations, however, do not result in twins but rather result in one degenerate embryo and one normal. Usually the right ovary ovulates during one heat period and the left ovary during the next.

The germ cell within the primary follicle is known as an oocyte. The process of meiosis begins in the primary oocytes during fetal development. The primary oocytes divide and become secondary oocytes (Figure 16–2). During adult life, all of the primary follicles contain primary oocytes. They have not yet halved in chromosome number. This last germ cell division occurs only when the follicle containing the cell begins to develop. The whole process is known as maturation of the ovum.

The ovaries are not located within the uterine tube, but just outside the end of it (Figure 16–3). The end of the uterine tube is somewhat funnel shaped with many small fingerlike projections called fimbria. The cells that line the uterine tube have tiny cilia and secrete mucus. The beating of the

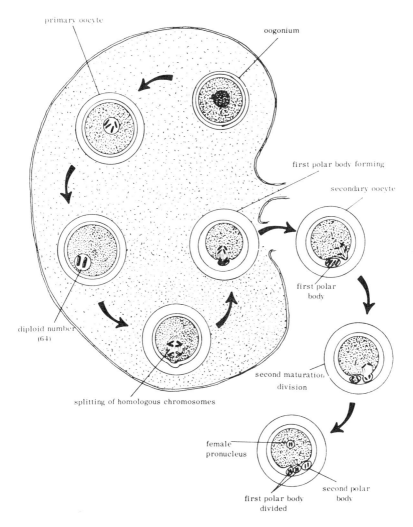

primary oocyte

oogonium

first polar body forming

secondary oocyte

first polar body

diploid number
(64)

second maturation
division

splitting of homologous chromosomes

female
pronucleus

first polar body
divided

second polar
body

Figure 16–2.   A diagram of maturation of the ovum.

cilia, which are microscopic, hair-like projections, creates a current of mucus, which is evidently responsible for moving the ovum into the tube. The end of the tube usually partially surrounds the ovary so that this process is facilitated.

The uterine tubes each empty into a horn of the uterus. In the mare, the body of the uterus is quite large in comparison to the horns. The neck of the uterus is the cervix. It is actually an extremely long sphincter muscle, which remains tightly closed, usually sealed with a mucous plug, except during heat or at parturition. The opening of the cervix can be viewed with the use

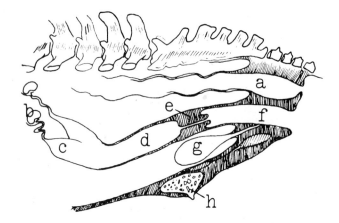

Figure 16-3.   The reproductive system of the mare: (a) rectum, (b) ovary, (c) uterine horn, (d) uterus body, (e) cervix, (f) vagina, (g) bladder, and (h) mammary gland.

of a vaginal speculum. Examination of the lining of the vagina and the opening of the cervix gives an indication of when the mare ovulates (Figure 16–4).

Many types of abnormalities of the female reproductive tract are found. Hermaphrodites occur about one in 13,000 births, according to Rauchback. In these mares, the entire uterus is commonly missing. Hafez has reported cases of intersexuality in horses in which the animal has some abnormal parts of both sexes. The specific genetic cause of this condition was covered in Chapter 8.

A hymen is occasionally found in the maiden mare. This is a membrane which blocks off the vagina near the cervix. It is rare and usually is not a complete blockage. It can be easily cut away before breeding and is not associated with other abnormalities.

Reproductive anatomical abnormalities of the mare generally attract little attention until breeding time. It is then found that the abnormality interferes with conception. Some of the more common abnormalities are:

1. Rectovaginal fistula. This opening between the rectum and vagina results in severe infection of the latter.
2. Tipped vulva. This condition results in "windsucking" and also causes vaginal infection.
3. Abnormally small uterus, or absence of uterus. Adonis claims that a small or absent uterus occurs because of a heritable factor. The result is usually sterility.
4. Small underdeveloped ovaries. This characteristic may be due to genetic factors, and may not respond to treatment.

Mother Nature has worked for eons to develop a complicated hormone system which causes foaling to occur in the spring when warm weather

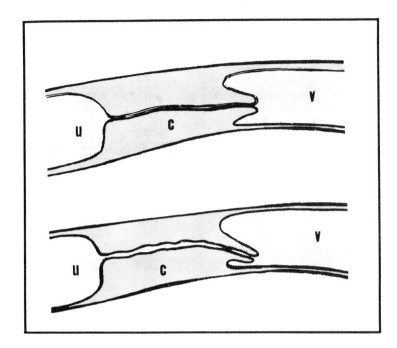

Figure 16–4.   During diestrus, the cervical opening into the vagina (v) is tight and rigidly held
in a central position. During estrus, the cervix (c) relaxes, allowing a continuous opening
from vagina to uterus (u).

begins to crowd out the winter chill. Man has been working at cross
purposes for the past century to turn this system around and make it function
abnormally so that foals are born in the difficult winter.

Man is ahead of the game. He has at last discovered the secret that allows
him to turn the tables, and now it can be pretty easy to fool Mother Nature.

Members of the Jockey Club in the early days were not aware of the
implications when they pronounced January 1 as the birthday of all
thoroughbred foals. They assumed that breeders would continue allowing
Mother Nature to have her way, bringing the foals in the spring, all within a
couple of months of each other, ensuring a rather evenly matched crop of
2-year-olds.

But year after year crafty breeders tried first one thing and then another to
get earlier foals and, as a result, a jump on the competition. Now breeders
must have January and February foals if they hope to be serious contenders
in competition later on.

The way Mother Nature went about her task was to gear the breeding
season with the length of the day. She built into the physiology of the mare a
mechanism that would stop the estrous cycle in the fall when the days got
shorter and hold the reproductive system in anestrus until the days were
longer in the spring.

The earth has been tilted on its axis for so long that plant and animal life on the surface has adjusted marvelously to the changing seasons the tilt brings. One of the most interesting adjustments is the estrous cycle of the mare. As the earth takes its 12-month journey around the sun, daylight is longer than 12 hours during half of the year and shorter than 12 hours during the other half. In the northern hemisphere the short days are from September 21 to March 21 and the long days are from March 21 to September 21. The opposite is true in the southern hemisphere.

The summer solstice is the longest day of the year. It occurs in the northern hemisphere on June 21. The winter solstice occurs on December 21. Thousands of years of evolutionary selection have programmed the mare's brain to recognize the length of day and begin her estrous periods only when the days are sufficiently long, that is, when spring has arrived. The actual length of day at any time of the year is also determined by the latitude. Therefore mares further north will begin coming in heat later in the season than those living closer to the equator.

The equinox is the time of the year when the day's length equals the length of the night. This day occurs twice a year: on March 21 and September 21. These dates are important to the horse breeder because they approximate the beginning and the end of the natural breeding season.

During the short days of winter, most mares enter a period of anestrus where they do not "cycle." The term "cycle" means that the mare passes through various stages of estrus: the first is estrus, or heat; the second is metestrus, the period of preparation for pregnancy; the third is diestrus, the period when the uterus is fully prepared to support the developing embryo; the last is proestrus, the time when the uterus recognizes that no embryo is present and prepares for another heat period (Figure 16–5).

Early in the season the cycle begins slowly, like a wheel lacking grease at the hub. It seems as though it has grown a bit rusty during the winter layoff. As a consequence the estrous cycle is much longer in the early season and ovulation (the process of the egg leaving the ovary) often does not occur, holding the mare in a prolonged heat. With no ovulation there can be no conception even if the mare is bred.

Dr. Robert Kenney, of the University of Pennsylvania, describes this time as the period of anovulatory receptivity—meaning the mare is receptive to the stallion, but does not ovulate. The percentage of mares showing estrus increases rapidly after the middle of February. At the beginning, only about 60 percent of those showing estrus actually ovulate. This percentage rises in April and May as the days grow longer. Even during the longest days, only 80 percent of cycling mares are ovulating, according to one study.

Thus, the conception rate is very low at the beginning of the season. It is only after the summer solstice that the conception rate climbs much above 50 percent. Figure 16–6 is compiled from information on a limited number of mares at similar latitudes. The percentages would vary with changes in

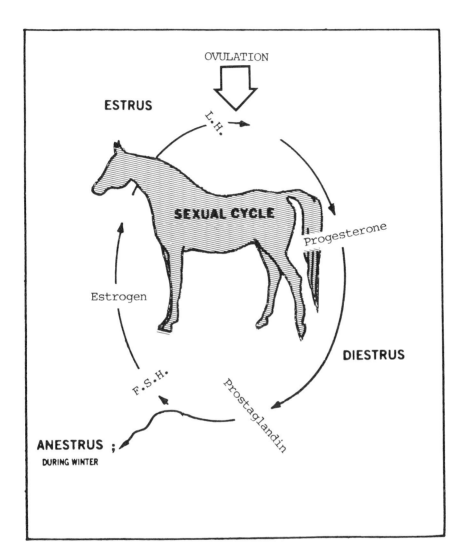

Figure 16–5.   The estrous cycle of the mare is commonly divided into two phases that blend one into another. External signs distinguish estrus from the balance of the cycle, which is lumped together and termed diestrus. Actually proestrus is the period during which the follicle is enlarging under the influence of follicle stimulating hormone (FSH). During this phase the vascularity of the uterine mucosa increases and the cilia lining the oviduct grow. Estrus is the period of sexual desire and acceptance of the male.

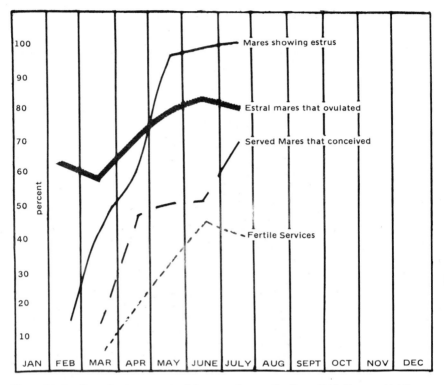

Figure 16–6. Reproductive aspects of the mare by month. (From R.M. Kenney: Vet Scope 19:16, 1975.)

latitude, with the percentage being higher earlier in the year when the mares are closer to the equator.

According to Dr. Kenney, both the stallion and the mare not only recognize day-length changes, but they respond to them. We do not know exactly how they recognize these changes and we do not whether it is the light or the darkness that they recognize, but it is certain that they do recognize the changes and that they do respond.

The season of fertility is described by Dr. Kenney as peaking in mid-July with a gradual increase up to that time and a gradual decrease after that as the days shorten again. Figure 16–7 shows the monthly incidence of ovulation in the northern hemisphere, and this factor roughly corresponds to fertility.

One question that arises is why so many mares are bred early in the season before the period of maximum fertility. Dr. Kenney explains:

> This is a result of the influence of the Universal birthday and many laws based on it. Until 1833, the Universal birthday of thoroughbreds and standardbreds was May 1, meaning actual breeding started June 15—the ideal time. But then the English

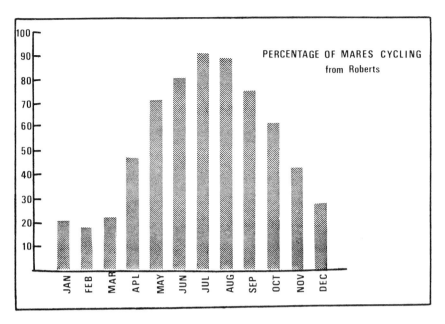

Figure 16–7.   Percentage of mares cycling each month during the year is shown. (Data from S.J. Roberts: Abortion and other diseases of gestation in the mare. *In* Current Therapy in Theriogenology. Philadelphia, W.B. Saunders Co., 1980.)

Jockey Club pushed back the birthday to January 1 where it has remained ever since—not only in England, but in the United States as well.

Early breeding has become necessary so that 2-year-olds will have maximum maturity and the greatest chance of winning in competition, be it racing or showing. This early breeding has brought a host of problems that would not occur if breeding were conducted in the natural season.

Dr. Kenney explains:

Many of the problems of early springtime breeding are related to long heat periods as well as to the lack of or delay in ovulation. However, there is a tendency to blame overcast weather, cold, lack of grass, and uterine infections for poor springtime breeding. Each year in February, March and into April, breeders say, "the mares are really slow coming around this year." In late April and May, however, they think the mares are "catching up." The mares are not slow and they are not catching up. The mares are only doing what they are supposed to do.

One of the first reports of successful early breeding in horses was by Dr. Wendell Cooper, in Michigan, nearly a decade ago. Figure 16–8 shows how he moved the peak of his breeding season from May and June to January

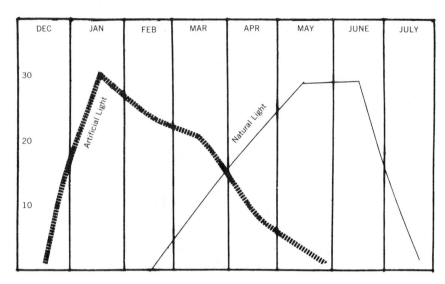

Figure 16–8.   Percentage of mares pregnant each week during the breeding season altered with artificial lights. (From W.L. Cooper: Wintertime Breeding of Mares. Symp. on Reprod. in Cattle and Horses. Am. Vet. Soc. for Study of Breeding Soundness. Michigan State University, 1972.)

with the use of artificial lights. Since that time, research regarding techniques of early breeding has continued.

Scientists at Colorado State University have been leaders in this study. Dr. Ed Squires, who has done much of the work there, says:

> Recently in our laboratory, a study was conducted in which nonlactating mares were exposed to 16 hours of light per day from December 1 to May 1, or until 50 days pregnant. Mares were maintained in a band outside during the day and in individual box stalls at night. Artificial light was provided by a 200-watt incandescent bulb 4 meters from the floor of each stall. After 60 days of exposure to artificial light, 32 of 34 mares had exhibited estrus. Based on data from our study, as well as other studies, we currently recommend the following artificial lighting program: 16 hours of light beginning December 1 to 15 and continuing until natural day length approaches 15 to 16 hours, or until mares are confirmed pregnant.
>
> Several questions still remain unanswered concerning the use of light for induction of estrus. For example, the exact intensity of light and type of light that is effective has not been adequately determined. Some breeders have reported successful induction of estrus in mares maintained in paddocks and exposed to either incandescent, vapor or fluorescent lighting.

There has been some evidence that the plane of nutrition plays a role in initiation of the breeding season. Mares should not be fat in midwinter. In fact, they should be a bit thin so that as the breeding season approaches, they can be gaining weight. One study showed that the onset of the breeding season was earlier for mares that gained weight between January 1 and May 22 than for mares that lost weight during this period.

Further research is needed to determine whether higher temperatures have an effect on initiation of the estrous cycle. Indications are that temperature may play a small role.

Many attempts have been made to discover a hormone, or a group of hormones, that will help to initiate the estrous cycle earlier in the year. Prostaglandin seems to work only after the mare has begun cycling. Colorado State University researchers tried using an oral progesterone-like (progestin) compound (Allyl Trenbolone), but found that it alone had no effect.

As Dr. Squires related:

> Since progestin alone was ineffective, we conducted an experiment combining the use of artificial light and progestin treatment. As a result of these experiments, it would appear that the effectiveness of Allyl Trenbolone for induction of estrus and ovulation in anestrus mares is dependent upon the amount of follicular activity at the time of initiation of treatment. The addition of artificial light in the study resulted in considerable follicular activity in 29 of 34 mares. Thus, in these mares the progestin was effective in regulating estrus.

There is a significant difference in mares that are in anestrus and those in a prolonged diestrus, although there may be no external differences. The diestrus mare needs the stimulus of increased length of day to cause her to cycle, while the mare with a prolonged diestrus has already begun cycling and hormones will usually help her to continue the cycle. Thus, it is important to put open mares under lights early in the winter. A Riverside California practitioner, Dr. Pierre Lieux says:

> It is only during the true breeding season that mares are expected to come in healthy, regular heats of 5 to 7 days, with a good ovulation toward the end. It is only at that time that one is able to formulate a true opinion of a mare's reproductive potential.

Dr. Lieux is in charge of breeding at several large thoroughbred farms. He has placed large lights in a paddock at these farms where those open mares that are to be bred early can receive at least 16 hours of light daily, beginning as early as November. He finds that many problem mares need to be examined and treated early in the season while they are cycling.

"When mares need to be examined prior to the breeding season," he says, "a true breeding season can be induced as early as December or January by keeping the mare under lights 1 or 2 months before the chosen date."

It seems too simple to be true; that the length of daylight controls the on-off switch to the mare's estrous cycle. Evidently, however, that is just about all there is to it. Horse breeders now have a way to solve many noninfectious breeding problems by simply putting mares under light about the first of December, for a total of 16 hours. A month later they can increase the mare's feed so that she begins to gain weight slightly. This practice is the secret to early breeding.

THE ESTROUS CYCLE

The estrous cycle varies slightly from one mare to another. When describing this cycle, it is necessary to speak in terms of averages. The average cycle in mid-season is 21 days.

Early in the spring the first few cycles are usually irregular and not typical of the more normal cycles later on. The pituitary gland produces follicle stimulating hormone (FSH) in the spring, the production of which is regulated by the length of day and other factors.

The cycle is traditionally described in relation to estrus, which is the "time of heat," the time of sexual attraction, easily identified by external signs. It takes an average of 3 days for the FSH to change the reproductive tract enough to begin signs of estrus; this is known as the period of proestrus. The signs of estrus are brought about by the hormone estrogen. This hormone is normally produced in the ovary, in a cyst-like follicle, which contains the ovum, later to be fertilized if pregnancy occurs. Other areas of the body also produce estrogen and if this occurs in a large amount, the mare may lapse into a "false heat," which can last for weeks or months. With some mares this false heat happens early in the season each year. It is not until later that these mares "settle down" and have normal cycles.

Estrus ends abruptly after an average of 6 to 8 days. The follicle ruptures from stimulation by luteinizing hormone (LH), which is produced in the pituitary gland. Early diestrus (or metestrus) lasts an average of 3 to 5 days. During this time the tissue in the part of the ovary that previously contained the follicle is organizing itself to develop a gland-like growth called the corpus luteum.

During diestrus, which averages 11 to 18 days, progesterone—sometimes called the hormone of pregnancy—is produced in abundance and prohibits the mare from continuing the cycle. If pregnancy has occurred at this time, the embryo begins to produce hormones of its own which inhibit the cycle and after about 40 days receives assistance from a hormone called pregnant mare serum gonadotropin (PMSG), which is produced by endometrial cups. PMSG will be described later.

If pregnancy has not occurred, a prostaglandin hormone is eventually produced to overcome the effects of progesterone and stimulate production of FSH that again initiates proestrus.

All of the hormones described have been synthesized and are commercially available. Their use at the proper time and in the appropriate dose can help regulate an abnormal cycle, making pregnancy a more probable situation. Whereas the cycle can get stuck at any stage, a hormonal boost can start it moving again.

One of the most celebrated hormones to be used recently in regulating the cycle is prostaglandin ($PGF_{2\alpha}$). A Kentucky equine veterinarian, one of many from all over the United States who used the prostaglandin drug when it was first available, said,

> I've used it in all categories of mares, including chronic barren, just-off-the-track, and maiden mares with different degrees of success, but I'd definitely say it's a good, useful tool. I'd call Prostin the most dramatic step in breeding management since I've been in practice, and that's 27 years.

Upjohn, the first company to come out with the drug, called its product Prostin. This pharmaceutical was followed closely by a similar one, Synchrocept, produced by Diamond Laboratories. These drugs can be used in three types of situations: (1) in mares with normal estrous cycles to bring about heat at the exact time desired; (2) in mares after foaling to bring on heat if the foal heat is missed; and (3) in apparent anestrus to bring on the initial heat in mares.

A California practitioner who used the drug on over 100 mares remarked: "I used it on those mares which were bred but didn't settle, and it did a good job bringing those empty 40-day mares into estrus." Another vet explained, "I use it to schedule mares to the stallion. I can breed more mares this way. . . earlier, too, and I've had good luck using it following foal heat."

Research on prostaglandins has been in progress for many years. They were first extracted from sheep vesicular glands in 1935 by Von Euler. Chemically, prostaglandins are long-chain fatty acids. A variety of types occur with many different physiologic actions. They affect blood pressure, cause diuresis, increase gastric motility, and have other effects. The uterine prostaglandin is called $PGF_{2\alpha}$.

During the breeding season, normal mares, cycling regularly, are bred 6 days after injection with the prostaglandin drug. An 80 to 85 percent conception rate results from the first breeding. This high percentage certainly simplifies breeding for both the busy breeding farm and the one-mare owner. It reduces the need for teasing, and eliminates much unnecessary boarding at the stallion farm, just waiting for the proper breeding time. It is advisable to examine mares for pregnancy before giving the drug if they have been exposed to the stallion since foaling. The drug may initiate abortion.

Breeding on the 9-day foal heat is a common method of "catching up" mares that are foaling late. This practice brings a foal almost a month earlier the following year. The problem is that the conception rate is only about 33 percent at the foal heat. Research has shown that mares injected with the prostaglandin drug about a week after the foal heat and then bred 6 days later will have a 65 percent conception rate. It is ineffective when administered prior to 5 days after ovulation.

Controlled breeding is now possible. With planning and calculated management the majority of foals will arrive in January and be at the head of their class. The wise breeder can take advantage of these new methods of fooling Mother Nature.

In considering the long-term genetic consequences of hormone therapy, horse breeders may be reluctant to use artificial means. Mother Nature moves swiftly to eliminate reproduction in those mares with hormone malfunction. These malfunctions are often the result of environmental factors, but may also be due to inherited genetic problems, which will only be passed on to the offspring when reproduction is made possible by hormone therapy. The genetic purist is therefore against the use of hormone therapy except in cases where environmental factors are known to be the complete cause of the hormone problem. Looking ahead, he sees a time when horse breeding will be possible only with the use of hormone therapy, and the horse will thus become completely dependent on man.

Normally, accumulation of estrogens in the system will inhibit FSH from the anterior pituitary. As estrogens increase, they also stimulate production of LH, which causes ovulation and development of the corpus luteum. This area in turn produces progesterone. It causes further favorable growth and development of the uterus lining, preparing it in such a way that it can produce "uterine milk" for nourishment of the early embryo, called a blastocyst. The early embryonic attachments are weak and frail.

The latest frontier of equine endocrinology research involves the hormonal interplay between the mare and early embryo. If a blastocyst is present at day 15, but then is removed non-surgically, the next ovulation is delayed, while if removal of the blastocyst occurs on day 14 or earlier, the normal interovulatory period occurs. The reason for this seems to be associated with estradiol production by the blastocyst. Estradiol evidently inhibits the release of prostaglandin from the endometrium that would otherwise destroy the corpus luteum. Some data show that estradiol administration to mares on days 15 and 16 will prolong the life span of the corpus luteum.

As the corpus luteum continues beyond day 16 and production continues, histotrophe and pregnancy-specific uterine proteins are produced. Data show that blastocyst estrogen has a synergistic effect on this action.

Irvine and Evans have been involved in research which shows that FSH surges and follicular development continue during pregnancy but cease with the onset of winter. Before day 35 occasional ovulations occur but few corpora lutea are formed. Fetal cells invade the uterus soon after day 35 and develop into endometrial cups, which produce pregnant mare serum gonadotropin (PMSG). This substance provides an LH-like effect causing

FSH-primed follicles to undergo final maturation, luteinization (usually without ovulation), and establishment and maintenance of the secondary corpora lutea. It may be that one role of PMSG is the formation of an immunoprotective barrier around endometrial-cup cells to prevent recognition and rejection of paternal non-histocompatible antigens on the surface of fetal membranes.

FSH production causes marked ovarian development between days 45 and 70, some ovaries becoming as large as 15 cm in diameter because of follicles and corpora lutea. These begin to regress after day 70 because of the luteotrophic role of PMSG at that time. From day 70 onward, progesterone production is taken over by the placenta, as shown by studies in ovariectomized mares. The placenta also produces increasingly large amounts of estrogens. Irvine and Evans state that maintenance of pregnancy is highly dependent on progesterone production from, first, the primary corpus luteum in one ovary, then the secondary corpora lutea in both ovaries, and finally the placenta. During the first few weeks of pregnancy, 150 mg progesterone per day is required to maintain pregnancy, and in later pregnancy 500 mg progesterone per day is required.

THE TIME OF BREEDING

With an understanding of the function of hormones and the sex cycle, one can begin to determine the most opportune time to breed the mare in such a manner that live spermatozoa are available in the uterine tube at the time of ovulation. No one has been able to prove just how long sperm deposited in the uterus will stay alive. However, Day reported from his research that spermatozoa will definitely survive over 2 days in the mare and usually 4 days or longer. This period may vary from stallion to stallion, but one with good hardy spermatozoa may be able to settle a mare even if bred 4 days prior to ovulation.

Every breeder in the business has a set time to begin breeding, a set time to quit, and a certain number of times to breed during the heat period. He always feels that his procedure is good because he has had good luck settling mares with it. If several mares are booked to one stallion, some consideration must be given to the number of services he performs. A wise breeder will develop a system which will settle the most mares with the fewest services possible.

If a good, young, healthy stallion is scheduled to breed only 15 to 20 mares during the season, one need not worry about when to breed. Breeding can take place safely every other day all through the heat. If conservation of the stallion is to be considered, the heat cycle of each mare should be noted before breeding. The length of the entire cycle should be determined (the time from the beginning of one heat to the beginning of the next).

Not all mares have "read the book" when it comes to showing signs of heat. Just like variations in the length of heat, mares will manifest heat in different ways. Some will be very characteristic, with all of the classic signs, while others will give little indication that they are in heat. Voss and Picket, who have done considerable research on breeding mares, claim that inadequate or improper teasing methods are one of the major causes of poor reproductive performance in mares. In their tests, mares were teased with two stallions each day. They checked nine criteria in determining whether a mare was in heat (Table 16–1). These figures represent the average of a 2-year study. As expected, not all mares showed positive in all categories when in heat. The signs were correlated with true heat, and some were found to be more constant than others. The negative coefficient of correlation indicates the extent to which a sign is absent. A coefficient of correlation of 1.0 would mean that all mares showed the signs when in heat.

Day found that almost 80 percent of the mares he palpated would ovulate within 2 days of the end of the heat period. Andrews and McKenzie reported that 77 percent of the mares they palpated ovulated within 24 hours of the end of heat. However, others, such as Trum, found that only 40 percent had ovulated within 24 hours of the end of heat. Of those mares, however, 77 percent did ovulate within 3 days of the end of heat. Some of the variation in these reports may be attributed to the use of mares with a wide variation in the lengths of their heat. Crowhurst, who has had much experience in breeding thoroughbreds, recommends that mares be bred on the second or third day of heat. He believes that a second breeding during heat is necessary only if the mare is still in heat 4 days after the first breeding. Caslick also supports this procedure when trying to conserve the stallion. In a group of 615 normal mares, he found that those bred on the second or third day of heat gave the greatest percentage of conception. All of these mares were bred only one day during the heat. Trum has gone so far as to prove that in 1543 cases he attended, breeding more than once during the heat period increased conception by only 3 percent when the mares were bred on the second or third day.

Table 16–1. The Signs of Heat in Mares. Some signs are more common than others as shown in descending order. A negative coefficient of correlation means a lack of that sign.

Sign	Coeff. of Corr.
Winking	.85
Squatting	.79
Urinating	.77
Tail Raising	.75
Kicking	−.73
Ears Back	−.68
Squealing	−.49
Fence Pushing	.37
Striking	−.34

From Picket and Voss

Consideration must also be given to the owner of the mare in that every effort must be made to settle the mare as soon as possible. However, unless the owner knows the length of heat for his particular mare, it might be well to pass the first heat without breeding. In so doing, one can learn the length of heat which is normal for that mare. Once this information is obtained, the mare should be bred 4 days before the scheduled end of heat. A mare with a heat period longer than 8 days who does not conceive from the first breeding should be examined by a veterinarian, who can determine ovulation by a rectal examination. Hormone therapy may be indicated to bring the cycle into normal limits.

Stocking has reported a high percentage of conception with a "one-service" breeding technique. From a group of 1063 mares, 878 conceived. This figure is close to most results with pasture breeding. Those mares who come in heat again after the third service should be given special consideration. (This topic will be covered later in the infertility section.)

A South African company, Animal Breeding Service (ABS), is marketing an electronic instrument designed to detect ovulation in domestic animals. The instrument has a probe which is inserted in the vagina to measure the electrolyte and volume changes of the vaginal mucus. Visual changes in the vaginal and cervical mucosa have been a favorite method (along with palpation of the ovary) of determining the time of ovulation, i.e., the time for breeding.

THE MALE REPRODUCTIVE SYSTEM

The testicles are located outside the body cavity, permitting spermatozoa to develop in a temperature somewhat lower than the internal temperature of the body. Sperm development occurs in tiny threadlike structures called seminiferous tubules. Within these tubules, lying close to the inner lining, is the reservoir of spermatogonia, the male germ cell line. These reservoir cells reproduce themselves by division into two identical cells, a process known as mitosis (Chapter 8).

Some of the spermatogonia move away from the basement membrane (Figure 16-9) becoming primary spermatocytes, beginning the process known as meiosis (Chapter 8). Basically what happens is that after two cell divisions the total chromosome number is reduced to half—one of each pair. This splitting is necessary so that the germ cells of each sex can unite without doubling the chromosome number. (The domestic horse normally has 64 chromosomes in each cell, or 32 pairs.) The separation of the chromosome pairs occurs when the primary spermatocyte divides. However, since the chromosome material has previously doubled, the halving of the chromosome number does not occur until the secondary spermatocyte divides to form a spermatid. The spermatids, over a period of several weeks, mature into spermatozoa.

The tissue in between the seminiferous tubules contains specialized cells

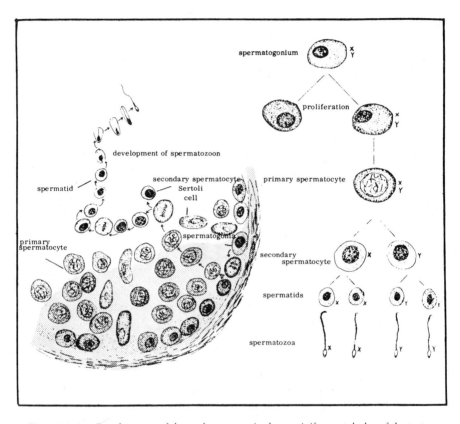

Figure 16–9.   Development of the male gametes in the seminiferous tubules of the testes.

which produce the male hormone, testosterone. Hormones are carried from their site of production to the rest of the body by means of the circulating blood. A deficiency of this hormone reduces sexual desire and, if severe enough, may inhibit the descent of the testicles. Testosterone regulates the growth and development of the external genitalia and the accessory sex glands. The production of this hormone is regulated by other hormones from the anterior pituitary gland, as discussed earlier.

Once the spermatozoa develop, they pass into the epididymis. This sac-like structure is located along the upper surface of the testicle.

Often one or both testicles fail to descend from the abdominal cavity into the scrotum as a colt matures. This condition is known as cryptorchidism. The testicles of the male fetus are located in the abdomen, in a position similar to the ovaries of the female. A ligament, the gubernaculum, is attached to the testicle at one end and the bottom of the scrotum at the other end. This ligament stretches down through the opening between the scrotum and the abdomen, known as the inguinal canal. Generally around the time of birth, shortly before or shortly thereafter, the gubernaculum contracts, pulling the testicle into the scrotum. This process is described further in Chapter 18. Occasionally the testicle may move back up again, for a short time, into the abdominal cavity. The descent of the testicles varies from individual to individual and from breed to breed.

The size of the penis, testicles, and other parts are determined by various genetic factors. The male reproductive system is diagrammed in Figure 16–10.

The epididymis acts mainly as a storage depot for the spermatozoa during the final stages of maturation. It takes about 3 weeks for completely mature spermatozoa to form. About one-third of this time is spent in the epididymis. During ejaculation spermatozoa are drawn from the epididymis and quickly passed to the exterior. The epididymis secretes a small amount of mucus which helps to move spermatozoa along the vas deferens. The bulk of the ejaculate is secreted by the accessory sex glands.

The ampulla is an enlargement of the vas deferens, which contains glandular cells. These cells add a small amount of secretion to the spermatozoa.

The vesicular glands (paired) secrete the albumin portion of the semen, which greatly increases the bulk of the ejaculate. There is a great difference in the amount of secretion from this gland in the various species as well as differences in the various breeds of the horse. The total semen volume varies from 50 to 300 ml; these differences are due mainly to the vesicular gland secretions. The lighter breeds of horses tend to produce a relatively small volume of semen.

The prostate gland produces a gelatinous secretion as the last portion of the ejaculate. Lieux suggests that it is noxious to living spermatozoa and that the lighter breeds secrete less of this product. The prostate is located at the junction of the two vas deferens and the urethra.

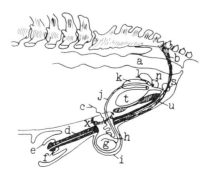

Figure 16–10.   The reproductive system of the stallion: (a) rectum, (b) retractor penis muscle, (c) spermatic vessels, (d) penis, (e) fold of the internal prepuce, (f) free part of the penis, (g) testicle, (h) epididymis, (i) scrotum, (j) vas deferens, (k) ampulla, (l) seminal vesicle, (n) prostate, (s) Cowper's gland, (t) bladder, (u) urethra, and (x) inguinal ring.

The bulbourethral glands (paired) produce a thin grayish fluid, which is the first portion of the semen to come out. It may act to lubricate the urethra for the rest of the semen to follow.

Upon erection, the penis may increase 50 percent or more in diameter and length. Erection is due to the filling of the corpus cavernosum and the spongiosum (cavernous blood sinuses in the penis) with blood. Control of the size of these sinuses is regulated by smooth muscle which surrounds them and the arterioles which deliver blood to them. Relaxation of these muscles causes filling of the sinuses and erection of the penis. The surface of the penis is richly supplied with sensory receptors which are extremely sensitive to touch. A stimulus from these receptors reflexes to the spinal cord and back to the muscle around the vessels of the penis, causing relaxation and subsequent filling with blood and erection. The same type of nerve stimulation can originate from the brain, where sight and smell may initiate an impulse. Thus penile erection can be initiated by either sensory receptors or the central nervous system.

Stallion puberty has been defined as the age at which an ejaculate reaches a concentration of 100 million spermatozoa and a 10 percent motility. Puberty will vary with environmental factors, but there are indications that inheritance also plays a role. One study found quarter horse puberty to be about 17 months. The breeding activities of a 2-year-old should be restricted, and probably eliminated entirely with a slow maturing breed such as the Arabian.

INFERTILITY

Sometimes it is fall before a breeder is willing to admit that he is not going to get his favorite mare settled this year. Although she has been bred many times throughout the season, after each breeding she has come back in heat. Since the stallion settled a dozen other mares this year, he has been ruled out as the source of the problem. Finally the breeder admits that it is time for a realistic evaluation of the breeding potential of his mare.

In discussing the matter with his veterinarian, the breeder discovers a number of possible reasons why his mare has not conceived this year. It could be a nutritional deficiency, a uterine infection, a hormone imbalance, a pathological change in the reproductive tract, or even a chromosomal problem.

It is wise for the breeder to analyze his breeding procedures before asking the veterinarian to conduct a clinical examination of his mare. He should ask himself, "Has the mare been regular in her heat cycle? Has she been bred at the proper time? Has teasing been adequate?"

The veterinarian will first discuss the breeding history of the mare in question. Previous foaling problems, estrous irregularities, or other abnormal reproductive irregularities offer clues to the nature of the current breeding problem. The veterinarian will want to know whether the mare had apparently "settled" earlier in the year and then returned "in season." A so-called silent heat can give the impression that a mare has settled when in reality she is cycling normally, but showing no outward signs. This type of mare will often begin to show observable signs during heat in late summer or early fall. The veterinarian will ask the owner about the mare's feed if she is thin or dull-coated. Poor nutrition can interfere with conception.

The veterinarian conducts both an external and internal physical examination. A general examination of all body systems will determine whether the mare can carry a fetus to term. Condition, attitude, conformation, and temperament are important considerations that may affect the breeding potential of a mare. The veterinarian looks at the shape, size, angle, and degree of closure of the vulva. He looks for scars or lesions on the vulva that might impair closure of the lips. He also looks for evidence of an external secretion from the vulva.

A mare vaginal speculum will be used for the internal examination. The veterinarian notes the color of the vaginal lining, the secretions present, any urine pooling, and the size and position of the vagina. At this stage of the examination the veterinarian may discover adhesions or abnormal openings or tears. He examines the cervix: the color, secretions, and swellings, or scars and adhesions.

An important part of the veterinarian's examination is rectal palpation. By inserting a gloved arm in the rectum, he can palpate the entire reproductive system through the rectal wall. The cervix, uterus, and ovaries are all carefully palpated to determine size, consistency, tone, and shape. First, he may discover that the mare is pregnant after all. He may find the uterus containing pus, cysts, or tumors. An experienced veterinarian can determine the shape and consistency of the ovaries, and can tell whether they contain a follicle or a corpus luteum which develops after ovulation. Also, ovarian tumors can be palpated.

Generally the veterinarian will take a uterine swab for culture to determine the type and sensitivity of infective organisms. He may also place a swab on a microscope slide for direct observation of bacteria, lymphocytes, and the nature of the epithelial cells. In difficult cases the veterinarian may take a uterine biopsy to determine the health of the uterine lining. Often an unhealthy uterine lining is the cause of sterility. Other checks that may be used in difficult problems are chromosome analysis and endocrine assays.

Before a final diagnosis is made, the veterinarian will want to gather an accurate record of details of the estrous cycle. This information reflects much about hormonal irregularities, which are the most common cause of sterility problems. Once he is able to pinpoint the exact cause of sterility, he can begin to correct the problem.

The breeding potential of a mare may be greater than a breeder believes. On the other hand, she may be hopeless. A thorough examination by a veterinarian will usually pinpoint the problem, and provide a means of evaluating her potential.

Many new techniques and diagnostic procedures available to the veterinarian make it possible to determine why a mare does not conceive and whether she even can be a good "economical" broodmare.

An important part of the evaluation will be determining the genetic soundness of the mare. Sterility problems do not always have simple solutions. Quite often there is more involved than simply diagnosing and treating a uterine infection. Often genetic factors are responsible for sterility, and if so, it may not be advisable to breed the mare, even if it is possible to get her to foal. Genetic poor breeders run in families. The daughters of these mares will often be genetic poor breeders. Genetics can affect the hormone system, causing imbalances that interfere with conception, or lethal genes may cause early death of the embryo. Chromosome abnormalities, which are now known to be more common than has previously been thought, cause a genetic sterility.

The practice of breeding mares with hormone problems and correcting their imbalances with synthetic hormones is leading to the production of more mares with hormone problems because we select for athletic ability rather than breeding soundness.

Another consideration related to the mare's genetics is whether she consistently produces weak or deficient foals that fail to develop and mature as successful athletes. Often a breeder is deluded into thinking that a mare is producing well for him, even though she produces only poor foals that die or have no potential. The breeder blames luck, nutrition, microorganisms and other factors, without considering that the foals have inherited a genetic weakness. Combined immunodeficiency is a genetic problem in the Arabian breed which results in the eventual death of foals that, at first, seem quite normal and healthy. This genetic problem went undetected for years because foal pneumonia and other diseases were always blamed for the death of the foals. Now attention is being focused on the production level of antibodies in colostrum. Some mares produce colostrum with an abnormally low amount of antibodies (a problem not confined to the Arabian breed), which increases the incidence of foal diseases. This problem may also be genetic.

Twinning is another genetic situation that leads to loss of foals and weak and undesirable offspring. The production of twins is known to be under genetic influence. Mares that produce twins are likely to do so again, and their daughters will be more prone to twin. Twin conception is the leading cause of abortion in horses. Twinning is on the increase in the thoroughbred breed because of the illogical practice of breeding back twin fillies because they are too small to make racehorses.

Conformation is the most important criterion determining whether a horse will be a good athlete. The breeder must determine this potential in selecting a broodmare. The veterinarian also looks at conformation as he conducts a thorough breeding examination. The mare is best suited for reproduction if she is large, with a good wide pelvis and a healthy mammary gland.

One of the most troublesome conformation defects in broodmares is the tilted vulva which leads to "windsucking." The labia of the vulva should be vertical on a normal mare; it should be on the same vertical plane as the anus. A deformed vulva and vagina have an immediate bearing on uterine health. Most mares cannot conceive until the problem is surgically corrected. Some mares with an abnormally shaped pelvis show this condition as early as 2 or 3 years of age, although it usually develops later in life after several pregnancies.

Poor muscle tone contributes to the "windsucking" problem. Often the membranous curtain which closes off the vagina to the exterior becomes so loose that air is aspirated through it when the mares moves. The vagina, and even the uterus, may become ballooned with air. This condition is known as pneumovagina, and provokes some type of chemical reaction which changes the chemistry of the mucosa, making it unhealthy for semen. The vagina and uterus become contaminated and infected. In extreme cases urine runs forward and pools in the vagina and sometimes runs into the uterus.

A Caslick operation is performed on mares with this abnormal conformation. The vulva is sewn shut from the top down, leaving a small opening at the bottom for urine. This stitching must be cut loose before breeding and foaling.

The condition of the cervix is an important indicator of fertility. The veterinarian will examine it carefully for scars and other abnormalities. Dr. Lieux says:

> The cervix is not only a vestibule through which the sperm gets to the uterus; it plays a very important role in the phenomenon of reproduction. This role is definitely related to its integrity. It has been our experience that whenever the cervix is not perfectly normal, but has scars, adhesions, cysts or other abnormalities, the broodmare is often unable to conceive.

Two other diagnostic procedures that have been developed within the last few years are the uterine biopsy and uterine endoscopy. When the veterinarian cannot find the source of the reproductive problem through the traditional examination, he may resort to one or both of these techniques.

The uterine biopsy is relatively simple to take. A long-handled instrument is inserted through the cervix into the uterus where a small chunk of uterine mucosa is pinched off, removed, and sent to a laboratory for analysis. A uterine infection is more consistently diagnosed with the biopsy than with a culture. Also, the ability of the uterus to support pregnancy can be assessed with a biopsy.

Dr. Lieux believes that the use of an endoscope to examine the uterus directly can sometimes reveal problems that would otherwise have been missed. With this technique he sometimes picks up conditions such as exudative endometritis, scars, cysts, and uterine adhesions.

Veterinary science has become efficient in the field of reproduction, but some problems still have no solution and some mares continue to be hopelessly sterile. The veterinarian can usually give a fairly accurate prognosis of the reproductive potential of a given mare after a careful and thorough series of examinations. The successful breeder will learn to depend on his veterinarian's reports.

BIBLIOGRAPHY

Andrews, F.N., and McKenzie, F.F.: Estrus, ovulation and related phenomena in the mare. Res. Bull. Mo. Agric. Exp. Sta. 329:117, 1941.

Baker, C.B., and Kenney, R.M.: Systematic approach to the diagnosis of the infertile or subfertile mare. In Current Therapy in Theriogenology. Philadelphia, W.B. Saunders, 1980.

Caslick, E.A.: The sexual cycle and its relation to ovulation with breeding records of the thoroughbred mare. Cornell Vet. 27(2):87, 1937.

Caslick, E.A.: The vulva and vulvovaginal orifice and its relation to genital health of the thoroughbred mare. Cornell Vet. 27:178, 1937.

Cooper, W.L.: Wintertime Breeding of Mares. Symp. on Reprod. in Cattle and Horses. Am. Vet. Soc. for Study of Breeding Soundness at Mich. State Univ, 1972.

Crowhurst, R.C., and Caslick, W.: Some observations on equine practice and its relation to the breeding of thoroughbred mares. North Am. Vet. 27:761–767, 1946.

Day, F.T.: Ovulation and the descent of the ovum in the fallopian tube of the mare after treatment with gonadotropic hormones. J. Agric. Sci. 29:459–469, 1939.

Day, F.T.: Survival of spermatozoa in the genital tract of the mare. J. Agric. Sci. 32:108–111, 1942.

Evans, M.J., and Irvine, C.H.G.: Serum concentrations of FSH, LH and progesterone during the estrus cycle and early pregnancy in the mare. J. Reprod. Fertil. (Suppl.) 23:193, 1975.

Ginther, O.J.: Occurrence of anestrus, estrus, diestrus and ovulation over a 12-month period in mares. Am. J. Vet. Res. 35:1173, 1974.

Greenhoff, G.R., and Kenney, R.M.: Evaluation of reproductive status of non-pregnant mares. J. Am. Vet. Med. Assoc. 167:449, 1975.

Hughes, J.P.: Clinical examination and abnormalities in the mare. In Current Therapy in Theriogenology. Philadelphia, W.B. Saunders, 1980.

Hughes, J.P., and Loy, R.G.: The relationship of infection to infertility in the mare and stallion. Equine Vet. J. 7:155, 1975.

Hughes, J.P., Stabenfeldt, G.H., and Evans, J.W.: The estrus cycle of the mare. J. Reprod. Fertil. (Suppl.) 23:161, 1975.

Hutton, C.A., and Meacham, T.N.: Reproductive efficiency on fourteen horse farms. J. Anim. Sci. 27:434–438, 1968.

Kenney, R.M.: Clinical Fertility Evaluation of the Stallion. AAEP Proc., 1975, p. 336.

Kenney, R.M.: Cyclic and pathological changes of the mare endometrium as detected by biopsy with a note on early embryonic death. J. Am. Vet. Med. Assoc. 172:241, 1978.

Kenney, R.M.: The Role of Endometrial Biopsy in Fertility Evaluation. AAEP Proc., 1978, p. 177.

Kenney, R.M., Ganjam, V.K., and Bergman, R.V.: Non-infectious breeding problems in mares. Vet. Scope 19:16, 1975.

Lieux, Pierre: Infertility in the mare. In Equine Medicine and Surgery. Santa Barbara, CA, American Veterinary Publications, 163.

Lose, M. Phyllis: Blessed are the Broodmares. New York, Macmillan Publishing Co., 1978.

Loy, R.G.: Effects of Artificial Lighting Regimes on Reproductive Patterns in Mares. AAEP Proc., 1968, p. 159.

Miller, W.C.: Practical Essentials in the Care and Management of Horses on Thoroughbred Studs. London, The Thoroughbred Breeders Assoc., 1965.

Osborne, V.E.: An analysis of the pattern of ovulation as it occurs in Australia. Aust. Vet. J. 42:149–154, 1966.

Oxender, W.D., Noden, P.A., and Hafs, H.D.: Estrus, ovulation and serum progesterone, estradiol and LH concentration in mares after an increased photoperiod during winter. Am. J. Vet. Res. 38:203, 1977.

Pickett, B.W., and Voss, J.L.: Abnormalities of mating behavior in domestic stallions. J. Reprod. Fertil. (Suppl.) 23:129, 1975.

Rauchback, K.: Hermaphroditism in the horse. Mh. Vet. Med. 16:220–221, 1961.

Roberts, S.J.: Abortion and other diseases of gestation in the mare. In Current Therapy in Theriogenology. Philadelphia, W.B. Saunders, 1980.

Roberts, S.J. (Ed.): Veterinary Obstretics and Genital Diseases, 2nd ed. Ann Arbor, MI, Edwards Brothers, 1971.

Rossdale, Peter D.: The Horse: from Conception to Maturity. Arcadia, CA, The California Thoroughbred Breeders Association, 1972.

Stocking, G.: Observations concerning conception in the mare. J. Am. Vet. Med. Assoc. 119:190–192, 1951.

Trum, BM The oestrous cycle of the mare. Cornell Vet. 40:17–23, 1950.

17

stallion management

Concern over the well-being of a stallion generally increases with the value of the animal. Although stallions are often syndicated for a million dollars or more and one stud fee can bring more than a year's wages for most men, it is difficult to locate good detailed information about stallion management (Figure 17–1).

A stallion requires a different kind of care from that given to mares and geldings. A handler who does not understand these differences may easily sour a stallion and thus reduce the animal's productivity. A temperamental stallion may be less able to settle a problem mare, or a soured stallion may hurt himself in some ridiculous manner simply because he is out of hand.

HANDLING THE STALLION

The devoted horse breeder with one or two stallions has spent much time and effort in procuring his animals. To him they are priceless. He naturally has a great amount of admiration and respect for them. These feelings will come through to the stallion and will tend to become mutual qualities. On the other hand, a high-priced syndicate often hires a temperamental handler who develops a fear or hatred of the stallion and who is intimidated by the great value of the animal. The stallion will also detect the attitude of this handler and will react in a negative way.

The relationship between stallion and handler is particularly important when the stallion is closely confined with little contact with other creatures. In this situation a harmonious relationship can develop between a stallion and his handler, and can be a basic part of the animal's happiness. The most

Figure 17–1.   Seattle Slew, a stallion syndicated for $12,000,000.

devoted handlers carry the relationship one step further, making the horse a true, and often affectionate friend. Then the stallion is content, enjoys human attentions, and will look forward to exercise and feeding time. In this situation, the "entire" (a term often used for the uncastrated male equine) has other things to look forward to than breeding. As a result he is easier to handle around mares.

For a handler to be affectionate and understanding it is not necessary, or even desirable, for him to be "kissey nice." A stallion must be handled with a firm hand (Figure 17–2). There should always be a quick reprimand when the horse is out of line. By the same token, the horse should be rewarded with a kind word and a pat whenever he deserves one. Gushy sentimentality has no place in handling a stallion. A stallion cannot become a pet in the context of a busy breeding farm.

By nature, a stallion is unpredictable. A handler must always be alert and attentive. He can be friendly and understanding without taking the animal too lightly. The stallion should always be handled with due respect for his quickness and power. A stallion can injure a person unintentionally. His full attention may be on a mare, causing him to plunge suddenly, whirl, or strike. If the handler is not wide awake, he may be in trouble rather quickly. A good handler is one step ahead of the stallion, in anticipation. He knows how stallions act and has had a great deal of experience in handling horses. Stallion management is no place for beginners.

Figure 17–2. A stallion must be handled with a firm hand. Sterling White has developed mutual respect with Gai Parada.

The best handlers do not believe in bullying a proud and noble stallion into submission. A stallion can be prancing excitedly at the end of a halter shank without being out of control. Heavy bitting and jerking, with much shouting and whipping, only shows how inadequate the handler is. This behavior shows a complete lack of trust between man and horse, and usually results in a soured animal.

The handler may have to be quite harsh with an ill-tempered stallion, but the key to success is skill and understanding. Even the good-natured stallion will show great spirit now and then, and may try to get his own way. This quality is an essential part of a stallion's personality; such behavior leads to the role of leader of the herd. It would be against the principles of good breeding to interfere excessively with this trait. Nevertheless, a handler must keep some sort of control. Potentially dangerous habits must be stopped at once, but it should be done on the basis of respect and esteem for the stallion.

HOUSING AND CARE

Keeping a stallion happy, healthy, and fit is the basic principle of successful management. A happy horse will make the most of his environment and will be easier to work with (Figure 17–3). Along the way his ability to settle mares will probably be enhanced. At least a healthy, happy horse

Figure 17–3.   Ken Crooks, of Van Vleet Arabians, admires and respects the stallions he manages, and as a result they are easy to work with.

will tend to have a longer life, thus increasing returns on the owner's investment.

Horses are as gregarious as humans. They cannot be happy in solitary confinement. The ideal situation for a stallion would be to run with a band of mares all year around. However, when dealing with expensive animals, this situation is not practical. There is too much danger to the stallion, to the mares, and to foals who might be with the mares. So the stallion is generally kept in his own paddock. To make him most at ease, he should be able to see other horses from his paddock (Figure 17–4). The nature of some stallions is such that it is necessary to keep them in a large box stall most of the time with only controlled exercise. Sometimes a goat, dog, or other animal can be a substitute companion for horses who crave constant company.

That the stallion needs a proper, balanced diet should go without saying. The danger is in letting the stallion get too fat during the inactive days of winter, and get too thin during the breeding season. The amount of feed necessary will vary tremendously between these periods and must be adjusted according to the need and condition of the horse. Vitamin

Figure 17–4.   A stallion should be able to see other horses from his paddock.

supplementation is a good safeguard. An adequate vitamin intake, especially vitamin E, will help keep a stallion's libido up.

A farm with expensive modern facilities will have provisions for a continual supply of fresh water. A struggling enterprise may be dependent upon a hand supply of water or filling a bucket or barrel with a hose. Under the latter circumstances it is easy to let down once in awhile and forget to keep a clean supply of water available at all times. The winter in the northern latitudes will bring freezing weather which adds to the problem of a continual supply of water. A good stallion manager will not let down his guard in this area.

An adequate amount of exercise is vital to the well-being of every stallion. It is unhealthy for any horse to spend his time in a box stall without an opportunity for daily exercise. A good workout on a hot-walker each day would be adequate. Horses in large paddocks will exercise enough on their own most of the year.

The ideal form of exercise is longeing by hand or by riding. This exercise gives the stallion contact with his handler and can greatly help in keeping him sane. Riding is especially good because it demands discipline along with exercise. A valuable stallion should be ridden only in a protected arena. Then if the rider becomes unseated, a valuable animal will not be galloping alone across the countryside.

Two months before the breeding season begins, the stallion should begin an especially rigorous schedule of exercise to put him in shape for the breeding season. Stallion managers throughout the world have a variety of favorite exercise schedules, most of which are quite effective. One of the most extravagant is walking the stallion on a lead shank for several miles each day. Another is galloping him around the training track.

The most common exercise method in the United States today is the hot-walker (Figure 17–5). The best procedure is to allow a stallion a short time in a paddock before placing him on the walker so that he can get the "steam" out of him, and settle down. At the onset of his pre-breeding-season exercise schedule, the horse should be walked a half hour a day. Some

Figure 17–5.   The hot-walker is a popular method of exercising horses.

managers advocate gradually increasing this time to as much as 2 hours a day just before the breeding season starts.

During the breeding season, exercise should continue, but at a somewhat reduced schedule. The stallion will be much less apt to hurt himself when in good physical condition while breeding. And if nutrition is adequate, some breeders feel that his good condition will result in greater fertility.

Grooming is an important aspect of year-round stallion management. Daily grooming serves several purposes. First, it will help to make the stallion feel better about himself and his handler. Grooming massages the skin and underlying muscles. It promotes better blood circulation and hair growth. Time spent at grooming gives the handler an opportunity to examine all areas of the stallion carefully for swellings and cuts and to observe the general condition of the animal. Grooming time provides more contact between stallion and handler, thus helping to build the rapport that comes with successful management.

Cross-tying is the most common method of restraint for grooming as it prevents the stallion from turning around or nipping. Grooming should be systematic. The handler should start at the same place with each tool and work over the body in the same order. This practice is not only more efficient, but gives the horse a sense of security that leads to greater cooperation. As an example, if feet are always cleaned in the same order, the stallion may eventually have the next foot picked up and ready at the proper time. Brushing and wiping should be done vigorously and swiftly, not halfheartedly.

The currycomb is used first to loosen dirt, scurf, and free hairs (a soft bendable rubber currycomb is most comfortable for the horse). The currycomb is used mostly on the body and neck, and only very gently on the legs and belly. The brush sweeps the coat free of dirt and loose hair. Hair should be brushed in the direction of growth.

After brushing, a large damp body sponge should be used over the face, ears, and lower legs. It is used also to clean out the lips, nostrils, and corners of the eyes. A separate sponge should be used to clean the sheath.

While cleaning the feet, the handler can check for signs of thrush, which is a smelly rotting of the frog. The hoof pick should be used from heel to toe. Caked mud and manure can be pried out in a large mass. The commissures and clefts of the frog should be carefully cleaned.

The final step in grooming is the use of a rub rag. It is rubbed from muzzle to tail to remove the last of the dust and to put the hair straight.

Cleaning of the sheath should not be neglected. A strong smelling black secretion called smegma combines with dirt and becomes uncomfortable to the stallion. A large accumulation of smegma and dirt may form in the blind pouch of the sheath, sometimes causing infection. A good cleansing of the sheath is necessary only about once a month in the off season. A thorough washing of penis and sheath before each breeding is often recommended.

Veterinary care should not be neglected. Routine vaccinations are important and should consist of protection against tetanus, encephalomyelitis, flu, and in some areas rhinopneumonitis and viral arteritis. The farrier should be called in on a regular basis—at least every 6 weeks. The teeth should be checked by a veterinarian at worming time. Most veterinarians recommend worming every other month.

TEASING

Teasing, or trying as some prefer to call it, is an important prerequisite to breeding. Stallions in the wild have a refined and effective teasing technique. They "ask" the mare whether she is ready and can read subtle answers as quickly as a wink.

Teasing, therefore, is based on natural tendencies, but it is in some ways contrived by man for his artificial breeding procedures. Teasing helps man to determine when a mare is ready to be bred, thus allowing fewer covers to obtain conception (Figure 17–6).

In the wild a stallion will approach a new mare from the front, with neck arched and nostrils distended. He will usually call loudly to her as he approaches head on. The stallion will, from experience, stand slightly off to one side of the mare's chest, just outside the range of her striking forefoot. Both animals will squeal and touch noses. The mare who shows encouraging subtle signs to the stallion will be investigated further. She will be standing rather rigidly as the stallion smells his way over her neck, shoulder, and back. The mare may swing her rear toward the stallion or just stand

Figure 17–6.   Much of the work is taken out of teasing on this breeding farm in Australia.

Figure 17–7.   This teasing barrier would be safer with padded corners and edges.

rigidly when in heat. All of the signs of heat seen in domestication serve a purpose in the wild.

The signs of heat have been explained in Chapter 16, along with information about the frequency and time of breeding. When dealing with expensive animals, a mare should be checked by a stallion across a barrier of some sort. This special-made teasing barrier can be constructed of solid wood with padded edges (Figure 17–7), a gate, or even a fence. The more

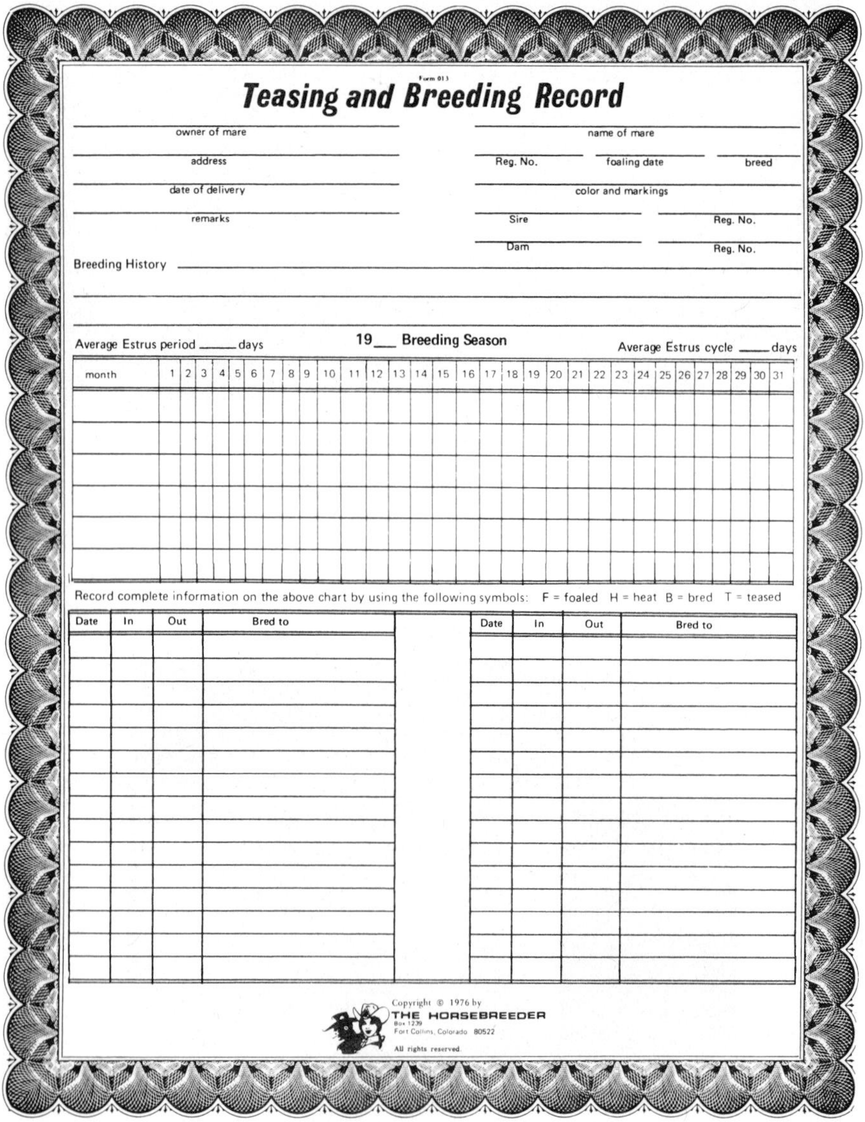

Form 013

Teasing and Breeding Record

owner of mare — name of mare

address — Reg. No. — foaling date — breed

date of delivery — color and markings

remarks — Sire — Reg. No.

Dam — Reg. No.

Breeding History

Average Estrus period ____ days — 19__ Breeding Season — Average Estrus cycle ____ days

month	1	2	3	4	5	6	7	8	9	10	11	12	13	14	15	16	17	18	19	20	21	22	23	24	25	26	27	28	29	30	31

Record complete information on the above chart by using the following symbols: F = foaled H = heat B = bred T = teased

Date	In	Out	Bred to

Date	In	Out	Bred to

Figure 17–8. A typical Teasing and Breeding Record.

expensive the stallion and mare, the safer the apparatus should be. A solid panel is most secure. This panel should be high enough to keep the stallion from placing his feet over it, and low enough for the stallion to reach the mare's flanks with his nose.

Generally, the first step in teasing is the touching of muzzles, with much snorting and sniffing. The handler should then allow the stallion to sniff along the mare's neck and withers, working his way toward the rear.

A mare not in season will not stand still for the stallion's advances, and will usually try to kick. She will give a loud angry squeal and strike out with a front foot. The experienced stallion who has been properly handled will lose all interest at this point.

A farm which has a large number of mares will require a special teasing stallion. The stallions who are used heavily for breeding will not be able to withstand the tiresome demands of teasing a large number of mares. The teaser should be aggressive and noisy. Often a pony stallion is the best because he is easier to handle. When the mares are not so expensive and are kept together in a large herd or pasture, the teaser can be placed in a pen next to them, or they can be moved each day into a small paddock next to the teaser. Those mares who are showing any signs of heat will come up next to the fence near the teaser. This method saves much time and manpower, and only occasionally will allow a mare in heat to be overlooked.

The purpose of teasing is to get a daily indication of the progression of each mare's estrous cycle (see Figure 16–5). Most good breeding farm managers keep a daily record on each mare (Figure 17–8).

BREEDING METHODS

Pasture breeding is the more natural method. It works well on a large ranch where the stallion is not to be used on outside mares. When spring arrives, the usual procedure is to place 10 to 20 mares in a large pasture with a stallion. The mares foal before being turned in with the stallion. Some may think that the practice is foolhardy to the foal during pasture breeding, but the foal's natural instincts keep him on the opposite side of his mother from the stallion and out of danger. Since pasture breeding is a time-honored method, registries accept it as legitimate. The breeding record simply indicates the total time that the mare was pastured with the stallion, rather than specific breeding dates.

Modern horse breeding is designed more to suit the needs of man, and economical considerations become paramount. Pasture breeding tradition-ally produces a higher conception rate. It does, however, have severe disadvantages for a modern system of operations. First, a group of expensive mares, most of which are strangers, will often injure each other during the course of a season. A mare owner who is paying a high breeding fee expects his expensive mare to come home in good shape. Thus, she must be kept in a small paddock or stall by herself and must be hand bred. Another factor is that with hand breeding a stallion can settle a much larger number of mares during a season. Also, an expensive stallion is less apt to get hurt when kept in a stall or paddock.

For the protection of the stallion, most farms use some sort of hobbles for the mare or a breeding chute. Sometimes "kicking boots" are put on the mare. These devices are put in place before the stallion is brought on the scene (Figure 17–9).

An unspoiled stallion will want to take his time at breeding. He should be allowed to tease the mare thoroughly for himself before covering her. When the handler and stallion are both convinced that the mare is ready, the stallion should be led up behind and to the near side of the mare, several feet away. At this point the stallion should be fully drawn; and if washing of the penis is to be done, it can be accomplished quickly at this time. The mare's external genitalia should have been washed prior to this time. Also the tail of the mare is usually wrapped to keep the long hairs from interfering with breeding (Figure 17–10).

The stallion is allowed to rear and mount. Some handlers try to assist the stallion at inserting the penis, but the best procedure is to allow conditions as close to pasture breeding as possible while eliminating as much chance for injury to the mare and stallion as possible. With this idea, some breeders refrain from washing either the mare or stallion routinely. Bacterial contaminants have always been a part of coitus, and there is little evidence that more than average cleanliness is necessary.

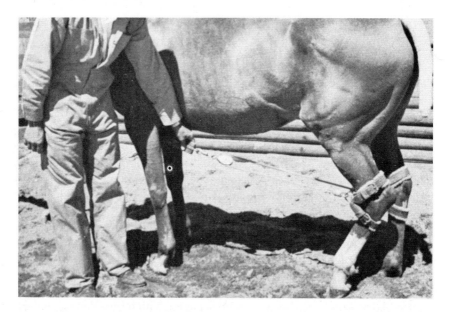

Figure 17–9.   Many mares need to be hobbled before breeding. (From O.R. Adams: *Lameness in Horses*, 3rd ed. Philadelphia, Lea & Febiger, 1974.)

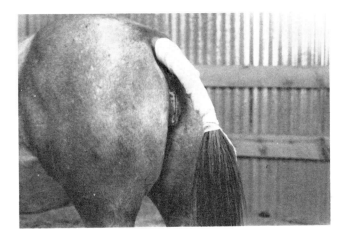

Figure 17–10.   The tail should be wrapped and the mare's genitalia should be washed before breeding.

After ejaculation, the stallion should be allowed to dismount on his own. If the mare is not in a breeding chute, she should be quickly turned after the stallion dismounts, so that she has no opportunity to kick at him. Also the stallion should be quickly led away from the mare, as he may be inclined to kick in excitement after covering.

A mare will generally have a tendency to squat and urinate after breeding. Some breeders feel that much of the ejaculate will be lost as the mare strains. As a consequence, the procedure has been to lead the mare quickly around the breeding area for a few minutes after the stallion dismounts. Scientific evidence has not shown that this practice is helpful, or that allowing the mare to urinate immediately reduces the conception rate.

Hand breeding does not always go smoothly. Various idiosyncrasies of mare or stallion can create problems. A common habit is for a spoiled stallion to lunge at the mare before she is ready. A bad stallion may stand on his hind feet a dozen yards from the mare and walk, squealing, toward her waving his front legs. Needless to say this action will scare even the toughest mare, causing her to lunge and kick at the stallion. If the stallion is not properly restrained at this time, he may injure the mare. Increased head restraint on such a stallion becomes necessary. Generally a stallion spoiled in this way is impossible to break of the lunging habit though he may improve. One way to teach him respect for mares is to let him experience pasture breeding. Under these conditions he will soon learn the reason for cautious teasing before mounting. The lunging habit develops because a young stallion has had no experience at natural breeding—all of his mares have been restrained.

Another bad habit is biting the mare. Some stallions make this habit a routine part of breeding. The mare owner objects, and so does the mare. An easy solution is to muzzle the stallion (Figure 17–11). Some breeders, instead, will place a driving harness collar or a rope necklace on the mare for the stallion to grip with his teeth.

Occasionally a stallion will become slow in interest, taking a long time to achieve an erection and may have difficulty maintaining one. In this case it may be helpful to warm the stallion up ahead of time with a good physical workout, such as cantering on the longe line for 5 or 10 minutes. Another method is to "tease" the stallion by bringing him to the mare, then taking him away, and repeating the process until an erection occurs.

Mares, too, may have bad breeding habits. Some will step forward during the breeding, throwing the stallion off balance. The use of a breeding chute will correct most undesirable actions by the mare.

To be successful at hand breeding, a breeder must keep in mind that pasture breeding is the ultimate. Procedures in hand breeding that deviate from pasture breeding must carry counterbalancing procedures. When free, the mare keeps the stallion honest and the stallion keeps the mare honest

Figure 17–11.   A muzzle as shown can be used to keep a stallion from biting. (From O.R. Adams: *Lameness in Horses*, 3rd ed. Philadelphia, Lea & Febiger, 1974.)

Figure 17–12.   Pasture breeding is the ultimate; the mare keeps the stallion honest, and the stallion keeps the mare honest.

(Figure 17–12). Thus, some breeders use as few artificial methods as possible.

ARTIFICIAL INSEMINATION

The earliest story of artificial insemination (AI) in the mare is dated 1322 A.D. It involved an Arabian who had a very good mare. He wanted to breed her to a certain prominent stallion, but his owner was unwilling and carefully guarded the stallion. One night while his mare was in heat, the Arabian saturated a cloth with the vaginal secretions, sneaked into the camp where the stallion was kept, and rubbed the secretions under the nostrils of the stallion. Evidently, this smell sexually excited the stallion enough to cause him to ejaculate. The Arabian caught the ejaculate in a clout, hurried back to his mare, and placed the clout in the vagina of his mare. It is almost inconceivable that such a procedure could result in a pregnancy, but according to the story it did.

Whether or not the story is entirely true, the idea of AI was evidently prevalent at that early date. Modern methods of AI employ more refined equipment but basically the same procedure.

Drs. B.W. Pickett and J.L. Voss, of Colorado State University, have been pioneers for over a decade in developing techniques for gaining maximum reproductive capacity from stallions, because of their commitment to the philosophy of AI. Even though AI is not permitted by the Jockey Club, the thoroughbred industry can still profit from the growing body of knowledge about semen production and evaluation that has come as a by-product of AI research.

The Colorado artificial vagina (AV) is an example of a development by Dr. Pickett and his group of a technique (and supporting special equipment)

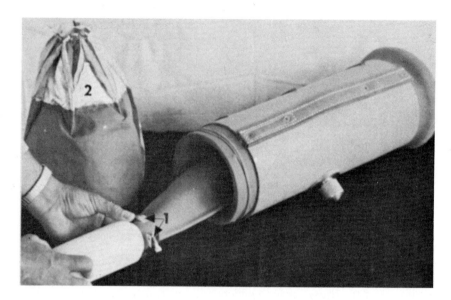

Figure 17–13.   The artificial vagina: (1) the hose clamp holding the collection bottle in place, and (2) the protective insulation jacket used to keep the semen warm. (From O.R. Adams: *Lameness in Horses*, 3rd ed. Philadelphia, Lea & Febiger, 1974.)

which makes the evaluation of a stallion more meaningful (Figure 17–13). After determining the necessary requirements for an AV, Dr. Pickett then set about developing the apparatus. He felt it must:

1. Elicit a favorable response from the stallion.
2. Maintain its temperature for at least 20 minutes regardless of the environmental temperature.
3. Be adjustable to accommodate stallions of various sizes.
4. Remove the gelatinous substance (gel) upon ejaculation.
5. Be constructed to maintain the semen in a viable condition after ejaculation.

Dr. Voss explains the Colorado AV at the short courses at Colorado State University, which are held several times each year:

> Precautions must be taken to protect the semen once it is collected. Although definite proof is lacking, it appears that the stallion's spermatozoa are more fragile than bull spermatozoa, which are highly susceptible to cold shock; therefore, precautions must be taken to prevent cold shock or sudden changes in temperature regardless of environmental temperatures.

The Colorado AV nylon duck jacket that fits over the collection bottle at the end of the AV provides adequate temperature control regardless of the external temperature.

Dr. Picket says, "An artificial vagina must be designed so that gelatinous material can be separated from the sperm-rich fraction during ejaculation. Approximately 75 percent of the spermatozoa are contained in the first three of six to nine spurts the stallion gives during ejaculation. Most of the gel is in the latter spurts. A milk filter is used in the AV to separate the gel from the spermatozoa."

A mare in heat is necessary for stallions that are not accustomed to regular collection with the AV. Dr. Voss explains the procedure:

> The first step is to twitch the mare. The plastic sleeve that is used to lubricate the artificial vagina is left in the AV to prevent loss of heat and the entry of dirt. The stallion is allowed to approach the mare in heat. When mounting occurs the stallion's penis is deflected and directed into the AV, after the sleeve is removed. The stallion handler has his hand behind the stallion's leg. This gives the stallion more support and reduces the possibility of injury to the collector.
>
> Once the stallion begins to thrust vigorously, the collector takes his right hand off the AV and places it on the shaft of the stallion's penis so that he can feel the urethra (Figure 17–14). After the first three spurts of semen are ejaculated the collector begins to lower the posterior end of the AV permitting the semen to run into the collection bottle. By the time stallion has ceased to ejaculate,

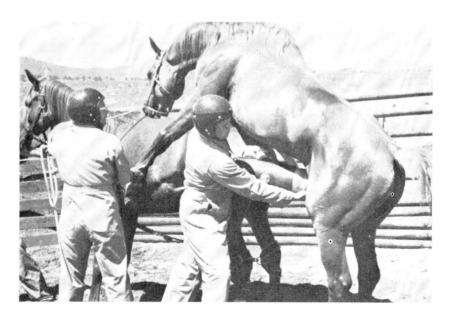

Figure 17–14.   Collecting semen from a stallion for a seminal evaluation and artificial insemination. (From O.R. Adams: *Lameness in Horses,* 3rd ed. Philadelphia, Lea & Febiger, 1974.)

the AV should be at about a 35 to 45 degree angle permitting separation of the sperm-rich fraction from the gel fraction. After ejaculation has ceased, the stallion's penis is stripped and examined for lesions, signs of infection, etc. At this time a culturette is used to take a sample from the urethra and another is used to take a sample from the prepuce to be cultured for the presence of pathogenic organisms.

The casing of the AV is heavy plastic varying in length from 21 to 35 inches and is 6½ inches in diameter. The anterior end is fitted with a heavy collar to prevent injury to the stallion. There is an outer leather carrying case and two rubber liners. Water at 135°F is placed between the two liners to keep the semen at body temperature. The goal is to maintain an internal temperature of 122°F.

SEMEN EVALUATION

Evaluation of the semen begins immediately after collection and upon entering the laboratory which nowadays has become a necessary part of a good stud facility. Since proper temperature is vital in the procedure, all the equipment that will contact the semen must be kept in an incubator at 100°F.

The semen is immediately poured into a graduated cylinder to be measured. The volume differs from stallion to stallion and between breeds, generally being about 50 to 100 milliliters (ml). More important than volume, however, is spermatozoal concentration.

Dr. Pickett and his group have determined that it takes from 100 to 500 million spermatozoa to give the maximum chance for conception. The average stallion gives much more than this in just 1 ml. "We recommend the Bausch and Lomb 'Spectronic 20' for determining concentration," says Dr. Pickett. "The sperm concentration per ml of gel-free semen is obtained from a chart prepared when the machine is calibrated."

Dr. Voss explains that not all spermatozoa are fertile. He has done some research aimed at showing which spermatozoa are most fertile. His most recent findings were reported at the 1980 meeting of the American Veterinary Medical Association, in Washington D.C. Traditionally, stallion spermatozoan morphology and motility have been the two most important criteria in determining fertility.

In summarizing his spermatozoa morphology study, Dr. Voss said, "Spermatozoa morphology may not be as valuable an indicator of fertility in the stallion as it is in the bull or ram" (Figure 17–15).

The established procedure for evaluating spermatozoan morphology is to smear a drop of semen on a microscope slide and stain the cells. The spermatozoan is made of a head, midpiece, and long thin tail. The tail normally rotates, propelling the head forward. Heads separated from tails

Figure 17–15.   Normal and abnormal stallion spermatozoa. A is normal; the others are abnormal with various types of structure. (From O.R. Adams: *Lameness in Horses*, 3rd ed. Philadelphia, Lea & Febiger, 1974.)

are undesirable. Some heads are abnormally shaped and it has been assumed that these are infertile.

After his recent studies, Dr. Voss says, "We are suspicious that only those spermatozoa that rotate with each lash of the tail and are progressing forward, without circling and jerking, are fertile."

Thus an ejaculate may have 50 percent motile spermatozoa, but with careful observation, it can be determined that only 20 percent are rotating and progressing normally. Under these circumstances, only 20 percent of the total number of spermatozoa should be considered in calculating the fertility level of the stallion. There are many types of stallion semen exam certificates or record forms (Figure 17–16). A permanent record of semen quality and quantity should be kept on every ejaculate.

Dr. Pickett believes strongly in the value of AI for horses. He says:

> Artificial insemination has been criticized or labeled as a means of reducing the accuracy of sire identification. Quite the contrary is

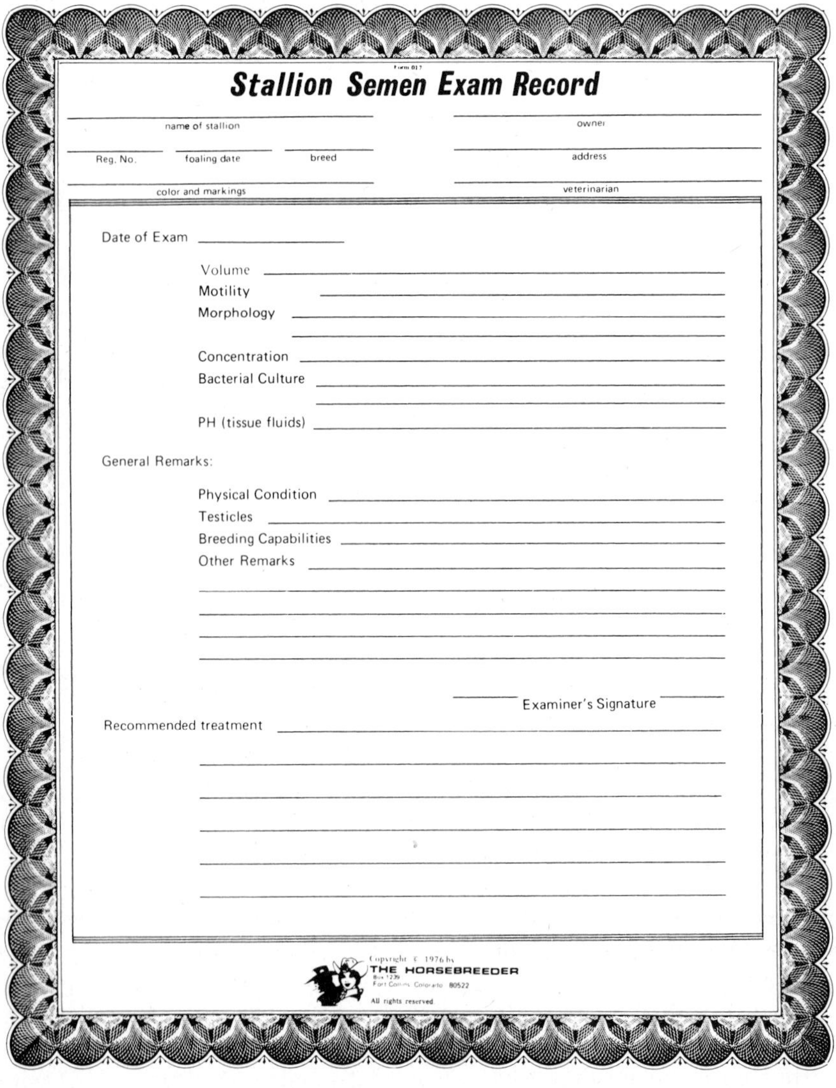

Form 017

Stallion Semen Exam Record

name of stallion ____ owner ____

Reg. No. ____ foaling date ____ breed ____ address ____

color and markings ____ veterinarian ____

Date of Exam ____

Volume ____

Motility ____

Morphology ____

Concentration ____

Bacterial Culture ____

PH (tissue fluids) ____

General Remarks:

Physical Condition ____

Testicles ____

Breeding Capabilities ____

Other Remarks ____

Examiner's Signature ____

Recommended treatment ____

Copyright © 1976 by
THE HORSEBREEDER
Box 1239
Fort Collins, Colorado 80522
All rights reserved

Figure 17–16. A typical Stallion Semen Examination Record.

true. Accurate records and keen observation are absolutely necessary in any successful AI program.

Artificial insemination allows utilization of the minimal number of spermatozoa for maximum reproductive efficiency. Small inseminate volumes, containing adequate numbers of spermatozoa for maximum fertility, are less likely to cause reaction or induce infection than are larger volumes. In a controlled experiment, a mean volume of only 1.5 ml of raw semen was used to inseminate 24 mares. A pregnancy rate of 75 percent was obtained during cycle one and 91.7 percent were pregnant after cycle two. The volume of raw semen necessary to obtain pregnancies in mares was less than expected. It is obvious that good pregnancy rates can be obtained with small volumes of semen, but care must be used during insemination of small volumes to insure the calculated amount of semen, containing the appropriate number of spermatozoa is placed in the body of the uterus. Appropriate semen handling techniques and sanitary procedures should always be followed.

Artificial insemination permits one to utilize a seminal extender. This is especially beneficial with stallions shedding pathogenic bacteria. The extended semen can be treated with the appropriate antibacterials, reducing the danger of contamination of the mare's uterus. The primary objective during the breeding season is to settle the maximum number of mares in the minimum amount of time. This would reduce labor and other expenses and provide earlier foals, thus shortening the foaling season, again reducing labor, etc.

Artificial insemination can aid in accomplishing these goals.

Dr. Pickett, believing wholeheartedly in artificial insemination, has said, "The registries should adopt a more enlightened attitude toward the use of AI since it has been clearly established that high conception rates can be obtained." He lists several advantages that AI holds over natural breeding.

When semen is diluted, as is the usual practice, the pathogenic organisms that might be there are reduced in number and can be destroyed with the use of antibiotics in the "extender." During the CEM outbreak a few years ago, AI became necessary to control the disease without a total cessation of breeding.

There is much less injury to valuable mares and stallions with the use of AI. Some stallions will occasionally object violently to a particular mare, and may injure her, the handler, or both. Gentle mares can be used for collection, or as is often done, the stallion can be trained to mount a "phantom" (a dummy made of wood and canvas) and ejaculate in the AV.

Stallions with chronic lamenesses or those who have been frightened and injured while breeding will be able to breed their normal share of mares by means of artificial insemination. Stallions have been trained to ejaculate into an AV without mounting a mare. Dr. Pickett relates, "Training to an AV was accomplished in at least one instance in which a stallion was unable to get an erection but showed excellent sex drive." Sometimes stallions that have been mismanaged during breeding will develop the habit of dismounting quickly while in the process of ejaculating, losing much of the semen. Collection of semen with an AV generally aids in correcting this habit.

Since a part of the artificial insemination procedure is to check the quality of semen with each ejaculate, minor changes in stallion fertility can be detected early and, it is to be hoped, corrected. Dr. Pickett says, "Many times AI is accused of being responsible for low fertility, when it only aided in pointing out the problem."

Studies have shown that one of the most common causes of infertility is overuse of the stallion. AI prevents this possibility. With use of the technique a full book of mares can be settled with collection of semen only every other day. "Some stallions can be used daily and occasionally twice or even three times per day without reducing spermatozoal output below the number needed for maximum fertility; while other stallions become infertile or even sterile after daily use," says Dr. Pickett.

High stud fees bring tremendous financial considerations of how soon and how often a visiting mare will be bred. Artificial insemination allows fair treatment of all mares by breeding them all every other day during estrus. When two or more mares need to be bred on a given day, AI is beneficial to prevent overuse of the stallion and still get the mares bred at the proper time.

"As stallions age, degenerative changes occur that reduce the total number of sperm produced, and in many cases those that are produced have a great number of morphological abnormalities which further reduce fertility," Dr. Pickett claims. Artificial insemination is a natural solution to this problem. In many cases AI may be the only means at hand to obtain more foals from an old, valuable stallion.

In the cattle industry, AI has been used to produce vast numbers of half-sibs, sometimes with accompanying genetic problems. This prospect is perhaps what scares many in the horse industry, causing them to take a position against AI. But artificial insemination is a much different procedure in the horse; it has been designed as an adjunct to natural breeding. The principal use of AI should be to equalize the productivity between stallions. If rules must be imposed against AI, let there be a limit set on numbers of offspring registered in any one year by a sire. Intelligent use of AI can only benefit horse breeding.

EXAMINING AND PREPARING THE STALLION

Many factors other than semen quality determine the breeding potential of a stallion. These other factors should be carefully evaluated prior to each breeding season to make sure that the stallion can withstand a vigorous breeding regimen.

First, the overall condition of the stallion should be noted. Poor flesh is a signal of improper nutrition or possibly some chronic disease. Generally the entire herd will show poor flesh if management and diet are a problem. With only one individual on the farm showing poor flesh, the problem is more likely an individual chronic condition which should be checked by a veterinarian. The veterinarian will want to know about any past illnesses, treatments, behavioral difficulties, past conception rates, and any problem with breeding performance.

Physical examination of the stallion should begin with a careful look at the back and the rear legs. Here most of the strain of breeding will be centered. Pain in the vertebral column or the rear legs can interfere with the stallion's desire to mount. Some conditions that may cause such a problem are bursitis, osteoarthritic conditions, neurological defects, spondylosis, and muscle atrophy in the hip and rear legs.

Psychological problems can sometimes interfere with the stallion's ability and desire to breed as much as a physical ailment. Ophthalmic problems, causing partial blindness, can affect a stallion's performance. Developing blindness can make a stallion apprehensive about approaching a mare and mounting.

The attitude of the stallion is all-important as he enters the breeding season. Those horses that are coming off the track need special consideration. They need to be wound down a bit at the track before being sent to the breeding farm. This cycling down process is generally achieved by decreasing the frequency and intensity of the gallops.

The usual procedure, once the stallion reaches the farm, is to remove his shoes and turn him into a large, safe paddock. The farm veterinarian will want to check him over when he arrives for some of the conditions described previously. Some farms routinely give a stallion off the track a tranquilizer for the first few days on the farm. Other techniques are also used to slow the stallion down the first few days. Some farm managers have the farrier trim the hoofs deliberately short when the shoes are pulled so that the horse is not apt to run and hurt himself. By the time the soreness is gone from the hoof, the horse is acclimated to his surroundings.

Some farms quarantine all new stallions for 21 days before allowing them to come into contact with other horses. George Atkins of *The Thoroughbred Record* describes the "breeding training" given some stallions just off the track:

> Many stallions are test-bred to mares (usually non-thoroughbred mares) before the actual start of the breeding season. This is done to give an inexperienced, shy, or fractious stallion a start on his new career. When the stallion actually covers a thoroughbred mare he will do so at less risk of injury, to himself or a valuable broodmare.

The role of learning in breeding has been covered at length by Dr. K.A. Houpt and D. Lein. They say:

There are many influences on sexual behavior. The hormonal influences and the external influences provided by the mare are important; no less important are learned factors. The hypothalamus is important in sexual behavior, but so too is the cerebral cortex. It is obvious from the discussion of sensory influences on sexual behavior that an experienced stallion needs very little external stimulation. A horse castrated as an adult after sexual experience may continue to show stallion-like behavior for years. Geldings may exhibit sexual behavior even if they were castrated as juveniles. Sexual behavior, it seems, is not totally dependent on hormones in male horses.

The learning of sexual behavior begins early in colts. Early experience is very important for normal sexual behavior and competence. Observations of play in foals has revealed that colts play much more than fillies and that much of colt play appears sexual in nature. There is a great deal of mounting interspersed with bouts of charging, nipping and other forms of mock-aggression. Full penile erection is often seen both during mounting attempts and mutual grooming bouts. This experience helps to perfect motor skills and familiarity with sexual postures. A colt raised alone with its dam will not have the benefit of this practice and experience; such colts often try to use their mother as a play partner. Owners often inquire about this behavior. It is not precocious and incestuous sexual behavior but rather an attempt at normal play. Colts deprived of play partners may, as adult stallions, be awkward and incompetent in their early breeding attempts. If possible, colts that will be used as sires should be pastured with other foals.

Houpt and Lein go on to say that too much early sexual experience is detrimental to normal libido when the foals mature. Also stallions that are used continually as teasers and are not allowed to breed may eventually show a loss of libido. There is evidence that stallions can learn to inhibit sexual behavior as easily as they learn to express it.

The physical surroundings where the stallion breeds the mare are important. A slippery floor may cause reluctance on the part of the stallion after a painful fall. A low roof where a stallion may strike his head can reduce his desire to mount next time.

The handler can be the greatest detriment of all if he uses inappropriate techniques. The handler should remain calm and communicate calmness to the stallion. The handler should know mare behavior and always hobble a mare if he is unsure of her tendency to kick. Many breeding farms require all mares to be hobbled during breeding.

Houpt and Lein explain the treatment of stallions with abnormal sexual behavior. "Patience and time are necessary with almost all cases. It is advisable to take advantage of the stallion's seasonal breeding pattern and institute behavioral therapy during the spring and summer. Advantage should also be taken of the stimulatory effect of the presence of other stallions."

They described how wild stallions tend to be more interested in copulation when other stallions are present. They claim, "Some stallions have mounted to breed mares only if another horse, even another mare, is present. The stallion may consider the second horse a competitor, or there may be another unknown reason."

Surprisingly, some stallions prefer one mare over another. If so, one can help a stallion recover from loss of libido by allowing him to mate with his favorite. Houpt and Lein describe how this is accomplished:

> The stallion can be teased with several mares to determine which he is most responsive to. A quiet mare may be necessary for a stallion that has been injured by another mare. A stallion that will not mount a mare may ejaculate into an artificial vagina. He may gain confidence and overcome his fear by this process and can later be induced to mount a mare.

Vicious behavior sometimes becomes a problem. It can manifest itself toward mares and/or handlers. Viciousness is sometimes caused from use of a stallion outside the normal breeding season. A vicious stallion may be less so in the midst of the breeding season. Overuse and rough handling can lead to viciousness.

Before the breeding season begins and periodically during the season, the stallion's penis and prepuce should be carefully examined for abnormalities. The procedure is explained by Dr. Dean Neely of Michigan State University:

> To examine the stallion's penis and prepuce, the animal is exposed to a mare long enough to stimulate let-down or the extension of the penis from the preputial sheath. Stallions may resent examination of their external genitalia and will kick straight out to the rear or sometimes "cow-kick" toward the examiner. Thus, the stallion should be examined in an open area where the examiner has freedom of movement and where the stallion will not kick solid objects, which cause leg injuries. The individual examining the stallion should protect himself by standing with his body beside the front quarters of the stallion and should gently slide his hand with a slow but steady movement down the chest wall and under the abdomen and then grasp the penile shaft.
>
> The penis and prepuce are examined by palpation and visual evaluation. The glans penis is examined with special attention to the urethral process. The urethral diverticulum is carefully evaluated for abnormalities and collections of smegma, often referred to as the bean. The examination is continued back over the shaft of the penis, observing closely for injuries and scars. While the penis remains

extended, both the internal and external folds of the prepuce will be visible and are evaluated for scars and vesicular or space-occupying lesions. Normally, a dense, brown-black, greasy smegma is present near the base of the penis but may be a deep violet color on lightly pigmented preputial skin.

Some of the most common penile and prepuce problems, in order of incidence of occurrence, are: lesions associated with trauma, tumors, penile paralysis, parasitic lesions, urinary calculi, and abnormal anatomy.

There is a rare equine genital disease known as equine coital exanthema, caused by a virus. Transmitted by coitus, the disease causes small lesions on the prepuce and shaft of the penis. These tiny water blisters soon develop into pustular papules. Within a week the papules have become encrusted. Small ulcers are evident when the scabs are removed from the papules.

The lesions cause only mild pain unless they become infected with bacteria, causing swelling of the prepuce and a thick, mucopurulent exudate over the penis. Without complications, the lesions heal in 7 to 9 days, leaving depigmented white scars on the dark-skinned areas. Sexual rest while the lesions are present helps prevent the spread of the disease.

A urethral bacterial culture is often a part of the examination for breeding soundness. However, since bacteria are natural inhabitants of the penis and urethra, it is difficult to interpret culture results. Dr. Neely says:

> Extreme care is needed in interpreting the results of such cultures, since the distal portion of the stallion's urethra usually harbors a wide range of microbial flora that do not cause an active disease of the stallion's genital system. An active infection of the stallion's genital system would be characterized by the presence of high numbers of inflammatory cells in the semen.

The scrotum should also be carefully examined while the testicles are palpated. Traumatic conditions of the scrotum can cause inflammation and increased testicular temperature, causing temporary sterility. A kick may cause a hematocele, a collection of blood within the testicular tissue. Dr. Neely describes it as:

> . . . a warm, fluctuant, somewhat painful swelling of the scrotum. As the hemorrhage organizes, it is palpable as a firm area; however, the majority of the scrotal pouch consists of a fluctuant, painless swelling because of seroma formation. Extensive adhesions often form within the tunics and can be detected by the failure of the testicles to slide easily within the scrotum. Unilateral castration of the affected testis is recommended if a decrease in sperm production occurs and lasts longer than 60 to 90 days, since any initial detrimental and temporary effect on spermatogenesis should have been reversed by this time.

Sometimes a generalized scrotal swelling or edema may be observed. This complication may be associated with some systemic diseases. Circulatory abnormalities may be a cause. If the edema does not disappear with increased exercise, a veterinarian should be consulted.

The testicles are palpated for size, shape, symmetry, and consistency. Testicular size is an important indication of the level of spermatozoa production. The normal testicle should be 8 to 12 centimeters long, 5 to 7 centimeters high, and 4.5 to 6 centimeters wide.

The epididymis can be palpated on the anterior-dorsal aspect of the testis. The body of the epididymis passes over the dorsal-lateral aspect of the testicle.

A testicular tumor, seminoma, is occasionally discovered by palpation. Usually only one testicle is involved. The affected testicle may be double in size and have a soft-to-fluctuant consistency.

Only after a thorough physical examination of a stallion in prime condition, will the veterinarian want to conduct a semen evaluation. Not only will this procedure involve several collections; it has meaning only when correlated with the physical examination.

BIBLIOGRAPHY

Dougall, Neil: Stallions: Their Management and Handling. London, J.A. Allen & Co., 1975.

Hughes, J.P., and Loy, R.G.: Artificial insemination in the equine: a comparison of natural breeding and artificial insemination of mares using semen from six stallions. Cornell Vet. 60:463, 1970.

Miller, W.C.: Practical Essentials in the Care and Management of Horses on Thoroughbred Studs. London, The Thoroughbred Breeders Assoc., 1965.

Neely, Dean P.: Physical examination and genital diseases of the stallion. In Current Therapy in Theriogenology. Philadelphia, W.B. Saunders, 1980.

Pickett, B.W.: Factors affecting stallion management. In Current Therapy in Theriogenology. Philadelphia, W.B. Saunders, 1980.

Pickett, B.W.: Use of artificial insemination in stallion management. In Current Therapy in Theriogenology. Philadelphia, W.B. Saunders, 1980.

Pickett, B.W., and Voss, J.L.: Abnormalities of mating behavior in domestic stallions. J. Reprod. Fertil. (Suppl.) 23:129, 1975.

Rossdale, P.D., and Ricketts, S.W.: The Practice of Equine Stud Medicine. Philadelphia, Lea & Febiger, 1980.

Terry, John M.: Artificial Insemination: Its Use as a Management Tool. AAEP Proc., 1978, p. 185.

Voss, J.L.: Stallion Management, Brood Farm Practices and Artificial Insemination. (Panel with Chris Cahill, B.W. Pickett, E.L. Squires and J.L. Voss) AAEP Proc., 1979, p. 31.

Wynmalen, Henry: Horse Breeding and Stud Management. London, J.A. Allen & Co., 1971.

18

pregnancy

The end product of horse breeding is pregnancy. Without it there is no need for concern about equine genetics. It begins with conception and ends with foaling, and the horse breeder must be attentive in the interim if all is to go well most of the time.

There is some variation in the length of gestation among breeds. The generally accepted gestation period is 340 to 342 days. A study of the Morgan horse showed a mean gestation period of 339.6 days. Records kept by Dr. Alan Purvis, a veterinarian studying induced foaling, mostly in Arabians, indicated an average gestation period of 338.4 days. Variations can be due to a number of factors such as sex of the foal, month of conception, size of foal, size of dam, and the plane of nutrition.

A foal crop of about 70 percent can be obtained with very little effort on the part of the breeder. The percentage can be increased only as the breeder becomes proficient in his business and knowledgeable about the biology of the horse.

FERTILIZATION

At the time of ovulation, the ovum resumes maturation which was begun earlier in embryonic development (Figure 16–2). It is thought that fertilization is an important aspect of the process. At least the ovum must be fertilized within 2 to 12 hours or the ovum degenerates and cannot develop into an embryo. Maturation is essentially a reduction of chromosome number to a haploid condition, as discussed in Chapter 8. After fertilization the ovum moves down the uterine tube slowly and does not reach the uterus for 5 to 6 days.

If the ovum is in the right horn, it may never enter the uterus, but may eventually implant in the lining of the horn. Research has shown that the many left ovary ova actually migrate to the right horn prior to day 70 and implant there.

Although it has been stated that it takes only one spermatozoon to fertilize the ovum, there is evidence that the chances of conception are greater in the presence of large numbers of spermatozoa. In one theory, an enzyme which is produced by the spermatozoa is needed in quantities obtained only from large numbers of spermatozoa. Therefore, one spermatozoon would not produce enough enzyme (hyaluronidase) to allow it to penetrate the layer of protective cells (corona radiata) around the ovum.

PREGNANCY DIAGNOSIS

Several techniques for pregnancy diagnosis involve either rectal palpation or hormone tests. The most rapid and the most accurate (when conducted by a competent and experienced individual) is rectal palpation.

The arm is covered with a rubber or plastic glove and inserted into the rectum to about elbow length. Since the uterus lies immediately below the rectum and the ovaries to each side (Figure 18–1), these structures can be readily palpated if the mare is sufficiently relaxed.

At 30 days the bulge in the uterine horn is no bigger than a walnut and is diagnosed more by the increased tone of the uterus than a bulge at the site of the embryo. Within the next 5 days the embryo enlarges 3 or 4 times. At about 40 days it is the size of an orange; at 2 months it is the size and shape

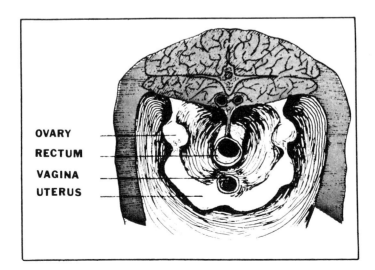

OVARY
RECTUM
VAGINA
UTERUS

Figure 18–1.   Cross section of the pelvic cavity in the mare, showing how rectal palpation can be used to identify other structures.

Table 18–1.   Data on Fetus and Uterus During Pregnancy

Days of Gestation	Size and Shape of the Chorionic Vesicle	Amount of Fetal Fluids in ml.	Weight of the Fetus	
16	1.8–2.2 cm. Pigeon's Egg			
20	2.6–3.2 cm. Bantam's Egg			
25	3.0–3.8 cm. Pullet's Egg	30–40		
30	4.2–4.5 cm. Small Hen's Egg	40–50	.02 g	Eye, mouth, and limb buds visible.
35	4.4–5.9 cm. Large Hen's Egg	60–90		
40	5.7–6.9 cm. Turkey Egg	100–150		Eyelids and pinnea have appeared.
45		150–200		
50	7.6 x 5 cm. Orange	200–350		

60	13 x 8.9 cm. Small Melon	300–500	10–20 g	Lips, nostrils, and beginning development of feet observed; eyelid partially closed.
90	23 x 14 cm. Small Football	120–3000	100–180 g	Villa of placenta present but without firm attachment, mammary nipples and hoofs visible.
120		3000–4000	700–1000 g	External genitalia formed but scrotum is empty.
150		5000–8000	1500–3000 g	May or may not have fine hair on orbital arch and tip of tail.
180		6000–10,000	3–5 kg	Hair on lips, orbital arch, nose, eyelashes, and fine hair on mane.
210		6000–10,000	7–10 kg	Hair on eyebrow, eyelids, edge of ear, tip of tail, and back.
240		6000–12,000	12–18 kg	Hair on distal portion of extremities.
270		8000–12,000	20–27 kg	Short fine hair over entire body.
300		10,000–20,000	25–40 kg	Prepuce developed; hair in mane and tail increased.
330		10,000–20,000	30–50 kg	Complete hair coat with final color.

of a football. More detailed data of the fetus and uterus are shown in Table 18–1.

Ponies are too small for successful rectal palpation; however, the hemagglutination-inhibition test can be used. Solomon has developed a simple, inexpensive method of testing for the gonadotropin hormone in pregnant mare serum. This hormone is present in the serum beginning just after the 40th day of pregnancy and lasting until the 120th day of pregnancy. The hemagglutination-inhibition test, extremely useful for routinely checking large numbers of mares, is quite accurate.

One commercial hemagglutination-inhibition pregnancy test kit detects equine placental gonadotropin in serum. This simple 2-hour test tube procedure has been shown to be 97 percent accurate when conducted 41 to 45 days after conception.

In the test, red blood cells coated with the hormone will agglutinate (clump together) and settle to form a mat of cells on the bottom of the reaction tube. When the gonadotropin in the serum of a mare is added to this system, inhibition of the combination of antibody with the gonadotropin-coated red cells takes place. The result, when using a round bottom test tube, is a "doughnut" pattern of cells equivalent to that seen in control tubes set up without anti-antibodies. This pattern constitutes a positive test. Serum from nonpregnant mares, which contains no gonadotropin, does not inhibit the hemagglutination reaction and thus shows a mat of cells with or without a faint ring. This pattern constitutes a negative test.

In the last few years the development of a technique that accurately measures hormone amounts in blood, milk, and other fluid has resulted in very early pregnancy diagnosis. After day 14, the nonpregnant mare will show less progesterone in her serum. On the other hand, a higher level of progesterone indicates pregnancy. These progesterone levels can also be measured in the milk of lactating mares.

EMBRYOLOGY

Within 24 hours after fertilization the ovum divides into two equal parts; within another 24 hours, each of these two parts divides again, giving rise to four parts. This process (cleavage) continues for the first week after fertilization, dividing the ovum (which is now termed a blastula) into 32 cells (Figure

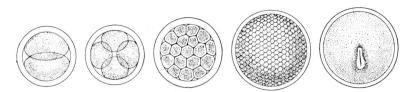

Figure 18–2.   The top figure is the first division of the ovum; it divides into many more cells, and finally takes on the first traces of an embryo.

18–2) by the sixth day. By the eighth day the embryo is approximately 1 ml in diameter. It doubles in size each day for a time and has been measured at 20 ml on day 14.

During the first 3 weeks the blastula develops a fluid-filled cavity, called a blastocyst. This cavity enlarges to about ½ inch in diameter. During this period of cellular growth, nourishment comes from "uterine milk," which is largely nutrients secreted by the cells lining the uterus. Roberts describes the first 15 days of pregnancy as the "period of the ovum." It produces estrogens and other substances that inhibit prostaglandin production.

The equine embryo begins the development of the circulatory system within the first few weeks—during the "period of the embryo," described as day 15 to day 60. A very primitive heart is formed, and vessels run out to the edges of the yolk sac and back. The embryo is recognizable as a horse by day 38, but is only about 2 cm long. The relationship of the yolk sac to the developing body of the embryo is illustrated in Figure 18–3. During the seventh week, the cells of the fetal membranes invade the endometrium and form "endometrial cups," which produce the PMSG detectable in the serum. The extent of the yolk sac at about seven weeks is diagrammed in Figure 18–4.

At this time the young embryo has developed a vascular system throughout the yolk sac, which carries nutrients picked up in the uterus to the developing body. The embryo is now floating in a sac of fluid called the

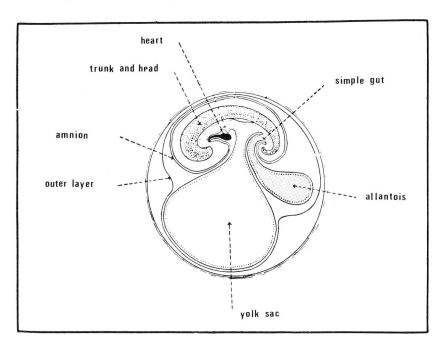

Figure 18–3.   Development of the embryo, day 21.

amnion. The yolk sac is much like the gut turned inside out. In this situation the secretions of the uterus are much the same as those inside the intestine of the developing embryo.

The change that occurs at about the seventh week of embryonic development can be envisioned by studying Figure 18–4. The allantoin then begins to take over the same function as the yolk sac. This time is extremely dangerous for the embryo. Its life blood comes from the vessels flowing to these extraembryonic membranes; and the changeover from choriovitelline placental circulation to the chorioallantoic type may kill the embryo owing to an insufficiency of oxygen. This lack, in turn, will cause an abortion. A 4- to 6-week abortion in mares is more common than most breeders suspect.

The association between the embryonic blood and the maternal blood is never very close in the mare. Oxygen and other nutrients must pass through the endothelium of the maternal blood vessels, the uterine epithelium, the lining to the allantoic cavity, and the endothelium of the fetal blood vessels. The histology of this association is illustrated in Figure 18–5. Very few other mammals have so much tissue between the two circulatory systems. This arrangement has both advantages and disadvantages. It leads to easy slipping of the membranes and subsequent abortion. On the other hand, it also makes for less trauma at birth. The extraembryonic membranes usually cause no trouble, and both mother and foal are ready to travel within minutes after parturition. While this association permits an easy separation of the fetal membranes from the lining of the uterus at parturition, thus affecting a quick delivery of the foal, a slow or difficult delivery is extremely

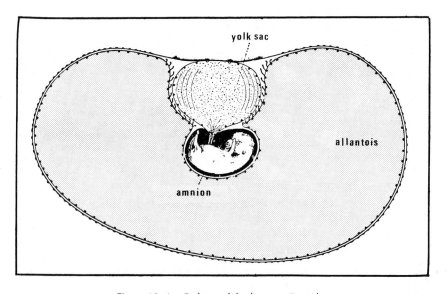

Figure 18–4.   Embryo of the horse at 7 weeks.

dangerous to the foal because of the possibility of a break in the maternal and fetal circulation.

The type of attachment illustrated in Figure 18–5 is not completely accomplished until about the 10th week. At this time the body length of the embryo is approximately 6 inches and the nutrients for the embryo are derived from the maternal circulation rather than from the "uterine milk."

The "period of the fetus" is described as occurring from day 55 or 60 to parturition. Clegg and co-workers believe that some of the gonadotropic hormone remains in small pouches in the allantois of the embryo after the uterine membrane stops producing the hormone. This hormone later serves to stimulate development of the gonads in the embryo at the proper time.

Of particular interest is the development of the gonads and genitalia. "Gonad" is a general term denoting either the male or female sex gland: the testicle, the ovary, or the primitive indifferent organ. In the beginning the gonad is the same in both sexes and is first intimately associated with the primitive kidney.

Eventually certain genes (probably on the Y chromosome) initiate changes in the cells of the male gonads so that they begin to produce some androgens. These hormones direct further development of the male reproductive system. If the embryo has no Y chromosome, the gonads begin to develop into ovaries. Female hormones (estrogens) are produced which stimulate development of the female genital system. If a testicle is to develop, the cells of the gonad begin to form cord-like masses which eventually differentiate into the seminiferous tubules. If an ovary is to develop, the cells become differentiated into many tiny primary follicles.

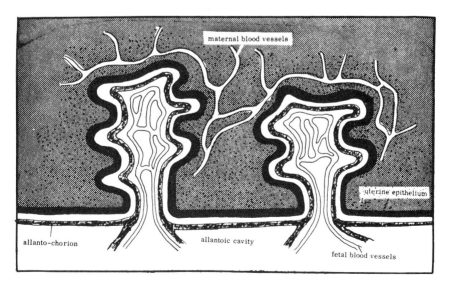

Figure 18–5. Nutrients in the equine placenta must pass through four layers of tissue from dam to fetus.

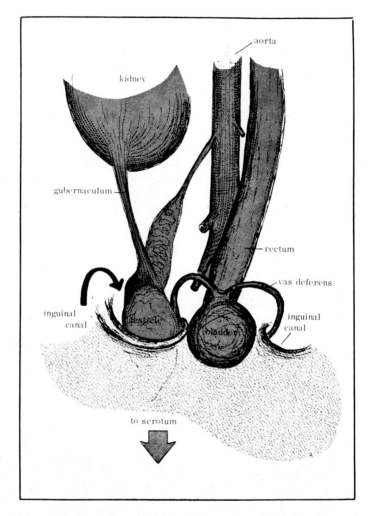

Figure 18–6.   The descent of the testicle in the horse.

As differentiation occurs, the gonads adjust their position. The ovaries gradually move back and to one side, and the developing uterine horns move with them. The testicles eventually move to a position within the scrotum. The position of the testicle before descent and as it passes through the inguinal canal is illustrated in Figure 18–6.

THE INTERSEX

Occasionally a foal is born in which the sex is impossible to determine. Such an animal is referred to as an intersex. A hermaphrodite would be an intersex in which gonads of both sexes are present, either separately or combined into one abnormal organ. More common in occurrence is the pseudohermaphrodite who has gonads of only one sex, but has reproductive organs with some of the characteristics of the opposite sex. Technically, sex classification in this case is on the basis of the type of gonad present.

A number of factors are involved in the normal development of sex in an animal; many of these are genetic in one way or another. Some sex abnormalities result from abnormal sex chromosome combinations, which are discussed in more detail in Chapter 8.

Cases of intersex cause one to wonder what initially determines the sex of a horse. In what manner do the chromosomes dictate the sex? Carl R. Moore has done some work with sex determination in the opossum. The opossum is

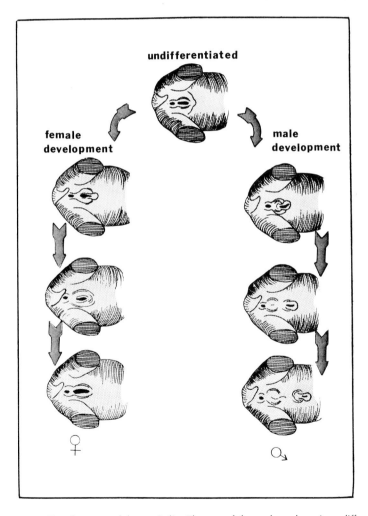

Figure 18–7.   Development of the genitalia. The sex of the early embryo is undifferentiated and has the potential to develop in either direction.

an ideal mammal on which to make such studies because the young are born at a very early stage of embryonic development and are, thereafter, kept in their mother's pouch. In this situation they are readily accessible for experimentation. Moore found that he could change the sex of an embryo by removing the gonads and administering the hormone of the opposite sex to embryos more than 63 days old. If he attempted this operation any earlier than 63 days, it had no effect in changing the sex of the embryo. From these studies he concluded that a hormone-like substance secreted by the body cells rather than by the gonads is responsible for the original sex determination in very young embryos. At a particular stage of development, depending upon the species, the sex influence is taken over by the hormones of the gonads. It may well be that this process occurs in the horse also.

The study of intersexes of various mammals indicates that the male hormones initiated by genes of the Y chromosome are strong enough to override the female hormones and cause development of a male. Basrur, in studying an intersex equine, found a testicle which contained a majority of cells with no Y chromosome. Evidently the few Y chromosomes which were present caused enough androgen production to produce a testicle.

The external genitalia also begin in the embryo in an undifferentiated state (Figure 18–7). The sex organs begin as a genital tubercle, a urogenital opening, and a pair of genital folds on either side. If ovaries begin to develop and produce estrogens, the genitalia develop toward those of the female. If the testicles begin to develop and produce androgens, the genitalia develop toward those of the male.

THE PREGNANT MARE

It seems that little contributes more to the production of strong, healthy foals than giving the mare a reasonable amount of exercise, although a certain element of risk does accompany the overwork of pregnant mares. With common care this risk is far outweighed by the benefits to the dam and offspring. When mares have judicious exercise and an adequate diet, foals dropped are not only bigger and stronger, but also resist the exposure to adverse influences, and thrive and grow much better than those from idle, ill-conditioned mares.

It is common for mares to be ridden or heavily worked right up to the time of parturition, especially when ranchers need the use of the mare. Pregnant mares as a general rule are much better off when ridden up to within 3 or 4 weeks of the time of foaling.

As pregnancy advances, and the calls of the growing fetus on the nutritive resources of the dam become greater, the amount of exercise should be diminished, and the food ration should be increased slightly.

In the stable, pregnant mares should be provided with plenty of room to lie down and stretch out on a good bed of soft litter. The floor of the stable should not slant too much in a backward direction. Broodmares do well when turned into a paddock at night with a dry shed for protection from the weather, and plenty of dry litter, if they get along with each other. With an adequate food supply, the open paddock is far more conducive to health than the atmosphere of the average stall, which is usually rather filthy.

When the weather permits, the mare can be turned to pasture for a few hours each day during the later weeks of pregnancy, without the risk attached to animals who have been more closely stabled. A bite of spring grass, before parturition, prepares mares for the more complete change of food which is shortly to take place and protects the foal from those severe attacks of diarrhea which result when mares are suddenly transferred from dry feed to pasture.

ABORTION

Peter Rossdale estimates early abortions from all causes at about 10 percent. This estimate is from records where a 40-day positive pregnancy test occurred and a subsequent check showed the mare not to be in foal. His figure seems much higher than should be expected on the average breeding farm, especially in light of the study he reported of the English Stud Book breeding records, which showed a national figure of about 3 percent for abortions occurring from the 40th day to term.

The fetus only attains a length of 8 or 9 inches by the fifth month, and it is, therefore, not surprising how few aborted fetuses are recovered. Because so few are found, there is a common belief that resorption of the dead foal often occurs. This is only speculation, however, and is not based on accurate information.

After the 300th day of pregnancy, a foal has some chance for survival, even though it is considered premature. Special care will make the difference. Foals born after 325 days are generally considered to be full term. The premature foal lacks complete development of lungs and some other organs, and is very thin and frail. Foals born before the 300th day will invariably die, even though there may be a heart beat.

There are four principal causes of abortion. The first and most common are the physiological problems such as hormone deficiencies and errors in development. Twin pregnancy usually results in abortion of one or both fetuses. Infections with bacteria, viruses, and other organisms can occur sporadically or in large numbers. Finally, abortions can occur in a number of ways from mismanagement.

Veterinarians have argued for years about just how much of a problem progesterone is in abortion. Improved techniques have made measurements of minute levels of blood hormones possible, and it appears that low levels of progesterone do cause abortion in many cases.

Progesterone can be supplied by the corpus luteum (CL) and/or the placenta. The primary CL is formed at the time of conception and remains present on the ovary until about 180 days of gestation. That is the first source of progesterone. The CL is formed on the ovary at the site of the follicle that has ruptured. Immediately after rupturing, the follicle fills with blood; but within a few days the cells have organized and changed to a type that secrete large amounts of progesterone which finds its way to the mare's blood stream.

Later on, the secondary CL is formed between about 40 and 140 days of gestation. Then after about 180 days, all of the CLs regress and the placenta becomes the final source of progesterone. So there are three sources of progesterone in the pregnant mare, which is an important consideration when one begins thinking about progesterone administration to avoid abortion.

Studies have shown that surgically removing the ovaries in pregnant mares early in gestation will cause them to abort; but after 120 days the pregnant mare doesn't appear to need the ovary. A research project at Colorado State University, headed by Dr. Ed Squires was aimed at learning the role of progesterone in pregnancy maintenance in mares. Based on this study, Dr. Squires now recommends the following schedule of treatment: 500 to 1000 mg repositol progesterone every four days or 250 to 500 mg of an oral progesterone every other day for mares that have an insufficient level of progesterone between day 30 and 140 of pregnancy.

Without the proper level of progesterone, the uterus will not respond to development of endometrial cups. These are made up of embryonic cells which move out from the embryo about day 35 of pregnancy. These endometrial cups begin to produce a hormone called pregnant mares serum gonadotropin (PMSG). Among other things, PMSG stimulates the development of follicles on the ovary which shortly transform into secondary CLs. These give a necessary boost to the mare's progesterone level.

The developing fetus may die from any one of a variety of genetic lethals (Chapter 10). These lethals cause the embryo to develop in such an abnormal way that it eventually becomes impossible for it to function in the uterine environment.

Russian researchers reported a study of 266 twin pregnancies. About 75 percent of them aborted both fetuses. Over half of those remaining were either born dead or died shortly thereafter, being too weak to survive. Thus, about 7 percent of the twin pregnancies resulted in a live foal, and in about half of these cases only a single foal was born alive or could be reared. Healthy twins are a rarity, as they have beaten tremendous odds. Other research indicates that multiple ovulations occur about 4 percent of the time with only one-fourth of these resulting in an observable twin pregnancy— dead or alive. Thus, the occurrence of twins is actually about 1 percent. Large mares, draft breeds, have a better chance of safely carrying twins to term, and the occurrence of twins is understandably higher. Some reports put the figure at 3 to 4 percent.

Infection can interfere with pregnancy by either interrupting the placental attachment or destroying the fetus by direct infection. A wide variety of bacteria can occasionally cause abortion: streptococci, *E. coli*, Klebsiella, staphylococci, and others. The most common occurrence of bacterial infection is when the uterus is infected at the time of breeding. The cervical plug generally keeps out bacterial invaders after pregnancy begins.

A study of early abortions, made by Dr. Tom Swerczek at the University of Kentucky, indicated that bacterial infection of the placenta may be a common cause of abortion in mares bred early in the season. He pointed out that during early pregnancy a subtle hormone cycle persists in the mare which periodically relaxes the cervix. This relaxing of the cervix, he claims, can allow bacteria to enter the pregnant uterus and will cause abortion.

Infectious organisms can be carried to the uterus through the blood, and settle there. This occurrence is common with several viral infections (see Chapter 22). The virus of rhinopneumonitis can be especially troublesome in this respect. Abortion occurs from about the 5th month on. Also, viral arteritis may cause abortion. Occasionally fungal infections occur, often brought about from excessive treatment with antibiotics.

Abortion as a result of poor management is probably less likely than commonly thought. The precautions that should be taken by the caretaker are covered in Chapter 20. Even poor nutrition does not generally result in abortion, although malnutrition during pregnancy can be damaging to the mare.

Dr. P.A. Doig claims that mares that are pregnant keep cycling on their ovaries as if they were not pregnant. With a 21-day pregnancy check, he expects to find follicles. They are normal. Dr. Doig describes two types of pseudopregnancy that occur which easily fool the mare, the veterinarian, and the owner into thinking that a pregnancy exists.

After ovulation, the corpus luteum (CL) is formed; it produces progesterone and in turn stimulates more follicle stimulating hormone (FSH) to begin again, in the diestral period, to get the mare ready for pregnancy or whatever. But, without development of pregnancy, prostaglandin is released from the uterus within 14 to 16 days. Its release normally causes luteolysis, or destruction of the CL and proestrus starts.

Some mares may lose the pregnancy at this stage, but for some reason they maintain the CL. Thus, the mare gives the outward impression that she is pregnant and because of the CL she does not come back in heat. Dr. Doig calls this condition "pseudopregnancy type one."

The CL is a small growth within the ovary that produces progesterone, the hormone of pregnancy. At this early stage the embryo is no bigger than the head of a pin and certainly cannot be palpated, but the progesterone causes changes in the uterus in preparation for pregnancy. The tone in the uterus is characteristic from day 15 to day 60. The uterus feels much like a radiator hose at this stage. After about day 27 or so, a small bump can be palpated just off center in either the left or right horns. This enlargement is ventral at first, but can be felt dorsally at about 45 days. It is the size of a large orange by 65 days.

On day 38 to 40 the mare develops the endometrial cups which produce the hormone PMSG. It continues to be produced until about the 150th day, regardless of whether the pregnancy continues. Some miscalculations in pregnancy diagnosis may result if based on the PMSG blood test. If a mare aborts prior to 150 days, the blood test will still be positive.

PMSG, among other functions, keeps mares from cycling regularly. We cannot override that with hormones. There is no known means to get a mare cycling with PMSG in her blood. This phenomenon is called pseudopregnancy type two.

Mares aborting after day 40 will not show external signs that they have aborted. They give a positive blood test and do not come back into heat until after about the 150th day. This picture is true of mares bred in March that seem to be pregnant until they come back into heat in September. Some people walk around the pasture looking for the aborted fetus, but it may have come out weeks earlier.

Why do these mares abort at this early stage? Probably the most common reason at this early time is a low-grade infection. How does a veterinarian determine whether a mare is infected? She may not have a positive culture. The mare's reproductive history is important in determining whether she is infected. An important piece of history here is whether she was pregnant at one time and then came up empty in the fall. An endometrial biopsy at this point will help determine infection.

One of the most controversial techniques is culturing. The uterus is the only structure that should be cultured. A cervical culture will not tell what is going on unless the mare is in heat and actually infected. Otherwise, a positive cervical culture is meaningless.

Many farms that use natural service will culture every mare presented for breeding in order to protect the stallion from infection by these mares. Farms that use artificial insemination may only culture barren mares and those that appear to have a problem. Dr. Doig describes his technique of taking a culture with a guarded swab. He inserts a finger in the cervix first and guides the swab past the finger into the uterus. He says that this procedure also helps to verify the patency of the cervix. Severe adhesions are found in about 10 percent of infertile mares checked by Dr. Doig.

Cultures should be correlated with cytology. Many veterinarians like to smear the swab on a slide, stain it, and look at the cells under a microscope. If neutrophils are found, the mare should be treated. Positive uterine culture should be expected in 30 to 35 percent of normal, healthy mares. Surveys also show that the same margin of error occurs on the other side, with many mares giving a positive culture when an endometrial biopsy shows no infection.

A study by Dr. Doig involved 92 mares that were biopsied and cultured at the same time. Of these, 28 had positive cultures, mainly Streptococcus and E. coli. Only five of these mares showed any signs of infection in the biopsy.

Five percent of the 64 mares with negative cultures showed evidence of infection on biopsy.

One of the most important clues to whether a mare may be infected is vulvar conformation. Some types of vulvar conformation encourage bacteria to enter by "windsucking." Years ago Dr. Caslick, a Kentucky veterinarian, described a simple operation which involved suturing the lips of the vulva nearly shut to reduce windsucking and subsequent infection. Although this operation is still commonly done, many veterinarians feel that it should be done more often. Dr. Doig claims that 80 percent of the infertile mares he sees should be sutured. Other veterinarians feel that mares should not be sutured until there is a pronounced forward angle at the top and a "falling in."

The vulva should be sutured down to the level of the floor of the pelvis. In a normal mare, there should be room to lay two fingers between the external ring of the anus and top of the vulva—about 4 cm. Many normal-looking vulvas that are set too high will suck air during estrus.

There would be far less problems with early abortion if more Caslick operations were performed. This procedure is as important as infusing the uterus. Any mare previously sutured should be sutured again after foaling.

Dr. Pierre Lieux, a Riverside, California veterinarian, believes that there is another important cause of early abortion in mares. He says:

> Many horse breeders assume that when a mare comes in heat and is bred at the proper time to a normally fertile stud, she should conceive. If she does not, it demonstrates, according to their thinking, that there is something wrong, such as a uterine infection, that a good veterinarian should be able to diagnose and correct. This is far from being the whole picture. Each mare is born with her own level of fertility or infertility as predetermined by her ancestry.

Lieux believes that some families of horses are genetically subfertile, or even infertile. He says:

> These poor-breeding performances shown by some families are most likely related to the existence of lethal genes. Lethal genes can destroy the embryo at the early phase of its development or prevent its formation by interfering with any mechanism of reproduction such as ovule formation, release, travel, fertilization, nidation, etc.

He feels that intense selection in horse breeding for performance, with no consideration for fertility, is causing equine fertility to degenerate.

> As an example, when I started practicing years ago, quarter horse mares were very easy to get in foal as compared to thoroughbred mares who had years and years of concentrated breeding behind them. An infected mare was then no problem to cure as it was with

thoroughbred mares. But now, because of the recent explosion of the quarter horse industry and mixing with thoroughbred there is no practical difference between the problems of the females of those two breeds.

Early abortion is now more often recognized in the mare because horse breeders are more knowledgeable and they work more closely with their veterinarians. When a mare comes up empty in the winter, they want to know why. As veterinary science progresses, more answers will become evident.

FOALING

A mare will generally foal at night. Studies have shown that 85 percent of all foaling occurs between 7:00 P.M. and 7:00 A.M. At this time she is more relaxed; and over the centuries foals born while the mare is relaxed tend to have a better chance of living, especially in the wild state. Therefore, one could say that this characteristic is genetic.

Foaling is initiated by hormones. A new hormone has been described and studied by veterinarians at the University of Guelph, Ontario. It is called dehydroepiandrosterone (DHA) and is produced by the fetal gonad (testicle or ovary). This hormone either causes production of another hormone called estrogen in the placenta or is somehow converted to estrogen by the placenta. More study should eventually reveal just what triggers the secretion of oxytocin from the posterior pituitary gland of the mare. This latter hormone causes uterine contractions and the letdown of milk in the udder.

The miracle of birth is how the mature fetus passes through the confining birth canal. This canal is composed of cervix, vagina, and vulva, which are soft structures and can stretch to whatever dimensions are necessary for the fetus to pass. However, these structures are surrounded by the bones of the pelvis, which have little "give." The shape and size of the pelvic canal dictate the way the fetus should pass through in order to avoid blockage.

The points of greatest mass within the fetus's body are the chest at the withers and the hips. The head and shoulders must pass out ahead of the chest or there will be too much mass for the opening. Passage of the withers is so tight that it is important for the backbone of the fetus to be up next to the mare's backbone for a perfect fit.

In the latter part of pregnancy while the fetus is maturing and gaining in weight, it normally lies on its back with its head directed toward the birth canal and its limbs flexed above it. As the processes of birth begin, the fetus rotates and the front limbs extend up to the nose for proper delivery.

For an unknown reason some foals develop in a backward position, and as the birth process begins, the rear legs extend, to come through the canal first—known as a breech presentation. Often the hips have difficulty coming through in this position.

The fluid surrounding the fetus in the amniotic and allantoic cavities make it possible for uterine contractions to force the fetus through the canal. This force is increased by contraction of the abdominal muscles to decrease the size of the abdominal cavity. The mare will usually lie down during parturition and this position puts extra force on the fetus by helping to reduce the size of the abdomen. Foaling is therefore much like squeezing toothpaste out of a tube.

During the first stage of labor the uterine muscles contract in waves which pass from the tip of the uterine horn to the cervix. This pressure moves the foal's front feet and nose up into the cervix and forces it to open wider. Continued contractions force a small part of the placenta out ahead of the fetus. It looks like a balloon filled with water.

The first stage of labor is recognized by how the mare acts. She is uneasy and may walk around in a circle and paw the ground and may even sweat. Sometimes the mare will retract the upper lip of her mouth at this stage. Milk may begin to squirt from the teats. The first stage varies in length from just a few minutes to several hours and may be related to the position of the fetus and the amount of oxytocin that is secreted.

The rupture of the placenta with the loss of some fluid, often called "breaking water," signals the end of the first stage of labor. Most mares lie down during the second stage of labor. They rise every once in a while and reposition themselves. Part of this activity is to help the foal to turn into the proper position for delivery.

Second stage labor averages 20 minutes in duration, varying from a few minutes to an hour. It ends when the fetus has passed through the birth canal.

Generally, a resting period occurs for both the mare and foal after the second stage of labor. The foal may simply lie quietly on his side with his rear feet still in the birth canal. The umbilical cord is still attached from the foal to the placenta at this time. Blood is draining from the vessels in the placenta to the foal. It is important at this time to leave the mare and foal undisturbed so that as much blood as possible can pass from the placenta to the foal. He will need all he can get. An anxious groom who rushes in at this point and pulls the foal free from the mare may deprive the little creature of as much as one-third of the blood he would receive otherwise.

Later during the third stage of labor the mare will expel the placenta. The foal soon attempts to rise or the mare gets up, and the umbilical cord breaks at a natural point about 1½ inches from the foal's navel.

Foaling goes well, with no complications, 90 percent of the time. Most mares never have a problem and are best left entirely alone. Other mares have a problem almost every time they foal. With expensive mares, an owner may not want to take his 10 percent chance and may call a veterinarian to "check the mare out" with the first signs of approaching parturition.

Since it is difficult for veterinarians to arrive at the exact time for foaling, the technique of artificial induction of foaling has become almost commonplace with some veterinarians. This practice allows the veterinarian to be present during foaling to alleviate any problems that may arise.

The veterinarian will initially ascertain the foal's position within the uterus. Before labor the foal is usually on its back. During the first stage and early second stage it rotates 180 degrees (as described previously), taking up the normal position for passage through the canal. Early in the process the foal may give the impression of being upside down.

As the second stage of labor begins, the veterinarian will insert his hand into the vagina and through the cervix to determine whether both front feet and the nose are coming out first. Then he will generally just stand back and observe, letting the mare foal on her own.

Dr. Alan R. Purvis, of Aldergrove, British Columbia, has routinely used induced labor for years. He says:

> In my practice, this technique has proven to be a safe, rational part of broodmare management. It has dramatically reduced the incidence of problems at parturition for both mares and foals. Dystocias (difficult delivery) due to fetal malpresentation are circumvented. Lacerations to the mare's colon, perineum, vagina and uterus are problems of the past. Retained placentas are virtually eliminated. I have never observed a so-called stillborn foal following an induction, or a foal that has died of asphyxiation or aspiration of amniotic fluid. Emergency calls to a dystocia at 4:00 A.M. are also a thing of the past.

Veterinarians use a commercially available injection of oxytocin to begin the induction, but as Dr. Purvis emphasizes, "it is not a procedure to be used lightly." A mare should be induced only if three important criteria are present. Dr. Purvis says:

> To satisfy the first criterion of an adquate gestation period, I believe that the mare must have at least 330 days of elapsed time from the last date of breeding. This rule of thumb has consistently resulted in excellent results for me and should be strictly adhered to.
>
> The second criterion for induction is satisfied with the arrival of adequate amounts of true milk in the udder. A drop or two of white milk is not significant. The presence of true milk occurs, in nearly all mares, hours or days before there is sufficient intramammary pressure to cause the typical preparturient waxing. Of course we have all observed those mares which foal naturally without waxing.
>
> The third criterion for induction is established on the basis of cervical softening. On arrival at the farm I make it a firm policy

before the cervical examination to reconfirm with the owner the last breeding date. I also examine the udder to confirm the presence of adequate amounts of true milk.

The attending veterinarian should also be very much aware that a widely dilated cervix may be observed in some pregnant mares as early as the ninth month of pregnancy and probably earlier. I have observed this type of mare to go to term and foal naturally with no problem.

Most mares foal within 30 to 40 minutes after injection with oxytocin. The total time varies from about 15 minutes to an hour. Dr. Purvis claims, "The delivery time appears to be more closely associated with nearness of the mare to natural delivery than with the amount of oxytocin administered."

It usually takes a foal nearly an hour to gain the standing position, after several unsuccessful attempts. The position of the foal's ears will indicate whether he has gained enough strength to accomplish the task. Right after birth the foal's ears will lie back against his neck. With time, as he gains strength, the foal will rise to an erect position.

The foal is attracted to the mare once he is on his feet. He will not immediately run up and start suckling but will make efforts to nuzzle the mare's chest and legs. Within 2 hours the foal should have found the teats and suckled. Some who have studied foal behavior have proposed that the new foal is attracted to suckle in an area of darkness that exists between the mare's hindlegs. Since most foals are born at night, this does not help in understanding how foals find the udder in the dark.

Changes are taking place in the lining of the foal's stomach after birth. Within the first 24 hours large protein molecules can be absorbed directly into the foal's blood without being broken down by the usual enzymatic process of digestion. This direct absorption allows the foal to benefit from the large immunoglobulin molecules in the colostrum. Suckling within the first 12 hours is extremely important if the foal is to gain benefit of the antibodies (immunoglobulins) in the colostrum (first milk). After 12 hours the ability to absorb them diminishes rapidly until after about 18 to 24 hours, very few of the antibodies are absorbed.

After a foal has first suckled, it will, with increasing sureness, find the udder at regular intervals and suckle more. Studies have shown that during the first week of life a foal will suckle, on the average, every 15 minutes.

One of the most common worries to the breeder is asphyxia in the foal. There is a certain amount of asphyxial stress on the fetus as it comes through the birth canal, but only if the umbilical cord becomes compressed between the foal and mare's pelvis will there be any real danger.

During birth the foal's oxygen-acquiring mechanism experiences a changeover from receiving oxygen across the placental tissue from the mare's blood to breathing oxygen into the lungs and absorbing it into the blood vessels of the lungs. Before birth the pulmonary blood vessels carry little blood. In the heart the blood is shunted past the pulmonary artery and directly into the aorta. When the changeover occurs, the shunt (the passage diverting blood) is closed and the blood is forced into the lungs. This situation happens by reflex action when asphyxia begins to develop in the foal.

The exact mechanism for this changeover is incompletely understood. Since this response is partly due to the gaseous expansion of the lungs, it is wise for the attendant, if possible, to clear the amnion from the foal's nose right after it comes out, to help the foal in his attempt to fill his lungs.

As the foal gains strength and suckles regularly through the critical first 24 hours, the attendant should be watching for a bowel movement. Until the foal passes the meconium (dung which the fetus stores in pellet form within the cecum, colon, and rectum), the first milk cannot pass through the entire length of the gut.

Retained meconium will cause mild colic and will need to be relieved with an enema. Veterinarians often use a commercial enema preparation that comes in a large plastic syringe with a tip made for insertion in the rectum. This simplification of the enema procedure has made it a routine practice for most veterinarians in attendance at foaling.

With good preparation and careful attention, the foal and mare should be healthy and fiesty the day after foaling. The owner can at that point focus his attention on the development of champion capabilities in the foal.

DIFFICULT PARTURITION

Though parturition is generally an easy and prompt act in the mare, in some instances it is extremely complicated and difficult. Many of these cases have a rapidly fatal termination, hence the great need for careful observation of the mare at foaling. When the foal presents itself in the genital passage in an unfavorable position or abnormal attitude, it will need help immediately. The normal presentation is shown in Figure 18–8.

Unlike the cow, the mare quickly becomes excited and restless, and even furious unless soon delivered. All veterinarians who have had to deal with cases of abnormal birth in mares are well aware of the difficult task that lies before them. If the foal is not born soon after straining commences, an examination is imperative, and if the cause of obstruction cannot be discovered and corrected, then the veterinarian ought to be called as soon as possible. Every minute's delay increases the gravity of the case.

A skilled attendant can first make an examination to learn the cause of obstruction to delivery. Should he find the foal in a favorable position, with the forelegs presenting and the head forward or resting upon them with sufficient room for the foal to pass through the canal, he should wait for the labor pains to produce its expulsion. If, however, the position of the foal is not favorable for a speedy birth, it must be corrected.

When the foal itself is the cause of obstruction, this difficulty may be due to the position of the limbs, body, or head. The forelimbs are perhaps most

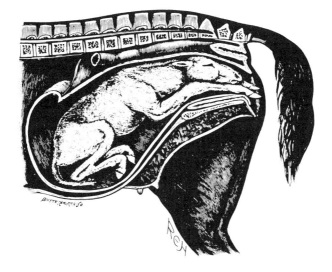

Figure 18–8.   The normal presentation of the foal.

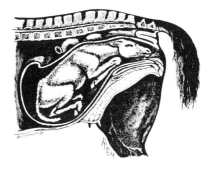

Figure 18–9.   A fetus with head presented but knees doubled back.

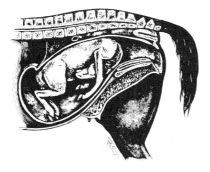

Figure 18–10.   A neck presentation.

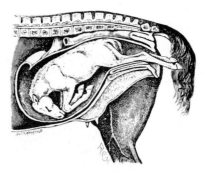

Figure 18–11.   A breech presentation.

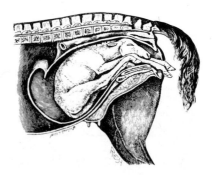

Figure 18–12.   A fetus with nose and all four feet presented.

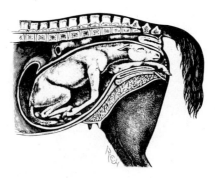

Figure 18–13.   A fetus with one forelimb displaced backward.

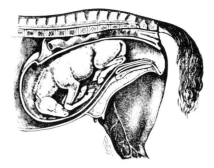

Figure 18–14.   A breech presentation with legs forward.

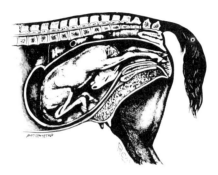

Figure 18–15.   A fetus with forelegs presented and head back between them.

often at fault; one or both are involved through a doubling-back at the knees (Figure 18–9). A similar flexion of the hindlimbs may occur at the hocks and be a cause of difficult parturition. The head, instead of being placed nose forward and between the forelimbs, may be bent downward. The neck will be thrown upward and backward, or toward the side of the foal's body (Figure 18–10). Instead of the head and forelimb coming first, it may be the hindlimbs. Only the tail and buttocks may be presented while the body may be reversed; instead of the back being toward the mare's back, the foal is more or less on its back with the legs upward.

Besides these malpositions there is the difficulty of twins and the occurrence of monstrosities. Deformity or diseases in the mare that cause narrowing of the genital passage may also be a cause of foaling problems. In cases of difficult parturition in the mare, much skill, patience, and physical strength are required.

Extensive knowledge and experience are necessary to correct the varied presentations shown in Figures 18–11 through 18–15. The amateur is likely to do more harm than good. He may even convert a comparatively simple

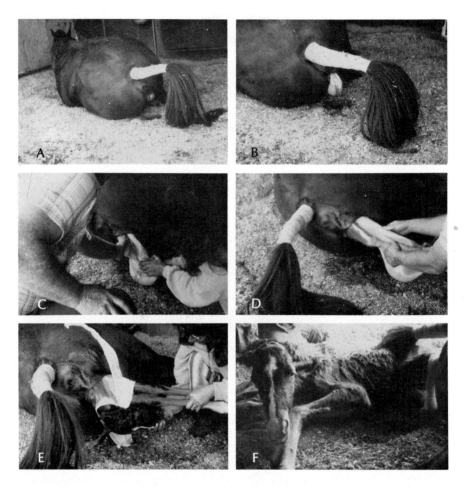

Figure 18–16.   The foaling experience.

case for an expert into a most difficult one if he is not competent in what he is doing (Figure 18–16).

EMBRYO TRANSFER

During the last decade, the proliferation of a myriad of cattle breeds in the United States has been made possible by embryo transfer. Importing purebred cattle is extremely expensive. The development of a large purebred foundation of so-called "exotic" breeds would have been very slow had it not been for this technique.

Japanese and British scientists modified the technique for horses originally. Now it has become commercially possible to transfer embryos from a valuable mare to various "incubator" mares that carry the embryo to term and act as "foster mother" (Figure 18–17).

Talk to a bovine geneticist who is involved in embryo transfer and he will tell you that the procedure is a long way from being practicable in horses. When interest in equine embryo transfer first began to develop in the United States, Dr. R.H. Douglas spoke to the International Embryo Transfer Society saying:

> Since the equine does not represent a major source of protein for human consumption and since the equine industry tends to be traditionally more conservative in regard to breeding policies

Figure 18–17.   The first embryo transplant foal born in the United States, Miss T.

applicable to propagation of purebred animals; little impetus has been present to investigate this area, aside from that arising from a purely academic interest.

He went on to say:

> When one considers the numerous benefits that such techniques offer the cattle, sheep and swine industry (multiplication of exotic breeds, multiple progeny from genetically superior females, building herds from a few base females, a mechanism to progeny test females, etc.), it becomes obvious that few, if any, apply to the equine industry as it presently exists. In my opinion the most likely areas of application of embryo transfer in the near future are (1) in obtaining offspring from proven subfertile or infertile females and (2) in managing valuable, older broodmares.

Since then, Dr. Douglas' words have proved to be prophetic on the one hand but too pessimistic on the other. He evidently knew of the transfer foal that had been produced at Texas A & M by a graduate student, Stephen Volgelsang, and considered it of purely academic interest. But Dr. Dwane Kraemer, the student's faculty advisor, owned the mare and proceeded to put the registration of the foal to test with the American Quarter Horse Association (AQHA). As Dr. Douglas had predicted—or following his prediction—the AQHA approved registration of future transfer foals under certain conditions, namely if the donor mare had proven to be subfertile (had failed to produce a live foal after *bona fide* attempts to achieve pregnancy during three consecutive calendar years) or the donor mare were over 15 years old.

One aspect of the AQHA ruling which is destined to stimulate much interest in embryo transfer is the provision that, "only one foal will be registered each year from each genetic dam. The first application received in the AQHA office shall determine which foal from any specific year will be the one registered."

While this wording may sound conservative, it is a major concession. The AQHA, in effect, recognizes that more than one (maybe a dozen?) foals may be produced in any one year from a donor mare. Since registration can be delayed until the foals are yearlings or even 2-year-olds, a breeder can have a group of sibs (half or full) in training—all with the potential of being registered. The best one of the lot can be registered to go on to the races. What happens to the others with potential? With tongue in cheek, many breeders say, "I'll find a place for them."

The AQHA has set the pattern for other registries. By 1981 a half dozen registries had approved embryo transfer, generally with the same qualifications as the AQHA. These registries were: Arabian, Appaloosa, paint, Canadian standardbred, and others. Dr. Ed Squires said, at his first annual Embryo Transfer Short Course, at Colorado State University:

Hopefully, in the future the breed registries will develop less restrictive guidelines for equine embryo transfer. This in turn will alter the type of mare that is available for an equine embryo transfer program. For example, if the majority of breed registries banned the age requirement for donor mares, this would change the type of mare entering into the program considerably. In addition, if the registries saw fit to allow two foals per year (per donor mare), this would certainly stimulate the equine embryo transfer business. One of the changes that has been made just recently by the AQHA is the legalization of movement of the equine embryo. At the convention in 1980, the AQHA passed a ruling that the donor mare could be flushed at one location and the embryo transported to another location for transfer. The transfer into the recipient mare must be completed within 30 days of recovery of the embryo.

It is a relatively simple procedure to collect an 8- or 9-day-old embryo (Figure 18–18) from a donor mare. A special catheter is inserted through the cervix into the uterus, guided by rectal palpation. The catheter has an inflatable balloon attached which is filled with air after it is inside. This confines the area to be flushed to the uterus. A liter of fluid will generally fill the uterus. When the fluid is siphoned out, it usually carries the early embryo with it. If not, a second or third flush will bring it to the exterior if it is in the uterus.

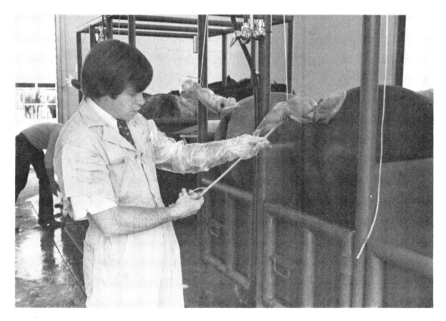

Figure 18–18.   Dr. Ed Squires, Colorado State University, collects an embryo from a donor mare. (Courtesy of George Jones.)

The embryo in the fluid is caught in sterilized glass or plastic containers and placed temporarily into culture media in an incubator. As soon as the recipient mare is ready, she can be inseminated with the embryo in much the same manner that semen is placed in the uterus during AI.

Kathy Imel, who pioneered the technique with Dr. Squires as a graduate student at Colorado State University, says, "In approximately 75 percent of the recovery attempts, the embryo can be seen as soon as it is poured into the dish. In fact, most of the day-9 embryos can even be seen in the cylinder prior to siphoning (from the uterus)."

Embryos collected on day 8 or 9 post-ovulation are in the blastocyst stage of development. They appear as fluid-filled spheres of slightly opaque cells (Figure 18–19). Occasionally, the embryo may be ovoid rather than spherical. Embryos are handled using sterile glass pipettes. Pipettes of various internal diameters are made and then fire-polished to eliminate any rough edges that might damage the embryo. For any given embryo, a pipette is selected that has an internal diameter large enough to allow free passage of the embryo into the pipette.

There has been much trial and error in developing the best technique for placing the embryo successfully in the recipient mare. The major problem with non-surgical transfer (insertion of the embryo through the cervix) has been that the recipient is in diestrus when the transfer occurs. Manipulation and opening of the cervix at this time cause production of prostaglandin by the uterus and the mare begins cycling as she cannot recognize that pregnancy has occurred. Any mare will lose a pregnancy in the same manner during that first diestrus if the cervix is forced open. The secret to successful nonsurgical transfer seems to be in giving a drug that will block production of prostaglandin.

The major stumbling block to large commercial equine transfer centers, such as those for cattle, is the inability to use frozen semen and artificial insemination. Hard-headed registry officials may be a long time in conced-

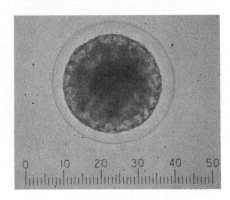

Figure 18–19.   An equine embryo at 6 days. (Courtesy of Rio Vista Farms, San Antonio, Texas.)

ing to this technique. As a result some veterinarians and breeders are beginning to think in terms of smaller, more individualized transfer facilities, located at many breeding farms.

Complications arise when a veterinarian attempts to coordinate a transfer at some facility a thousand miles away. The donor mare must be bred under close supervision with frequent palpations to make sure that ovulation has occurred. The mare must then be transported a long distance only to have the veterinarian on the other end fail in an attempt to recover the embryo. Back again to the stallion, the mare is bred, then returned to the transfer facility.

As long as the only purpose of embryo transfer is to salvage old, infertile mares, the technique will have a low success rate. For some reason, many of the problem breeders will not have fertilized ova (embryos) after breeding. Whether this situation is due to a blockage of the oviducts or a loss of the ova before they enter the oviduct is not known. This question is currently a major area of investigation by those interested in embryo transfer in horses.

Will embryo transfer eventually become accepted by the registries for use in other than problem breeders? Serious horse breeders who are attempting to improve their stock—even if the primary drive is to win at the races—deserve more from their registries than political double-talk about the economic implications of "wide-open" embryo transfer. Embryo transfer has the potential to improve any breed if conducted intelligently, and all people associated with it can be better off.

What moral right has a registry to deny registration of a transfer foal? Has it forgotten its original intent, to improve the breed and serve the breeders? Every registry began humbly, accepting the word of the breeders as to which sire and which dam were responsible for foundation stock. Are registries now so far above the breeder and tangled in so much red tape that there is no way to profit from a scientific breeding breakthrough that can improve their breed probably more than any other factor?

The key to improving the collection percentage in problem breeders seems to lie in careful evaluation of the specific problem from which the mare suffers. Many of these mares carry so much exudate in the uterus that an embryo cannot survive the initial 8 days that are necessary for the procedure. Some mares may not be ovulating properly; others may have blocked oviducts.

The veterinarian who expects to use embryo transfer in horses successfully will have to be an expert in sterility problems of the mare. He will have to know each mare, work with her for an extended period of time, and prepare her for eventual embryo collection. When he does work with the mare as described, collection percentages should improve dramatically.

BIBLIOGRAPHY

Betteridge, K.J.: Embryo transfer procedures and results obtainable in horses. *In* Current Therapy in Theriogenology. Philadelphia, W.B. Saunders, 1980.

Gerneke, W.H., and Coubrough, R.I.: Intersexuality in the Horse. Anderstepoort J. Vet. Res. 37:211, 1970.

Hillman, R.B.: Induction of parturition in the mare. *In* Current Therapy in Theriogenology. Philadelphia, W.B. Saunders, 1980.

Lose, M. Phyllis: Blessed Are the Broodmares. New York, Macmillan Publishing Co., 1978.

Males, Ron, and Males, Val: Foaling: Broodmare and Foal Management. London, J.A. Allen & Co., 1979.

Miller, W.C.: Practical Essentials in the Care and Management of Horses on Thoroughbred Studs. London, The Thoroughbred Breeders Assoc., 1965.

Roberts, S.J.: Gestation and pregnancy diagnosis in the mare. *In* Current Therapy in Theriogenology. Philadelphia, W.B. Saunders, 1980.

Rossdale, Peter: The Horse: From Conception to Maturity. Arcadia, CA, The California Thoroughbred Breeders Association, 1972.

19

the foal

The production of foals is what it is all about for the horse breeder. He lives for the coming of the foals in the spring. All of his work the year round culminates as the foals are dropped. Genetic planning, thoughtful care, and continual concern all become evident when an outstanding foal crop arrives.

Something about a good foal inspires dreaming and optimism. During the first few weeks of life the balance and style of the individual reflect his greatest potential. From 6 months of age and for the next few years there will be a string of awkward periods when the head is too short, the croup is too high, the neck is too short, or his conformation in general seems ungainly. But, during the first summer, the devoted horse breeder finds himself often just leaning on the fence gazing at his foals, drinking up their beauty, and pondering their potential.

CARE OF THE NEWBORN

When a foal is born, the attendant or veterinarian should conduct a thorough examination of the young animal in order to determine abnormalities immediately. These congenital abnormalities are often from genetic causes (Figure 19–1). If the foal should die in a few days as a result of a genetic problem, it is helpful to have a record to go back to later. The more facts at hand, the better a breeder can reduce the incidence of genetic defects in his herd.

The nostrils and mouth should be carefully checked to see whether all is normal. The eyes should be observed for cataracts or other ocular abnor-

Figure 19–1.   Congenital abnormalities may become immediately evident at birth. This animal has the white foal syndrome, and thus cannot live a normal life.

malities. The anus should be checked, and the occurrence of a bowel movement noticed. An enema is indicated if no meconium (fetal bowel content) has passed. The penis should be observed in colts and the vagina in fillies. Various genetic abnormalities are described in Chapter 10.

The bowels of the foal at birth contain a considerable amount of fecal matter, consisting of the solid remains of bile, and other secretions thrown out by the mucous membrane of the intestines during development. Usually the bowel contents are discharged soon after birth as a soft greenish or yellowish brown substance known as meconium. In some cases, this material becomes hard and dry, and cannot be discharged unless an enema is administered.

The foal may be straining; if so, warm soapy water may be injected into the bowel by means of a soft flexible rubber tube, two or three times during the day. Should this procedure fail to effect removal, the breeder should call a veterinarian without delay. If constipation becomes habitual in the foal, the dam must be allowed an extra supply of carrots or green food, and the addition of bran to her daily grain.

The raw edges of the broken umbilical cord will serve as an entrance area for infectious organisms into the system of the foal if it is not properly disinfected immediately. A strong tincture of iodine (7 percent) is generally painted over the umbilicus. Unless this procedure is followed routinely, navel or joint ill may result, and will seriously hamper the growth and development of the foal.

Figure 19–2.   Apparently misshapen legs at birth generally come into normal limits with time and care.

At the time of birth and for a week or two afterward, foals often present an unshapely and awkward appearance. Their hocks or knees, or both, are acutely flexed, and their fetlocks may almost touch the ground (Figure 19–2). The limbs give the impression of being incapable of supporting the weight of the body. It should, however, be remembered that where there is no bending of bones, or shortening of ligaments or tendons, the foal invariably "straightens up," and the deformity gradually disappears as growth proceeds. In those cases where the bending of the joints is due to contraction of the tendons, the defect may be remedied by mechanically stretching or dividing the latter using a surgical procedure.

THE PREMATURE FOAL

Foals, when born before the full term of gestation, are sometimes still enveloped in the fetal membranes or afterbirth. As they are then disconnected from the dam, respiration is only possible by exposure to the external air, so the membranes must be removed promptly. Then, breathing may be set in motion by providing a little artificial respiration, or by shaking the foal in the air by its hind legs.

It is important that the afterbirth be promptly removed from the box and buried in some unfrequented place and sufficiently deep to guard against its being uncovered and eaten by dogs. Burial of the afterbirth should occur only after a careful check to make sure that the membrane is intact since missing parts may later cause trouble in the uterus.

The milk of mares which foal prematurely is always scant and of indifferent quality for the first 2 or 3 days, and may require supplementation by milk from another mare or Foal-Lac. An adequate supply of antibodies at this time should also be of concern, as will be explained later in the chapter.

Hand-feeding will be necessary night and day, every hour or so, until the foal is strong enough to feed itself, but after the first 36 hours the period between meals may be gradually extended. When it has acquired sufficient strength to support itself, the foal may be returned to the dam. How it will be received by her is a question which must not be overlooked, and the attendant should stand by until the mare has settled down to her offspring and shows a desire to nurse it.

ISOERYTHROLYSIS

The foremost authority on equine blood typing (see Chapter 25), Dr. Clyde Stormont, said, "The indiscriminant practice of transfusing plasma into filly foals is often priming these future mamas for development of neonatal isoerythrolysis in their foals later in life." What he meant was that blood transfusions should be made only when the blood of recipient and donor are known to be relatively compatible.

Neonatal isoerythrolysis, commonly referred to as NI, seems to be on the increase these days because of the increasing practice of blood transfusion in horses (Figure 19–3). Dr. Ann Bolling, of the Serology Laboratory, University of California, Davis, describes NI:

> Neonatal isoerythrolysis (NI) of horses is an acute hemolytic disease of newborn foals caused by immunologically mediated red cell lysis. The culprit antibody molecules which combine with specific antigens on the red cells of an NI foal are passively acquired by the foal as it absorbs antibodies from the colostrum it has sucked from its dam.

Foals are healthy when born; NI does not develop until the foal gets colostrum. In 12 to 36 hours symptoms develop: the foal stops nursing; becomes dull, sluggish, and weak; and icterus becomes apparent with eventual development of yellowish urine. The red cells of the foal are destroyed by antibodies that the mare has developed against the foal's blood.

How can this situation happen? It is known that proteins occurring naturally in an animal's body will cause antibodies to be produced against that foreign protein. The first exposure to the foreign protein works like a vaccination by setting up the potential for production of the specific antibody. After a few weeks the immune system is ready for massive production of that specific antibody upon subsequent exposure. The immune system treats foreign red cell antigens as if they were invading bacteria or viruses.

Figure 19–3. As the foal first begins to nurse, there may be concern for isoerythrolysis.

A large group of factors involved in blood typing are known as alloanti-gens, which are variations in protein structures on the red cells. Sixteen types of alloantigens have been identified during blood typing of one horse. These variations in protein structure are reflections of eight genes which regulate production of these proteins (see Chapter 25). The genes are active in the bone marrow cells on chromosomes of the nucleus.

Dr. Stormont has called these red cell antigens "blood factors." He has divided them into eight genetic systems. Each one of the genetic systems has two or more alleles that can be identified. Each horse carries two alleles for each genetic system. These alleles are inherited from that horse's parents and limit which blood factors that horse can pass on to its progeny.

As an example, a horse may have alleles a^{11} and a. It may have two a alleles or two a^{11} alleles, or any combination of two alleles in the group of the A system. It cannot have two a alleles unless both sire and dam had at least one a allele. This principle forms the basis for parentage analysis.

These blood factors are detected by alloantibodies which must be produced and stored in the laboratory. An alloantibody exists for each of the antigenic factors except those designated by a single lowercase letter (a, c, k, p, q, t, and u). These antibodies are produced by injecting the various blood types into experimental animals such as rabbits, guinea pigs, or even other horses. The foreign blood factors cause the experimental animals to produce antibodies against whatever blood factor was injected. Serum from these animals is then called antiserum.

A drop of antiserum is placed on a slide with a drop of blood to be tested. The specific antibody in the antiserum identifies a specific antigenic factor when the red cells are destroyed by hemolysis (a bursting of the cells) or agglutination (a clumping together).

Some antigenic factors are stronger in their ability to cause alloantibody production in experimental animals. As an example a^{41} causes a very potent alloantibody to be produced against it. This alloantibody is the same one that a mare may produce against a^{41} alloantigens on the foal's red blood cells. Because most of the other antigenic factors of the A system are relatively weak antigens, a horse that does not carry an allele for a^{41} is said to be A negative.

Since there is much variation in the antigenic potency of the various blood factors, only a few of them pose a danger in possible NI cases or in blood transfusion. The q^{a} factor is also a strong antigen. A mare that does not have a q^{a} allele is known as Q negative and will produce strong alloantibodies against q^{a} blood cells. Dr. Stormont estimates that 90 percent of all blood incompatibilities in horses arise from a^{41} and q^{a} problems, so one should mainly consider these two factors when a blood transfusion is given.

This knowledge can help the horseman prevent NI problems. Blood typing will identify A and Q negative mares, which are the source of most NI problems. A survey of 407 thoroughbreds by Stormont and his co-workers showed that over 92 percent were A positive and over 74 percent were Q positive. This combination leaves only a small percentage of mares that can cause NI in their foals and then only under certain specific conditions.

A negative mare must first be sensitized with A or Q positive blood. This sensitizing may occur when she is a foal if for some reason she does not get enough colostrum and the veterinarian gives her a blood transfusion. Plasma is used for a transfusion and even though it seems to contain no red blood cells, enough are always present to cause an antigenic reaction. If blood types were not checked when the transfusion occurred, chances are greatest that the blood was A and Q positive. In that case, the negative filly is set up for later alloantibody production against the cells of a positive foal.

NI does not always occur when conditions are proper or not always to the same degree. The reasons are not clear, but it may be due to the variations in ability of mares to produce good amounts of antibody over long periods of time. The placenta through which nutrients pass from mare to fetus is designed to keep out the red blood cells and antibodies. In most cases the placenta acts so effectively that a negative mare may carry a positive foal and not have her system come in contact with the positive antigens of the fetus's red blood cells. Generally no NI will occur in this case unless the mare carries some alloantibodies from a previous sensitization to positive blood (a transfusion or a previous positive pregnancy in which the placental barrier was compromised). If this sensitization occurred several years previously, the antibody production will usually be so low that the foal can survive with only mild anemia developing, or no symptoms at all.

How can the knowledge of blood types help the farm manager guard against NI? First it would seem to be good practice to have all broodmares blood-typed with a record of A and Q negative mares. These mares should then be checked 3 weeks before foaling for development of alloantibodies. The Serology Laboratory at the University of California, Davis (the same place that does the blood typing for The Jockey Club) will analyze the serum of the mare for alloantibodies which may affect the foal. The price is reasonable, and results are available by foaling time. There is no danger until foaling, since the only way a foal can receive the dangerous antibodies is through the mare's colostrum.

One cross-matching test uses the red blood cells of the foal and the plasma of the dam. They are placed together in vitro and if the foal's cells become clumped together because of the dam's alloantibodies, there is danger of NI.

When impending NI is discovered, it can be avoided by simply withholding the mare's colostrum for 36 hours. By that time the foal's intestinal lining will have matured so that the alloantibodies cannot be absorbed. During that period the foal will need an alternate source of colostrum or a blood transfusion.

Every large breeding farm should have a blood donor horse or two available for transfusions. Most veterinarians will be happy to work with the manager in checking the blood donors, first to see that the donor is A and Q negative and periodically to see that the donor has not accidentally become sensitized to positive blood. Only the veterinarian should give injections or draw blood from the donor as it is imperative that the equipment be sterile, with no traces of blood from other animals.

The value of a negative donor is that the blood transfused into the recipient will not contain positive alloantigens which would sensitize a negative recipient. The danger of a negative donor is that should it become sensitized, its plasma could contain alloantibodies against the red blood cells of a negative recipient.

PEDIATRICS

One of the most rapidly expanding areas of knowledge in equine medicine is pediatrics. Concentrated research in the last few years has brought forth many details about the changeovers in physiology at birth of the foal. This knowledge helps us to handle the newborn more effectively.

Among the more important changes that occur at birth are (1) the initiation of respiration, (2) a change in the circulation, and (3) a change in the absorptive and digestive capacity of the intestinal tract. The best care and treatment of the newborn foal must take into consideration the immaturity of various systems which, in the past, were sometimes assumed to function in the same capacity at birth as in the adult. The areas that deserve special consideration are: the immune system, the persistence of some fetal hemoglobin, renal function deficiency, the inadequacy of the so-called blood-brain barrier in preventing the diffusion of many drugs into the brain, and inadequacy of drug polarization and elimination.

In reviewing his experience in pediatrics a few years ago, Dr. William R. McGee told those attending a Stud Manager's Course, in Lexington, Kentucky:

> When we think of the complexity of respiration and the many hazards present in the birth process, it is amazing how smoothly this adjustment is accomplished. Likewise, it is not surprising that respiratory difficulties occupy such a dominant place as a cause of death in the newborn.

He went on to say:

> The average newborn begins to breathe spontaneously almost immediately after birth. It is a most urgent adjustment, that of establishing pulmonary respiration to take over the functions of gaseous exchange previously carried on through the placenta.

An excellent review of the initiation of respiration in foals was presented by Dr. Ronald J. Martens at the 1978 American Association of Equine Practitioners (AAEP) Convention. He said:

> In a normal vaginal delivery, the thoracic cage is maximally compressed after delivery of the head. The upper airway is thus exposed to the atmosphere which has a lower pressure than that inside the thorax. The fluid produced by and contained in the fetal lungs is thereby partially expelled. Following delivery of the thorax there is an immediate elastic recoil which tends to suck air into the upper respiratory tract; the diaphragm rhythmically contracts causing a negative intrathoracic pressure and air partially inflates the lungs.

This information comes from human studies.

Dr. Martens went on to say:

> Respiration is initiated by many factors. The immersion reflex which inhibits breathing is no longer present. The respiratory and associated centers of the previously insulated newborn are assailed by sounds, light, gravity, temperature variations, tactile stimuli and odors; the limbs and neck also flex which stimulates stretch receptors. At the same time, the tissues are using up oxygen without replenishment and adding CO_2 and H ions without elimination. All of these stimuli, and others, are important in initiating the vital first breath. As the lungs aerate, the pulmonary flow normally increases

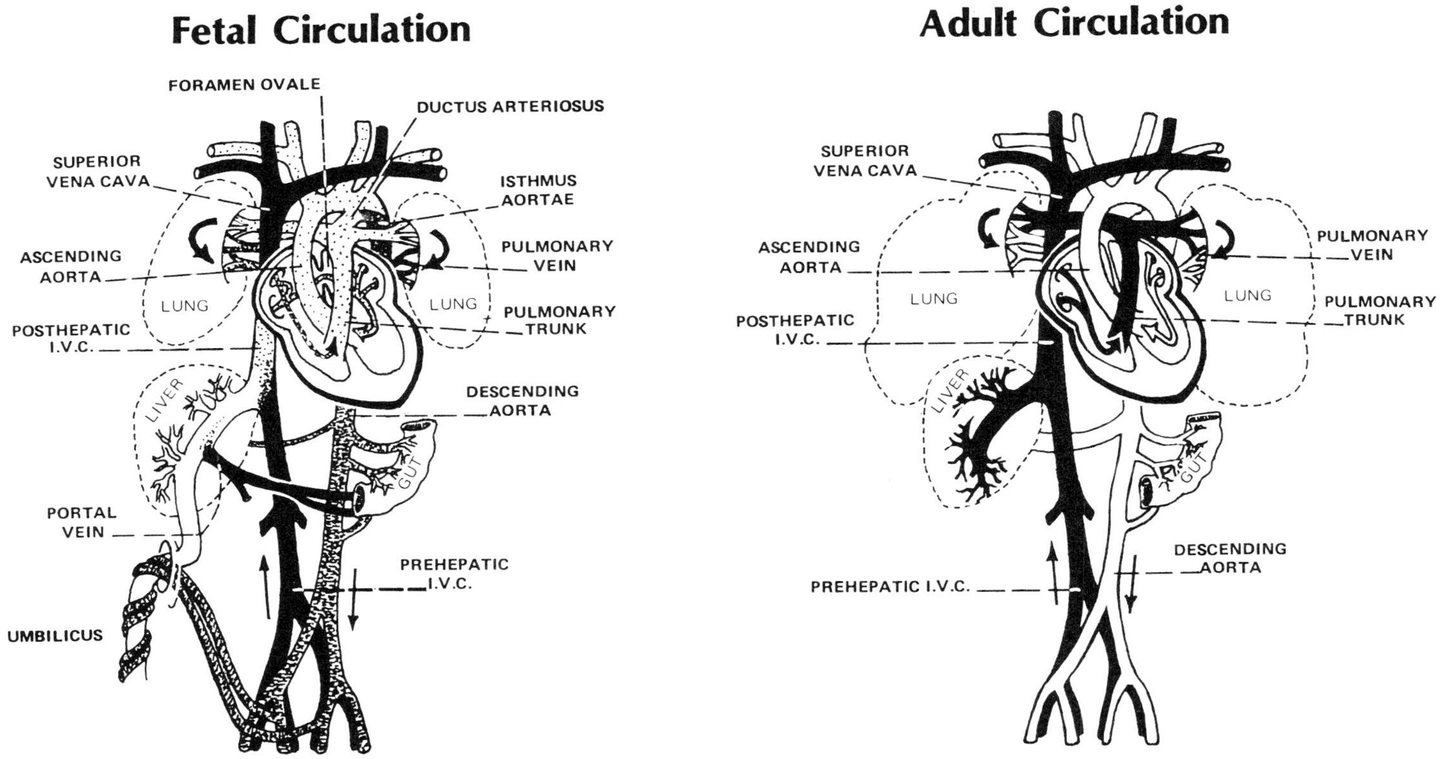

Fetal Circulation

FORAMEN OVALE

DUCTUS ARTERIOSUS

ISTHMUS AORTAE

PULMONARY VEIN

PULMONARY TRUNK

LUNG

DESCENDING AORTA

GUT

PREHEPATIC I.V.C.

SUPERIOR VENA CAVA

ASCENDING AORTA

POSTHEPATIC I.V.C.

LUNG

LIVER

PORTAL VEIN

UMBILICUS

Adult Circulation

PULMONARY VEIN

PULMONARY TRUNK

LUNG

GUT

DESCENDING AORTA

SUPERIOR VENA CAVA

ASCENDING AORTA

POSTHEPATIC I.V.C.

LUNG

LIVER

PREHEPATIC I.V.C.

Figure 19–4.   Fetal and adult circulation in the horse.

due to a decrease in the pulmonary vascular resistance and reduction in the pulmonary arterial pressure (Figure 19–4).

The pulmonary arterioles of the fetus and neonate have a thick muscular medial layer. These vessels possess very active vasomotor tone and are capable of constriction which can markedly increase the pulmonary vascular resistance.

He describes a histamine-mediated response of hypoxic pulmonary hypertension, and says:

This response is beneficial in the fetus as it diverts blood away from the lungs, through the ductus arteriosus and foramen ovale, and back to the placenta where respiration occurs. In the newborn this response can be fatal. The ductus arteriosus is still anatomically patent and the foramen ovale is held closed by an increased left atrial pressure. With the initiation of pulmonary arterial constriction, right heart pressure may predominate as in the fetus, shunting blood away from the lungs and back toward the placenta which is no longer present. This causes the foal to utilize anaerobic glycolytic pathways which result in increased lactic acid production and a metabolic acidosis. Consequently, a vicious cycle follows.

The fetus has right ventricular dominance, which is primarily due to the high vascular resistance of the unexpanded lung. The elevated pressures in the right atrium force open the flap-like foramen ovale (between the right side and left side of the heart), which in the foal is a thin delicate structure with multiple fenestrations. This allows the highly oxygenated blood returning from the placenta a more direct access to the left heart, thus reducing the degree of admixture with low oxygenated blood which would result if complete mixing occurred.

Dr. Martens claims that it may take 2 to 3 hours to achieve a relatively normal acid base balance in the foal, and 12 hours or more to achieve adult levels of homeostasis.

A neonatal maladjustment syndrome has been reported by P.D. Rossdale, a British veterinarian. This respiratory distress syndrome in the foal is due to a delay in the normal physiologic maturing of the foal and tends to return the cardiopulmonary system to the fetal state. If the condition is prolonged, it leads to secondary complications or death. Some of the causes of this respiratory stress, known as neonatal asphyxia, are: (1) partial or premature placental separation, (2) umbilical cord compression, (3) cesarean section, (4) prolonged uterine contractions, (5) drugs, and (6) septicemia.

While the foal is still involved in its cardiopulmonary adjustment, it must be up and around enough to obtain an adequate supply of colostrum. Dr. Lance Perryman of Washington State University feels that at least a pint of good colostrum is necessary to bring the foal's immunoglobulin G to the desired level. He says that this protein must be taken orally within the first 18 hours and preferably before 12 hours have passed. The epithelial lining of the intestinal tract rapidly matures between 12 and 24 hours so that the large immunoglobulin molecules cannot be absorbed. Serum from a healthy donor horse is recommended for those foals that do not obtain an adequate amount of colostrum.

Care of the navel stump is the first item of consideration once respiration is established, says Dr. William McGee. "It can be said that the efficiency of this treatment decreases in direct proportion to the time it takes to do it!" He explained that the objective is to treat the navel before exposure to infection, not to attempt to kill infection which has been present for a time. He likes tincture of iodine.

Also an initial consideration should be routine administration of penicillin at recommended doses, according to Dr. McGee. He warns that too much streptomycin can cause eye damage; he claims that this precautionary injection can delay rapid progress of certain conditions until more specific treatment can be administered.

A third initial consideration is the administration of a warm, mild, soapy water, or glycerine in water, enema shortly after the foal is up and around. He recommends use of a 1- or 2-quart enema bag or can and a soft rubber tube. "The tube should never be inserted more than 2 to 4 inches into the rectum," he says.

In discussing initial nursing, Dr. McGee says that many normal foals are unable to get up and nurse for 2 to 6 hours because of muscular incoordination. He recommends that the foal receive help if he has not nursed on his own within a couple of hours, or that the mare be milked out and the foal fed from a bottle.

Through many years of experience, Dr. McGee has observed a pattern of good and bad years as far as incidence of foal septicemia is concerned. He says some foals are born with infections and fail to gain the strength to get up and nurse. Others, even though able to get up, often present the signs associated with the "dummy foal." They cannot nurse, and wander aimlessly about the stall, bumping into the wall or other objects with no apparent concern. Dr. McGee feels that the more common "foalhood infections" are caused by streptococci, staphylococci, *Shigella equi,* and *Escherichia coli,* either individually or in combination.

Dr. McGee says these infected foals soon become dehydrated:

> Failure to recognize this and supply the necessary types and amounts of fluid to correct the fluid balance at the beginning, in addition to antibiotics, will result in poor response to the overall treatment.

When the septicemia progresses to joint ill, it is usually indicative of a streptococci infection, claims Dr. McGee.

> The degree of tissue damage is extremely important and indicates more than anything else the type of treatment to be used. If recognized early enough, it is entirely possible to clear such a joint by local poultice application and systemic antibiotic administration. When there is only moderate soreness, without appreciable swelling or significant rise in temperature, I prefer to try such treatment.

Dr. McGee emphasized that "a day or so without treatment will change this mild case to one showing lameness, acute soreness and swelling around the joint, and two or three degrees of fever." He says treatment at this stage must be considerably more energetic and must include high levels of more potent antibiotics.

According to Dr. Humphrey D. Knight of the University of California:

> Conditions that weaken the foal and favor infection are chronic endometritis, dystocia, mishandling of the foal at birth, overcrowding and poor ventilation. Conditions that modify the environment so that select microbes can proliferate are poor hygiene and sanitation, repeated use of a foaling stall, and the selective pressure placed on the microbial environment by the overzealous use of antibiotics.
>
> In my opinion, the prophylactic injection of an antibiotic into a healthy neonate is of unproven value and unnecessary. Typically, invasion of the foal by the organisms occurs before or soon after birth; bacteremia results and the foal is quickly overcome. The umbilicus is usually considered the major portal of entry; however, respiratory, nasopharyngeal and gastrointestinal sites may be of equal importance.

Dr. Martens offers the following list of approaches to the treatment of septicemia and, in particular, pneumonia:

1. Minimize stress in the neonatal period.
   a. assure a patent airway and ventilate if necessary.
   b. correct metabolic acidosis with sodium bicarbonate.
   c. give intravenous glucose to replenish metabolic reserves.
   d. protect against hypothermia.
   e. minimize handling and excitement.
   f. provide a clean environment.
2. Assure an optimal immunological status.
   a. immunization, parasite control, and proper nutrition of the dam during gestation.
   b. adequate colostral or serum substitute intake by the foal.
3. Minimize ventilation/perfusion abnormalities.
   a. acid-base balance.
   b. fluid therapy.
   c. bronchodilatation.
   d. mobilization of obstructing secretions.
   e. prevention of hypostatic congestion.

4. Provide relative comfort.
   a. antipyretics.
   b. cough suppressants in paroxysmal coughing but not agents that totally inhibit coughing.
5. Antimicrobial therapy.
   a. determine specific causative agent if possible and treat accordingly.
   b. use adequate dosages and frequencies for sufficient lengths of time to assure adequate blood levels.

Dr. McGee believes that rupture of the bladder occasionally occurs when the structure is full at birth. "The compression of the foal's abdominal cavity and contents during passage through the birth canal simply causes a 'blowout' through the wall of the bladder," he says. Or the rupture can result from a sudden violent jerk on the navel cord, such as when the mare foals standing up, he maintains.

The initial signs of this condition can be easily overlooked, Dr. McGee says:

> The foal appears quite normal to the casual observer for 2 or 3 days. About the third day it appears a little dull and uncomfortable. The temperature is generally near normal, respiration somewhat shallow and a little faster than normal, and membranes are slightly injected, often with a slight yellowish cast. While not showing colicky symptoms as such, the foal will be seen to strain occasionally with tail lifted. Quite often this is not too disturbing to the owner or manager and he administers a dose of combiotic and some milk of magnesia. Perhaps an enema is given also.
>
> By the following day all of the signs have increased in intensity. The temperature is now around 102 degrees; the respiration faster, although more shallow. The foal is definitely more uncomfortable and presents a rather anxious appearance. Closer examination finds the abdomen distended, with the feel somewhat of a beach ball full of water. This is the picture when the veterinarian arrives.

The diagnosis can be confirmed by inserting a hypodermic needle into the abdominal cavity and obtaining free-flowing urine. "Prompt surgical repair is indicated," he says, "and when done promptly can be expected to result in rapid and complete recovery."

Dr. H.K. Knight discussed "Infectious Diseases of the Newborn Foal" at the 1978 AAEP Convention. He said of equine herpesvirus 1, "Usually infection by this agent results in respiratory disease in young horses, or abortion of the fetus. Recently, Bryans and co-workers described a clinical syndrome in neonates resulting from infection by equine herpesvirus 1 (EHV1) during the prenatal period. Some foals were ill at birth and some appeared weak; however, the majority were born healthy and well developed. Weakness, lethargy, and failure to nurse, the common signalment of neonatal disease, developed during the first week of life. Respiratory distress and diarrhea ensued. The foals contracted a variety of secondary bacterial infections which failed to respond to therapy and they succumbed after a few days to 2 weeks of life. Lesions common to all included interstitial pneumonia and underdevelopment of the spleen, thymus, and adrenal cortex."

Dr. Knight said the frequency of meningitis in foals seems to be lower than in infant humans. He explains:

> Perhaps newborn foals with septicemia succumb too rapidly for the signs of meningitis to be noticed. When meningitis is present during the initial septicemia, early death is the usual outcome. However, in cases where the initial bacteremia is contained but meningitis develops secondary to a chronic focus of infection (e.g., lung abscess), it may be well worth treating even though the prognosis must still be considered extremely poor.

A recently described liver infection that occurs in foals is known as Tyzzer's disease. Dr. Knight says that in the liver:

> . . . the damage consists of foci of necrosis and inflammation containing bundles of aligned bacilli. The disease strikes foals usually during their first month of life and has a peracute course. Foals may die suddenly without evidence of previous illness or they may have a short period of semi-coma preceding death. The course is usually too short for successful therapy, and recommendations for prevention and prophylaxis have been difficult to make because of incomplete knowledge of the epidemiology of the disease. Isolation and careful attention to sanitation is always beneficial and may minimize outbreaks. Prophylactic use of antibiotics may be of some value but antibiotic susceptibility tests have given varied results. Tetracycline and penicillin have been recommended.

An infectious diarrhea may occur in foals often accompanying an overwhelming infection. "Initial therapy should be directed toward correcting the defects in homeostasis," days Dr. Knight, "and only after this should attention be directed toward antimicrobial therapy."

Corynebacterium equi pneumonia is one of the most insidious and devastating in foals, says Dr. Knight.

> The history is almost always the same: the foal was born normally and seemed strong for the first 4 to 6 weeks. Then, maybe the owner noticed some vague signs of lethargy and failure to thrive, or possibly the foal began to cough but still appeared alert and continued to nurse. The clinical picture is of a somewhat underweight, but otherwise normally developed foal showing dyspnea to a varying degree. It is still an active animal but definitely distressed. On auscultation of the lung, an increased harshness of breath

sounds and, invariably, adventitial sounds are heard. The temperature usually fluctuates during the day and may reach 104°F in the afternoon and drop to 102°F in the morning. In treated cases there is usually some improvement initially followed by a progressive deterioration. Seldom does an animal survive more than 2 weeks once severe dyspnea develops.

Fortunately, we can do something for streptococcal pneumonia, says Dr. Knight. "Most of our cases occur in the spring and early summer, most often in foals 4 to 8 months old. However, we also see the disease in younger animals and adults."

Close attention to young foals and the application of the vast amount of new knowledge about equine pediatrics in the past decade should greatly increase the survival rate of foals.

EARLY TRAINING

An important detail in the successful raising of foals is routine early training. If the foal does not quickly become accustomed to man and allow itself to be handled easily, it will probably be injured the first time it gets into trouble. Almost every foal will get a cut or two before weaning or will be in need of veterinary treatment of some sort.

The easiest procedure is to work with the young foal a little every day or so. Petting, talking, and rubbing the baby horse helps to gentle him. Within a few weeks a small well-fitted halter should be put on the foal and he should be taught what it means to be led.

Within 6 to 8 weeks the foal can be easily taught to lead without much danger that he will hurt himself. At the same time the feet should be routinely handled so that a battle does not ensue every time a trim is needed.

This early training will make future handling of all kinds much easier, and will lead to a gentler disposition in the long run. A short grooming session once a week will help the foal to retain what he has learned. The breeder who has taken the trouble to provide this early training will fare much better when an emergency arises and cooperation of the foal is needed.

IMMUNITY

The modern horse breeder must concern himself with immunity, the response of his horse's body to foreign invaders. The horse breeder, through selection, has a great deal of eventual control over the immune system of his horses. Ia Tizard describes it in *An Introduction to Veterinary Immunology*:

> The most important of the general factors that influence disease resistance are genetic. Under natural conditions, spread of disease through a population may therefore initially eliminate all susceptible animals but leave a resistant residue to multiply and make use of

the newly available resources such as food. By appropriate breeding programs it is therefore possible to develop strains of animals that are either highly resistant or susceptible to a specific disease.

Another way that a horse breeder has influence over immunity is through the feed he provides. A poor level of nutrition reduces disease resistance.

Immunity to invading organisms comes about by two basic mechanisms; the first is a neutralization of toxins produced by the organisms, and the second is destruction of the organisms. These immunities are brought about by cells of the body which engulf the organisms and by antibodies, which are large "active" protein molecules that attach to and destroy the power of the foreign invaders.

The immune system of the newborn foal is not yet completely developed. The cells which engulf foreign invaders have not yet matured and are not fully active. The antibodies circulating in the blood of the dam are too large to pass through the placental barrier of the uterus. Immunological protection comes from the colostrum received by the foal during the first few hours of nursing. The "first milk," which contains a large supply of antibodies produced by the mare, circulates throughout the system of the foal after ingestion. Normally this colostrum feeding gives enough protection to last the foal a few weeks until his own system begins to mature.

Genetic immunodeficiencies, discussed in Chapter 10, will not become evident until the antibodies of the colostrum are depleted from the foal's system. After birth the normal foal starts to develop a light active immunization against the organisms it contacts in its environment. This ability increases until 6 or 8 months of age. Immunization injections prior to about 4 or 5 months usually have little lasting effect. When discussing immunity, the words "generally" and "usually" are common because great natural variation exists among individuals as to the development of the system and the strength of the system at maturity.

The strength of the immune response has been an active selective factor in nature for millions of years. The unknowing horse breeder, with syringe in hand, often fails to select against horses with a weak immune system, thus allowing it to perpetuate itself and increase the horse's dependency upon man and his antibiotics.

FAILURE OF PASSIVE IMMUNITY

It has become alarmingly clear how badly the foal needs the help of its mother's antibodies during the risky time of life after birth, before it can build its own defense system. Most foal deaths during the first weeks of life occur because of a lack of these vital antibodies. Research has shown that about 25 percent of all foals have some degree of passive (colostral) antibody transfer failure. About the third month, the foal's self-produced active immune system affords significant protection (Figure 19–5).

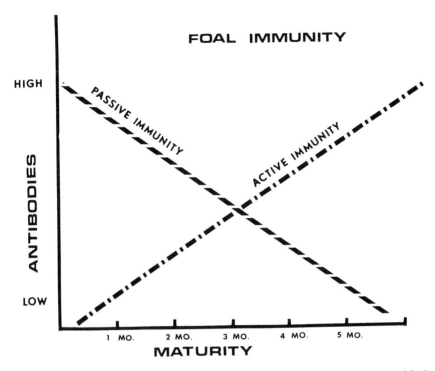

Figure 19–5.   The changing dominance of passive versus active immunity in a normal foal.

The colostral antibodies have been identified as immunoglobulin G (IgG), and its measurement is expressed in mg/dl. When the level of IgG is below 200 mg/dl, 78 percent of such foals die, and 25 percent of foals will die with levels between 200 and 400 mg/dl. Normal serum level of IgG is approximately 1500 mg/dl.

A research study showed that all foals dying of infection during the first week of life have insufficient blood levels of antibodies. This characteristic is true in all breeds. They die of such conditions as hepatitis, pneumonia, arthritis, nephritis, pleuritis, and omphalophlebitis. Such foals are unable to cope with the normal bacterial invaders ever-present in the barnyard environment.

Passive transfer of immunity to foals may fail for several reasons. One cause is the mare that bags up prematurely, then drips the precious colostrum away before the foal is born. When this begins to occur, the horseman should "milk out" the colostrum and save it in the freezer, to be given to the foal within the first few hours of birth. Another solution is to call in a veterinarian who can induce labor with a hormone shot if the mare is otherwise ready to foal.

Some mares produce colostrum which is, for reasons unknown, deficient in antibodies; thus, their foals do not have adequate immunity. A study of 11 mares which produced antibody-deficient foals was conducted by Dr. Jim Crawford of Washington State University. Dr. Crawford showed that about half of them had insufficient levels of antibodies in their colostrum. Through a complicated biological process, these antibodies pass from the mare's blood to her milk, through the tissues lining the mammary gland. This process is somehow interfered with in some mares. It may be a genetic defect, or it may be a result of certain types of drug therapy at the wrong time.

Another cause of failure is the foal's suckling pattern. Sometimes, even with assistance, a foal will not suck. The foal seems to lack a desire to nurse and often will appear rather weak. In this case the colostrum can be milked from the mare and administered by stomach tube. This is a job for the veterinarian, however, since it is easy to get the tube down the trachea to the lungs.

Occasionally a foal which swallows adequate amounts of good colostrum will for some reason be unable to absorb the antibodies through the intestinal lining. Normally the foal's gut is able to absorb these large molecules during the first 24 hours of life. Certain drugs, such as steroids, administered during this time can block the absorption from the gut. Within 24 hours the gut lining matures to the point that antibodies cannot be absorbed intact; digestive enzymes destroy these immunoglobulins.

Breeders who produce a large number of foals each year continually run into problems with failure of colostral antibody transfer resulting in early foal death. To be prepared, a bit of colostrum should be "milked out" and saved from each mare that has an abundant supply. A couple of quarts of frozen colostrum will go a long way toward solving colostrum failure problems and will result in a more healthy foal crop.

Modern techniques offer the opportunity to evaluate the serum antibody level of young foals. A simple test has been perfected by veterinarians at the University of California, Davis, to determine the relative level of antibodies in a foal. All foals which appear to be weak should be checked.

After the initial 24-hour period, antibodies can be administered to a deficient foal by means of blood serum transfusion. Many veterinarians maintain a supply of frozen serum to be used in such cases. Some veterinarians feel that it is more advisable to use fresh serum from a gelding on the farm where the foal is kept to ensure that the serum contains specific antibodies against all the microbial invaders on that particular premise. Normally 1 or 2 liters of serum gives an adequate level of antibodies for the foal.

THE ORPHAN FOAL

Sometimes the udder of the dam is functionally destroyed, or so far damaged as to be incapable of producing a supply of wholesome milk, or the dam may die and leave the offspring to be reared by foster mothers or by hand.

To procure a foster mother is always a difficult task. Occasionally, a mare will lose her foal, and a foal will lose its mother, about the same time. In this case it is an advantage to bring the survivors together. Although the problem is alleviated somewhat, the foal may develop digestive disorders and diarrhea at first, especially if the foal has not received the first laxative milk of its dam. Whether hand rearing is desirable depends on the age, character, and breeding of the offspring. The rewards of what can amount to a full-time job the first week or so with a new orphan foal can be great indeed as the little creature begins to respond to his human "mother." The younger it is when deprived of its natural parent, the greater amount of trouble it will give. With no foster mother available, another source of food supply is the cow. The most suitable milk for this purpose will be obtained from a heifer a week after calving. Once having begun with the milk of a particular animal, it is best not to change.

Although the following figures show that the same constituents are found in the milk of the cow as that of the mare, the actual and relative proportions of these constituents differ to a considerable extent. To have this milk more closely approximate the composition of mare's milk, water must be added to reduce the proportions of casein and fat. Also the deficiency of sugar must be made up. At first the proportion of water to cow's milk should be one part of the former to two of the latter, but as time goes on, one part to three will be more suitable. Later, water may be excluded altogether.

	Cow's Milk	Mare's Milk
Water	87.0	88.0
Fat	4.6	1.0
Casein	4.0	1.6
Sugar	3.8	8.9
Salts	0.6	0.5

Borden Chemical Company has developed, through years of experimentation and research, a product called Foal-Lac (Table 19–1), which contains all of the essential ingredients of mare's milk. To assure the foal of extra nutritional support, the level of vitamins and minerals in Foal-Lac is higher than in mare's milk.

Guidelines offered in *The Borden Guide to the Care and Feeding of Orphan and Early Weaned Foals* are: The foal should be fed four times per day (every 6 hours if possible) for the first week as an orphan. If the foal adapts well to the feeding schedule and handles the increases without difficulty, it may be changed to a schedule of three times per day (every 8 hours if possible) during the second week. If the foal does not readily consume the increases in feeding, it should be continued on four feedings

Table 19–1. The Ingredients of Foal-Lac and Mare's Milk Compared

Ingredient	Mare's Milk	Foal-Lac
	(Solids Basis)	
Protein (%)	19.6 %	20.2%
Fat	14.2%	14.4%
Fiber	0.0%	0.2%
Milk Sugar (Lactose)	53.4%	52.6%
Ash	3.6%	7.2%
Calcium	0.9%	0.9%
Phosphorus	0.5%	0.8%
Cobalt	——	1.1 mg/lb
Copper	——	14.6 mg/lb
Iodine	——	0.8 mg/lb
Iron	——	57.8 mg/lb
Magnesium	400.0 mg/lb	501.0 mg/lb
Manganese	——	24.0 mg/lb
Potassium	0.5%	1.6%
Sodium	0.18%	0.8%
Zinc	——	20.3 mg/lb
Vitamin A	1818.0 USP Units/lb	10,500.0 USP Units/lb
Vitamin D_2	varies	2,100.0 USP Units/lb
Vitamin E	——	2.5 Int. Units/lb
Vitamin K	——	2.0 mg/lb
Thiamine	1.2 mg/lb	2.4 mg/lb
Riboflavin	0.8 mg/lb	9.2 mg/lb
Pantothenic Acid	12.1 mg/lb	17.0 mg/lb
Pyridoxine	1.2 mg/lb	2.4 mg/lb
Niacin	2.0 mg/lb	5.2 mg/lb
Choline	——	636.0 mg/lb
Folic Acid	4.0 mcg/lb	266.0 mcg/lb
Biotin	——	13.0 mg/lb
Vitamin B_{12}	12.1 mcg/lb	11.6 mcg/lb
Ascorbic Acid	404.0 mg/lb	454.0 mg/lb

per day. Three feedings per day should be sufficient for the third week, and can be reduced to two feedings per day during the fourth week. Finally the feedings should be reduced to one per day for a couple of days prior to discontinuing liquid feeding.

Directions on the container outline two different feeding schedules. One of these, which should be used with a ration formulated for young foals, contains a minimum of 16 percent protein supplemented with vitamins and minerals. The second one should be used with a simple grain mixture of a low-protein horse feed. These feeding schedules, including the introduction of Foal-Lac pellets and the grain ration, should be self-explanatory so they are discussed only briefly.

Feeding liquid Foal-Lac in an open bucket will probably bring the most success. To start the foal, rub a little on its mouth and place its nose in the milk. It sometimes helps to hold the bucket for the foal until it starts to drink. Then lead it to the permanent feeding area with the bucket. The bucket should be placed about the height of the foal's shoulder and tipped slightly toward the foal so that all the milk can be readily consumed. The foal may require 1 or 2 hours to drink the milk during the first couple of days.

Wash the bucket in hot, soapy water after each feeding. Rinse thoroughly and invert the bucket so that it dries before the next feeding.

Consult a veterinarian if there are any signs of disease, fever, parasites, or other conditions. A worming program should be established even for a young foal to keep it free from parasites. Foals less than 1 week old should be out of the weather, especially at night. A heat lamp in the stall is desirable if the foal is orphaned during the first 3 days of life and especially if foaling occurred in cold weather. Place the lamp in such a position that the foal can get away if it becomes uncomfortable. If it is doing well, the lamp can be removed after the first week.

At the start of the second week, some grain or a special complete foal ration should be offered. It may be desirable to place a couple of handfuls of good-quality, heavy rolled oats in the feeding box along with the grain mixture or complete foal ration. Foals generally like rolled oats, and the oats will encourage them to start eating the grain ration. If the grain is not consumed in 2 days, remove it from the box and offer fresh food. The foals can be allowed to eat all the grain ration they want if there are no problems with stool consistency. They may be eating 1 to 2 pounds per day by the time they are 1 month old.

During the third week Foal-Lac pellets should be introduced and mixed with the grain ration. Just add a small handful at first until the foals eat the pellets readily. As the Foal-Lac powder is decreased in the 30-day liquid schedule, Foal-Lac pellets should be increased so that by the time liquid feeding is discontinued, or shortly thereafter, the foals are receiving 2 pounds of Foal-Lac pellets per day. If the 8-week liquid feeding schedule is followed, Foal-Lac pellets should be fed at the rate of 1 pound per day until the 50th day. Then increase Foal-Lac pellets to replace the powder as it is reduced so that the foal receives 2 pounds of pellets per day by the 60th day.

Good-quality, dust-free alfalfa hay should be made available at about 3 weeks of age, particularly if pasture is not available or if alfalfa is not included in grain pellets.

Provide fresh water starting the second week. A trace mineral salt block may be used, particularly if a home-mixed grain ration is given. A complete foal feed will include adequate salt.

Orphan foals should be exercised daily. After they are 1 to 2 weeks old, depending on their condition and the weather, they can have free access to pasture as long as there is a protected feeding area and a place they can go for shelter from wind and rain.

Table 19–2.   The Use of Foal-Lac

Schedule 1

Age of Foal	No. of Feedings Day of Liquid Foal-Lac**	Warm Water Pints	Total Daily Foal-Lac Powder, Lbs*	Total Daily Foal-Lac Pellets, Lbs
0–3 days	4 (4)	4 (4)	1 (1)	0
4 days	4 (4)	6 (6)	1.5 (1.5)	0
5 days	4 (4)	6 (6)	1.5 (1.5)	0
6 days	4 (4)	8 (8)	2.0 (2.0)	0
7 days	4 (4)	8 (8)	2.0 (2.0)	0
2nd week ***	3 (3)	8 (8)	2.0 (2.0)	0
3rd week	3 (3)	8 (8)	2.0 (2.0)	¼ (¼)
4th week	2 (3)	6 (10)	1.5 (2.5)	½ (½)
29 days	1 (2)	4 (10)	1.0 (2.5)	1 (1)
30 days	0 (2)	0 (12)	0 (3.0)	1 (1)
31–35 days	0 (2)	0 (12)	0 (3.0)	1½ (1)
6th week	0 (1)	0 (12)	0 (3.0)	2 (1)
7th week	0 (1)	0 (8)	0 (2.0)	2 (1)
8th week	0 (1)	0 (4)	0 (1.0)	2 (2)
9th–12th week	0 (0)	0 (0)	0 (0)	2 (2½)
13th–16th week	0 (0)	0 (0)	0 (0)	1 (2½)
17th–20th week	0 (0)	0 (0)	0 (0)	½ (1)
21st–24th week	0 (0)	0 (0)	0 (0)	½ (1)
Total Foal-Lac consumed in 6 months			50 (113)	169 (240)

*Three level cupfuls equal one pound

**Divide the total daily water and Foal-Lac powder allowance equally into the number of feedings indicated.

***Start offering grain ration in small amounts. Fresh water should be made available at this time. Provide a small quantity of alfalfa hay unless the roughage is provided in the mixed grain ration or the foals are allowed access to pasture. A salt block should be provided with a home-mixed grain or low-protein horse feed program.

Foals orphaned during the second or third week of life will be 20 to 30 pounds heavier than foals orphaned during the first week unless the former have not received sufficient milk from the mare or the mare's milk has distressed the foal. If the foal is in a weakened condition, follow the directions for younger foals outlined previously. Otherwise start the foal on a 1-pound level of Foal-Lac powder for 1 or 2 days. If the foal readily consumes this amount and shows no digestive disturbances due to diet change (from mare's milk to Foal-Lac), the foal can be brought up to 1.5 pounds and then 2 pounds of Foal-Lac powder over a period of a couple of days. The schedule can then be picked up for a 3-week-old foal, as shown in the detailed feeding schedule (Table 19–2). These foals can start right off on a feeding frequency of three times per day.

Foals that have been with the mare as long as 2 weeks are often more difficult to get on Foal-Lac. One should not be concerned if they do not

Table 19–2.   The Use of Foal-Lac (Continued)

Schedule 2

Age of Foal

Age	
0–1 week	Normal program for mare with foal.
2nd week	Introduce foal to grain ration by tying foal with the mare at the feed trough. Fresh water should be available.
3rd week	If foal is eating, tie away from mare during feeding and introduce the foal's own grain ration. Provide ¼ to ½ pound of Foal-Lac pellets per day.
4th week	Continue to encourage dry feeding during mare's feeding time. If possible this should attain 1 pound of Foal-Lac pellets and 1 pound of grain or complete feed per day.
5th week	Foals can be weaned anytime after they are 4 weeks old and in good health. Note previous discussion on handling the separation of mares and foals and other management suggestions. To ease the transition, powdered Foal-Lac mixed with water may be tried for 1 week. Some foals may not readily accept the liquid feed. Feed liquid 2 times per day. Limit alfalfa-brome hay to 1 pound per day.

	Amount of Warm Water (pints)	Amount of Foal-Lac Powder/day (pounds)	Amount of Foal-Lac Pellets/day (pounds)	Amount of Grain Ration/day (pounds)
5th week	4	1	1	1
6th week	0	0	1½	1½
7th week	0	0	2	2
8th week	0	0	2½ (3)	2½
9th week	0	0	2½ (3)	2½ (3)
10th week	0	0	2 (3)	3½ (3½)
11th week	0	0	1½ (2½)	5 (5)
12th week	0	0	1 (2)	7 (7)
5th and 6th month	—	—	0–½ (½–1)	Full Feed
After 6 months	—	—	0–½ (0–½)	Full Feed
Total Foal-Lac pellets			100–150 (150–200)	

The quantities of Foal-Lac pellets recommended depend on the quality of the grain ration. Figures in parentheses are suggestions for use with a simple unfortified grain mixture. The other figures are for use with a ration formulated for foals containing a minimum of 16% protein and fortified with vitamins and minerals.

After 6 months of age the protein content of the grain ration might be reduced to 14% protein.

The continued use of Foal-Lac after 6 months will help provide extra condition and bone development. Some research has shown continued growth stimulation from this low level Foal-Lac feeding after 6 months, but some research has shown no increase when a 20% protein ration was continued after 6 months.

readily consume all the Foal-Lac liquid, as they will consume up to 2 pounds of powder per day.

Foals orphaned after 1 month with the mare or intentionally weaned from the mare at this age are sometimes reluctant to drink liquid Foal-Lac. The foals should be offered liquid Foal-Lac at a level of 1 pound of Foal-Lac powder per day divided into two feedings. If the foal readily consumes this amount, it can be increased to 2 pounds of powder after 1 or 2 days and continued for 7 to 10 days or until the foal has adjusted to living without the mare. The liquid feeding at this age serves to bridge the gap from the mare to dry feed, although it is not required for the production of a strong, healthy animal.

During this time the orphaned foal should be encouraged to start eating the grain ration, Foal-Lac pellets, and hay so that it can be on a dry feed program within the week or 10 days mentioned previously.

WEANING FOALS

For several reasons it may be desirable to wean the foal from the mare as early as 1 month of age. Some of these are (1) a need to show the mare, (2) fewer problems with returning the mare to the stable for personal or commercial riding, (3) sale of mare or foal without the sale of the other, and (4) ease of shipping the mare (without foal) for breeding purposes. If a breeder plans to wean early, the foal feeding program should be designed so that it can be accomplished with a minimum of effort.

Early weaning should be accompanied by the use of Foal-Lac, and special procedures are necessary. After the foal is 1 week old, it should be tied at the feeding trough with the mare. This practice encourages the foal to eat the grain ration. Again, some rolled oats may aid the consumption of the mixed feed. Place some feed in the foal's mouth, so it can become acquainted with this new experience. Starting during the second week, put a small quantity of Foal-Lac pellets in the feeding area where the foal is tied so it will also be familiar with this feed.

Foals will vary greatly in their aggressiveness at the feed trough. Some will start eating dry food at the first exposure while others may not eat much even by the time they are 4 weeks old. If a foal does start to eat dry feed readily, it will help the weaning process to tie the foal away from the mare at feeding time and give it access to its own feed.

When the time comes for early weaning, say, 4 to 6 weeks of age, the break should be complete. Do not try to bring the mare and foal together for periodic nursing or the foal will not eat readily and both the mare and foal will be nervous. The mare and foal should be separated from both sight and hearing of each other if at all possible. They will try to get together for a couple of days, especially if they can see and hear each other. This situation could result in management problems.

Make sure that the early weaned foal has fresh water. It is desirable to feed a special foal ration with a minimum of 16 percent protein and fortified with vitamins and minerals. The milk nutrients will be provided by the pelleted Foal-Lac. The Foal-Lac should be brought up to 2 pounds per day as soon as possible. The feeding of liquid Foal-Lac as discussed previously will ease the transition.

Foals that are weaned at 4 weeks or older and occasionally foals orphaned after they are 2 or 3 weeks old will become nervous and may not eat well because they miss the company of the mare or other horses. Chances are that only one foal is orphaned or has been early weaned. To overcome this problem, it is well to place another farm animal or a gentle dog with the foal for companionship.

As the Foal-Lac liquid is increased, a temporary stool looseness may appear. If this stool gets very watery and persists for 24 hours, the Foal-Lac should be reduced to the original level for 2 days and then increased, perhaps more slowly than indicated in the feeding directions. If the condition does not correct itself, a veterinarian should be consulted.

The hair around the foal's muzzle may be lost after a couple of weeks because of contact with the liquid; the hair returns normally when the foal is off liquid feed.

These feeding directions will result in good sound foals. Some may prefer to feed liquid Foal-Lac at higher levels and for longer periods than indicated, particularly if the foals are doing well and extra condition is desired. Some have fed 6 to 8 pounds of Foal-Lac powder mixed with the appropriate amount of water per day to foals that are 2 to 3 months old. These animals take on a fine condition and are excellent for sale or display. Be careful to avoid developing epiphysitis at these higher levels. Foal-Lac powder mixed with water can be continued through 4 to 6 months in place of pellets if the owner prefers this approach.

The chart shown in Table 19–2 gives the Borden schedule for feeding of light horses (900 to 1200 pounds mature weight). Ponies will require about one-half these quantities. Note that there are two sets of directions; the choice depends on the quality of the grain ration used.

MARES AND FOALS

Data on the behavior of mare and foal are beginning to surface. Recent studies by K.A. Houpt and T.R. Wolski at Cornell indicate that a foal recognizes its dam by means of vision, smell, and other factors. Their data are reported in *Applied Animal Ethology*.

Their study showed that visual recognition is important to the mare in identifying her own foal. They used foals varying in age from 3 weeks to 6 months. The mares had learned to recognize their foals previously by sight. They tested the role of smelling by blocking olfaction with mentholated ointment in the nose of the mares and on the foals. Although the mare's identification was slower, they presumably used other means and were successful in identifying their own.

Sound is another factor that mares were found to use to identify their foals. The Cornell veterinarians made tape recordings of the foal's neighs and played them to the mares and found that the mares neighed more often in response to their own foal.

These authors mention the theory that the initial identification of a foal is entirely olfactory—smell that comes from the placenta and amniotic fluid in which the foal has developed. They described how occasionally a mare will prefer to stay in the vicinity of the amniotic soaked bedding rather than follow her foal that is led away. They say that the initial licking and sniffing are very important in establishing the foal's identity to the mare.

It must be that the sight and sound eventually take over as means of identification. Drs. Houpt and Wolski claim that mares pastured with other mares and foals sniff their foals more often than do mares in stalls. They report that if only her own foal is present, she has little need to check that the correct foal is nursing.

In contrived test situations, it took an average of 30 seconds for a foal to distinguish its dam from another. With vision blocked, it took twice as long. With smell blocked by mentholated ointment, it took a foal 3 times as long, on the average, to recognize its dam. With both vision and smell blocked, foals were still able to recognize their dams, but it took more than 3 minutes on the average.

Foals with quiet mothers were consistently slower at finding them. The researchers found that the foals quite regularly approached the wrong mare if she were neighing. Tests with tape recordings of dam's neighs showed that foals cannot readily distinguish one mare's neigh from another.

Also a part of the study was collection of data about suckling and how carefully the mare looked after the foal. They found that a foal will suckle, on the average, four times every hour during the first week and this drops to an average of once an hour by 5 months. The data showed that a mare will stay within 5 yards of her foal 95 percent of the time during the first week. By 8 months the mare will remain at this close distance only 20 percent of the time.

It is difficult to identify specific reasons why a mare will reject her own foal. The Cornell team postulates that equine maternal behavior probably depends on the rising progesterone and falling estrogen levels seen in the parturient mare. They site evidence in rat experiments where blood-borne factors cross-transfused can induce maternal behavior. Also they tell of combinations of estrogen and progesterone injections producing maternal behavior in rodents.

Genetic factors, as yet unrecognized, must also play a part in maternal behavior since mares that are good mothers tend to have daughters that are good mothers, the authors claim. This result could be accomplished through inheritance of proper glandular function. They also claim that maternal behavior is somewhat learned, citing evidence that foal rejection occurs more frequently in first-time mothers.

Imprinting is a phenomenon that helps in understanding a mare's behavior at foaling. The young equine becomes imprinted with the idea of a large moving object and seeks to follow it. This imprint concept helps to explain why the mare tends to keep all other large animals, including people, away from the foal during the first few days of life. Eventually the foal learns to recognize its mother by means of sight and smell.

What is initially a blessing, as the normal mare develops an obsession with close proximity to the foal, eventually becomes a curse at weaning time. Either the mare or foal, or both, often fret themselves into losing weight and sometimes injuring themselves if improperly handled.

A note in The Chronicle of the Horse told of a unique method of weaning on some thoroughbred farms in Kentucky. The technique is called "interval weaning." It is described as follows:

> In this method, mares and foals from the barn are turned out together in a large field. They are allowed to run together long enough to develop a herd sense, in that they move together, and have a definite pecking order established. Then, one or two at a time, the mothers of the oldest, strongest foals, now four to six months old, are removed from the pasture. This should be done quickly and quietly, preferably while the foals are occupied somewhere else, so as to reduce the disturbance to the herd. The mares must then be removed to an area of the farm completely out of sight and hearing of the weanling herd. The foals, of course, miss their mothers and are upset for a time, but they quickly notice that their contemporaries are still calmly munching grain from the creep-feeders or lying about in the sun-warmed grass. They may band together in their loneliness, and will even seem to set up their own herd structure within the larger herd. As more and more mares are removed, over a period of three to possibly seven weeks, depending on the age spread of the foals, the band of broodmares is quietly turned into a weanling herd, with its own leaders and pecking order established.

The writer of the note warns that the entire operation should be carefully supervised and that the pasture fences must be safe and strong. It is reported that most foals quickly adjust to the herd of familiar horses, without their mothers.

Those who have conducted "interval weaning" say that weanlings are calmer, have less growth interruption and a lower injury rate. They also mention the absence of the ear-shattering sounds of a typical weaning day using conventional methods.

BIBLIOGRAPHY

Arthur, G.H.: Veterinary Reproduction and Obstetrics, 4th ed. London, Bailliere and Tindall, 1975.

Barty, K.J.: Observations and procedures at foaling on a thoroughbred stud. Aust. Vet. J. 50:553, 1974.

Becht, James L., and Page, E.H.: Neonatal Isoerythrolysis in the Foal: An Evaluation of Predictive and Diagnostic Field Tests. AAEP Proc., 1979, p. 247.

Bergin, W.C.: A Survey of Embryonic and Prenatal Losses in the Horse. AAEP Proc., 1969, p. 121.

Davis, Lloyd E.: Drug Therapy in the Neonatal Animal. AAEP Proc., 1978, p. 449.

Knight, Humphrey D.: Infectious Diseases of the Newborn Foal. AAEP Proc., 1978, p. 433.

Lose, M. Phyllis: Blessed are the Broodmares. New York, Macmillan Publishing Co., 1978.

Martens, Ronald J.: Perinatal Physiology and Pathobiology of the Foal. AAEP Proc., 1978, p. 411.

McGee, W.R.: Barren Mares, Foaling Mares, Maiden Mares (panel with J. Fred Arnold, Walter C. Kaufman, James D. Smith, Don Witherspoon and W.R. McGee). AAEP Proc., 1979, p. 349.

Miller, W.C.: Practical Essentials in the Care and Management of Horses on Thoroughbred Studs. London, The Thoroughbred Breeders Assoc., 1965.

Pittenger, Peggy Jett: The Backyard Foal. New York, Arco Publishing Co., 1973.

Platt, H.: Etiological aspects of perinatal mortality in the thoroughbred. Equine Vet. J. 5:116, 1973.

Purvis, A.D.: Elective Induction of Labor and Parturition in the Mare. AAEP Proc., 1972, p. 113.

Roberts, S.J.: Veterinary Obstetrics and Genital Diseases, 2nd ed. Ann Arbor, MI, Edwards Bros., 1971.

Rogers, Tex: Mare Owner's Handbook. Houston, TX, Cordovan Corp., 1971.

Rossdale, P.D.: A clinician's view of prematurity and dysmaturity in thoroughbred foals. Proc. R. Soc. Med. 69:631, 1976.

Rossdale, P.D.: Blood gas tensions and pH values in the normal thoroughbred foal at birth and in the following 42 hours. Biol. Neonate 13:18, 1968.

Rossdale, P.D.: Neonatal problems in the horse. In Current Therapy in Theriogenology. Philadelphia, W.B. Saunders, 1980.

Rossdale, P.D.: Some parameters of respiratory function in normal and abnormal newborn foals with special reference to levels of PaO_2 during air and oxygen inhalation. Res. Vet. Sci. 11:270, 1970.

Rossdale, P.D.: The Horse: From Conception to Maturity. Arcadia, CA, The Calif. Thoroughbred Breeders Assoc., 1972.

Rossdale, P.D., and Leadon, D.: Equine neonatal disease: a review. J. Reprod. Fertil. (Suppl.) 23:685, 1975.

Rossdale, P.D., and Mahaffey, L.W.: Parturition in the thoroughbred mare with particular reference to blood deprivation in the newborn. Vet. Rec. 70:142, 1958.

Rumbaugh, G.E., and Ardans, A.A.: Neonatal Equine Immunity: Its Acquisition, Assessment and Disorders. AAEP Proc., 1978, p. 425.

Steven, D.H., and Samuel, C.A.: Anatomy of the placental barrier in the mare. J. Reprod. Fertil. (Suppl.) 23:579, 1975.

Tizard, Ian R.: An Introduction to Veterinary Immunology. Philadelphia, W.B. Saunders, 1977.

Traver, D.S.: Foal Diarrhea Syndrome (panel with R.H. Martens, A.M. Merritt, Lance E. Perryman, W.L. Scrutchfield and D.S. Traver). AAEP Proc., 1979, p. 195.

20

farm management

The most successful farm manager is one with a vast amount of experience in raising and breeding horses and one who knows the importance of keeping up to date with new concepts and ways of doing his work. A good manager will keep a daily log as a permanent record of the many happenings on a busy horse farm (Figure 20–1). The proper use of a log book can be of tremendous assistance from year to year. Notes can be reminders throughout the year that certain activities should be done. Comments about various horses, appraisals, and other data become invaluable as time dims memory.

MANAGEMENT BY AGE

Many managers have found their jobs much easier when horses can be managed as units, with group feeding and with attention focused on the whole group rather than on individuals. The broodmares are one of the primary considerations and will be discussed first.

During the early months of pregnancy the mare demands no special care beyond that included in the term "good management." Mares do best at pasture during this time and need only be quickly observed daily to see that no injuries have occurred. A good manager will take time from his busy schedule to sit among the mares at pasture at least once a week and observe each mare carefully as to condition. He should analyze her in respect to the weeks before in order to detect subtle changes that may signal a problem. This aspect of management cannot be left to temporary or inexperienced help.

EQUINE MANAGER'S DAILY LOG

Sheet No. _____
Day of Week _____
Day of Month _____

_____
Name of Manager

_____
Name of Farm

General Remarks _____
_____
_____
_____
_____
_____
_____

Supplies Needed:
Check when ordered

☐ _____
☐ _____
☐ _____
☐ _____
☐ _____
☐ _____

Mares Foaled:

_____
_____
_____
_____

Horses Sold:
Check when records
are completed

☐ _____
☐ _____
☐ _____
☐ _____
☐ _____

Horses Purchased:
Check when records
are completed

☐ _____
☐ _____
☐ _____
☐ _____
☐ _____

Supplies Received:
from previous days orders

_____
_____
_____
_____
_____
_____

Bookings Requested
Check when contract
completed

☐ _____
☐ _____
☐ _____

Horses Put in
Training

_____
_____
_____

Horses Taken Out of
Training

_____
_____
_____

Mares on Night Foal-
ing Watch

_____
_____
_____
_____

Horses Departing	Reason
_____	_____
_____	_____
_____	_____
_____	_____
_____	_____
_____	_____

Horses Arriving	Reason
_____	_____
_____	_____
_____	_____
_____	_____
_____	_____

Stallions Bred	To
_____	_____
_____	_____
_____	_____
_____	_____
_____	_____

Mares Serviced	By
_____	_____
_____	_____
_____	_____
_____	_____
_____	_____

Figure 20–1.   A typical manager's daily log record.

As fall approaches and the pasture stops growing, the pregnant mare will be about 6 months along and will begin to require more careful treatment and a more watchful eye. At this time it is advisable to move the mares to paddocks where they can be put into small compatible groups. On some farms mares are moved to box stalls for the winter. However, a paddock with a large open-sided shed allows for more exercise and much less individual care. In the winter, there is generally little or no grazing in the paddock so that a feeding system must be established. Feeding is simplified if no more than four or five mares are kept together in a single paddock and if care has been taken to see that the mares get along well with each other. There should be plenty of room at the hay bunks and there should be more feed buckets available for grain and pellets than there are horses in the paddock because the mares lowest in the pecking order will be continually pushed away from their feed. They soon learn to rotate to the additional feed bucket, but there seems to be less rotating and confusion if there is an extra bucket or two and if they are placed far enough apart. Mares that need a special diet can be kept together in one paddock where each can be caught and tied at feeding time. Larger farms may be able to divide the diet varieties into three or four categories and shift mares according to their needs, thus avoiding the daily chore of catching mares that need a special diet.

Mares, when in foal, and especially when near foaling time, have a greater tendency to roll. If there are hollows and open ditches, they may become caught and injure themselves or cause the death of the fetus. The ground in the paddock, therefore, should be level. There should be no enclosed sheds with single doorway openings; the entire side of the sheds should be open. Pregnant mares, like other horses, will sometimes try to pass through a doorway in one solid group. The severe bumps that occur in this situation could also cause the death of a fetus.

Adequate records give a clue as to when foaling time will occur for each mare. Several weeks before she is due, each mare should be carefully observed daily. Then about 2 weeks before the scheduled event, the mare should be moved to a foaling stall. This large loose-box stall or temporary shed should offer plenty of room for the mare to move about, with protection from inclement weather and freedom from drafts. The stall should be a minimum of 8 by 10 feet, and larger if possible. For less spacious facilities or during the heaviest part of the foaling season, it may be necessary to leave two or three mares together in a large paddock during their foaling time. This practice is safe if the mares are compatible and the weather is not severe.

Within a few days after foaling, mares can again be grouped together in paddocks with their foals. As soon as the weather permits in the spring, the mares and foals should be placed out in the larger pasture, so the foals have the chance they need to exercise.

Abortion and premature birth are the most serious accidents that can happen to pregnant mares. Though both terms are often applied indiscriminately, "slipping the foal" is the term generally employed when the young creature is expelled at any time before it is fully developed. "Premature birth" signifies that the young creature has been born ahead of schedule, yet with all of its organs sufficiently formed to enable it to live for at least some time in the external world. When an abortion occurs at a very early period of pregnancy, it may not produce any appreciable disturbance in the mare's health. The embryo usually escapes intact and often unnoticed.

Abortion has many causes (see Chapter 18). Improper nutrition or toxins may be a cause. Accidents may occasionally cause abortion, although not as commonly as once thought. Genetic lethals, as described in Chapter 10, may be a cause. When several cases follow each other quickly in a breeding farm, the question of a contagious infection arises, and veterinary advice is necessary. In fact, it is well to consult a veterinarian with every abortion. The careful manager will find an aborted fetus soon after the abortion and often it will be in good enough condition for a laboratory to diagnose the problem.

The safest procedure after abortion is to isolate the mare. If she has been in a stall, it should be cleaned, disinfected, and allowed to air out thoroughly. The bedding should be burned or buried. Even if the mare has been in a paddock, there should be an attempt at a thorough clean-up. The fetus and all the membranes should be placed in a plastic sack, in a freezer, for the veterinarian.

When abortion takes place, the flanks fall in a little, unless it happens early in gestation, and the abdomen may descend. The vulva and vagina dilate slightly and often a gelatinous, red-tinged fluid can be found on the buttocks. The mammary glands may develop rapidly and even produce milk. Before a later-stage abortion the mare may suddenly appear dull and dejected, or she may be restless and uneasy.

After weaning in the fall the foals will require a considerable amount of special care. They can be kept in rather large groups, depending on the size of the paddock. Weaning and shortly thereafter is often an excellent time to place the young animals in a loose-box stall for the first time. The keeper must then spend some time with the animals. Special attention during this period can pay high dividends. In a stall the individual eating habits of the foal can be observed, and the daily handling for exercise will serve to train the foal to more intense human contact. The foals should have been halter trained prior to weaning. Daily grooming will gentle them considerably more at this point.

The weaned foal should be wormed and vaccinated according to the schedule provided in Chapter 22. The hooves should be trimmed and the teeth should be checked to see that they are all wearing evenly.

After a short time in a stall the foal will be eating well on his own and accustomed to the world of human beings. At this time the youngster can be turned into a large paddock with others of its kind. As with horses of any age group, adequate feeder space should be provided, although weanlings tend to fight less than the older horses.

During their first winter, weanlings should be caught up about every month and handled some so that they do not forget their previous training. The sexes should be separated the first winter; otherwise, a precocious female may turn up bred. They should be gaining well by early spring so that they can be the first animals on the farm to try out the spring pastures. Early, fast-growing pasture is less nutritious than later pasture and, therefore, must be supplemented with other feed until late spring. It works well to have a small paddock available next to the pasture so that the yearlings can come in for daily feedings during this time. Before they are finally turned out completely to pasture in late spring, they will need another round of vaccination, worming, and hoof trimming.

As winter approaches, the coming 2-year-olds will require more attention than any other group on the farm. It is at this time that their serious training should begin, regardless of whether they are to become racehorses, show horses, or just pleasure-riding horses. Because of the heavy demands of this age group, many breeders prefer to sell most of their stock as yearlings, or send them to a training stable where they can receive adequate attention.

Those 2-year-olds that remain on the farm can be alternated between paddock and stall and trained as help is available and weather permits. At this age, the youngsters should not be heavily ridden, but many other procedures are available to occupy their attention. They can become

Figure 20–2.   Two-year-olds should become accustomed to the bit, and learn about ground driving, although they should not be heavily ridden.

accustomed to the bit, and learn about ground driving (Figure 20–2). They should be taught to load into a trailer. They should become used to a saddle on their back. Perhaps most important, they should learn to enjoy stall life and the feeling of being well groomed. Stall confinement should not be overdone, however, at this age because they need heavy exercise to strengthen bones, tendons and hoofs.

FEEDING

Proper feeding is the most difficult part of equine care, especially in the winter when there is no pasture. During the cold weather it is easy to become slipshod and neglectful. Feeding may become irregular—a feast or a famine. Generally this situation does not result from willful neglect, but ignorance on the part of the horse keeper.

Horses have individual needs, some requiring much more feed in winter to hold their weight while others can get by on very little and still remain fat. When temperatures dip, a bit more energy will be required. Those horses on a heavy working schedule need more nutrients than those just standing in the paddock. Heavily parasitized horses may require much more feed than after they have been wormed. Growing horses need more feed per pound of body weight.

Hay continues to be the main staple of the majority of all horses. The amount of hay needed per day per horse depends upon the total digestible nutrients (TDN) it contains. A good rule of thumb to remember is that hay contains about 50 percent TDN, as opposed to 65 to 85 percent TDN in grain.

The quality of hay will determine the amount of TDN. In general, legume hay (e.g., alfalfa and clover) contains more TDN than grass hay. The age of the hay is important since hay loses its nutrient value with time. The coarseness can also affect the TDN. The best hay is leafy and contains fewer large stems. A good quality alfalfa with no mold is excellent horse feed. Mold in alfalfa hay must be carefully avoided; for this reason many horsemen prefer to feed grass hay. Moldy hay can cause colic and may lead to heaves.

How much TDN does a mature horse need? A good round figure is three-quarters of a pound of TDN per 100 pounds of body weight of the horse; thus a 1000-pound horse would require 15 pounds of average quality hay per day. A 75-pound bale of hay should last 5 days under these conditions. Up to twice as much poor-quality hay may be necessary to give the same nutrition.

Grains contain more TDN per pound than hay. Corn has the highest percentage, about 80 percent; good barley will contain about 75 percent. Oats has been a traditional grain for horses; the TDN of oats can vary from about 55 percent to 70 percent, depending on the amount of fiber in the oats.

The average horse at rest will maintain adequate condition on good quality alfalfa hay without grain supplement. Good alfalfa hay will contain all the vitamins needed as well as an adequate level of protein. Mineral content of hay varies somewhat depending on the soil conditions where the hay was grown. Trace minerals fed free choice is usually a good idea.

The feeding of breeding animals should receive special attention during the winter, which is the time to consider the proper condition for those mares to be bred early in the spring. Mares conceive more readily if they are gaining weight. Difficulty may be encountered, however, if the mare is too fat. Mares should not receive increased feed until just prior to the breeding season. During the winter the stallion should be fed "to effect," keeping him in good shape without excess fat. It is better for the stallion to be a trifle fat going into the breeding season than underweight.

A common misconception is that mares need significantly more feed during early gestation. Studies have shown that the nutrient needs of the pregnant mare do not increase until the latter one-third of pregnancy. During this time, requirements will increase to about 1 pound of TDN from hay per 100 pounds of weight along with ¼ to ½ pound of grain per 100 pounds of body weight. Save the extra feed for the lactation period, when the mare's nutrient requirements will increase dramatically. A mare may produce as much as 500 quarts of milk monthly, and at this time she may need as much as 1¾ pounds of grain per 100 pounds of weight along with her regular 2 pounds of hay per 100 pounds of body weight.

Probably the first consideration of proper feeding is water. Dr. Lon Lewis of Colorado State University emphasizes this aspect when advising about feeding horses. Although he recommends that good-quality water be given free choice, he says that one gallon of water per 100 pounds of body weight per day is the average amount consumed. This quantity can vary depending on weather, the horse's activity, and the type of feed consumed.

Water can be dangerous to the horse following heavy exercise, causing colic and founder. "What I like to do after cooling the horse out a bit," says Lewis, "is to go ahead and allow the horse roughage—hay or pasture—for a good hour before giving water free choice." He describes this as the procedure when physical activity is discontinued. During exercise the horse should be watered as frequently as possible and be allowed to drink as much as possible.

Keep the water and feed bunk far enough apart so that the water will remain clear of feed. "It seems that contractors building stables love to put the water container right next to the feed bunk," he says.

Trace mineralized salt should also be available free-choice for the horse. This costs little more than the plain salt block and eliminates trace mineral deficiencies that occur in some areas. Trace mineralized salt does not contain calcium and phosphorus. Minerals such as manganese, cobalt, iodine, iron, and zinc may be present in ample quantity in the feedstuff; in fact they generally are, so the feeding of trace mineralized salt is simply a low-cost insurance against trace mineral deficiencies.

The only nutrients usually of concern in the horse ration are energy, protein, calcium, phosphorus, and vitamin A. The feedstuff used to provide these nutrients are roughages, concentrates, and commercial rations, which may be a mixture of varying amounts of roughages and concentrates.

Roughages are either legumes or grasses. The major legume fed in most areas is alfalfa, although clover, bird's-foot trefoil, lespedesa, and similar chops are fed in other areas. A variety of grasses are fed and are quite good for the horse, e.g., timothy, brome, crested wheat, bluegrass, and orchard grass.

Green alfalfa is excellent pasture for the horse. (Bloat is a problem only in the ruminants.) However, a mixture of grass and alfalfa provides the best pasture for horses because it offers the greatest total quantity of nutrients and the longest grazing season. "In the past, many horsemen have considered grass hay to be the best hay for the horse," admits Lewis, "and if you're talking about poor-quality hay, this is true." Poor-quality alfalfa tends to be very stemmy with few leaves. The leaves contain two-thirds of the energy and three-fourths of the protein of the hay. When the hay is mishandled, most of the leaves can be lost—and most of the value of the hay. In contrast, the grass hay leaves are more firmly attached to the stem and are larger and longer, making loss more difficult. Also, alfalfa is more prone to dustiness and moldiness than is grass hay.

Since the horse is susceptible to the ill-effects of moldy, dusty feed, this type of hay should be avoided. Fungal spores develop easily in alfalfa hay that is not handled properly, and fungal spores are the major cause of chronic respiratory problems in the horse. Chronic pulmonary emphysema, chronic bronchitis, chronic cough, heaves, and bleeders result from these spores. Also microtoxins are present in moldy hay and since the horse is particularly susceptible to these, moldy hay should be avoided.

Never feed moldy hay to a horse. If dusty hay must be fed, sprinkle it down with water before feeding. Regardless of the type of hay, this practice of sprinkling may be necessary for the horse with chronic respiratory problems. "My first choice for the horse with chronic respiratory problems is to put him out on pasture," says Lewis. "If that isn't possible, then your next best choice is to go to hay cubes or wafers."

When it comes to average- to good-quality hay, alfalfa is much better, nutritionally, than a grass hay. Grass and alfalfa are usually similar in total digestible nutrients (TDN), but alfalfa is generally about 10 percent higher in energy, has on the average about twice as much protein, about three times more calcium, and about five times more keratin (vitamin A precursor). Much more important than the type of hay is the quality of the hay.

Several factors determine the quality of hay. First, it must be cut in the early stage of maturity. For legumes, the ideal is at, or just before, the time of flowering; for grass hay, when the head just begins to protrude. Allowing hay to stand after the first flowering, or after the head protrudes, results in a decrease in the crude protein and an increase in the crude fiber content at the rate of a half percent per day. It also results in a decrease in the digestible energy content at the rate of three-fourths of a percent per day.

This may not sound like very much, but as Lewis points out, "Twenty days after the first flowering the crude protein is reduced from a normal 15 percent to 5 percent, digestible energy decreases from a normal 65 percent to 50 percent and the crude fiber content will increase by 10 percent. The potential of the hay is changed from good to very poor in that 20-day period of time before cutting."

The second factor of hay quality is that it be free of mold, dust, and weeds. The third factor is that it not be exposed to extensive weathering, which is evidenced by the absence of the green coloring. The fourth factor is the leafiness of the hay.

"I think probably the best criterion for judging the quality of the hay," says Lewis, "is whether or not you would like to sleep on it. If it is nice and soft and clean, that's good hay to feed your horse."

The first cutting of hay may contain more weeds and trash, which can decrease the nutritional value of this cutting. Also it is often more difficult to get the first cutting put up before a rainfall.

The period of the fastest growth often comes during the second growth, midsummer, when the temperature is highest. Hot weather and adequate water cause this rapid growth. As a result, the stems grow faster than the leaves and the quality of the hay is reduced. For this reason, the third cutting, or the cutting after the temperature is lower, gives the best quality of hay.

The stage at which the hay is cut and the handling of the hay after it is cut are much more important than which cutting it is. When the hay is cut a little late, it will be poor quality no matter how it is handled. If the hay is cut at just the right time, it can still lose much quality by being rained on and having to lie in the field longer to dry. Extra raking and turning will cause it to lose valuable leaves. These factors are more important to hay quality than which cutting it was.

Other sources for roughage for horses are silage or haylage and cubes or wafers. Silage and haylage can be fed to the horse successfully, but most horses do not like the fermented taste until they become accustomed to it. However, once they get used to it, silage and haylage become a good source of roughage, provided they do not contain mold or microtoxins.

Hay cubes or wafers have a number of advantages as well as disadvantages compared to long-stem hay. Any of a variety of different pellet sizes are fine for the horse. Those about the size of the end of the little finger are best. If, however, the pellets are fed on the ground, a much larger pellet is desirable. These pellets should be well formed—not soft and crumbly.

Cubes or pellets have a number of advantages (Figure 20–3). The horse can take in about 130 percent more feed than with hay. Waste is usually decreased by as much as 50 to 100 percent with the use of pellets. Pellets

Figure 20–3.   Pelleted hay and grain mixtures are successfully used by many horsemen as the only feed.

and cubes may decrease feed storage space by as much as 300 percent and transportation costs as much as 25 percent. Cubes and pellets are not dusty as hay can be.

Some horses may look better on cubes and pellets because they have less intestinal fill. Supplements of vitamins, minerals, or anything else can be easily added. Pelleted feeds can be fed with less ingestion of manure.

Pellets and cubes also have disadvantages. Generally they are more costly. But taking all factors into consideration—storage, wastage, transportation—they may not be more expensive in the long run when the feed is not homegrown. Some horses may choke on pelleted feeds. This hazard can often be overcome by placing several large rocks among the pellets to eliminate excessive gulping of the pellets.

Sometimes the fine material that falls from the pellets and accumulates in the bottom of the feed bin can be troublesome. This problem occurs less often if the pellets are kept and fed dry. The amount of molasses in the pellets will determine how crumbly they become. They will be soft and chewy if there is too much molasses. The best ratio is about 1½ to 2 percent molasses.

With some setups, pellets and cubes may be harder to feed, but with a mechanized setup they can save labor.

Another important disadvantage of pellets and cubes is that it is not possible to assess the quality as it is with hay. For this reason, pellets and cubes have occasionally gotten a bad name, which is undeserved. If good-quality feed goes into the pellets, they will be of good quality also.

A study by Dr. Harold Hintz at Cornell showed that feeding pellets increased wood chewing. He determined that with a conventional hay-grain ration, the experimental group of horses was eating 81 grams per day of new pine two-by-fours. When changed to pelleted feeds, they ate four times that amount. To reduce this problem, it is well to feed 4 to 5 pounds of long-stem hay with a pellet ration. Also increasing the frequency of feeding will reduce the amount of wood chewing.

The major cause of wood chewing is not nutrition. The horse eats wood because he is idle and has nothing else to do. Providing the horse with companionship and more room helps to reduce wood chewing.

Roughages by themselves may not provide adequate amounts of nutrients to meet the horse's requirements during the last 3 months of pregnancy, during lactation, and during growth and work. At these times the horse needs a concentrate mix. Such a mix for the horse may contain protein, vitamins, and mineral supplements in varying amounts; however, the major ingredient of concentrates is cereal grains. Any of a number of different cereal grains can be fed to the horse.

Oats is by far the most popular although on an energy basis, oats is the most expensive cereal grain. Energy needs make up 70 to 90 percent of all feed consumed. This need is the major reason that cereal grain is fed—it is a good source of energy. Therefore, it makes sense to compare cost on an energy basis.

If oats, which contain about 70 percent TDN, can be purchased for $8.00 per hundred pounds, and corn, which contains 90 percent TDN, can be purchased for $9.00 per hundred pounds, a little calculating shows that corn is the best buy.

The way to make this determination is to divide the cost by the nutrient content, in this case, energy. If protein is the concern and a protein supplement is the item of choice, divide the cost by the percentage of protein. When buying a phosphorus supplement, divide the cost of the supplement by the percentage of phosphorus in it.

In the energy example, comparing oats with corn, the energy in oats will cost $11.43 per hundred pounds of TDN. The corn will cost $10.00 per 100 pounds of TDN. In this case corn is a much better buy.

Oats is not only more costly, but more variable in quality because it contains differing amounts of hulls. Good oats contain 30 percent or less hulls. Poor oats may contain over 50 percent hulls. De-hulled oats, or racehorse oats, can be purchased, but if figuring out the cost on an energy basis, it is pretty hard to justify the cost.

Oats continues to be the most popular grain for horses partly because of its low energy content. Because of that lower energy content, it is harder to overfeed or founder a horse on oats; it just takes more to do it, according to Lewis.

"I think the major reason oats is the most popular grain among horses, however, "says Lewis, "is a matter of habit." Also the horse owner is not generally as cost-conscious as other livestock producers.

Because corn is the cheapest grain in terms of energy, it makes up the bulk of many supplements. One gallon of corn contains the same amount of energy as 2 gallons of oats. "This is where people get in trouble feeding corn," says Lewis. "You'll commonly hear people say, 'Don't feed the horse corn, it'll make them too high, too spirited.' And this certainly is true on a volume for volume basis—or even a weight for weight basis."

But by feeding on an equal energy basis, there is no more chance for excessive weight gains, health problems, and other complications than with oats. "The horse's system doesn't care where it gets the energy. The amount of energy is the factor determining problems, not the source."

Often you hear that corn is a "hot" feed, a heating feed. Some people say, "Feed it during the winter; it will keep the animal warm." Dr. Lewis says, "Yes, it is a good feed during the winter, but not because it is a hot feed."

Actually, the amount of heat produced in the digestion, absorption, and utilization of corn is one-third less than it is for oats. The greatest amount of heat produced is in the utilization of roughages. The best feedstuff to keep the animal warm during the winter is roughages. If the animal cannot take in enough roughages to meet his energy needs, then grain will be needed.

Barley is a good cereal grain for the horse. It is intermediate between oats and corn with respect to energy content. Wheat and milo are also good cereal grains for the horse. Milo and wheat have small hard kernels and need to be broken down.

Cracking, rolling, and grinding of wheat and milo increase the feeding value by about 15 percent. Cracking or grinding corn increases its feeding value by only about 5 to 7 percent, and rolling oats or barley, both of which have a fairly soft kernel, increases their value by only 2 to 5 percent. These figures determine what the horseman can afford to pay for processing grain. If the cost of rolled oats is more than 5 percent higher than whole oats, it is not worth the extra cost.

Rye can also be fed to the horse. It is not very palatable to the horse and should not make up more than one-third of the concentrate mix. Wheat should also not make up more than one-third of the concentrate mix. Wheat contains the protein called wheatglutin, which gives wheat flour its sticky, doughy consistency when wet. It tends to wad up in the horse's stomach and cause problems.

Grains are often obtained in concentrate mixes sold by various feed companies or as a part of complete feeds in pellet form. The complete feeds contain about 10 to 25 percent concentrate and generally utilize alfalfa as the roughage. They are intended to be fed as the sole ration for the horse.

Alfalfa is an excellent source of lysine, a specific amino acid of protein needed by the growing horse. Soybean meal and milk-based protein also contain adequate amounts of lysine for the growing horse. Calf Manna and Start To Finish contain the proper protein for the growing horse.

Urea is commonly fed to ruminants, but it is of little benefit as a protein supplement for horses. Most of the urea fed is absorbed from the small intestine intact and excreted in the urine. The horse can tolerate higher levels of urea in the feed than can cattle, but there is little value in it for him.

FEEDING FOALS

According to Dr. Lewis, the most common mistake made by horsemen is overfeeding foals. He emphasizes that the most practical procedure is to feed alfalfa hay free-choice to the pregnant mare and the foal.

The broodmare will eat 2.5 to 3.0 pounds of hay per 100 pounds of body weight per day, and this quantity will supply her total digestive nutrient needs. When alfalfa hay is fed, the broodmare will exceed her protein requirements and may need a high-phosphorus salt mix supplement, which can be fed free-choice.

Dr. Lewis emphasizes that most common forages, including alfalfa and cereal grains, contain only about 0.25 percent phosphorus, while equine requirements are close to 0.35 percent. Because of this disparity, he advocates use of a free-choice high-phosphorus salt mix. The formulation of the high-phosphorus salt mineral mix is shown below:

High-Phosphorus Salt Mineral Mixes
1. Equal parts trace mineral salt and dicalcium and monophosphate.
2. Two parts Purina's 12:12 or Co-op's Perfect 36, with one part monophosphate.
3. All—Co-op's op-T-Min.
All of these mixes give 8 percent calcium, 16 percent phosphorus, and trace mineral salt.

The normal, healthy, lactating mare will give enough milk to supply 100 percent of the foal's requirements for the first 2 to 3 months. Therefore the diet of the mare during this time must be adequate if the foal is to receive all of its needs.

Good lush pasture or good-quality hay and a mineral mix supply 80 to 120 percent of the lactating mare's protein and calcium and phosphorus needs. If any legumes are in the pasture or alfalfa hay is fed free-choice, she will get plenty of protein and calcium, but will be somewhat deficient in phosphorus. Dr. Lewis therefore recommends supplementing the lactating mare with a balanced calcium-phosphorus salt mineral mix, which can be fed free choice, or if grain is being supplemented also because of poor roughage quality, three ounces per day of the balanced calcium-phosphorus salt mineral mix should be fed.

Balanced Ca-P Salt Mineral Mixes
1. One part trace mineral salt and one part dical.
2. Purina's 12:12 Mineral.
3. Co-Op's Perfect 36
All of these mixes give 10 to 12 percent calcium, 10 to 12 percent phosphorus, and trace mineral salt.

Dry, mature, grass pasture or grass hay will not supply enough protein, calcium, or phosphorus. A concentrate mix or grain will have to be fed to bring a lactating mare up to her requirements. She will need 0.5 to 1.0 pound of concentrate per 100 pounds body weight, per day.

The feeding program is often dictated by the price of hay versus grain. If hay is expensive in relation to grain, feed 1.5 pounds per 100 pounds of body weight, per day, of both hay and concentrate. If grain is more expensive than hay, which is more often the case, the mare should be on free-choice roughage. If roughage is alfalfa, no concentrate is needed, nor is a concentrate needed if the mare is on good, lush, green pasture. But with both of these a balanced calcium-phosphorus mineral mix will be needed.

If the roughage is of questionable quality, or if the mare is a hard keeper, she will need to be fed some concentrate during lactation and this concentrate should be supplemented with 3 ounces per day of the balanced calcium-phosphorus salt mineral mix.

After the 12th week of lactation, milk production will begin to fall. A common mistake at this time is to continue feeding the mare the same amount, causing her to gain excessive weight. Usually a foal will begin to eat with the mare as his body requirements increase. This means that the foal's normal energy requirements are met up to and until weaning. Dr. Lewis says that by weaning time the foal should be eating 0.5 to 0.75 pound per 100 pounds body weight, per day, of concentrate. He emphasizes, "Don't ever creep feed foals, free choice! This can lead to epiphysitis, contracted tendons, and other problems."

If the mare and foal are in a pen at night, the mare can be tied while the foal eats his supplement. If on pasture, a creep-feeder can be used so that the foal can get away from the mare to eat. Dr. Lewis recommends starting controlled creep feeding (not free choice) at 3 months of age. When a number of mares with foals are on pasture together, the amount of concentrate can be calculated so that each receives no more than 0.75 pound per 100 pounds body weight, per day. As an example, if 10 foals averaging 400 pounds in weight are using the creep-feeder, 30 pounds of concentrate are needed daily. It is best to divide this ration, 15 pounds in the morning and 15 pounds in the evening.

A foal should be weaned at 4 to 5 months of age, says Dr. Lewis. Waiting longer pulls the mare down too much without benefitting the foal substantially. Several days before weaning, stop all concentrate to the mare and decrease the amount of roughage by half. This practice helps decrease the milk production and reduces the excess distention of the mammary gland at weaning. Dr. Lewis says, "Don't milk the mare out. On the day of weaning remove the mare from the familiar pen, if possible, rather than the foal. Make the separation abrupt, complete and final."

One reason that Dr. Lewis recommends weaning at 4 to 5 months is that it is harder to feed the large foal when he is still on the mare. Weaning before 4 months is too early as it requires feeding too much concentrate, which predisposes to growth and bone problems.

After weaning, the concentrate should be increased to 0.1 to 1.5 pounds per 100 pounds body weight, per day, for the foal. Also roughage, preferably alfalfa, should be available free-choice. By 12 months of age, the foal will have two-thirds of his mature weight and about 90 percent of his height.

Yearlings need only 0.5 to 1.0 pound per 100 pounds body weight, per day, of concentrate with free-choice roughage. Dr. Lewis emphasizes that overfeeding of concentrate is the most common error made in feeding the growing horse.

EPIPHYSITIS

Veterinarians have been diagnosing epiphysitis for several years without knowing exactly what the condition entails. Meanwhile, equine nutritionists have been trying to determine exactly what is involved in the condition and what causes it. Although research continues, enough facts are at hand to describe the malady as a syndrome, with various manifestations.

Dr. Luis Medina described epiphysitis as, "an inflammation of the end of a long bone. However, the affected anatomical structure is actually the growth plate." This veterinarian says:

> Epiphysitis is common in the heavily fed, rapidly gaining, or over-fat, older foal and yearling. It is also seen in the young thoroughbred subjected to hard training. Excessive body weight or stress crushes and damages vulnerable cells of the immature growth plate. The bone responds by laying down new bone in an attempt to repair the injury. This is the bony enlargement (of the fetlock) that one sees.

Even before epiphysitis becomes visually evident, it causes pain for the foal and results in contracted tendons, according to Dr. James Rooney. He explains:

> The limb joints have a normal, "usual," standing angle or position in which nerve endings are not stimulated, are "silent." Whenever the joint leaves that position, proprioceptive signals are

generated which induce muscle contraction to return the joint to the normal, neurologically silent position. Instability of the joint, for any reason, and/or continuous displacement of the joint from the normal position, will initiate proprioceptive signals and muscle response. If repeated continuously or with high frequency, a vicious cycle may eventually be set up. Once contracture begins, instability of the joints (particularly fetlock and coffin) will add to the fixing of the vicious cycle. Implicit in this hypothesis is the point that it is muscle contraction and contracture leading to shortening, probably, of intermuscular connective tissue, which causes contracture and not shortening, relative or absolute, of the tendon. . .

Dr. Lon Lewis has called it the epiphysitis syndrome and described its many manifestations as being enlarged fetlocks, dropped hocks, splay feet, hip dysplasia, angular leg deformities, contracted flexor tendons, the wobbler syndrome, and osteochondritis dissecans. He said that signs of impending epiphysitis are observation of an "open knee," enlarged fetlocks, and/or excessively upright pasterns. Recently there is a trend to call the condition osteochondrosis.

Dr. Lewis said that the severity and form of manifestation of the syndrome are determined by the interaction of at least four causes: (1) rapid growth rate, (2) trauma to epiphysis, (3) nutritional imbalances, and (4) genetic predisposition.

While a rapid growth rate usually has a genetic cause, it is expressed through nutrition. A high energy intake can allow the maximum growth rate, which may be excessive. Early stunting sometimes occurs; then later compensatory growth occurs when the youngster is on a normal diet.

Trauma often occurs because of an excessively heavy body putting extra force on a bone that is too fine and delicate. Under these circumstances, normal activity can provide the necessary trauma for epiphysitis to begin. A wire cut, or other injury on one leg, causing its disuse, can lead to excess weight on the good leg and development of epiphysitis, which, in this case, often leads to angular deformity as well. Trauma brings damage and pain, with the pain causing continued reflex contraction of tendons. Since the flexor tendons are stronger in the horse, these overpower the extensor tendons, resulting in an upright pastern and sometimes eventual knuckling over at the fetlock. Dr. Lewis says that this explanation of the cause of acquired contracted tendons, which originated with Dr. Rooney, is more plausible than the commonly used explanation that the bones are growing faster than the tendons.

Nutritional imbalances that help in bringing on the epiphysitis syndrome are associated with calcium and phosphorus imbalances and protein and energy excesses. A calcium deficiency is the usual picture, says Dr. Lewis, not phosphorus deficiency. The excess calcium can tie up phosphorus to a certain extent. The reverse is more dramatic—excess phosphorus tying up calcium. Thus, the calcium-phosphorus ratio is extremely important. The mature horse can get by with a ratio in the range of 8 : 1 to 0.5 : 1 (Ca : P), but it is safer if the range is 6 : 1 to 0.8 : 1. The growing horse must have a closer range in the calcium-phosphorus ratio, 3 : 1 to 0.8 : 1.

While it has not been definitely proven that excess protein can cause epiphysitis, it can be observed clinically. Since protein accounts for 20 percent of bone, excess protein seems to be a logical factor. Moreover, excess protein can cause problems in two ways. First, protein is a source of energy, and excess energy is known to cause epiphysitis. Second, protein tends to increase urinary calcium excretion, which could cause a deficiency. Research in pigs and dogs has proven that excessive energy in the diet leads to such things as hip dysplasia and crooked legs.

The genetic influence in epiphysitis is expressed in many ways. Rapid growth has a genetic base. Heavy bodies are genetic, and fine bones are genetic. An excessively straight conformation, which begins as a genetic trait, can increase trauma to the joints. Other genetic factors, as yet undefined, cause certain family lines to be much more prone to epiphysitis than others.

Treatment of epiphysitis should begin with diet correction. First, take all concentrates away from the animal and feed roughage, free choice. This change will decrease the energy intake, slow the growth rate, and decrease the phosphate intake. Generally all of these are needed before the problem can be solved. If alfalfa is fed, add a high-phosphorus salt-mineral mix. If the weanling will not eat the mineral mix, it can be fed in 1 pound of grain daily—add 4 ounces of high-phosphorus salt-mineral mix.

Beyond nutritional changes, treatment will depend on the manifestation of the syndrome. Exercise is detrimental with epiphysitis, swollen fetlocks, and angular deformities (crooked legs). Therefore, Dr. Lewis recommends complete stall rest, that is, 24 hours a day for 4 to 6 weeks. Some improvement should be observed by then. If contracted tendons is the principal manifestation, exercise is needed. Phenylbutazone is indicated with contracted tendons. It should be used early and continued until the pasterns improve in their angle. The reason is that if pain is the cause of the contracted tendons, reducing the pain should reverse the process. Phenylbutazone is contraindicated if epiphysitis occurs without the upright pasterns. It tends to slow healing of epiphysitis.

SPECIALTY SUPPLEMENTS

So much is printed about the virtues of different supplements that no wonder everyone thinks that a performance horse cannot get by on hay and grain. "Most of my clients have already made up their mind on what additives and feeding programs work for them." said Dr. P.B. Altemuehle, of

Hebron, Kentucky recently in a survey of veterinary opinion about supplements. "I find it easier to argue politics and religion than to try to change their minds on feeding horses."

Another veterinarian, Dr. T.E. White of Natchitoches, Louisiana said, "I am not a big believer in the use of feed supplements, additives or hormones in horses. A good-quality grain ration with the proper amount of protein and good grass or alfalfa hay can optimize performance. In my opinion, the use of supplements and additives probably does more for the owners and trainers than it does for the horses."

Dr. Lon Lewis, the veterinary nutritionist mentioned earlier, said that the only nutrients the horseman has to be generally concerned about are energy, protein, calcium, phosphorus, and vitamin A. "The feedstuff used to provide these nutrients are roughages, concentrates, and commercial rations, which may be a mixture of varying amounts of roughages and concentrates," he said.

Several factors determine quality of hay, as covered previously, but as long as it is good quality, clean and green, high-powered vitamin and mineral supplements will not be needed. Dr. Lewis recommends that a trace-mineral salt block be available to every horse just to ensure that these minerals are supplied in adequate amounts. Moreover, since alfalfa is so high in calcium, he also recommends a high-phosphorus mineral supplement. These supplements are not expensive and are the only ones needed unless a horse is sick or debilitated.

Dr. Atlemuehle said, "Some of my clients are spending money foolishly for unnecessary supplements. When a client approaches me for an honest answer, I point out what is available, the costs and whether all the ingredients are actually needed. I encourage everyone to feed and treat each horse as an individual."

Protein requirements of the horse run between 8 and 16 percent. Cereal grains contain 8 to 12 percent protein; grass hay or pasture contains 5 to 12 percent protein; and the legumes from 12 to 20 percent protein. Therefore, depending on the type of feed and the particular requirements of the horse being fed, additional protein may be needed. Adding a protein supplement to the concentrate is a ready solution.

Major protein supplements for the horse are cottonseed meal, linseed meal, soybean meal, and milk-based protein. Protein quality is not important for the mature horse. Any amino acids can make up the protein in the feed. The growing horse needs at least 0.7 percent of the amino acid lysine in the ration.

Alfalfa is an excellent source of lysine. Soybean meal and milk-based protein also contain adequate amounts of lysine for the growing horse. Dr. S.S. Gutman of Long Valley, New Jersey, said, "I rely on the use of Calf Manna (Carnation) at ½ pound in the morning and evening feedings."

These milk-based supplements can be overdone, however. Dr. White said, "I believe that overfeeding foals is a main cause of contracted flexor tendons and that overfeeding vitamins A and D causes skin and skeletal problems."

"Many of the high-priced feed additives contain large amounts of unsaturated fatty acids to give the hair coat a shine," said Dr. Lewis. "But the effect can be produced at a fraction of the cost by adding cooking oil to the ration. One to 2 ounces of cooking oil added to the ration twice a day will put more bloom on the coat than most of your high-priced feed additives. This may also help the horse shed off earlier in the spring."

Another veterinarian, Dr. B.P. Kelly of Lexington, Kentucky, admonished:

> Additives and supplements do little, if anything, for horses free of disease, on a good parasite control program and receiving a balanced diet. The key to good performance is a balanced diet. In my practice area there is a great deal of tradition in the horse business. Many people think because Granddaddy raised a stakes winner on straight oats, that's all that's necessary today. This is obviously untrue considering what we know regarding the nutritional requirements of horses. If an improper diet is consistently fed, almost any supplement will help. Peak performance cannot be achieved until the diet is balanced. There is no supplement on the market that can balance a poor diet.
>
> If the diet is complete, supplementation is probably unnecessary. I sometimes recommend a pelleted vitamin supplement at half the recommended level to ensure adequate vitamin intake when pelleted feeds are used. Perhaps I'm fooling myself, but I think I can detect improvement occasionally. Many horses on the track and in the show ring become run down, stale, and anemic. It is fairly common practice to begin sticking them with vitamin B12-liver-iron preparations, and adding tonics to their feed. This does wonders for the veterinarian's wallet but nothing for the animals. A balanced diet and rest, back on the farm away from the stress of training, brings these animals back into shape, although it may take many months.

Equipoise, a Squibb product, is a common supplement that many horsemen swear by. Chemically it is boldenone, a hormone that affects the reproductive capacity of stallions. It is related to testosterone and increases competitiveness and muscle tone. Dr. J.F. Thompson of Spring City, Pennsylvania, said, "I use boldenone to speed up conditioning and stimulate the appetite, but good-quality hay is required for the Equipoise to work."

Feeding boldenone is not done without danger to the horse, however. Colorado State University scientists, Drs. Pickett, Voss, and Squires, gave the substance to stallions in an experiment and measured the effect on sperm output and testicular size. They found that the product drastically reduced the reproductive capacity of stallions. No research has yet been done to show the effects on mares. As Dr. Kelly said, "The use of hormones to optimize performance is a waste of time and effort. I can think of no case in which the use of hormones is justified to aid performance. I have had some success with Equipoise to increase aggressiveness in some animals. However, I can't say if their performance is improved of if they just become alert losers."

As Dr. Kelly admits, veterinarians sometimes dispense supplements because of client pressure. A Lee's Summit, Missouri veterinarian, Dr. R.E. Hertzog said:

> We use and dispense a wide variety of commercial feed supplements in our equine practice. In our pleasure-horse practice we probably use supplements because of a lack of good basic nutrition. Many of our clients do not understand the need for the basic requirements and, due to summer overgrazing and winter feeding of poor-quality grass hay, many horses lack protein and total digestible nutrients. Our first approach is to have the owners supply good-quality alfalfa hay, and grain and vitamin supplements if needed. Many horses on a poor plane of nutrition also suffer from other management problems, including parasitism. Anemic horses often have a dramatic rise in packed cell volume after deworming, diet balancing, and the administration of Vi-Natura or Lixotinic. We have seen only a minimal response to the administration of anabolic steroids.

The average adult performance horse needs no high-powered supplements, according to the experts. Good-quality hay supplies almost everything this horse needs when idle. As he begins to work, his energy needs go up. This extra energy can be supplied by feeding grain.

Feeding management is as essential as the amount and type of feed. Proper feeding technique is an art in which the individuality of each animal must be considered. Feeding should be on schedule, regardless of how hard it might be snowing or raining. A horse will require less feed, and waste less feed, if fed on time. Make changes in feed type gradually to avoid colic. Clean fresh water should always be available—an especially difficult problem in the winter, when ice must be taken care of routinely. The feeders and buckets should also be kept clean.

TRAINING AND EXERCISE

The science of horse breeding does not encompass training of horses for performance. To be done properly, this area of expertise should be left to those who specialize in it. A breeder who does not sell all of his stock as young untrained animals will generally hire a full time trainer, or send the promising individuals to a professional trainer. However, a good horse breeder will not leave his young stock completely undisciplined. A young horse must learn certain lessons to reduce the likelihood of being injured. Also adequate exercise is essential to proper physical development of a potential athlete, and without some schooling a horse cannot be properly exercised.

Foals should be handled gently every day the first week of life and then at least once a week up to weaning time. The foal learns quickly to respect the pressure of a lead rope if he is taught while quite young. By the time the foal is several months old, he should know how to stand quietly when tied, and should not fight when his feet are picked up for trimming. This is all the discipline needed prior to weaning, but it is absolutely necessary if the foal is to be handled safely when injury occurs, shots are to be given, and worming is necessary.

At weaning time the foal's early training can be reinforced and the youngster can be further gentled with daily grooming for a week or two. At this time the young horse should begin to learn about loading into a horse trailer. At this age the foal can easily be pushed in. After the foal has been fed in the trailer a few times, he will feel at home getting into any trailer.

One other discipline that a weanling must learn to submit to is to stand to be caught. Many young horses have a tendency to run to the far end of the paddock and "turn tail" when a person tries to catch them. This habit can become frustrating for both horse and handler and will sooner or later lead to injury. Most weanlings will learn the lesson with minimal stress if left in a small enclosure for a few days, dragging a lead rope that is about 6 feet long. The rope should be of a soft cotton in case the foal becomes tangled in it. A soft rope will not cause rope burns. During this procedure the handler should make it a point to "catch up" the weanling several times a day. Each time the youngster "turns tail," the lead rope can be picked up and firmly pulled to show the youngster that he should be facing the person who approaches.

With these lessons firmly in mind, the unruly foal has become a pleasant animal to deal with. If the lessons have been given with a gentle hand, the weanling will look favorably upon further contact with people and his potential will be greatly enhanced. At this point the weanling can be turned into a paddock with others of its kind for the winter.

The following spring the youngsters will be a much better lot because of the discipline encountered the previous fall. They will be ready to display their best side in the sale ring or to begin a light training program with good attitudes.

Before training of the yearling begins in earnest, some consideration of conditioning and exercise is important. Colts and fillies in the paddock seldom exert themselves long enough to develop a good muscle tone. Lack of proper exercise usually becomes a real problem when the yearling is first brought into a stall to begin training. Often the feed is increased without sufficient exercise and too much fat is laid down. This procedure neglects preparation of a horse's muscles, legs, and wind for the strenuous work of racing or other performance activities. The horse will tire easily and will huff and puff when asked to exert itself.

In the beginning the yearling should be started on the hot-walker for 30 to 40 minutes at a time. As the exercise is increased, it is best to break it up into at least two periods a day. The most effective exercise gives the horse variety, sometimes being ponied, sometimes being longed, and sometimes being put on the hot-walker. Plenty of exercise can be obtained by working the horse at liberty in a small, round corral, urging the horse with voice and whip signals as in longeing.

GROOMING

Good grooming makes the stabled horse healthier as well as enhancing its appearance. Proper grooming massages the skin and underlying muscle, thus increasing the circulation and promoting healthy hair growth.

The rubber or plastic currycomb is used first on a dirty horse. This loosens caked dirt and brings loose hairs to the surface. In addition, the currycomb massages the hide and muscle underneath. Grooming should be systematic (Figure 20–4). One should start with the same tool at the same place each

Figure 20–4. A typical supply of grooming equipment.

day, and work over the body in the same order so that the horse gains a sense of security and becomes accustomed to the procedure more quickly.

A stiff brush is used after the currycomb. Vigorous strokes will get between the hair and down to the skin. The groom should brush in the direction of the hair growth. A soft brush follows the stiff brush and is used to smooth down the coat and to remove the scurf brought to the surface by the stiff brush. Again the brush is used in the same direction as the hair grows. The stiff brush and currycomb are not used below the hock or knees of the horse. A rub rag is used after the brushes to bring a shine to the coat of the horse.

Daily grooming should include care of the feet. A hoof pick can be used to pry out the manure and mud from the foot (Figure 20–5). The frog and the cleft should be checked for the beginning of thrush. This fungus infection occurs in the foot of a horse and is generally caused by damp, dirty stalls.

When the weather warms up in the spring, the yearlings should get a bath. Until the horse is accustomed to a hose, several buckets of warm water will suffice. A large sponge can be used to wet the horse down. A generous amount of soap or non-detergent shampoo should be used with as much water as possible. The rubber currycomb can be used to massage the skin and loosen the dirt. A first bath may require two or more rinses after sudsing to get all the dirt. The mane and tail should also be washed carefully. The tail can be dunked into a bucket. After the final rinse, the horse is scraped as dry as possible. If the weather is not very warm, the horse should be covered with a blanket. It is best to walk the horse until he is completely dry.

Figure 20–5.   Cleaning the feet should be a regular task when horses are stabled. The foot shown is in need of trimming.

Figure 20–6.   When preparing a yearling for sale, clipping the long hair is important. Trimming of the coronet, pastern and fetlock is shown.

Yearlings that are prepared for sale will look much better if the shaggy winter hair is trimmed. Horses from extremely cold climates may look better with a body clip, but generally just a shortening of unwanted hair without clipping it off close to the skin is all that is necessary. A completely body-clipped horse must be blanketed to protect him from cold and drafts.

The hair around the coronet band should be clipped short as should most of the hair of the fetlock and the posterior part of the cannon (Figure 20–6).

The long hair of the head and face is best clipped as shown in Figure 20–7. The hair that protrudes from the ears should be trimmed to an even edge. A bridle path cut in the mane just behind the ears gives a refined look to the head and makes the halter fit more comfortably.

Figure 20–7.   A diagram showing where to clip the hair of the head and face.

Later in the year after workouts become more strenuous for the yearling, there will be a great amount of stress on the legs, especially the front legs from the knees down. It is a common practice to brace and rub the legs after a vigorous workout. A liniment is rubbed into the tendons and ligaments and then often the legs are bandaged. The rubbing and liniment increase circulation and reduce swelling that may result from stress and strain.

Although the routine use of leg bandages after bracing and rubbing is not as commonplace as it once was, there is still a place for them when increased heat in the tendons indicates a threatening bow or bucked shin. The method of applying a stable bandage is shown in Figure 20–8. First a cotton pad is placed evenly around the leg. The bandage is then started at the fetlock and wrapped downward over the fetlock and pastern. The bandage can be streched tightly under the back of the fetlock and then wrapped up the cannon bone. Self-adhesive bandages work fine and stay in place well.

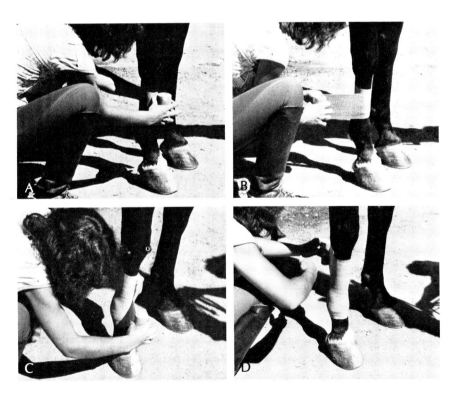

Figure 20–8. The leg wrap should begin in the center of the cannon bone (A). The self-adhesive bandage sticks to itself as it is wrapped down toward the fetlock (B). After it completely enwraps the fetlock, the wrap continues back toward the knee (C). The wrap is finished just below the fetlock (D), and no pin is necessary to hold it in place. (Courtesy of 3M Company.)

BIBLIOGRAPHY

Barbalace, Roberta Crowell: Light Horse Management. Minneapolis, MN, Burgess Publishing Co., 1974.

Bradbury, M., and Werk, S.: Horse Nutrition Handbook. Houston, TX, Cordovan Corp., 1974.

Breuer, L.H., Bullard, T.L., and Yeates, B.F.: A Suggested Schedule for Management for a Horse Breeding Farm. Proc 10th Annual Horse Short Course, Texas A & M Univ., 1970.

Davies, J.: Fire protection for horses is different. Thoroughbred Record *193*:556–557, 560, 1971.

Ensminger, M.E.: Horses and Horsemanship. Danville, IL, Interstate Printers, 1978.

Ensminger, M.E.: Stud Managers Handbook Annual (beginning with 1965). Edited by Agriservices Foundation, Clovis, CA.

Finnery, Humphrey S.: A Stud Farm Diary. London, J.A. Allen & Co., 1959.

Hintz, H.F.: Growth Rate of Horses. AAEP Proc., 1978, p. 455.

Kays, John M.: The Horse. New York, Arco Publishing Co., 1969.

Kronfeld, D.S.: Feeding on Horse Breeding Farms. AAEP. Proc., 1978, p. 461.

Lewis, Lon D.: Nutrition for the Broodmare and Growing Horse and Its Role in Epiphysitis. AAEP Proc., 1979, p. 267.

Miller, W.C.V.: Practical Essentials in the Care and Management of Horses on Thoroughbred Studs. London, The Thoroughbred Breeders Assoc., 1965.

National Academy of Sciences: Nutrient Requirements of Horses, 4th ed. Washington, D.C., National Academy of Sciences, 1977.

Posey, Jeanne K.: The Horse Keeper's Handbook. New York, Winchester Press, 1974.

Schryver, H.F., and Hintz, H.F.: Factors Affecting Calcium and Phosphorus Nutrition in the Horse. AAEP Proc., 1978, p. 499.

Schryver, H.F., and Hintz, H.F.: Nutrition and reproduction in the horse. *In* Current Therapy in Theriogenology. Philadelphia, W.B. Saunders, 1980.

Swidler, David T.: All About Thoroughbred Horse Racing. Miami, FL, Hialeah Guild Publishers, 1967.

Ulmer, D.E., and Juergenson, E.M.: Approved Practices in Raising and Handling Horses. Danville, IL, Interstate Printers, 1974.

Widmer, Jack: Practical Horse Breeding and Training. New York, Charles Scribner's Sons, 1942.

Willoughby, David P.: Growth and Nutrition of the Horse. San Diego, CA, A.S. Barnes, 1975.

part v

practical horse breeding

21

facilities

In most horse breeding operations there is a larger cash outlay for facilities than for the horses themselves. This is especially true for the more successful farms. The land and the improvements on it are important factors in whether a horse breeding farm can fulfill its potential. Often a breeder is so intent on acquiring good stock that facilities are neglected at the beginning. Many breeders come to realize that they have backed into their location and are using barns and fences designed for another type of operation.

FARM SITE

Most breeding farms begin as part-time endeavors with the owner engaged in business or working at a job in a specific locality. Thus, the site selection is limited to a local farm. Few breeders have the opportunity to analyze a large geographical area for the best site. When possible, a breeder should take advantage of establishing the farm in traditional horse country. Many like the bluegrass area of Kentucky and Virginia or the gently rolling land of Pennsylvania and Ohio. The glassland area of central Florida has proved to be a good horse breeding country. Climate can be such an important factor in horse breeding that many full-time breeders eventually seek the southern latitudes and choose a site in Southern California, Arizona, or Texas.

Today with continually spiralling land prices, some horse breeders have re-evaluated the importance of raising their own hay. They have decided that it is most economical to buy the best hay available and design their

Figure 21–1.   Horses seem to thrive best with mild weather, well-drained pastures, plenty of shade, and a good supply of fresh water.

facility with paddocks which produce only browsing grass. In this way, they can have smaller farms than would otherwise be necessary.

Better land for less money is often found a long distance from the city, and at first consideration that may be the place to establish the farm. However, the availability of necessary services is important also and may have a greater economic impact over the years than the cost of the land. Is there a good veterinarian nearby? Is there a source of part-time help available? What about a farrier? Many horse breeding farms have been established in remote areas that seemed perfect in every respect, but eventually were relocated nearer to town so as to be more convenient to services.

Horses seem to thrive best with mild weather, well-drained pastures, plenty of shade, and a good supply of fresh water (Figure 21–1). These items are important in selecting the best site for the farm. Selecting the best location is no idle decision; many factors must be weighed carefully in order to choose a place that will be suitable for years to come.

PREVENTING ACCIDENTS

Some horse farms seem to have all the "bad luck." At least it seems like bad luck since they have most of the problems when it comes to injured horses; whereas other farms, even with a large number of horses, may go for months without any serious veterinary problems. What makes the difference? Usually it is that the accident-prone farm does not have a manager who makes a conscious effort at accident prevention (Figure 21–2). Conscious effort in this area eventually becomes an unconscious part of overall good management, and is truly one of the basic secrets that make the difference between good and poor managers. This effort is important in many aspects of management, such as maintaining proper facilities, feeding regularly, and keeping the horses contented.

Accidents are by far the most common type of equine veterinary problem. The majority of these could well have been predicted by the veterinarian

Figure 21–2. Poor facilities are generally "an accident waiting to happen." An example of dangerous facilities at a small quarter horse racetrack.

Figure 21–3. Protruding nails and board ends should all be eliminated.

had he seen the facilities before the accident happened. Poorly constructed fences or fences in ill-repair can contribute to the accident problem. Fences do not have to be expensive to be safe. A good tight wire fence is not as dangerous as some people think, especially if the posts are not too far apart.

Gate openings should have no rough edges or projections. Nails halfway out and rails projecting a few inches past the post can cause nasty wounds if a horse crowds the side as he passes through the opening. The same is true of stall door openings. The edges should be constructed in anticipation that the horse is going to scrape it every time he goes by (Figure 21–3).

The average farm will tend to accumulate a certain amount of junk. This buildup must be watched closely where horses are raised. Pieces of wire, boards with partially embedded nails, metal fragments, and the like should be loaded up and hauled away on a regular basis.

Everyone has seen that run-down farm with a couple of strands of barbed wire hanging between shaky posts and a broken panel for a gate. Occasionally such a farm gets by with as few accidents as any other. Why? It happens only because the horses have grown accustomed to the place and no new animals are brought in to create excitement. Contentment is the secret. This factor is very important in accident prevention, even on a farm with sturdy, clean, well-kept facilities.

Practically all accidents happen at a time when the horses are upset, excited, or discontented. The wise manager is alert to this reaction and is quick to adjust the situation so that horses "settle down" quickly. At times it is unavoidable or undesirable to keep the horses perfectly calm. They become upset because things are changed. Friends are separated, a mare is in heat, or a new horse is put into a group. Even small things can be

Figure 21–4. Good facilities can make better management easier.

upsetting, such as changing the order of feeding, or making the glutton wait until last. The manager with good facilities has more flexibility in handling the horses in the way he thinks necessary because he does not have to worry about one of them getting upset and breaking through that shaky gate.

On the other hand, regardless of the facilities, the good manager does his best to see that all the horses remain as content as possible. This care greatly reduces the chances of an accident. He does not make the glutton wait until last to be fed just for spite. Still, if there is reason for him to be last, it becomes a regular thing. If a horse is to be separated after being accustomed to his friends, he is placed in a safe enclosure. New horses are given the opportunity to become acquainted over a stout, safe fence a few days before being placed in the same enclosure with others. The good manager is regular in feeding habits, being consistent in the feeding time, amount, and nature of feed. The manager of a large farm will keep horses of unlike age and type separate. Geldings will be kept together, mares in foal together, weanling foals together, and so forth. Good facilities can make management easier (Figure 21–4).

FENCING

Fences should be designed and built with safety in mind. The smaller the enclosure, the more the fence will be tested and the safer it should be. Woven wire fences can be more dangerous than simple smooth wire unless the weave is tight and no openings are large enough for a hoof to slip through (Figures 21–5, 21–6). One rail or board along the top of the fence is helpful in warning the new horses or foals that the wire fence is actually a barrier. Poles make safe fences as long as they are high enough to discourage jumping (Figure 21–7). Fences of 1-inch boards can be dangerous as they are easily broken, leaving pointed and splintery ends. Some helpful information about various types of fence construction is provided in Table 21–1 and Figures 21–8 through 21–10.

The gate is the most important consideration of the fence as far as accidents are concerned. The horse becomes accustomed to going through the gate opening, and he is far more apt to test the strength of the closed gate, when he becomes excited, than the fence itself. Gates must be strong enough to withstand the full force of a horse once in a while. A horse that breaks a gate one time is far more inclined to try it again, and will probably be injured in the process (Figure 21–11).

Fence corners are important considerations also. If a horse is going to challenge a fence, it is likely to be at the corner. Rails 4 to 5 feet high should run out about 10 feet from each corner to make a wire fence a safer proposition.

Important factors to consider in fence layout are accessibility, flexibility, and efficient use of labor. Roads and walkways should be designed so that all pastures and paddocks are easily accessible. Planning should include the

Figure 21–5.   A good woven wire fence.

Figure 21–6.   Smooth wire, instead of barbed wire, makes an inexpensive fence.

Table 21-1. Safety Requirements for Various Fence Types

Material	Material Specifications			Construction Details		Comments
	Post	Line fence or rails	Fence Height	Number & Spacing Rails or Mesh of Wire	Distance between post on centers	
Steel or aluminum	7½'	10' or 20' rail	60"	3 rails; 20" centers	10'	Because of the strength of most metal rails, fewer rails and posts are necessary than where wood is used.
	7½'	10' or 20' rail	60"	4 rails; 15" centers	10'	
	8½'	10' or 20' rail	72"	4 rails; 18" centers	10'	
Board	7½', 4"–8" diameter	2" x 6", or 2" x 8"	60"	4 boards	8'	
	8½', 4"–8" diameter	2" x 6", or 2" x 8"	72"	5 boards	8'	
Poles	7½', 4"–8" diameter	4"–6" diameter	60"	4 poles	8'	
	8½', 4"–8" diameter	4"–6" diameter	72"	5 poles	8'	
Woven wire	7½', 4"–8" diameter	9 or 11 gauge stay wire	55"–58"	12" mesh	12'	Woven wire is satisfactory for larger areas where the concentration of animals is not too great. But it is not recommended for corrals, paddocks, or small pastures. Use 1 or 2 strands of barbed wire (with points 3" to 4" apart) on top.

Figure 21–7.   Poles make a safe fence.

Figure 21–8.   Chain link fences are expensive, but durable, safe, and eye appealing.

Figure 21–9.   Pipe fences are durable and safe.

probability of bad weather, heavy snows, and flooding, and should provide for normal management under adverse conditions. Good drainage is an important prerequisite for the location of paddocks and stalls (Figure 21–12).

It is wise to keep the possibility of future expansion in mind when planning the fence layout. Paddocks and barns should be located so that more can be placed adjacent to them as the need arises. The overall layout or farm plan should probably include many more facilities than will be needed at first. With such forethought little effort and material will be wasted when the operation expands.

Figure 21–10.   Rubber-nylon fencing is attractive and safe.

Figure 21–11.   A well-made pole gate makes one of the sturdiest and most attractive gates available.

Labor costs are often a neglected expense at a breeding farm. Poor layout, long distances to walk, mud, and difficult gate latches can all lead to severe increases in the amount of time needed to care for the horses. A successful farm will be designed with some thought for the personnel who work there. The efficient use of labor and an attractive atmosphere lead to contented help and will keep costs down.

Figure 21–12.   Good drainage is important for areas that will have heavy use.

Many types of fences have been used successfully for horse farms. Most farms use a combination of fence types depending upon the size of the enclosure and the number and type of animals kept inside.

Electric fences are often used as a temporary device to contain horses in a pasture when no time or money is available to build a permanent fence. Once horses have become accustomed to where the wire is located, it works quite well. Usually small cloth flags are tied on the wire a few feet apart so that it can be more easily seen. Sometimes an electric wire is used just inside another fence to keep horses from "riding the fence."

Wire fences are most often used to enclose larger paddocks and pastures. A 2 × 4 inch V-mesh horse wire is the best type to use. A board running along the top of the wire makes the fence much more attractive and affords a more visible barrier, which can be very important at times. Twisted smooth wire can be obtained as cheaply as barbed wire and often can serve as a more economical fence. At least four strands, tightly stretched, are necessary. Wire fences should be used with wooden posts or round metal posts with smooth tops. The T-angle metal posts that are driven in the ground can be very dangerous for horses who tend to rear up and come down on top of the post, opening up terrible wounds.

Board fences have traditionally been the fence of choice for horses (Figure 21–13). They are easily seen and can be quite strong. However, they are expensive for large enclosures and are generally chewed and easily damaged in smaller enclosures. They require a lot of maintenance. Post and rail fences, although expensive, are becoming quite popular for horses. They tend to show the effects of cribbing less and require less maintenance. Pressure-treated posts and rails are the most desirable, as they usually prevent chewing and ground rotting.

Chain link fences are being used more often in construction of new horse farms (Figure 21–8). Their price has become more competitive in recent years as the price of lumber has increased. Chain link makes an excellent enclosure for stallions and for the keeping of visiting mares, whose safety is extremely important. The top of the fence should be smooth and protected by a board or pipe. Boards can be used quite successfully with chain link fences. Placed outside the wire, they eliminate the cribbing problem.

Figure 21–13.   Board fences have traditionally been the fence of choice for horses.

Pipe fences are popular for small enclosures (Figure 21–9). They are extremely durable and safe. They are prohibitively expensive unless some sort of inexpensive used pipe can be obtained. An inventive horseman has marketed a plastic pipe fence which is surprisingly strong and adds an extra safety factor because of its flexibility.

Rubber-nylon fencing has recently been marketed (Figure 21–10). It is available in widths from 2 to 4 inches, and is quickly installed. Its greatest value may be in conjunction with woven wire to give a visible barrier.

Gates should be strong and well hung. A gate should not be placed over a low spot where one must stand in a mud hole to open it. Metal gates seem to be the most durable. Many aluminum gates are in use and are quite acceptable. In some parts of the West, sturdy pole gates are available. These gates are both safe and attractive (Figure 21–11).

BARNS AND STALLS

The type of buildings necessary on a breeding farm depends upon the climate and the specialized needs of the farm. Breeders in the cold or wet areas that have long periods of inclement weather may be more interested in a large structure which covers stalls, indoor arena, tack room, and feed storage. Breeders in milder climates with many days of sunshine may think more in terms of stalls set in rows with separate tack room and feed storage barns, and exercise facilities completely out in the open.

A breeding farm with little or no reason to train will need a different type of barn than a training stable. The horse breeder is concerned about housing for several types of horses: the stallion, visiting mares, dry mares, mares in foal, foaling mares, weanlings, and yearlings.

In all but the most severe climates, breeding stock get along quite well when several of similar types are kept in a large paddock with a single open-sided shed. A shed similar to the one shown in Figure 21–14 is suitable

Figure 21–14.   Open-sided sheds make excellent shelter on a breeding farm.

for mares and young stock. The opening should face away from the direction of prevailing winds and preferably open to the south or east to take advantage of the winter sun. This type of housing offers a minimum of work; there are no stalls to clean or horses to exercise. On those farms where controlled winter lighting is used to help bring mares into heat earlier, it will be necessary to keep the mares in an enclosed barn or in stalls, which substantially increases the amount of work required.

Four or five large paddocks, each with an open-sided shed, should be adequate for the small breeder. Preferably these paddocks should be rather pie-shaped, offering easy access from a central area. Feed storage, tack rooms, and foaling stalls should be located in the central area. The stallion quarters and visiting mare paddocks should probably also be in this central area on small breeding farms.

Large farms with more diversified operations will generally be better served with at least three separate working areas on the premises, not necessarily located adjacent to one another. The horse raising operation can be arranged as described previously. A separate breeding center will be helpful when several stallions stand at one farm and when a large number of visiting mares are expected. Quarters for these "outside" mares should be isolated, not only for their own protection, but to avoid bringing in a continual string of diseases to the regular farm horses. Most breeding farms offer a choice of box stalls or small private paddocks for visiting mares. An economical way of building these paddocks is to place them side-by-side with a long, open shed across one end. The shed supplies shelter to all the paddocks, but keeps the mares separated. A third working area on the farm would be for horses in training.

Most breeding farms have a breeding barn close to the area where the stallions are kept, with a breeding chute and a veterinary room (Figures 21–15, 21–16) so that breeding can continue uninterrupted during bad

Figure 21–15. A good breeding barn is an important part of the facility.

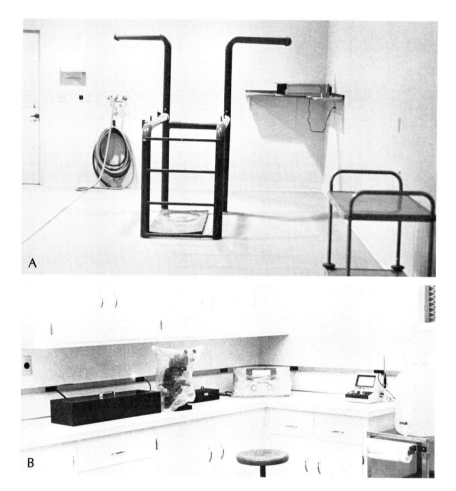

Figure 21–16.   A veterinary treatment room can be very useful on a busy breeding farm. It
should include treatment stocks and cabinet space.

weather and the veterinarian has a clean, efficient place to "set up" when
checking and treating mares. On smaller farms that want to consolidate, it
may be advantageous to have the foaling stalls adjacent to the breeding barn
and veterinary room. The breeding area of this barn should be at least 20 ×
20 feet and 12 to 15 feet high. Some sort of dust free, slip-proof floor would
be a helpful feature.

The training area will require individual stalls and a training arena or track
(Figure 21–17). A hot-walker is also necessary in this area as stalled horses
are in need of daily exercise. A modern technique of exercise is the equine
swimming pool (Figure 21–18). Using this form of exercise builds leg
strength while avoiding damaging stress. Swimming is an important aid in
recuperation from certain types of leg injuries. The well-planned training

Figure 21–17.   The training area will require individual stalls. The "show place" may have expensive permanent structures (above); the new metal structures are just as durable and useful and are less expensive (below).

area will have places for manure disposal and feed and bedding storage near the stalls. Tack rooms and a grooming area should also be located adjacent to the stalls.

Peter C. Smith lists six major needs of a stable in his book, *The Design and Construction of Stables*: dryness, warmth, adequate ventilation but freedom from drafts, good drainage, good lighting, both daylight and artificial, and adequate and suitable water supply. In line with these basics, an infinite variety of farm and stable layouts can be developed depending on the needs and desires of individual owners. The farm layout should provide facilities for a well-rounded operation situated in a useful arrangement.

A breeder who is in the enviable position of building a facility from the ground up may find it well worth his time to consider some modern

Figure 21–18.   The equine swimming pool builds muscle strength without the damaging stress. It is good exercise for horses recuperating from lameness.

mechanisms which make life easier for horses and hired help. One of these is a system of automatic waterers. Water pipes can be run underground to each stall and paddock easily before structures are erected, to ensure a continual supply of fresh water to every spot on the farm. Another system worth considering is an automatic fly sprayer. Several types of electrically operated aerosol dispensers are available. These systems pipe a liquid insecticide to any area and periodically spray the premises against flies. Generally a short spray about every 15 minutes will substantially reduce the number of flies present. This fly sprayer more than pays for itself in the long run by reducing wasted energy from horses fighting and worrying about flies. Sometimes the weight loss from bothersome flies can be tremendous. Breeders who plan a large amount of pelleted feed might consider some sort of automatic feeding system which reaches all stalls and paddocks and operates on a time clock. Such a device can save many dollars in labor costs.

PASTURES AND HAYING

After all of the paddocks, arenas, living and working areas have been plotted out on the farm, the remaining acreage can be allotted to pasture for hay production. Those farms that have a large number of horses not in training can utilize pasture and can benefit from more pasturage than hay fields, although most hayfields can double as pasture when needed.

The best pasture is one with high-quality forage. The more use that can be made of good pasture, the less expensive feeding will be. Good pasture forage is higher in digestible nutrients than is comparable hay; it is harvested by the horse, requires no storage, and requires no manual labor for utilization.

Shade is important in all pastures. If no trees are available, open sheds can be provided. Fresh clean water is also a necessity. A flowing stream is a real plus factor for any pasture. Mineral and salt blocks will add the finishing touches to a perfect equine nutritional environment.

The geographical location will determine the type of pasture that can be utilized. Also the amount of pasture required to support a given number of horses will vary as to forage type, soil fertility, and amount of moisture. The average mature horse on pasture will eat and trample about 1000 pounds of forage per month. As an example, if a good irrigated field in an area that is moderate in climate will produce 3 tons of hay per acre during a 6-month growing season, then during that same period each acre should support one mature horse for 6 months. Probably the ideal situation would be to rotate pasture use and, in the preceding situation, allow grazing only about half the time in any given pasture, and harvest 1½ tons per acre. With this type of management, 100 acres of hay-pasture would handle 50 mature grazing horses for 6 months and also produce 150 tons of hay for winter feeding. Of course, a better area might produce twice this amount. Most counties of the United States have extension agents that can tell a newcomer what he can expect from his land.

The county agent can also advise the newcomer on which kinds of pasture and hay grasses can be grown in the area (Figure 21–19). Certain species will grow best while others, not well suited for the region, will grow little or not at all.

Dry land pasture may have to be cleared of sage and shrubs to make maximum profit from it. If the native grass is abundant and rainfall is slight, it might be advisable to leave the terrain alone. Crested wheat grass, intermediate wheat grass, or Russian wild rye make excellent dryland pasture. In some areas dryland pasture will be more productive with some Orenberg alfalfa or Ladak mixed in.

Some farms can benefit from leveling and grading before planting pasture hay, or even dryland pasture. The county agent will advise that the field be plowed, reseeded, harrowed, and packed with a roller if necessary to make a firm seedbed for reseeding. The seedbed should be free from old sod and weeds because old plants will compete with new seedlings if they are not completely removed.

The county agent may also advise a soil test to determine what type of fertilizer will be needed. Fertilization can improve plant tolerance to dry weather and increase the protein content, as well as making an overall healthier plant.

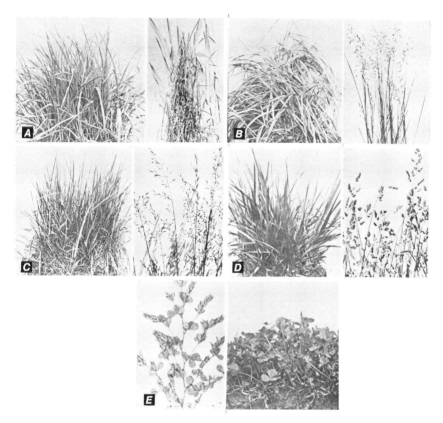

Figure 21–19. Varieties of grasses available for pasture: (A) timothy, (B) orchardgrass, (C) bluegrass, (D) fescue, (E) clover.

If pasture is to produce hay also, some consideration should be given to a forage type that will make good hay. The legumes are generally higher in protein and minerals than grass hay, but tend to have more problem with moldiness. A mixture of legumes and grass is quite good. Most of the commonly grown leafy grasses and legumes make high-quality hay. Weeds and undesirable plants lower hay quality by adding woody material, which causes bad tastes and odors, thus lowering acceptability and digestibility.

The growing conditions of hay can affect its quality. A lack of water will reduce the leafiness. On the other hand, too much water can also diminish leaf numbers through disease. The soil fertility will also affect leafiness.

Harvesting procedures drastically affect hay quality. Unless the hay can be put up properly, it is better left to grazing. The stage of plant growth at the time of cutting is important. The protein content becomes lower as the plant advances from the vegetative stage to the reproductive stage. The digestibility is also reduced as plants mature because of increased stem thickness.

Figure 21-20.   A farm layout which provides facilities for a well-rounded operation situated in a useful arrangement. At the left is a large barn for the farm-owned mares. Individual paddocks can be added in the area to the left if desired. The barn on the right can be used for outside mares. The breeding barn is central, and the covered training area upper central. The office is at the far left, as well as individual paddocks for breeding animals, each with separate shelters.

Legumes should be cut just as a few flowers begin to show, and grasses just as the seed heads begin to appear. Field-cured hay generally loses more quality than artificially cured hay. However, hay must be sufficiently dried before being baled; otherwise, it develops a musty moldy odor which can be toxic to horses. New machinery can crush and "condition" stems at the time of mowing. Then the hay can go immediately into windrows for final drying. The result is a better quality hay.

Seven factors are important in achieving high-quality hay. The first is species of plant. The higher the percentage of legume, the higher the hay will be in feed value. The stage of maturity when cut is a second factor. A third factor is the percentage of leaves that remain. Leafiness increases the quality. The amount of green color is also a factor. A bright green color indicates a minimum of bleaching and leaching losses of vitamins and nutrients. The aroma or fragrance is a factor in hay quality. It should have a clean "crop" odor, not a moldy dusty smell. Stemminess is important to consider. Large numerous stems reduce the quality. Finally the amount of foreign material present is important. Such items as stubble, weeds, sticks, and dirt make bad hay. All aspects of the facilities must be considered in an effort to improve breeding operations (Figure 21–20).

BIBLIOGRAPHY

Barbalace, Roberta Crowell: Light Horse Management. Minneapolis, MN, Burgess Publishing Co., 1974.

Blickle, J.D., et al.: Horse Handbook: Housing and Equipment. Ames, IA, Iowa State Univ., 1971.

Dunn, Norman: Horse Handling Facilities. Stud Managers Handbook, Vol. 8. Clovis, CA, Agriservices Foundation, 1972.

Ensminger, M.E.: Horses and Horsemanship. Danville, IL, Interstate Printers, 1978.

Guenther, H.R.: Horse Pastures, Horse Science Series, Cooperative Extension Service C151. Reno, Univ. of Nev., 1974.

Hollingsworth, Kent: A Barn Well Filled. In The Blood-Horse, 1971.

Jeffries, N.W.: Better Pastures for Horses and Ponies. Montana Cooperative Extension Circular 294 (May). Bozeman, Montana State Univ., 1970.

Langer, Lawrence: Know Stable Design and Management. Phoenix, Farnam Horse Library, 1973.

Miller, W.C.: Practical Essentials in the Care and Management of Horses on Thoroughbred Studs. London, The Thoroughbred Breeders Assoc., 1965.

O'Dea, Joseph C.: Veterinarian's View of Farm Layout. The Blood-Horse, Dec. 1966, p. 3790.

Smith, P.C.: Design and Construction of Stables. London, J.A. Allen, 1967.

Ulmer, Donald E., and Juergenson, E.M.: Approved Practices in Raising and Handling Horses. Danville, IL, Interstate Publishers, 1974.

Willis, Larryann C.: The Horse Breeding Farm. San Diego, CA, A.S. Barnes, 1973.

22

veterinary care

$\bigvee$eterinarians who manage a horse breeding farm as a full-time endeavor are the first to point out that horse health should not be accepted simply as a lack of disease. They know from experience that it is better described as a state of maximum economic production.

A complete animal health program involves much more than prevention, detection, and treatment of disease. It starts with knowledge of past reproduction performance of the herd based on complete records. Good records will indicate disease patterns and show the level of performance that has been obtained. In order to establish an effective horse health program, one must have knowledge of past management practices.

A systematic yearly schedule of events on a horse farm leads to better horse health. Each part of the total operation has its proper time. Changes in feed type and amount should vary with the season. Working within the framework of a schedule reduces procrastination or the neglect of some rather important procedures during the year. A breeder might start with an annual schedule of events as shown in Figure 22–1. After careful considera-tion a breeder might want to modify the schedule to suit his particular situation.

Activity on a horse breeding farm is low early in the year, so this is the most convenient time to vaccinate the mares. Choose a warm day and do them all at once, or if special arrangements do not have to be made with a veterinarian, vaccinate one or two each day as they are fed. Immunization at this time is especially helpful as it will give passive immunity to the foals through the colostrum they receive at birth. A bit of precaution is necessary

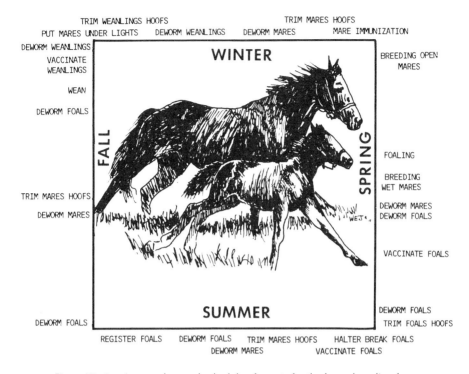

TRIM WEANLINGS HOOFS   TRIM MARES HOOFS
PUT MARES UNDER LIGHTS   DEWORM WEANLINGS   DEWORM MARES   MARE IMMUNIZATION

DEWORM WEANLINGS   BREEDING OPEN
    VACCINATE   WINTER   MARES
    WEANLINGS

    WEAN

DEWORM FOALS

FALL   SPRING   FOALING

                                                                   BREEDING
                                                                   WET MARES

TRIM MARES HOOFS   DEWORM MARES
DEWORM MARES   DEWORM FOALS

                                                                   VACCINATE FOALS

                        SUMMER   DEWORM FOALS
DEWORM FOALS   TRIM FOALS HOOFS

        REGISTER FOALS   DEWORM FOALS   TRIM MARES HOOFS   HALTER BREAK FOALS
                            DEWORM MARES   VACCINATE FOALS

Figure 22–1.   A general annual schedule of events for the horse breeding farm.

here in that some vaccines can harm a fetus at certain times of pregnancy. Checking with the veterinarian can be important.

During the latter stages of pregnancy each mare should be carefully observed from day to day. The grain ration is best fed separately so that each mare gets the proper amount to keep in good flesh. This rationing requires constant attention as often a mare will begin to lose weight slowly so that someone seeing her everyday may not notice the subtle change. At this time special heed should be given to the vitamin and mineral content of the ration.

Foaling time is extremely critical. Certain routine procedures should be followed in order to give the foal its best chance for survival. The navel should be disinfected and tetanus antitoxin given. If there is any reason to suspect infection, a precautionary dose of long-acting penicillin should be given to the foal. Generally an enema at birth is a good idea. A pint of warm soapy water serves well.

It is good to wash the mare's udder before the foal first nurses, and keep it clean with a mild soapy solution. If the mare does not drop the afterbirth within 12 hours, a veterinarian should be consulted. Give the mare a meal of bran mash immediately after foaling as it is a helpful laxative. The amount of feed must be increased drastically after foaling because of milk production (see Chapter 20).

Early in the year mares that have foaled should not be bred on the foal heat. Waiting for the 30-day heat will not set them back in their foaling time, and in the long run will ensure a greater conception rate. Dry mares should have a thorough veterinary examination before the breeding season begins. The heat cycle should be charted as early as it begins and breeding begun only when it seems regular.

Mares that have not been wormed since early fall should be as soon as they recover from foaling. Except under unusual circumstances foals need not be wormed until 2 months of age. When kept in crowded conditions, foals may need worming as often as every month during their first year. Foot trimming for the foals should begin early, but not before halter-breaking. A monthly hoof trim not only helps legs to grow straight, but also teaches foals to allow their feet to be lifted. Complete gentling and training of the foal are very important, not only to avoid later injuries, but also to facilitate quick and proper treatment in case of disease or injury.

Vaccination of the foals should begin at 2 to 3 months of age. At this time they are beginning to lose their passive immunity received from the mare. Immunization should not be too early, however, as the young system is not capable of building complete immunity for some time (see Chapter 19). Creep feeding the foals at about 3 months of age helps to ensure adequate nutrients. Some breeders prefer to feed foals daily in separate stalls, and this practice will generally produce better results.

Late summer is the ideal time to brand the foals or register them. Getting the paperwork done early can eliminate headaches later. Castration of colts can begin even while they are nursing in later summer. This is less traumatic than castration shortly after weaning. Early castration is easier on everyone, including the colts. It helps to develop more even dispositions, and more rapid learning. There is less tendency for cresty necks to develop, giving flexibility for good performance later. In very hot climates it may be advisable to postpone castration until cool weather in late fall.

It is not advisable to wean all foals at once. Some will have been born later or will be retarded in development. Weaning in pairs is helpful over a period of a month or two in the fall.

The year's activities are concluded with a pregnancy examination for the mares. Even if they have been previously checked in foal, it is wise to check for subsequent abortions that may have occurred. This determination is important so that the mare can receive proper winter care for her condition.

Most of the previously described yearly activities are normally thought of as management rather than health considerations. However, to the professional horse breeder, complete horse health means maximum production. This health program can come only with exacting vigilance throughout the year to see that all potential problems are prepared for and all horses are given every bit of protection possible.

DISEASE

The first question that should come to mind when a horse breeder is confronted with a disease problem is whether it is contagious, that is, caused by an organism and capable of transmission from one animal to another. The diagnosis of a contagious disease signals concern for all of the other horses on the premises.

In general, the horse is a remarkably healthy animal. It has tremendous natural capabilities to fight infection, and usually does so without ever raising concern from its owner. Most organisms are rather specific in their infection. As examples, the human cold virus, the dog distemper virus, and the tuberculosis bacterium generally find themselves defenseless against the natural immune system of the horse. There are a few exceptions, such as the anthrax bacterium, which can bring about sudden death in almost every domestic animal and man. The facts that horses have an excellent auto-immune system and that most organisms are species specific limit the danger from infectious organisms. The list of common contagious organisms of the horse is rather short. Also many of the contagious diseases are confined to young horses, or to breeding animals. This characteristic narrows the possibilities when a diagnosis is attempted.

It is convenient to think of contagious organisms as falling into two groups: viruses and larger organisms. The viruses do not respond to antibiotic therapy while the second group does. Vaccines are used to control the viruses. Virus diseases reduce the white blood cell count, a fact that is extremely helpful in diagnosis. Bacterial infections raise the white blood cell count.

Equine Infectious Anemia (EIA) is caused by a virus. The government program of Coggins testing and elimination of carriers of this disease has caused much controversy. In the acute stages of the disease, the horse has a high temperature, severe depression, and loss of weight. The virus causes a rapid destruction of red blood cells and results in anemia from which the disease gets its name. More commonly the disease is less noticeable at the onset and eventually causes the horse to be unthrifty over a long period of time. The exact methods of transmission are not known but blood-sucking flies do carry the disease from infected horses to others. EIA is not considered a common disease problem in most areas, and would not be the first consideration when a horse comes down with a virus infection. There is no effective vaccine available for this disease.

Equine Encephalomyelitis has received much public attention in the last few years because it can also be contracted by man. It is a virus infection of the brain carried by mosquitoes and other blood-sucking insects from wild birds which act as a reservoir. A large outbreak occurred in the United States in 1971, but since then the disease has been considered rather rare. Annual vaccination is always recommended because of the high death rate and incapacitation when it does occur.

Equine Influenza is probably the most common disease of the horse. The incidence varies from year to year and from area to area, but in bad years 50 to 90 percent of all horses in some areas will be affected. Symptoms are quite similar to the disease in man—an initial high temperature, depressed appetite, and a watery nasal discharge. Treatment is of little value, but rest and good care help to keep the disease under control. Debilitated animals, especially those that get damp and cold, may develop such a lowered resistance that pneumonia sets in. Influenza viral strains seem to drift from one type to another over the years, making it difficult for vaccine manufacturers to keep their products highly successful without constant vigil. Consequently some years the vaccine is more effective than other years. Currently it is believed that vaccination will last only about 3 months, necessitating a quarterly vaccination program in order to keep all horses protected.

Equine Viral Rhinopneumonitis can be easily confused with influenza. It starts with a rise in temperature and a watery nasal discharge which often develops into a persistent cough. Estimates are that 80 percent of all weanlings and other previously unexposed or unvaccinated young horses in some areas will show symptoms of the disease during the months of August through December. A profuse nasal discharge has led to the nickname of "snots." A more important consideration is the effect of the virus on pregnant mares. It causes abortion; and one estimate is that fully 20 percent of all equine abortions are attributable to "rhino." It can strike just a few animals, or can create an "abortion storm," severely setting back a breeding program of many years standing. The "rhino" virus can live for months in bedding, equipment, and surroundings. In areas with a considerable number of broodmares, "rhino" can become increasingly more frequent as a result of a contagious cycle. Mares may recover, then be re-infected by sick animals, and in turn pass the disease to healthy animals resulting in up to 90 percent abortion in a herd. Affected pregnant mares may not show visible symptoms of "rhino." There is, however, a viremia in which the organism penetrates the fetus and multiplies. Although the mare's defense system, through production of antibodies, eliminates the virus from the rest of the mare's body, these antibodies cannot penetrate the fetal membranes to protect the fetus against the disease. Abortion occurs when the fetus can no longer function. Sometimes the virus doesn't kill the fetus, only weakens it. The infection persists and results in a weak foal that will often die. Prevention is important with "rhino" since no effective treatment has been established. For primary immunization, two injections given 4 to 8 weeks apart is recommended. Then a booster annually with a single dose will keep up the immunity.

Viral Arteritis is a third disease affecting the upper respiratory tract. It is easily confused with influenza or "rhino." It also may cause abortion. Generally, however, it is a much milder disease than the other two and recovery is usually prompt. A pregnant mare contracting the disease will have a 50 to 80 percent chance of abortion. A vaccine is available, but because the disease usually occurs in a mild form, it is not in common use.

In general, bacterial infections of horses seem to center in specific areas of the body and are not as highly contagious as the virus diseases. *Contagious equine metritis* (CEM) is an example. The bacteria that cause this infection settle in the reproductive organs of the mare and the stallion. *Strangles* or equine distemper is another bacterial disease that becomes isolated in the lymph glands of the head and neck. Only occasionally will it become systemic, and then there is a great danger of a fatality. Otherwise strangles is a bothersome problem that slows a horse down for awhile with subsequent recovery. Occasionally the organism can be very persistent causing throat abscesses that remain for months. Bacterial infections are not as threatening since the veterinarian has such a wide variety of antibiotics that help in treatment.

Tetanus is a bacterial disease of an unusual type. It is not contagious in that it is not passed from one animal to another. The organism is ever-present in barnyard surroundings. When it becomes introduced into the tissue of a horse by means of a wound, the organism can proliferate and produce a neurotoxin which generally will cause death of the animal. Tetanus toxoid is given as a routine prevention to foals and to others as an annual booster. Unprotected horses can be given temporary protection, when a wound is discovered, with tetanus antitoxin.

Many veterinary problems that occur in horses are metabolic in nature or are due to injuries, for example, colic, cancer, diabetes, blindness, and lameness. The average horse breeder is not equipped to diagnose and treat the variety of problems that might come along. The professional horseman learns to use his veterinarian and to trust his judgment.

FOAL DIARRHEA

Foal diarrhea can be serious or inconsequential. The knowledgeable groom should be able to handle the simple diarrheas and should know when to call the veterinarian for the serious cases. There are many causes for foal diarrhea and the different kinds vary in seriousness.

A loose stool during the first few days of life is a condition that should have immediate attention before it leads to serious systemic changes in the foal's fluid and electrolyte balance. At this early age, the most common cause is dietary.

Because the mare's udder is smaller than the cow's, we often imagine that the equine mother does not produce a large quantity of milk, but for the first few weeks many mares will produce 2.5 percent of their body weight daily, that is, 25 pounds of milk from the average 1000-pound mare. A 100-pound foal may have some trouble handling one-quarter of this weight daily in milk. It works best for him when he gets a little every 10 to 15 minutes.

When too much milk is put into the stomach at once, not enough enzyme action occurs before it balls up in the stomach and passes undigested into the intestine. This heavy concentration of milk in the intestine causes an osmotic attraction of fluid into the intestine and a loss of blood electrolytes with it. Then an abnormal bacterial fermentation occurs producing organic acids and carbon dioxide. These interfere with fluid passing out of the intestine into the blood stream. The result is a diarrhea. If allowed to continue, the foal becomes dehydrated. Death could result within a few days as complications continue to mount.

The best thing for such a foal is to go without milk for 12 to 24 hours. The mare should be milked out so that she is not uncomfortable and so that her milk production will not diminish during the procedure. A popular product for use in humans, Kao-pectate, is also beneficial for early foal diarrheas.

A "foal heat" diarrhea often develops from unknown causes at about 6 to 12 days of age. This timing usually corresponds to the mare's estrous period, which comes on at that time and lasts only 2 to 3 days. Many causes for this diarrhea have been proposed. Hormonal changes in the mare have been implicated most commonly, with the assumption that these are reflected in the milk supply. Another thought has been that the estrous secretions from the vagina contaminate the udder and cause the problem in the foal. Still another proposal is that the nature and quality of milk change during estrus, upsetting the foal enough to cause the diarrhea.

At about the time of the "foal heat," strongyloid larvae can be seen in the milk of most mares. One theory proposes that these are taken in by the foal and a diarrhea develops from it. Just why the Strongyloides are not present in the milk earlier is not known.

Regardless of the cause of the "foal heat" diarrhea, it should be of little concern. If it is truly a "foal heat" diarrhea, it will persist only for a day or so and will clear up on its own.

Diarrheas developing after about the third week of life are more commonly caused by microorganisms. Recent studies have shown that most infectious diarrheas occur in foals that have a deficiency of immunoglobulins. This problem occurs more commonly than has been thought in the past—nearly 25 percent of all foals suffer from such a deficiency. When this condition occurs, the normal microorganisms of the environment gain a foothold and multiply rapidly within the foal's intestine. Viruses, bacteria, and parasites—all can cause diarrhea under such conditions.

A recently discovered viral cause is known as rotavirus. Infections in foals were first described in England a few years ago and then later found to occur in Kentucky. Several large outbreaks were described in Texas. Now the organism can be found just about anywhere. This infection occurs when the foal is between 3 and 8 weeks of age. The first sign is lack of nursing. In the Texas outbreak, therapy that was usually effective was oral electrolytes, Pepto-bismol, and spectinomycin. The antibiotic was effective because it reduced bacterial complications. After recovery from rotavirus diarrhea, most foals were stunted, and some had occasional relapses of diarrhea. Only two foals out of 70 affected, died. Death was from severe dehydration.

Bacteria that commonly cause foal diarrhea are *Escherichia coli, Clostridium perfringens,* and Salmonella. The most troublesome of these is Salmonella. This organism can become resistant to many antibiotics, and it can remain as a chronic infection in the intestine of foals and adult horses. Some horses become Salmonella carriers, continually reinfecting other foals. Salmonellosis is a serious condition requiring careful treatment by a veterinarian.

Ascarids, large roundworms, are an important parasite in foals. They can grow to large numbers quickly within the intestine of a foal. Diarrhea may be one result of their presence. Sometimes these worms become so numerous that they completely obstruct the intestine and cause death.

Strongyle larvae are present in the environment of just about every foal. They are swallowed with feed or water. They molt in the intestine and then begin to migrate through the intestinal wall and along the lining of the blood vessels. The resulting damage can cause diarrhea. Deworming at an early age can alleviate the parasite problem. Deworming should be done every 6 to 8 weeks to keep down the parasite numbers.

There are other causes of diarrhea in foals. Some foals eat large quantities of sand, which can lead to intestinal upsets. Some foals eat the mare's feces. Sudden or drastic changes in the hay or concentrate diet of young horses can kill off the beneficial intestinal microflora and cause temporary diarrhea. Indiscriminate use of oral antibiotics can alter the intestinal microflora also. Antibiotics in the intestine will kill off the "normal" bacteria and allow the pathogens a free rein. Continued use of oral antibiotics will result in a buildup of resistant pathogens that can eventually lead to the death of the foal.

When a diarrhea occurs, the groom should watch the foal for signs of dehydration and reduce the milk intake if possible. The foal will normally stop drinking milk on its own. The presence of increased body temperature may signal an infectious diarrhea. When the diarrhea persists for 48 hours, and the mare is not in estrus, a veterinarian should be consulted.

DISEASE DETECTION

The experienced veterinarian can certainly see trouble at a glance when it comes to most disease conditions in most animals, including the horse. The astute horseman who remembers and gains from experience can also become skilled at detecting disease.

Sometimes just a glance may cause a "red light" to flash in the veterinarian's head, warning that something is wrong; but it takes more than that simple glance to apprehend the true and complete nature of the problem. The good veterinarian will conduct a thorough and methodical examination of a sick horse in order to arrive at the correct diagnosis every time. The horseman, although often lacking in extensive experience with every disease condition, can learn much and solve many of his own problems by following a predetermined procedure when examining a horse for disease.

Although the veterinarian may use a variety of diagnostic tools, such as a stethoscope or a hoof tester, he gains most of his information from simple observation and palpation with hands and fingers. Remember that some diagnostic tools (a thermometer, for example) are useful in the hands of the average horseman as well.

A number of so-called casual observations will often produce the clue needed to determine what is wrong with the sick horse. Without ever laying a hand on the horse, and before approaching him, one can note a few general conditions. Is the horse eating; if not, how long has he been off feed? Does he seem a bit apprehensive, or is he unusually dead-headed? Has he been sweating? Has he been lying down or maybe rolling? Does he stand with one leg at rest more than he should? Has he been stamping his feet, pawing, or switching his tail? Has he lost any hair from rubbing certain spots? Are there any fresh feces in the stall or corral where the horse is kept? If so, are they of normal consistency? Are they the proper color?

The normal body temperature for a horse is 100°F. When it gets over 102°F, one should begin to suspect an infectious disease. The thermometer should be shaken down to below 100°F before being inserted into the rectum of the horse. The correct procedure is to stand as close to the horse as possible facing toward his rear with one arm over the hip. The tail is raised with this hand and the thermometer inserted with the other. The forward end of the thermometer should rest next to the wall of the rectum for best results. It is best to have a string tied to the end of the thermometer, as it can be sucked in and lost or defecated out and lost.

The horse loses his appetite with most infectious diseases but may not with injuries or lameness. Colic will cause the horse to act uneasy. He may turn around to look at his belly, lie down and roll; he may paw, kick at his belly, or stamp his feet. Sweating may be an indication of severe colic. The lack of a bowel movement may indicate an intestinal obstruction.

The pulse rate can offer important information. Normally it varies from 28 to 42 beats per minute. The pulse can be counted by placing the sensitive part of the longest finger over the submaxillary artery in the inner aspect of the lower jawbone. The artery here is about the size of the ink cartridge in a ball-point pen. It is easy to find. The pulse can readily be felt and counted. Although exercise naturally increases the pulse rate, an increased pulse rate at rest is an indication of some sort of circulatory disturbance (Figure 22–2).

The breathing rate is important with respiratory problems. It may be the telltale clue with pneumonia or heaves. A 6-year-old horse will normally have 8 to 12 inspirations per minute at rest. The rate will be a bit faster in younger animals. Factors other than rate are important with respiration. Is

Figure 22–2.   The equine pulse is taken from the submaxillary artery.

the breathing mainly in the chest, or is it mostly further back in the abdomen? Strong abdominal breathing will eventually result in a distinct furrow between the rib cartilages and the lower border of the abdomen. A painful chest condition can cause this kind of breathing.

Careful examination of the head and neck are important in detecting the cause of some diseases. Both eyes should be looked at carefully. Is there a discharge? Are the eyes bright or dull? The eyelids can be pulled up and back to get a good look at the white portion. With a little experience subtle color changes can be recognized. The white can take on a bluish color in some conditions, such as recent founder. It can become yellowish with other diseases. Is there an exudate from the eye? One eye, or both? This fact can determine the difference between a systemic disease and a local injury.

The nostrils may tell a story. Is the delicate lining reddened? With certain conditions, small pinpoint hemorrhages occur in the nostrils. Any nasal discharge should be noted, and whether it is on one side or both. Sinus infections, abscessed teeth, and tumors are problems that make close examination of the head of vital importance. Often swellings about the head are missed without carefully directed attention. Troublesome tumors may be found, or swollen lymph glands, which indicate infection. Annual floating of the teeth of older horses will help reduce the occurrence of sores on the tongue and inside surface of the cheeks from the sharp edges of the teeth. Floating is the filing of the sharp edges of the teeth, as shown in Figure 22–3.

Figure 22–3.   Floating the teeth.

Sometimes the odor of the breath may be characteristic of a disease problem. Teeth should be checked carefully. Are teeth broken or are the edges unusually sharp? If so, these teeth may be causing sores.

The entire body should be observed for swellings or cuts and sores. Muscle atrophy in a shoulder, hip, or leg will indicate a lack of use, and point to a problem area. A look down the sides, behind the elbows, and under the breast is important.

Just the overall stance of the horse may tell a secret, such as with choking. When something is lodged in the esophagus, the horse may stand with extended neck.

Careful observation may reveal swellings about the knees or fetlocks. Are there hock enlargements? Are there obvious splints, or thickened tendons? Sometimes leg swellings can be deceiving. Comparison with the opposite leg will often show up swellings that might have been missed otherwise.

Much will be learned by carefully looking over the horse at rest. Afterward, only the novice will walk away thinking that he has seen everything. As often as not, the horse must move out before the source of his troubles can be determined.

The observer should have someone else lead the horse out and away. Does the horse have proper balance? Does he hold his head off to one side? Herein may be clues to a disease of the brain. Is the horse reluctant to move, acting as if he were carefully walking on eggs? This is a sign of founder. Is he stiff? If so, does he warm out of it?

The good veterinarian, or horseman with years of experience, will rapidly make all of these observations. The casual observer may not realize that the veterinarian is looking at so many places or asking himself so many questions. The veterinarian who seems always to "home-in" on the problem correctly has his systematic examination procedure down pat. It is automatic with him. In this way he has developed a trained eye. If the horseman wants to become proficient at disease detection, he also needs a trained eye.

PARASITISM

Internal parasites (worms) have adapted themselves quite well to a comfortable existence within the body of the horse. They occur in many varieties and in vast numbers within an individual animal. Parasitism is insidious in its damage. It robs nutrition, and it damages tissue throughout the body. Many fine horses with tremendous genetic background fail to reach their potential every season because they have been stunted, crippled, or outright killed by internal parasites (Figures 22–4, 22–5, 22–6).

Worms are probably more of a menace to horses than they used to be; they seem to be a hardier bunch and present in greater numbers. The high price of real estate has forced many horses together into small pens, an ideal situation for the perpetuation of internal parasites.

Recent discoveries have pinpointed some fallacies in traditional deworming procedures. The danger to foals has, in the past, been underestimated. It is now known that many foals become severely infested with roundworms within 30 to 60 days. Research has shown piperazine to be an effective anthelmintic against these worms. It is the treatment of choice for these parasites during the second month of the foal's life. Within 60 days after deworming the ascarid parasite population has again built up to dangerous levels.

By weaning time the most dangerous internal parasite is the large strongyle or the bloodworms. An effective anthelmintic against them is thiabendazole or chemicals of the same family such as cambendazole, fenbendazole, and mebendazole. Almost every drug company produces one of these kinds of anthelmintics if they are in the deworming business. As a result these anthelmintics are available in paste, powder, granules, suspensions and pellets. Any form works fine as long as the horse gets a full dose within 24 hours.

By midsummer in most areas of the country the bot flies are out in large numbers. They lay their eggs on the hairs of the horse. The eggs soon hatch and find their way into the stomach of the horse where they can do great harm. The drugs listed previously will not eliminate bot larvae. One of several other anthelmintics must be used to fight these parasites. The two most popular are dichlorvos and trichlorfon. For many years carbon disulfide was the only effective anthelmintic and was used routinely by all veterinarians against bots, in spite of the dangers associated with its use.

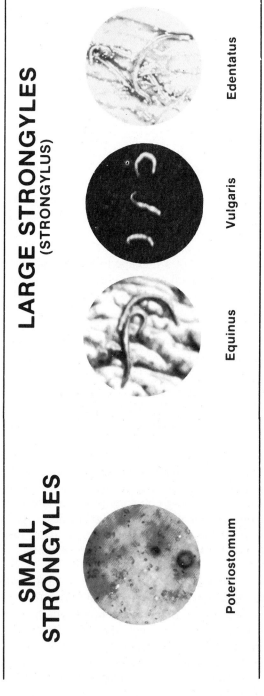

Figure 22-4.   Strongyles, as viewed in the intestine. (Courtesy of Farnam Companies.)

ASCARIDS

Parascaris

Figure 22–5.   Ascarids, as viewed in the intestine. (Courtesy of Farnam Companies.)

BOTS
(GASTROPHILUS)

Intestinalis

Figure 22–6.   Bots, as viewed in the intestine. (Courtesy of Farnam Companies.)

Some forms of dichlorvos and trichlorfon are effective against parasites other than bots. For this reason these drugs are good anthelmintics to use routinely in rotation with the other types of anthelmintics. Rotation of type is important because parasites sometimes build a resistance to one anthelmintic if it is used consistently every 2 months. Pyrantel is a good anthelmintic to rotate with the benzimidazoles. Recently, an injectable anthelmintic, Ivermectin, has become available. It has a broad spectrum of action and its effects last longer. Ask your veterinarian.

Table 22–1 shows most of the brand names under which you can purchase a dewormer, including those used by the veterinarian. It also shows the generic name.

Table 22–1.   Dewormers

GENERIC NAME	TRADE NAME	COMPANY
Thiabendazole	Equivet Tz	Farnam
	Equizole	Merck
	Top Form Performance	Top Form
	Equizole A (includes piperazine)	Merck
Cambendazole	Camet	Merck
Fenbendazole	Panacur	National
Mebendazole	Talmin	Pitman-Moore
Febantel	Rintal	Haver-Lockhart
	Cutter Paste Wormer	Cutter
Pyrantel	Imanthal	Beecham
	Pyraminth	Beecham
	Strongid-T	Pfizer
	Banminth	Pfizer
Piperazine	Foal Wormer	Farnam
	Equivet Jr.	Farnam
Dichlorvos	Equigard	Shell
	Equigel	Shell
	Tridex	Fort Dodge
Trichlorfon	Anthon	Cutter
	Combot	Haver-Lockhart
	Dyrex	Fort Dodge
	EquiBot Tc	Farnam
	Negabot	Cutter
Avermectin	Ivermectin	Merck

Administering dewormers by stomach tube has been a traditional procedure. Thousand of dollars and thousands of hours have been spent on this practice. The new paste wormers and injectable Ivermectin are changing things (Figure 22–7). They have brought the cost of deworming down enough so that a horseman can afford to do it every other month, thus reducing the danger of death from colic.

Today every horseman should be aware of what is available for deworming and should have a regular deworming schedule for his horses. Few horses do not need deworming. If serious doubts exist, take a feces sample to a veterinarian and have him check it with a microscope.

The most lethal parasite is *Strongylus vulgaris,* the well-known bloodworm. This tiny killer parasite is estimated to be the cause of most colics in horses where effective parasitic control measures are not practiced.

The adult *S. vulgaris* lives in the intestinal tract of the horse. His ancestors have been plaguing horses for thousands of years. These parasites have probably killed more horses over the centuries than bacteria, viruses, starvation, bad weather, or any other cause.

These small (12 to 25 mm in length) worms attach themselves to the lining of the cecum and the large colon where they suck blood from their host. The

Figure 22–7.   Paste wormers are effective and easy to administer.

irritation causes small nodules in the intestinal lining, which can become infected with bacteria. The worms also lay eggs which pass out and fall to the pasture, paddock, or stall floor. The eggs are well-protected against adverse weather conditions, but when the proper temperature and moisture come about (around 80 percent relative humidity and 26°C), the eggs hatch into larvae. This stage can happen as early as 1 week after they have passed out of the horse.

The infective larvae are very active. They crawl considerable distances. In the pasture they move up onto blades of grass where the horse has a better chance of ingesting them. In the stall they crawl onto the bedding or onto food that may have fallen onto the floor. In dirty paddocks they may be accidentally washed into the water container.

These infective larvae play the waiting game. They may crawl up and down a blade of grass a hundred times before being eaten. They are not as active in the winter. Cold weather does slow their development, but they survive even subzero temperatures by forming a protective sheath over themselves.

When the larvae are finally swallowed by an unsuspecting horse, they withstand the digestive juices within the stomach and penetrate through the intestinal wall, mainly in the posterior small intestine, cecum, and ventral colon, and search out the tiny arterioles that are abundant in the wall of the intestine. They like it there just under the lining of the small arteries. They migrate through the tissue, causing damage along the way and progressing upstream against the flow, finally entering the lining of larger and larger arteries. In about 2 weeks the average larva has reached the anterior mesenteric artery. Some settle into other nearby intestinal arteries.

Then the larvae set up housekeeping. They seem to like it in the arteries. The horse's tissue begins to react against them, calling in various types of cells to attempt to fight the foreign invader and wall it off so that its toxic waste products will not bring damage to other tissue. This action results in a sac, called an aneurysm. This sac is formed by the dilatation of the walls of the artery, which tends to fill with blood. This blood then clots up, causing problems. It soon begins to form a plug in the artery, which is known as a thrombus. Parts of this may become dislodged or even break off and float downstream to where the diameter is smaller and there form another plug, which is known as an embolus.

As these arteries are blocked off, they may cause what is known as "thromboembolic colic." Colic can come from any type of intestinal pain, but this trouble is caused by a lack of blood supply in the areas that would have otherwise been supplied by the blocked arteries. In severe cases a portion of the intestine dies because of the lack of circulation. Within a day or two this portion of the intestine develops a necrotic hole allowing the contaminated intestinal content to enter the abdominal cavity. This contaminated material causes peritonitis and death.

Equine abdominal arteries have been fighting a battle with these bloodworms since before recorded history. As a result the equine has evolved a mechanism that causes collateral circulation to develop rapidly in the abdominal area of crises. In this manner, many would-be fatal blockages are circumvented by the quick formation of new vessels (within a few days).

Cornell veterinarians have been studying means of combating these aneurysms. They use a new technique, called arteriography, in their studies. By injecting a substance in the bloodstream which will show up on the roentgenogram and outline the abdominal arteries, they have been able to monitor the development or regression of aneurysms. This capability helps them to determine the effectiveness of any particular treatment.

Research has shown that small numbers of S. vulgaris larvae can cause harm and even death. In experiments, researchers have given controlled doses of the larvae to young horses. They report that 2500 larvae will cause death in about 89 percent of worm-free foals, 3 to 10 months old. The course of the infection is characterized by a marked increase in temperature, loss of appetite, rapid loss of weight, mental depression, loss of physical activity, abdominal stress, constipation or diarrhea, and death in 14 to 21 days. The average barnyard infection is not so dramatic, but even a dozen or so larvae will cause a thickening of the lining of the abdominal arteries and possible death as a result.

"Verminous arteritis (swollen arteries due to bloodworm larvae) has been incriminated as a cause of death in 10 to 33 percent of abdominal crises," Dr. J.R. Georgi told a group of veterinarians recently at Cornell University.

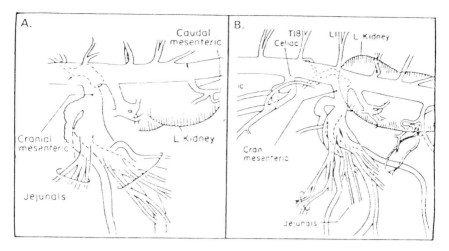

Figure 22–8.   A diagram showing the relative difference in size of the anterior mesenteric artery before treatment (A) and 30 days after treatment (B).

"It has been recognized as a cause of weight loss and ill-thrift and generally regarded as a serious and widespread obstacle to the health and optimum development of your horses." Dr. Georgi said that necropsies of 130 horses showed 86 to 95 percent had lesions in the cranial mesenteric artery and 50 to 65 percent had lesions in the cecal and colic arteries.

The traditional method of attacking bloodworm menace has been with the oral anthelmintics given in a single dose. Thiabendazole is an effective drug which eliminates the adult parasite in the intestinal tract. The procedure has been to use thiabendazole and other anthelmintics on a regular basis the year around in an attempt to keep the adult population at the lowest possible level so that fewer larvae can be produced.

Dr. Georgi described the research in which he used repeated doses (twice a day for 2 days) of a new drug, albendazole, to kill the larvae in the tissue. A single dose of anthelmintic has not been effective in the past. Now Ivermectin is used to kill larvae in the tissue. He took roentgenograms with the special technique which allows visualization of the abdominal arteries. Using young horses with already thickened arteries, he took x-ray studies before and after treatment for comparison (Figure 22–8). Albendazole, given in the repeated dosage of 50 mg/kg, reduced the thickening of the arteries to near-normal within 60 days.

FOOT CARE

The foal is born with a very soft pad of horny material covering the bottom of the foot. This soft material protects the dam's uterus from what could be a rather sharp-edged hoof. After birth, as the foal walks around, this soft pad is worn away. As the days go by, the foal begins to develop the mature hoof anatomy with distinct frog, sole, and hoof wall.

Some foals will be born with terribly crooked legs, but such leg deformities may straighten up by themselves within a few weeks. Other times early crookedness only gets worse without some sort of artificial leg angle correction. Some foals will be born with rather straight legs that become crooked as the weeks pass.

Nutrition determines the composition of leg bones and cartilages, which often have a great deal to do with a developing crookedness, or a natural straightening. Another factor is the overall leg conformation of the foal. If the legs are basically straight in all aspects, the feet tend to land square and level when the horse is in motion. Legs that are not basically straight show deviations in the flight pattern of the feet, which do not land square and level.

During the first two months of the foal's life, the hoofs begin to wear unevenly if the legs are not basically straight. A hoof worn unevenly tends to exaggerate the crookedness of the foal's leg because of the softness of the bones. By the same token, corrective trimming at an early age can create subtle pressure on the legs causing them to straighten under some circumstances.

In corrective trimming the secret is to trim down the area of the foot that has grown too long because poor conformation keeps the foot from landing flat. When the wall is too long on one side, it is impossible for the foal to set this foot down level. After trimming, each step will force the leg toward a more normal position. As this subtle force continues, the crooked leg begins to grow straight.

As the foal gets older, portions of the hoof wall may begin to flare out at the bottom as a result of poor conformation. Rasping this portion of the wall down on the outside to reduce weight-bearing in this area can help in straightening the leg. If the wall and ground surface of the foal's hoof is rasped regularly and consistently, the hoof wall will tend to assume a normal shape.

If the breeder neglects a foal's hooves, years of planned breeding can be for naught. The most well-bred horse with crooked legs might just as well be a cur. Therefore, foot care of the foal is a very important aspect of horse breeding.

It is best to keep a record of foot care on all foals, especially those in need of corrective trimming. Best results can be obtained if the hoof is rasped every other week. This short interval may not be long enough to determine where the excessive foot wear is occurring. A glance at the record will indicate which area of the hoof needs to be rasped.

Straight-legged foals should be trimmed about once a month, or at least checked to see whether wear is even and whether the wall has grown too long. About ¼ inch of wall should project below the sole all the way around

Figure 22–9.   Hoof trimming equipment: (a) rasp, (b) nippers, (c) hoof knife, (d) soft cotton
rope.

the hoof. The sharp outside edges of the wall should be rounded slightly with
the rasp to reduce chances of a crack developing.

On the average, the hoof grows at a rate such that new horn at the coronet
reaches the lower border in about 10 months. The actual rate of growth
varies depending on the condition of the horse. In most instances a monthly
trim should continue until the youngster is ready to be trained and shod.
Broodmares can usually get by with a trim every other month. A simple set of
trimming tools should be used regularly on every breeding farm (Figure
22–9).

TRAILERING

Trailering horses is part of our modern age. The busy horse breeder has
occasion to haul mares to other studs, young horses to trainers, yearlings to a
sale, or sold horses to their new homes. A good horseman should be aware
of the dangers involved in travel and the ways to handle them.

Pulling horses is such an easy procedure that one often tends to neglect
good management practices. Following these few simple rules should

alleviate most of the problems that do occasionally arise: (1) vaccinate in advance; (2) wrap the legs; (3) blanket in cool weather; (4) carefully watch water consumption; (5) feed conservatively; and (6) keep a detailed travel log.

Preparation for the haul should begin a few weeks in advance. Hauling to a new area with exposure to new horses and new bugs can often result in contagious disease. A horse should always be vaccinated for flu about 2 weeks before leaving. Under some conditions vaccination for strangles at the same time should be done. If other routine vaccinations such as equine encephalomyelitis and tetanus have been neglected, now is the time to get them done.

Also 2 weeks prior to the trip, a veterinarian should be consulted about whether a Coggins test will be needed. It often takes 2 weeks to get the official results on this blood test. The veterinarian will also need to issue a health certificate for travel. In many western states, a brand inspection is necessary before a horse can legally be hauled across county lines.

When loading for the trip, it is well to stop a moment and consider the horse as an individual. If one is unfamiliar with how this particular horse hauls, it is advisable to take some special precautions. It pays to wrap the legs from the fetlocks to the knees with a wool bandage. This procedure helps to prevent stocking up, and reduces the incidence of cuts and bruises, which are so common when the horse becomes fractious. A horse that has

Figure 22–10.   Electrolyte feed additives help reduce dehydration.

Date	Odometer Reading	Location	Gas, Gallons	Gas, Cost	Other Expenses		Condition of Horses	Feed	Water	Feces	Remarks	Time

Figure 22–11.   A typical travel log.

never been hauled before or one that continues to be nervous may require a head bumper, a padded hood-like apparatus with holes for the ears. In cool weather it is advisable to have horses blanketed while hauling. Although most horses haul well without special consideration, some take it so hard that a tranquilizer becomes necessary.

Hauling horses sets up conditions that often lead to colic and other digestive problems. Horses sometimes will fail to drink enough water and will nervously swallow large bites of hay or grain, which cause choke. It is very important to watch the water intake to be sure that the horse drinks regularly. If the horse does not drink, it should not be allowed to eat. Long trips with little water consumption and nervous sweating can lead to dehydration. In these cases, an electrolyte imbalance occurs, the expression in the eyes of the horse changes, and a quick dullness of the hair develops. An electrolyte feed additive is helpful in this instance (Figure 22–10).

It is always a good idea to bring plenty of the horse's regular feed along. Colic is less likely if the horse can continue on the familiar feed. A horse with a history of colic problems while traveling should be given a dose of mineral oil before departing. The addition of bran to the grain ration on a trip can help in keeping the stool loose. Periodically during the trip, the stool should be observed carefully; the hardness of the biscuit, and the smell of it, can be an indication of whether colic is about to develop. Mucous covered feces or no feces at all are danger signals.

At rest stops it pays to take note of the bedding used. Some horses not accustomed to straw may consume large quantities, causing colic. Terribly dusty stalls can cause pulmonary emphysema. Never use spoiled or rotten hay for bedding.

Horses can travel much farther without a rest stop than most people realize. They seem to get along well for 24 hours at a stretch when necessary. Frequent unloading for exercise (every few hours) is not important and may even tend to upset the horse, making it reluctant to re-enter the trailer.

When traveling, it is wise to keep a detailed log of happenings (Figure 22–11). There should be a place for the typical financial information necessary for tax purposes, but also columns for detailed information as to care of horses and equipment on the trip. Each stop can be recorded, noting time of day and whether water and feed were given. There should be a place for remarks on the condition and attitude of the horse and space to make notes as to bowel movements. Accurate observation requires that the feces be removed quite often. This is a good idea anyway. It does the horse's feet no good to stand hours on end in wet feces.

FIRST AID

Every breeding farm should have a veterinary supply cabinet (Figure 22–12). Typically it will contain the following items, and more: (1)

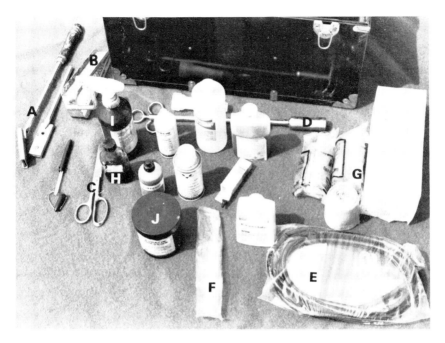

Figure 22–12.   The typical first aid kit has various tools and instruments, depending upon the
expertise of the groom. Here are teeth floats (A), dental wedge (B), scissors (C), balling
gun (D), and nasogastric tube (E). It is important to have a sterile syringe and needle (F).
Various types of bandages (G) are important. A bottle of 7% iodine (H) is perhaps the
most vital drug in the kit. Tamed iodine in a spray-top bottle (I) is handy for use on foal
navels. Furacin dressing (J) or a similar antiseptic ointment is important. Other drugs in
the kit, as shown in the center of the photograph, vary depending on the type of horse
farm involved.

thermometers; (2) scissors; (3) disinfectants; (4) bactericidal salves, such as
Furacin ointment; (5) liniments, braces, and lotions; (6) antibiotic eyedrops;
(7) petroleum jelly; (8) cotton, bandages, and tape; (9) leg wraps; (10)
alcohol; (11) iodine; (12) purple lotion for cuts; and (13) scarlet oil.

First aid is necessary in many instances when the horse should have
immediate attention and the veterinarian cannot come for awhile. Some of
the more critical conditions are described next, with suggestions for aid.

Colic can be a deadly serious condition and usually requires prompt
attention. Generally the first signs will be sweating and rolling. The horse
will act uncomfortable and will look extremely worried. If colic is suspected,
be sure to tell the veterinarian with the first call; then he will know it is
urgent. An attendant should stay with the horse and should not let him roll
as the gut can be ruptured. Walking the horse helps to disperse the gas in the
intestine. Walking to the point of exhaustion, however, only makes things
worse.

Wounds are generally more alarming to the horseman than to the horse.
As long as bleeding is not profuse, a wound can wait a few hours for

treatment. It is best to leave a wound open without treatment of any sort until the veterinarian arrives. The veterinarian may want to suture the wound, and healing may be retarded with some types of treatment. Bleeding on the legs can be controlled with a tourniquet placed above the wound. Cuts on the heel often bleed badly and when the veterinarian is not available shortly, the entire foot should be bandaged tightly.

Acute lameness in all legs can be due to such things as founder and azoturia. Both conditions require a veterinarian's immediate attention. While waiting, all feed should be removed from the horse. Keep the animal warm and quiet.

Choke can be serious, but often the horse will get over it without help. In some cases a green foam may come out of the nostrils. The horse will look extremely worried and will hang his head, and may alternately lie down and get up, but will probably not try to roll. Feed and water should be removed. Sometimes massaging the esophagus along the neck will break up the wad of food that is causing the problem. Massaging toward the head works best.

Strangles can be an emergency situation when an abscess forces off the air intake in the throat. In such instances it is important to explain the critical aspect of the situation to the veterinarian on the phone so that he will act quickly. Meanwhile keep the horse quiet. Do not try to move him or work with him. If he goes down and no veterinarian is available, make a slit in the windpipe about 6 inches below the throat. This "last resort" procedure should be done only if one is quite sure where the windpipe is located. It lies just under the skin and, if done properly, the incision will induce very little bleeding.

BIBLIOGRAPHY

Butler, Doug: The Principles of Horseshoeing. Published by the author, Alpine, TX, 1974.

Harris, Susan E.: Grooming to Win. New York, Charles Scribner's Sons, 1977.

Hayes, M. Horace: Veterinary Notes for Horse Owners. New York, Arco Publishing Co., 1968.

Kirk, Robert, and Bistner, Stephen: Handbook of Veterinary Procedures and Emergency Treatment. Philadelphia, W.B. Saunders Company, 1978.

Mitchell, D.: Equine herpesvirus I abortion. In Current Therapy in Theriogenology. Philadelphia, W.B. Saunders, 1980.

Rooney, James R.: The Sick Horse: Causes, Symptoms and Treatment. Cranbury, NJ, A.S. Barnes, 1977.

Seiden, Rudolph: Livestock Health Encyclopedia. New York, Springer Publishing Co., 1968.

The Illustrated Veterinary Encyclopedia for Horsemen. Various editors. Grapevine, TX, Equine Research Publications, 1975.

keeping records

As emphasized throughout this book, the keeping of accurate records often results in better care and management of the horses. Poor management can negate much of the good of excellent breeding; and if a profit is to be realized from a horse breeding endeavor, accurate records are a necessity.

HEALTH AND CARE RECORDS

The most basic of records are those that pertain to the health and care of the individual horses on the farm. A record should be initiated when each foal is born and each horse is purchased for the farm. This individual horse record (Figure 23–1) should contain space for identification of the horse, the feeding history, veterinary care, hoof care, and a breeding record.

A more extensive veterinary record may be desired. Sometimes it is helpful to have specific areas to note the vaccinations given over the lifetime of the horse (Figure 23–2). Another helpful aspect of a veterinary record is a special section for the listing of each veterinary visit, with room to describe the nature of the problem and the diagnosis. A complete worming record is also important.

Breeding Records

Permanent breeding records are essential for many reasons. A record of a mare's heat cycle will help in efficient scheduling of a busy stallion. Once a mare is settled, an accurate record will help in determining the future foaling

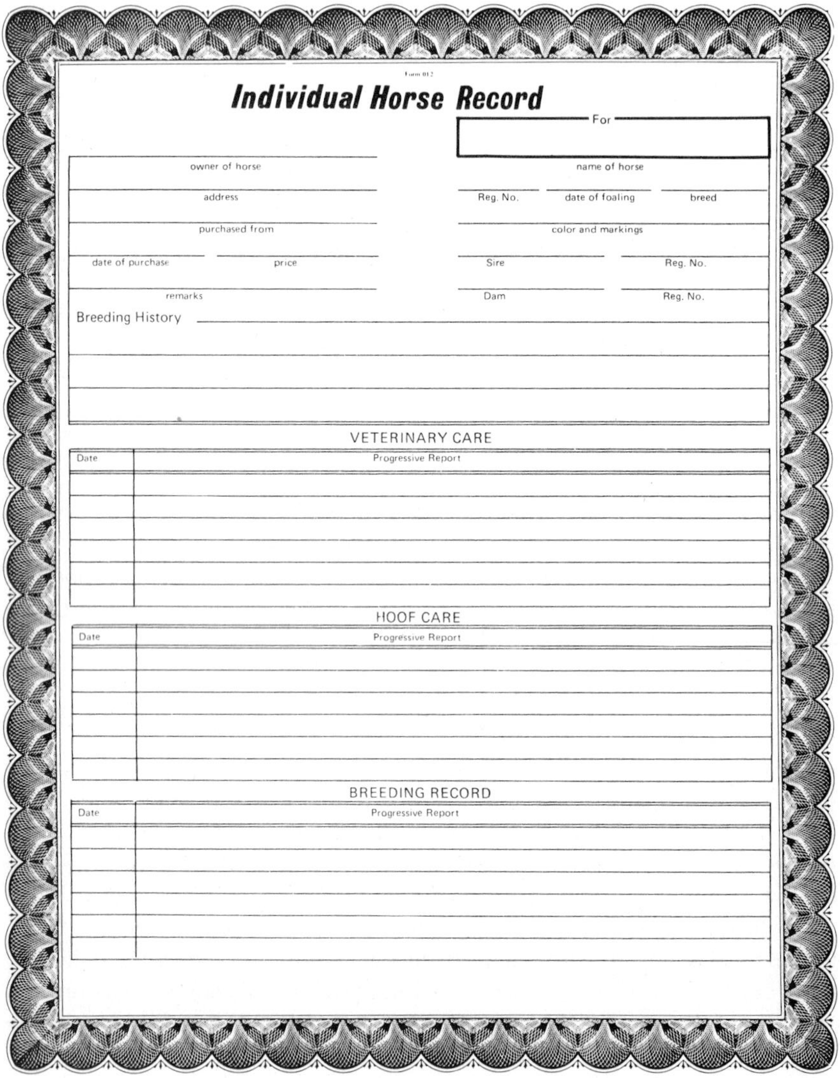

Individual Horse Record

For

owner of horse | name of horse

address | Reg. No. | date of foaling | breed

purchased from | color and markings

date of purchase | price | Sire | Reg. No.

remarks | Dam | Reg. No.

Breeding History

VETERINARY CARE

Date	Progressive Report

HOOF CARE

Date	Progressive Report

BREEDING RECORD

Date	Progressive Report

Figure 23–1. A typical individual horse record.

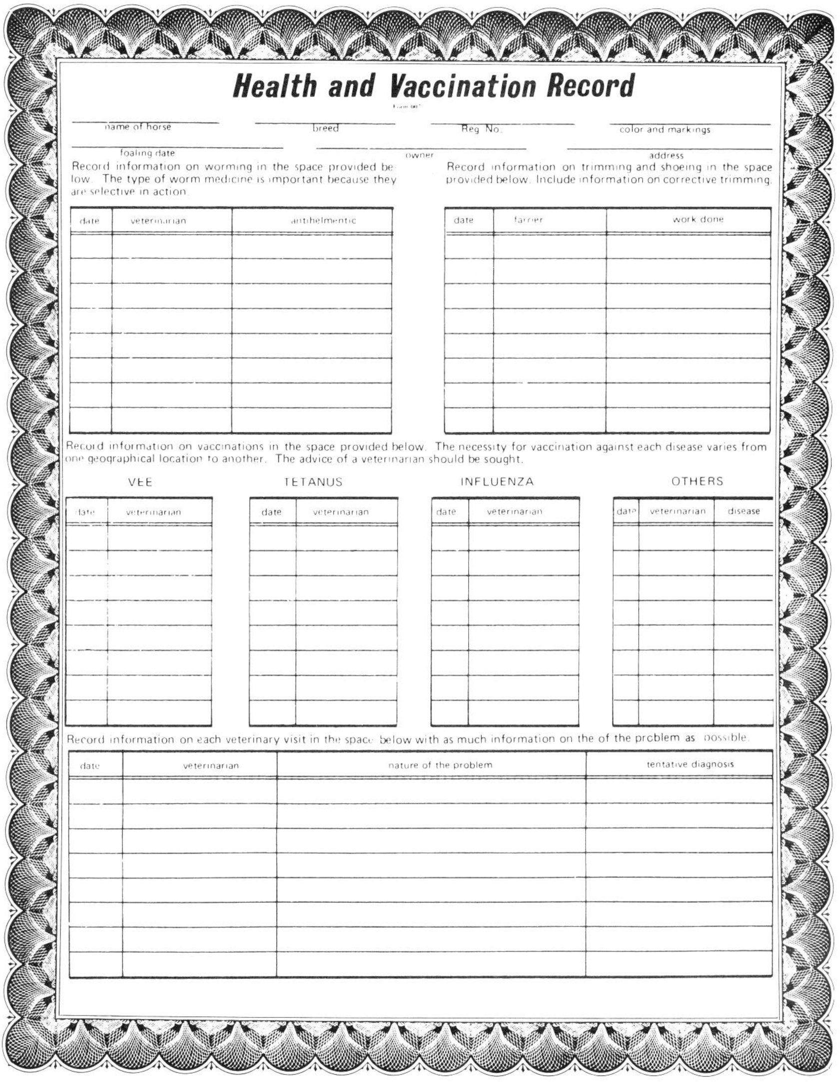

Health and Vaccination Record

name of horse	breed	Reg. No.	color and markings
foaling date		owner	address

Record information on worming in the space provided below. The type of worm medicine is important because they are selective in action.

Record information on trimming and shoeing in the space provided below. Include information on corrective trimming.

date	veterinarian	antihelmentic		date	farrier	work done

Record information on vaccinations in the space provided below. The necessity for vaccination against each disease varies from one geographical location to another. The advice of a veterinarian should be sought.

VEE		TETANUS		INFLUENZA		OTHERS		
date	veterinarian	date	veterinarian	date	veterinarian	date	veterinarian	disease

Record information on each veterinary visit in the space below with as much information on the of the problem as possible.

date	veterinarian	nature of the problem	tentative diagnosis

Figure 23–2.   A health and vaccination record.

Stallion Breeding Record

Form 001

An official record of the mares bred during the 19___ season to:

_____   _____   _____
        name of stallion   breed   Reg. No.

_____   _____
        year foaled   color and markings

The following mares were bred to this stallion as shown below:

	name of mare	Reg. No.	owner of mare	dates mare was bred - if pasture bred give dates in and out of pasture	year
1					
2					
3					
4					
5					
6					
7					
8					
9					
10					
11					
12					
13					
14					
15					
16					
17					
18					
19					
20					

CERTIFICATION: I do hereby certify that the above named mares were bred to this stallion on the day(s) shown above.

_____   _____
            date   signature of stallion owner or manager

                                        _____
                                                    address

Figure 23–3.   A stallion breeding record.

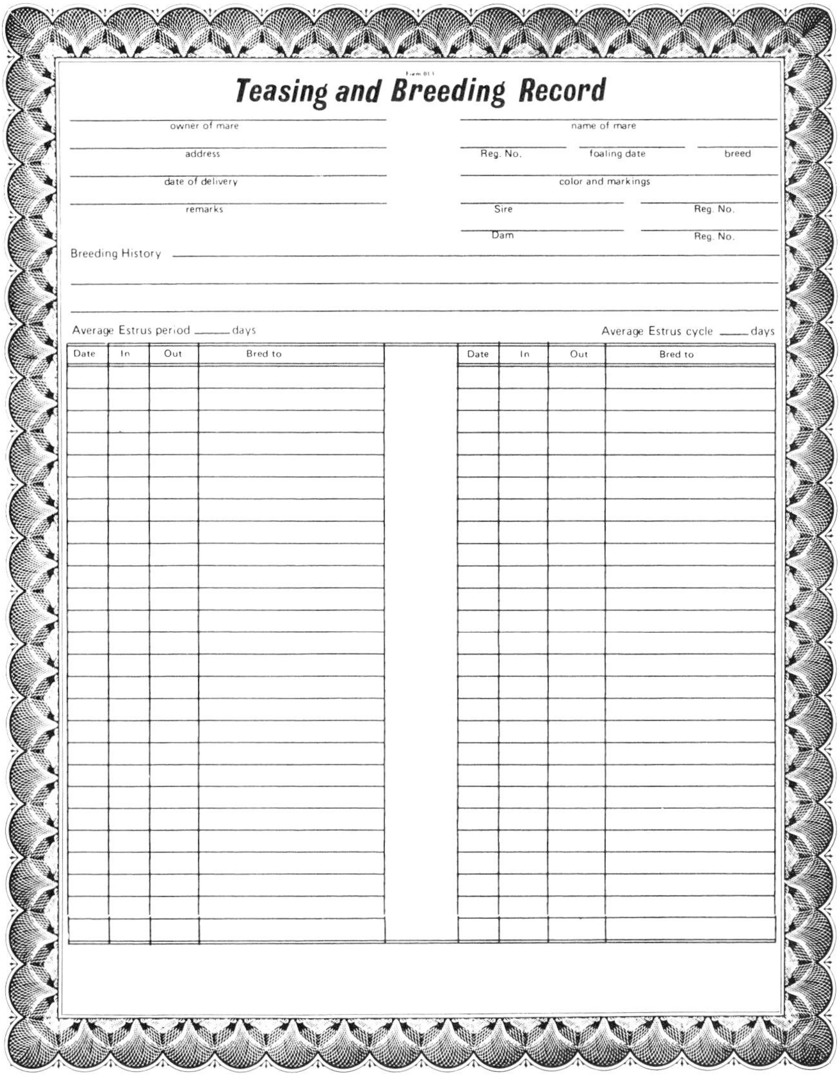

Teasing and Breeding Record

Form 811

owner of mare		name of mare	
address	Reg. No.	foaling date	breed
date of delivery		color and markings	
remarks	Sire		Reg. No.
	Dam		Reg. No.

Breeding History _____

Average Estrus period _____ days   Average Estrus cycle _____ days

Date	In	Out	Bred to		Date	In	Out	Bred to

Figure 23–4. A teasing and breeding record.

date. When a farm has more than one stallion, it is added insurance to the memory if records show to which stallion a mare is bred. Registries demand that this sort of record be annually compiled in a stallion service report. This report can be made from a stallion breeding record (Figure 23–3) kept daily during the breeding season. A record should be kept for each individual broodmare to show at a glance her breeding record over the years (Figure 23–4). This record should include space to describe any breeding problems treated by the veterinarian.

A separate record should be kept for each broodmare being bred (Figure 23–5). It should include space to record the position of the estrous cycle each day of the breeding season.

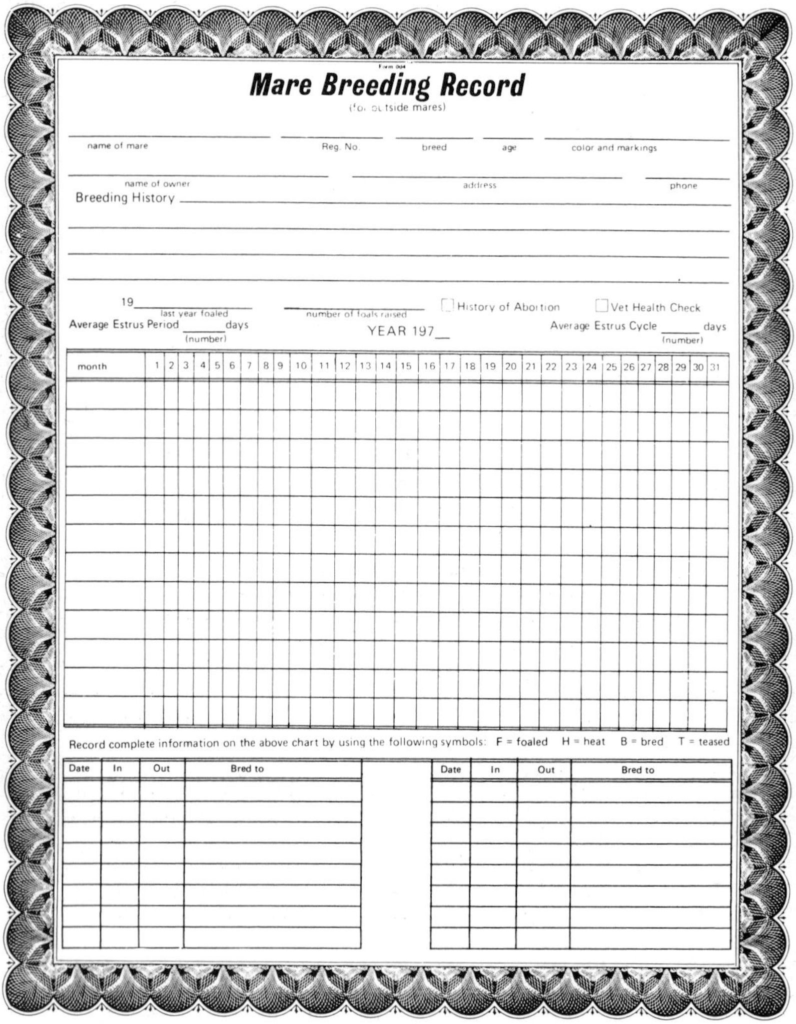
Form 004

Mare Breeding Record

(for outside mares)

name of mare ______ Reg. No. ______ breed ______ age ______ color and markings ______

name of owner ______ address ______ phone ______

Breeding History ______

19______ last year foaled ______ number of foals raised ☐ History of Abortion ☐ Vet Health Check

Average Estrus Period ______ days (number) YEAR 197__ Average Estrus Cycle ______ days (number)

month	1	2	3	4	5	6	7	8	9	10	11	12	13	14	15	16	17	18	19	20	21	22	23	24	25	26	27	28	29	30	31

Record complete information on the above chart by using the following symbols: F = foaled H = heat B = bred T = teased

Date	In	Out	Bred to	Date	In	Out	Bred to

Figure 23–5. A mare breeding record.

Performance Records

For those horses not sold as yearlings, a performance record will eventually be needed. The simplest of racing records should include complete information on every race entered (Figure 23–6). It is important to have space for the date, track, distance, time, jockey, trainer, placing, amount won, and general remarks.

A show horse record should reflect information pertinent to the type of showing done. A record of halter showing must include date, show name,

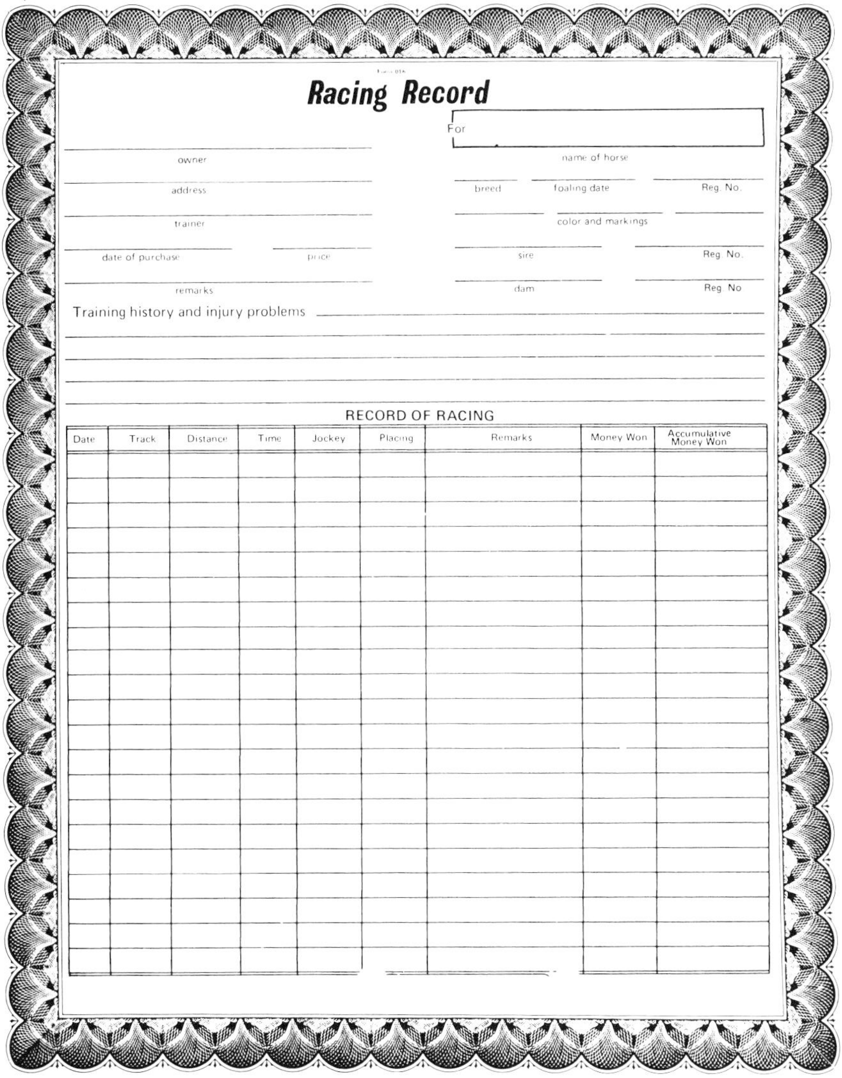

Racing Record

For _____

owner					name of horse		
address				breed	foaling date		Reg. No.
trainer					color and markings		
date of purchase		price			sire		Reg. No.
remarks					dam		Reg. No

Training history and injury problems _____

RECORD OF RACING

Date	Track	Distance	Time	Jockey	Placing	Remarks	Money Won	Accumulative Money Won

Figure 23–6.   A racing record.

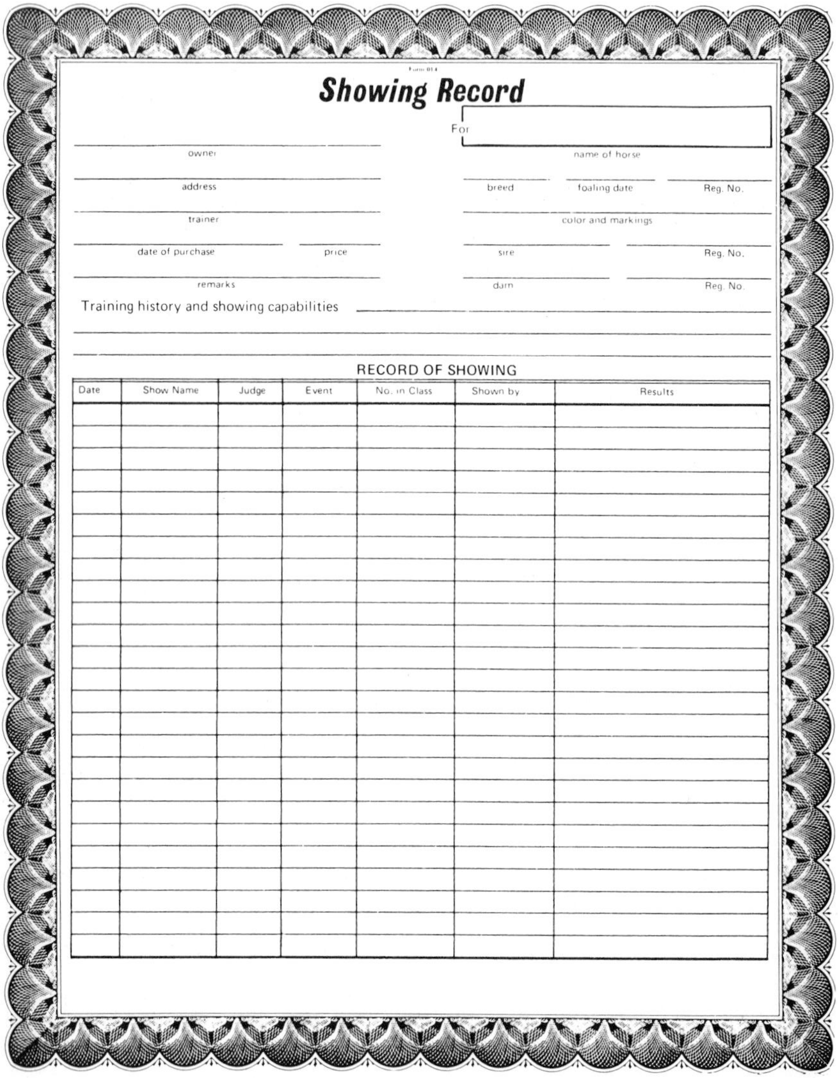

Showing Record

For _____

owner	name of horse
address	breed foaling date Reg. No.
trainer	color and markings
date of purchase price	sire Reg. No.
remarks	dam Reg. No.

Training history and showing capabilities _____

RECORD OF SHOWING

Date	Show Name	Judge	Event	No. in Class	Shown by	Results

Figure 23–7. A showing record.

judge, event, number in class, whom shown by, and the results (Figure 23–7).

MANAGER'S RECORDS

The professional manager will keep not only a daily log (Figure 23–8) but also a variety of tally records for future reference. A good daily log will have space for general remarks, which can serve as a reminder of things to do or of unusual happenings, which could later become a valuable reference. If the daily log is rather complete, it can serve as a reference when filling out more complete records about individual items. Foaling records, for example (Figure 23–9), should be kept as a permanent record, although the wise manager will note foalings daily in his log as this time of year is often busy and the permanent record keeping may be left to the weekend. The mare record of produce (Figure 23–10) should also be a permanent record kept by the manager. This information is helpful when later applying for registration.

When a large number of stalls are occupied, it can be helpful to the manager to have the stable boys keep a stall record (Figure 23–11) for each stall. In this way, the boys know what is to be done in each stall, and the manager can keep tabs on what is occurring on a daily basis.

The manager may find it helpful to keep an equine veterinary record (Figure 23–12), with space to record each visit by a veterinarian to each horse. An important part of this record is notes on follow-up recom-

EQUINE MANAGER'S DAILY LOG

Month ________ Day of Month ____

Day of Week ________

________ Name of Farm

________ Name of Manager

General Remarks ________

Supplies Needed: Check when ordered

☐ ____
☐ ____
☐ ____
☐ ____

Stallions Bred	To	Bookings Requested Check when contract completed	Horses Arriving	Reason
		☐		
		☐		
		☐		

Mares on Night Foaling Watch	Mares Foaled:	Horses Departing	Reason

Figure 23–8. A typical equine manager's daily log.

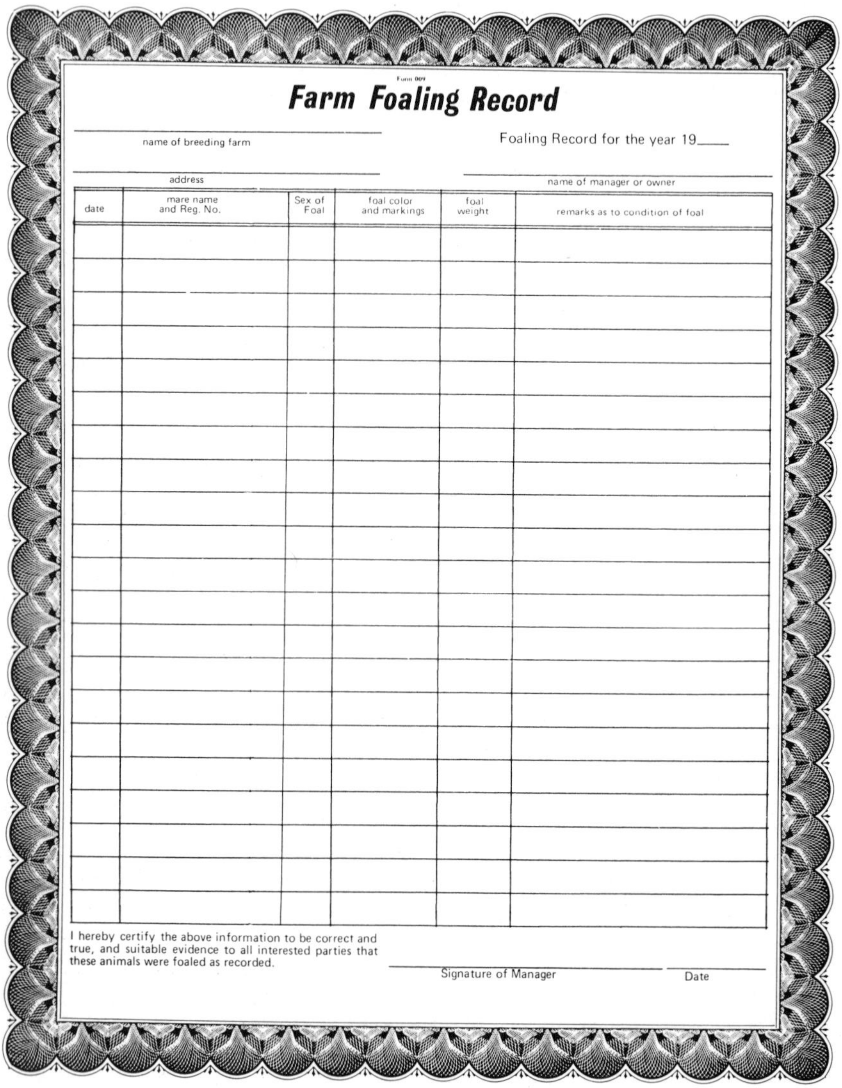

Form 009

Farm Foaling Record

name of breeding farm

Foaling Record for the year 19____

address

name of manager or owner

date	mare name and Reg. No.	Sex of Foal	foal color and markings	foal weight	remarks as to condition of foal

I hereby certify the above information to be correct and true, and suitable evidence to all interested parties that these animals were foaled as recorded.

Signature of Manager

Date

Figure 23–9. A farm foaling record.

Mare Record of Produce

name of mare   breeding farm

breed   Reg. No.   address

foaling date   color and markings   owner

purchased from   price

date bred	sire	date foaled	color and markings	name and Reg. No.	remarks

Record of Sales:

date of sale	age	price	sold to	remarks

Figure 23–10.   A mare record of produce.

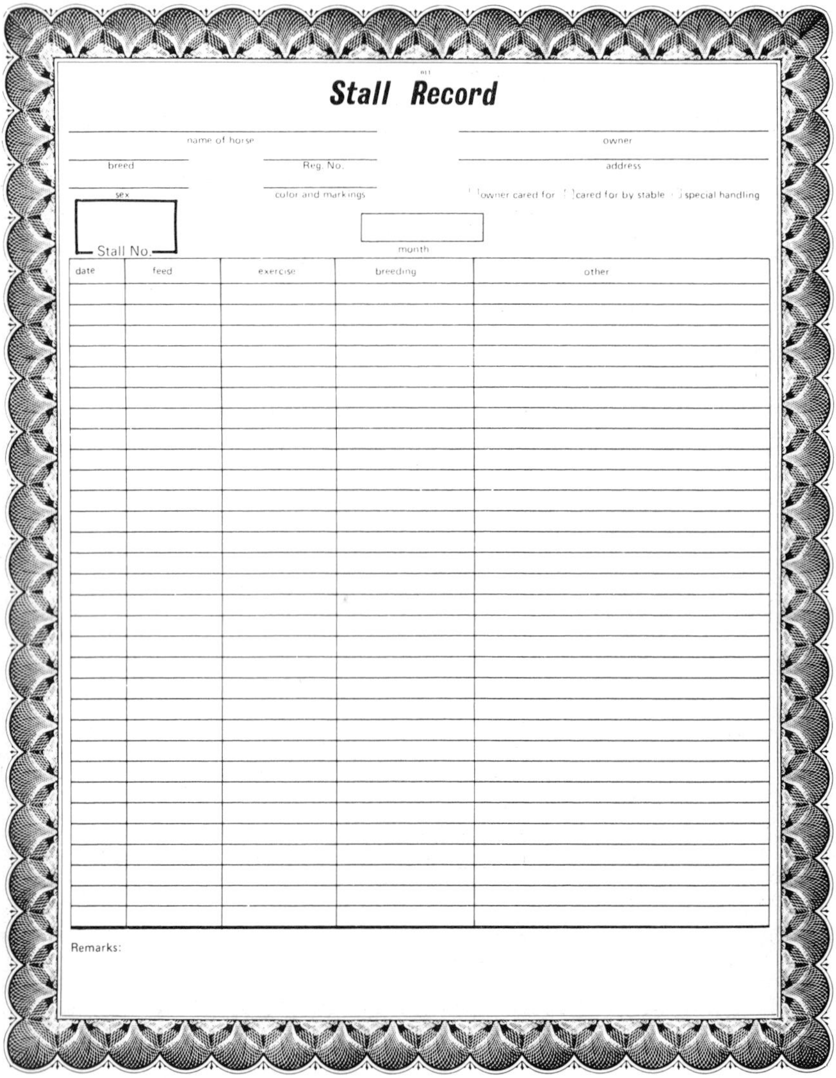

Stall Record

name of horse | owner

breed | Reg. No. | address

sex | color and markings | ☐ owner cared for ☐ cared for by stable ☐ special handling

Stall No.

month

date	feed	exercise	breeding	other

Remarks:

Figure 23–11. A stall record.

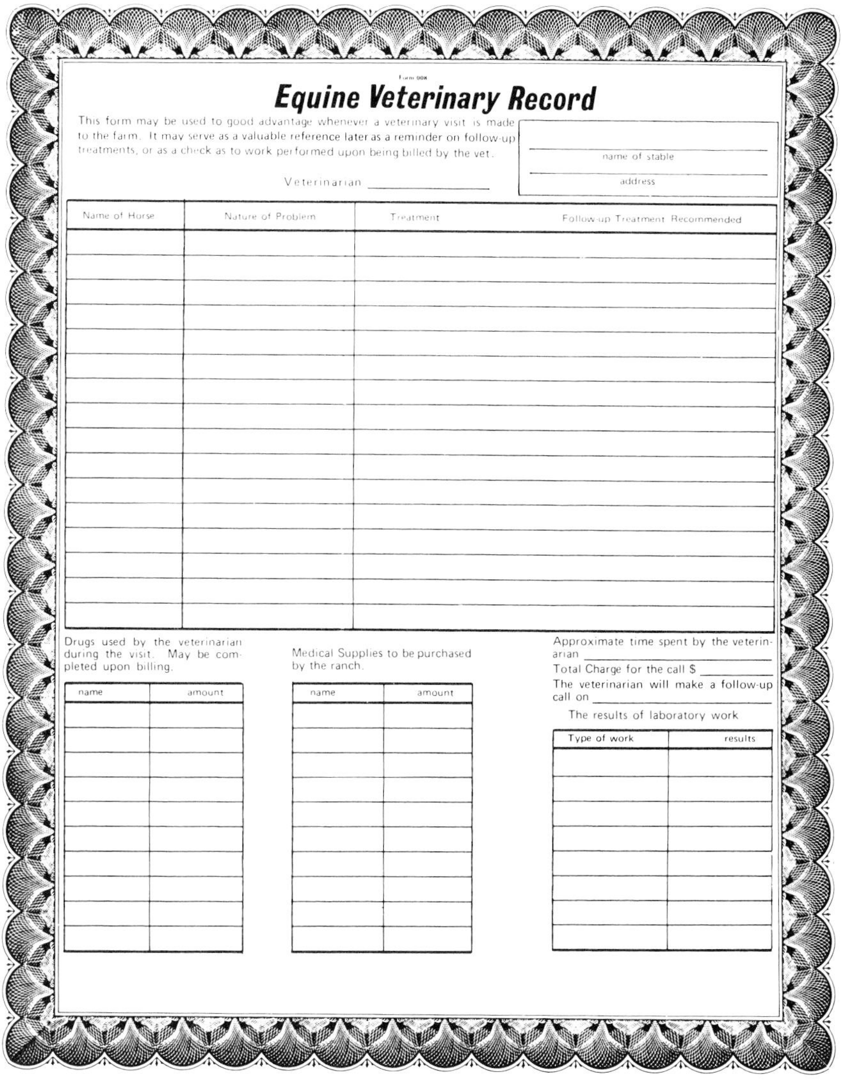

Equine Veterinary Record

Form 504

This form may be used to good advantage whenever a veterinary visit is made to the farm. It may serve as a valuable reference later as a reminder on follow-up treatments, or as a check as to work performed upon being billed by the vet.

name of stable

address

Veterinarian _____

Name of Horse	Nature of Problem	Treatment	Follow-up Treatment Recommended

Drugs used by the veterinarian during the visit. May be completed upon billing.

name	amount

Medical Supplies to be purchased by the ranch.

name	amount

Approximate time spent by the veterinarian _____
Total Charge for the call $ _____
The veterinarian will make a follow-up call on _____
The results of laboratory work

Type of work	results

Figure 23–12.   An equine veterinary record.

mendations. All of the forms illustrated in this chapter can be obtained from *The Printed Horse,* P.O. Box 1908, Fort Collins, CO 80522.

REGISTRATION RECORDS

A manager must be familiar with the procedures and requirements of the registry he deals with, whether it be the Jockey Club, The American Quarter Horse Association, or whatever. Each registry does things a little differently, and most of them put out an instruction or rule book. They usually supply necessary forms free of charge. The farm manager should keep a supply of registry forms on hand.

FINANCIAL RECORDS

Although owners of most large breeding farms have their books and tax returns prepared and kept by an accountant, basic records must be compiled at the farm on a daily or weekly basis.

The keeping of good records is probably more important to financial success in the horse business than in any other. Not only will complete and accurate records help in showing what the business has done and point the way to where the business is headed, but they are essential in proving to the IRS that the enterprise is truly worthy of being classified as a business. Without records, a horse business is only a hobby in the eyes of the tax man. Good records can be helpful in preparing credit applications and can help in securing maximum social security benefits.

A complete record system should contain enough information to ensure that deductible or nontaxable expenses can be explained. It should identify the source of receipts. All cash and property received should be completely identified as to source; otherwise there may be a later dispute as to a nontaxable status on some items. It should prevent omission of deductible expenses. Cash expenses should include as much detail as possible in order to help substantiate their validity to the IRS. Neglecting to note a $25.00 expense could result in increased income taxes of as much as $3.50 or more. It should determine depreciation allowances. Only accurate records can allow a horse breeder to determine an acceptable depreciation schedule for tax purposes. It should be designed to take advantage of capital gain and loss provisions of the IRS. Inadequate records in this respect may make it impossible to postpone paying certain gains, or to deduct certain losses that would otherwise be allowed by the IRS. It should establish reportable earnings for self-employment social security taxes. Since later social security benefits to the self-employed depend upon prior earnings, the record should be accurate. It should provide additional detail on items reported in the income tax return. Completeness in this regard may help avoid increased tax assessment in case of an IRS audit.

WEEKLY EXPENSE RECORD

Farm or
Stable name _____   Week from _____ to _____ Month of _____ 19 ____

		DISTRIBUTION OF DISBURSMENTS												DEPRECIABLE		
Day	Description	Feed	Farrier & Vet	Farm Suppls.	Trans-porta-tion	Train-ing	Labor	Adver-tising	Taxes Lics. Fees	Stud Fees	Misc.	Loan Pmts.	Real Estate Pmts.	Equip.	Horses	Farm Main.
Total for this week																
Total for the month																

Figure 23–13.   A weekly expense record.

										Long Term Capital Gain		
Day	Description	Stud Fees	Boarding	Winnings	Training	Livestock Sales	Misc. Produce	Misc.		Regular Horses	Real Est.	Horses	Other

MONTHLY INCOME RECORD

Farm or stable name _____ Month of_____ 19 ____

Total for this page													
Total for the month													

Figure 23–14.   A monthly income record.

A bookkeeping system begins with a weekly expense record (Figure 23–13). It is best to purchase a special bookkeeping system designed for a horse breeding business. If one cannot be found, it will work as well to buy prepared sheets with blank columns and fill in the proper headings. It is important to have proper headings under distribution of disbursements so that budgeting and annual tallying can be done accurately. There is some advantage in keeping depreciable items separate from other expenses. Some of the items that warrant a special column are: feed, farrier, veterinarian, supplies, transportation, training, labor, advertising, taxes/licenses/fees, and stud fees.

Since income items generally occur in a few large sums, a monthly income record is adequate (Figure 23–14). It should also contain appropriately labeled columns so that an annual tally will reveal the various sources of income accurately. The major sources of income on the breeding farm are: stud fees, boarding, winnings, training, livestock sales (other than horses), miscellaneous produce (generally hay), and regular horse sales. The sale of horses, even breeding stock since 1969, is counted as ordinary income. It is best to keep long-term capital gain income in separate columns.

Since not all income is collected immediately, it is helpful to keep a record of monthly charged business and bills collected (Figure 23–15). This list serves as a quick reference for determining the amount of uncollected income for the month and also helps in figuring whether ends are meeting financially each month. Even though a business may be on a cash basis, with income only counting when collected, it is helpful to know the total picture.

A monthly summary record (Figure 23–16) should have a space for the total of each of the income and expense columns kept for the month. As the year progresses, previous months' totals should be tallied in order to keep a running tab on each column. A helpful aspect of a monthly summary record is a simple monthly statement which shows net income for the month and for the year to date. A part of such a statement should be a total business summary of paid business and charged business for the month and for the year to date. A summary of the cash position each month can be helpful, especially where cash flow is limited.

Those who are serious about making a profit in the horse business will be concerned about monthly budget and projections (Figure 23–17). A business on a cash basis has four sources of actual income for the month: cash on hand, cash sales, collections from credit accounts, and loans or other cash injections. This income must be balanced against the various items of expense, including the owner's cash withdrawals. Other items that are important in assessing the true financial picture each month are a comparison of accounts receivable and accounts payable. Both of these sometimes tend to get out of hand, creating financial problems. It is important to realize the feed and horse inventory each month. A rise in accounts payable may not be serious when there is an accompanying rise in feed and horse inventory.

MONTHLY CHARGED BUSINESS									© Caballus Corporation Form 551
Date Charged	Customer	Stud Fees	Boarding	Training	Livestock Sales	Misc. Produce		Misc.	Long Term
Totals for the month									

Farm or Stable name _____   Month of _____ 19_____

BILLS COLLECTED									
Date Paid	Date Charged	Customer	Stud Fees	Boarding	Training	Livestock Sales	Misc. Produce	Misc.	Long Term
Totals for the month									

Figure 23–15.   A record of charged business.

MONTHLY SUMMARY RECORD

Farm or
Stable name_____   Month of _____ 19 ____

INCOME SUMMARY	Stud Fees	Boarding	Winnings	Training	Livestock Sales	Misc. Produce	Misc.	Regular Income Horses	Horses	Real Estate	Other
Totals this month											
Totals previous months											
Totals to date this year											

EXPENSE SUMMARY	Feed	Farrier and Vet	Horses	Trans.	Training	Labor	Adv.	Taxes Licens. Fees	Stud Fees	Misc	Loan Pmts.	Real Estate Pmts.	Equip.	Horses	Farm Maint.
Totals this month															
Totals previous months															
To date this year															

Figure 23–16.   A monthly summary record.

A true financial statement shows the balance between assets and liabilities (Figure 23–18). A financial statement can be prepared each month and, when done on a regular basis, indicates the progress being made in the business. When a detailed financial statement, properly signed and dated, shows significant assets over liabilities, it can sometimes be used to obtain a loan to improve the cash flow of the business.

Complete accurate records kept each month will be a great help at the end of the year in compiling the annual records on the business. At the end of the year an annual summary record with totals of each category each month can be compiled (Figure 23–19). This breakdown is helpful in determining seasonal trends of various types of expenses and income.

The purchase of horses by a business is considered to be a capital outlay which will bring benefits for years to come. Therefore the cost must be prorated over the useful life of the animal. The depreciation for any one year is the only tax expense allowable. A number of methods of depreciation are possible. In some cases the animal has no value at the end of its useful life (such as when a broodmare is kept until death). In other cases there is a salvage value (such as when a mare held for racing purposes is later sold as a broodmare). Horses that are bought for resale need not be set up on a depreciation schedule as long as the animal is not kept beyond the end of the taxable year. One of the most simple methods of figuring depreciation is straight line depreciation with no salvage value figured. A horse with a 10-year useful life would in this case have a depreciation of $100 per year if the original cost of the animal was $1000.

Each taxpayer is allowed to decide the useful life of his horses as determined by his own set of circumstances, but whatever life he selects

MONTHLY BUDGET AND PROJECTION							
INCOME		at month's start or last month's totals		projected budget		actual	
Cash on hand							
Cash sales							
Collections from credit accounts							
Loan or other cash injections							
EXPENSES	Total cash available						
Feed							
Farrier and Vet							
Horses (non-depreciable)							
Training							
Labor							
Advertising							
Taxes, Licenses, Fees							
Stud fees							
Misc.							
Loan payments							
Real Estate payments							
Equipment							
Horses (depreciable)							
Farm Maintenance							
Reserve and or Escrow							
Owner's withdrawal							
OTHER ITEMS	Total cash paid out						
Cash Position							
Accounts Receivable							
Accounts Payable							
Feed inventory							
Horse inventory							
Bad Debts							
Depreciation							
Total sales volume (cash and charged)							

Figure 23–17.   A monthly budget and projection form.

may be revised by the IRS. Most guidelines allow from 10 to 15 years' useful life for breeding animals. Horses held strictly for racing would have a somewhat shorter useful life, and a higher salvage value, except for geldings.

The method of depreciation selected, whether it results in a large or small initial depreciation, can be very important. When first starting in business, it may not be as important, taxwise, to show a profit. It is best during this period to claim as much depreciation as allowable. In later years, when a profit is essential, less expenses will be shown in depreciation. In a business that shows a profit every year, it will not be so important when depreciation is taken. Every horse breeder should keep a permanent horse depreciation summary record (Figure 23–20). With this type of record the net value of each horse each year is known. This net value is subtracted from the gross sale price when a horse is sold to determine the net profit or loss on that individual horse. As an example, suppose that Queen Mary was purchased for $5000, and depreciation in prior years for Queen Mary has been $3000. This would leave a net value of $2000. Suppose that she was sold for $2500. Tax would be due on only $500 of the sale price.

A horse breeder may also want to keep an equipment and facilities depreciation summary record (Figure 23–21). This type of record works the same way in determining a net profit or loss when equipment or real estate is sold. Guidelines to acceptable lives of equipment and improvements are: machinery and equipment—8 years; non-mechanical equipment—10 years; buildings, block with floors and foundations—25 to 30 years; buildings, frame and pole—20 years.

Employees of a breeding farm business are required to have FICA and tax deductions withheld from their pay. The government requires that detailed records be kept on these deductions and that the money be put on deposit at a bank for the government. An employee withholding record should also be kept (Figure 23–22).

Capital gains income should be recorded on a separate summary sheet. Capital gains income allowances vary from year to year, and it is advisable to check with the IRS before arbitrarily assuming what qualifies toward capital gains. Gains of equine breeding stock since 1969 must be treated as normal income. A capital gains summary sheet is a helpful record to maintain (Figure 23–23).

A net income computation (Figure 23–24) determines the taxable income of the business. Accurate and complete records as outlined in this chapter will help in establishing that a breeding enterprise is a business rather than a hobby. To deduct horse losses against ordinary income, the endeavor must be classified by the IRS as a trade or business. Kenneth A. Wood, in *The Business of Horses*, outlines the factors considered by the IRS in determining business or hobby status:

> 1) How active the taxpayer has been in the enterprise; 2) How regular the taxpayer's activities have been; 3) Whether the taxpayer

MONTHLY STATEMENT

Total month's paid business	_____ line A
Total bills collected from previous months	_____ line B
Total month's income (A+B)	_____ line C
This month's disbursements—minus shaded Columns from weekly expense record	_____ line D
This month's net income	_____ line E
Previous month's net income	_____ line F
Net income to date this year	_____ line G

BUSINESS SUMMARY

Total month's paid business	_____ line A
Total month's charged business	_____ line B
Total month's business (A+B)	_____ line C
Total business previous months	_____ line D
Total business this year (C+D)	_____ line E

CASH POSITION

Cash on hand at month's beginning	_____ line A
Total month's paid business	_____ line B
Bills collected from previous months	_____ line C
Money borrowed	_____ line D
Total cash available this month (A+B+C+D)	_____ line E
Total monthly expenses	_____ line F
Money taken from business (personal)	_____ line G
Total money from business (F+G)	_____ line H
Cash position at month's end	_____ line I

A

Figure 23–18A.   A form for summarizing business volume and cash position.

FINANCIAL STATEMENT	© Caballus Corporation Form S53
ASSETS	**LIABILITIES**

ASSETS		LIABILITIES	
Cash on hand	$ _____	Notes payable (secured by signature) $ _____	
Cash in banks	$ _____	Notes payable on equipment	$ _____
Notes receivable	$ _____	Notes payable on horses	$ _____
Accounts receivable	$ _____	Notes payable on real estate	$ _____
Value of stored feed	$ _____		
Value of equipment	$ _____	Accounts payable:	
Value of tack	$ _____	Feed	$ _____
Value of horses	$ _____	Farrier and Vet	$ _____
Value of real estate (including improvements)	$ _____	Training	$ _____
		Advertising	$ _____
Prepaid expenses:		Taxes, licenses, Fees	$ _____
Utility deposits	$ _____	Balance of stud fees	$ _____
Stud fee advances	$ _____	Construction and Repair	$ _____
Futurity fees	$ _____	Other	$ _____
Entrance fees	$ _____		
Leases	$ _____	Notes receivable discounted	$ _____
Other	$ _____	Accrued taxes	$ _____
_____	$ _____	Accrued interest	$ _____
		Other liabilities:	
Stallion syndication	$ _____	_____	$ _____
Investment securities	$ _____	_____	$ _____
		_____	$ _____
Other assets:			
_____	$ _____		
_____	$ _____		
_____	$ _____	**Total Liabilities**	**$**
_____	$ _____		
_____	$ _____	**NET WORTH**	
		Excess of assets over liabilities	$ _____
Total Assets $			
		Total liabilities and net worth	**$**

Business Name _____

signature of Owner _____   date

B

Figure 23–18B.   A typical financial statement.

ANNUAL SUMMARY RECORD

Summary for the year from _____ 19_____ To _____ 19_____

INCOME

MO.	Total Months Income	Stud Fees	Boarding	Winnings	Training	Livestock Sales	Misc.		Reg. Income Horses	Long Term Capital Gain		
										Horses	Real Estate	Other
Jan.												
Feb.												
Mar.												
Apr.												
May												
June												
July												
Aug.												
Sept.												
Oct.												
Nov.												
Dec.												
Totals												

EXPENSES

MO.	Feed	Farrier & Vet	Horses	Trans.	Train-ing	Labor	Adv.	Taxes Licenses Fees	Stud Fees	Misc.	Loan Pmts.	Real Estate Pmts.	Equip.	Horses	Farm Maint.
Jan.															
Feb.															
Mar.															
Apr.															
May															
June															
July															
Aug.															
Sept.															
Oct.															
Nov.															
Dec.															
Totals															

Figure 23–19. An annual summary record.

HORSE DEPRECIATION SUMMARY							
HORSE	Date Purch'd	Life Expect'y	Actual Cost	Prior Depreciation	This year's Depreciation	Depreciation to this date	Net Value
TOTALS							

Figure 23–20.   A horse depreciation record.

EQUIPMENT AND FACILITIES DEPRECIATION SUMMARY							
EQUIPMENT ITEM	Date Purch'd	Life Expect'y	Actual Cost	Prior Depreciation	This year's Depreciation	Depreciation to this date	Net Value
TOTALS							
IMPROVEMENTS (item)	Date Acquired	Life Expect'y	Actual Cost	Prior Depreciation	This year's Depreciation	Depreciation to this date	Net Value
TOTALS							

Guideline Lives of ADR

Machinery and Equipment — 8 years

Non Mechanical Equpiment — 10 years

Building — Cement Block with floors and foundations — 25-30 years

Building — Frame and Pole Type — 20 years

Figure 23–21.   An equipment and facilities depreciation record.

EMPLOYEE WITHHOLDING RECORD						
Name	Social Sec. No.	No. of exept.	Married Single	Phone No.	Address	

Name								Hourly Wages $	
Quart-er	hrs. Worked		Total Earnings	Deductions					Net Pay
	reg.	o.t.		FICA	Fed. Tax	State Tax			
1									
2									
3									
4									
Total									

Name								Hourly Wages $	
Quart-er	hrs. Worked		Total Earnings	Deductions					Net Pay
	reg.	o.t.		FICA	Fed. Tax	State Tax			
1									
2									
3									
4									
Total									

Name								Hourly Wages $	
Quart-er	hrs. Worked		Total Earnings	Deductions					Net Pay
	reg.	o.t.		FICA	Fed. Tax	State Tax			
1									
2									
3									
4									
Total									

Name								Hourly Wages $	
Quart-er	hrs. Worked		Total Earnings	Deductions					Net Pay
	reg.	o.t.		FICA	Fed. Tax	State Tax			
1									
2									
3									
4									
Total									

Figure 23–22.   An employee withholding record.

CAPITAL GAINS SUMMARY SHEET						© Caballus Corporation Form S56	

REAL ESTATE SOLD

Description	Date Acquired	Date Sold	Original Cost	Prior Depreciation	Other Costs	Expense of Sale	Total Capital Gain
–		Totals					

OTHER CAPITAL GAINS ITEMS

		Totals					

GRAND TOTAL

Capital gains income allowances vary from year to year and it is advisable to check with the I.R.S. before arbitrarily assuming what qualifies towards capital gains. Gains on breeding stock since 1969 must be treated as normal income.

Figure 23–23. A capital gains summary sheet.

has a reasonable expectation of profit. Since the 1969 Tax Reform Act and the 1971 Revenue Law, if you can show a profit in 2 out of 7 years it is generally presumed a business is being conducted. Where losses run beyond a 5 year period the IRS may still allow a business classification if one or more of the following circumstances prevail: 1) Considerable capital appreciation has occurred; 2) changes have been made towards a more profitable operation; 3) the taxpayer's training, education, and experience is related to the horse business in a way that makes him something of an expert; 4) the taxpayer has spent much time and effort in the horse enterprise; 5) the taxpayer does not have substantial income from "other" sources; 6) the annual loss has been declining significantly over the years; and 7) the losses have been caused primarily by unforeseen circumstances that may not occur in the future. Complete and accurate records may be one of the most important factors in nailing down the tax position.

The figure on line 9 of Figure 23–24 is the one on which the federal and state taxes are based in regard to ordinary income. The figure on line 10 may be added with other long-term capital gain (from non-horse activities) to give

NET INCOME COMPUTATION

© Caballus Corporation Form S54

Year's Income, excluding long term capital gain . . . . $_____line 1
(total of all income columns on reverse side of sheet not shaded)

Year's Expenses, excluding depreciable items . . . . $_____line 2
(total of all expense columns on reverse side of shaded)

Depreciation:

    Real Estate Improvements (taken from depreciation sheet). . . . . $_____line 3

    Equipment (taken from depreciation sheet) . . . . $_____line 4

    Horses (taken from depreciation sheet–this year's column) . . . . $_____line 5
    (note–this should not include the non-depreciable horse)

Interest . . . . $_____line 6

Total Operating Costs for the Year . . . . $_____line 7
    (line 2 3 4 5 6)

Other Deductions (if any) . . . . $_____line 8

Net Income before Capital Gains . . . . $_____line 9
    (total of lines 7 and 8 minus line 1)

Long Term Capital Gains (from capital gains summary sheet) . . . . $_____line 10

Figure 23–24.   A form for net income computation.

a total capital gains figure. The records related previously will give all of the information necessary to make out a federal or state income tax report in those aspects relating to the horse enterprise.

BIBLIOGRAPHY

Dyke, Bill, and Jones, Bill: The Horse Business: An Investor's Guide. Fort Collins, CO, Caballus Publishers, 1975.
Oppenheimer, H.L.: Cowboy Arithmetic. Danville, IL, Interstate Press, 1971.
Willis, Larryann C.: The Horse Breeding Farm. San Diego, CA, A.S. Barnes, 1973.
Wood, Kenneth A.: The Business of Horses. Published by the author, Chula Vista, CA, 1973.

24

business aspects

Gone are the days when horse breeding could be a sideline endeavor conducted the way one does his racetrack betting. For many reasons one cannot merely put extra pocket change into horse breeding and, when a horse is sold, go whistling down the street with a fat bank roll.

Even the small-time owner with three or four broodmares in his backyard cannot be nonchalant about the enterprise. The price of well-bred horses is often astronomical; feed is expensive; veterinary and farrier fees are high; and interest on enough land to pasture a half-dozen horses represents almost a year's wage for a laboring man. When everything goes right, the profit margin is narrow, and it is not often that everything goes right. More often than not, the principal reward is esthetic. The average man simply could not engage in horse breeding if he were not allowed to deduct his expenses against his income. Only good business practices and a breakeven operation will keep the endeavor out of the hobby class tax status.

BUDGETING AND PLANNING

All of us with special interest in horses feel that the IRS has taken a rather narrow view of horse breeding, racing, and showing. They have placed a severe burden upon the horseman to show a respectable profit. Somehow they have missed the concept that the majority of small breeders, especially in the racing industry, have hopes that someday they will hit it big with the one-in-a-million longshot horse they have bred and raised themselves from stock that they obtained with limited means. Nevertheless, a breeder must live under the IRS rules, and his business must bend to the IRS dictates. Thus, intelligent budgeting and planning are paramount issues!

Some horse breeders use their farming business to pad the actual losses of the horse enterprise. When income and expenses of a general farm or ranch are all thrown in together in one system of accounting, it is impossible to see that the horse breeding enterprise alone may be losing money at the expense of cattle breeding, or at the expense of crop production.

The horse business does have several profitable aspects. The primary consideration of this book has been the breeding, raising, and selling of horses. A common adjunct to this area is the profit from stud fees for those who own a stallion. Many breeders wish to expand their enterprise to include racing—with the hope of making a profit. Some with more facilities than needed for their own horses want to train or board horses for others as a part of their enterprise. This variety of combinations offers a multitude of income expectations for those in the horse business.

Budgeting can begin with totalling of expected income over the first 5 years of the business (Table 24–1). It takes a while to get into full production, and only under special circumstances will the first year or two of horse breeding show a profit. It will be helpful from a budget standpoint, in the initial purchase of breeding stock, to buy some bred mares. Following this practice will allow yearling sales within the second year of business, and could prove to be one of the most economical ways of obtaining some top name breeding.

A breeder may find himself, at this point, with several years of unprofitable horse breeding behind him, only now realizing that something must be done quickly to turn a profit for the IRS. If so, one solution would be to sell enough stock this year to put the operation in the black, for once. This decision is difficult for a breeder who has been in business only a few years. Generally inexperience has caused him to have more money in his horses than their quality justifies. Selling at what may seem low prices is painful, but it will raise the capital to continue, perhaps in a better direction. It will also allow a few more years of "building up" with slim or nonexistent profits allowed by the IRS.

Planning 5 years in advance may be a new experience for some breeders. It will be especially difficult for the individual with a small operation and limited resources because the small breeder is unable to show how ends will meet without a miracle occurring, such as some of his colts winning to give his breeding stock a good name, or some millionaire becoming enthralled with his breeding and offering to buy everything he can produce.

As an example of a 5-year budget (Table 24–1), assume that the breeder begins with a dozen well-bred mares in foal to some promising stallions. Opportunities exist at every price level. Some may consider a well-bred mare worth $1000 and another may consider a well-bred mare worth $20,000. For this example, they will be averaged at $5000 each. Twelve mares at this price will total $60,000.

Table 24–1. A Hypothetical Financial Projection for 5 Years, Typical of a Beginning Horse Breeding Operation

Item	Hypothetical 5 Year Projection				
			EXPENSES		
	1st year	2nd year	3rd year	4th year	5th year
Grain & Feed Additives (hay raised on farm)	$ 1,500	$ 2,500	$ 3,000	$ 3,500	$ 3,500
Farm Lease	1,000	1,000	1,000	1,000	1,000
Farrier and Veterinarian	1,500	2,000	2,500	2,500	2,500
Labor	1,500	2,500	5,000	8,000	10,000
Insurance	1,500	1,500	1,500	1,500	1,500
Farm Supplies & Tack	200	500	1,000	1,000	1,000
Registration Fees	300	300	300	300	300
Advertising	1,000	1,500	1,500	1,500	2,000
Stud Lease (for breeding mares at home)	10,000	10,000	10,000	10,000	10,000
Outside Stud Fees (2 mares)	2,000	2,000	5,000	5,000	10,000
Board (2 mares at stud)	600	600	600	600	600
Transportation (principally mares to stud)	600	800	1,000	1,000	1,000
Racing or Showing Costs			5,000	5,000	5,000
Miscellaneous	500	500	500	500	500
Depreciation	(6,000)	(6,000)	(6,000)	(6,000)	(6,000)
TOTAL	$28,200	$31,700	$43,800	$47,400	$54,900

	INCOME				
Outside mares (to leased stud)	1,500	2,000	2,500	3,000	3,500
Boarding outside mares	360	480	600	720	840
Sell 5 yearlings (avg. ½ mare value)		12,500	12,500	12,500	12,500
Sell 2 2 yr-olds (avg. value of mares)			10,000	10,000	10,000
Sell 1 outstanding prospect				10,000	10,000
Sell 1 winner					20,000
TOTAL	$ 1,860	$14,980	$25,600	$36,220	$56,840

START-UP COSTS

12 mares	$60,000
Farm Purchase	optional
New Improvements	10,000
Machinery	10,000
TOTAL	$80,000

Machinery, which might include a pickup, horse trailer, and a tractor, could cost from $10,000 to $20,000. The conservative investor in our example starts with used equipment costing $10,000.

The typical businessman who is accustomed to dealing with enterprises which have large gross incomes is always struggling to show every item of expense. The business that can borrow 100 percent of its capital has a tax advantage in that the interest is entirely deductible, helping to reduce the taxable profit. The horse breeder is in a different situation. While he needs to deduct every real expense, he does not want to contrive expenses. Since all interest can be deducted as a personal expense, after the horse business profit has been determined, it is advantageous for the horse breeder to assume that all of the start-up costs, including real estate, are paid for with out-of-pocket cash. This cash, of course, may be raised by putting a mortgage on the house and the mountain cabin.

In the example, $60,000 will be needed for horse purchases, $10,000 for equipment purchases, and also probably another $10,000 to renovate barns and fences on the new facility where the horses are to be kept. This $80,000, along with the projected first-year deficit of about $20,000, means that a little over $100,000 will be needed the first year to begin the enterprise. In addition the purchase of a small farm will generally be necessary.

The real estate part of the endeavor is best left as a personal enterprise and reported on the tax return as a real estate investment. An accountant should be consulted on details of this as tax laws and rulings change from year to year.

A farm suitable for a horse breeding business with a base of 12 mares might range from $50,000 to $250,000 depending on the area. Whatever the price, it can be justified as a real estate investment which has no taxable relationship with the horse business. Just to keep the record straight and the IRS happy, a small lease fee should be assessed to the horse business. In the example, $1000 has been shown as an annual lease fee.

The example cited is conservative in several aspects, as should be any projection. Expenses are figured at the maximum that should be expected, and income is figured at the minimum that might occur. It can be expected that purchases made the first year will reflect a breeder's inexperience and lack of contacts, regardless of how much "good" advice he gets from those who "know." A beginning breeder who pays $5000 for a mare the first year will often have opportunities in succeeding years to buy mares of comparable quality at half that amount; or he will have the chance to buy mares of twice the quality at the same price.

The wise beginner will plan on this happening and whenever possible will sell a mare for a bit less than he originally paid after she has produced a foal or two. This will give an opportunity to purchase better quality mares as time goes by. Also, as mares are replaced, the breeder can fudge a year or two on age, and thus keep the average age of his mares about the same over a long period of time.

However, the beginner should not allow himself to be slipshod and flippant in his purchases the first year. On the contrary, he will have to bring all of his intellect and self-control to bear when purchasing. He will have to eliminate emotional buying as completely as possible, and purchasing on sudden impulse. As a general rule, the beginner would be better off passing all purchases that seem to be worth the money, and yielding only when he seems to have found an outstanding bargain. More advice is given on purchasing later in this chapter. Unless the beginner uses exceptional skill and has average good luck, he will find it difficult to raise yearlings that will sell for $2500, as projected from $5000 mares.

As the years pass and the breeding stock improves, one can expect the average selling price of his yearlings to improve. Since it will take much longer than 5 years to bring this increased price, the first 5-year projection should not count on it. Often unrealistic projections will anticipate the production of quite a few winners within 5 years, even from a mediocre start. The projection, as shown in the example, has several forces at work to improve the quality of the stock so that the second 5-year projection can show a much higher percentage of winners. The selling and repurchasing of better mares with the same money is one way to upgrade. The practice of taking a couple of mares a year to good outside studs will upgrade the stock. The amount of the stud fee generally reflects the quality of the stud; therefore, as income improves, more can be invested in stud fees, which in turn increases the chances of raising a winner.

Another important aspect of the plan in the example given is the leasing of a stallion the first year. For the money spent, leasing over a 5-year period will provide much more quality than purchasing will. The example provides for a total of $50,000 for the leasing of a stallion during the 5-year projection. If the beginner (and many experienced breeders also) were to take $50,000 to purchase a stallion the first year, chances are he would not nearly approach the caliber of stud that could be leased for $10,000 a year. Even if comparable value could be obtained in a purchase, the breeder may find his ideas changing so much after the first few years that he would want to try a different bloodline. Leasing offers much more flexibility than purchasing.

Joining a stallion syndicate offers another option. Syndication memberships are often sold with one breeding per member. A breeder purchasing 10 shares would then have the right to breed 10 mares a year. In some ways this practice is as undesirable for the new breeder as buying, and it prohibits the breeder from selling fees to other breeders to help defray the stallion's expenses. In the example, a few outside mares are bred each year to the leased stallion. Publicizing the stallion is one reason for the advertising expenses shown.

LIABILITY

Any business carries with it a certain amount of personal liability. This aspect is also true of the horse business. Hiring employees seems to carry a greater liability each year, when one considers minimum wages, withholding taxes, FICA, unemployment insurance, and workers compensation insurance. Liability drives up the cost of doing business because it necessitates more insurance.

Many breeders have elected to incorporate their breeding enterprises to reduce the personal liability. Incorporation has both advantages and disadvantages, which can best be explained by a corporate lawyer. Besides reducing liability, incorporating can be an effective means of inheritance transfer. Among the disadvantages is the fact that profits receive a much higher taxation.

With U.S. courts awarding ever-increasing amounts of money for injury and death, it becomes important for any businessman to consider liability insurance. Liability insurance can be purchased to protect individuals or corporations. Ordinary home owners' insurance will not cover losses arising from a business, such as horse breeding.

CONTRACTS AND LEASES

The large amounts of money involved in the horse business demand that agreements be specific and contracts be comprehensive and binding. Oral and informal written contracts can be binding, but they are not effective when dispute or controversy arises.

One of the most commonly used contracts is the breeding contract (Figure 24–1) or, as it is often called, "The Stallion Service Agreement." The essential parts include: the stallion and mare identification; the name of the two principal parties involved; an explanation of just what is agreed to in terms of breeding; the fee, and how it should be paid; the type of guarantee that the mare will foal; the type of care that will be provided for the mare; the assurances that the mare is healthy and free from infection; and specifically the way in which the breeding will be documented so that the resultant foal can be registered. Disputes often arise owing to injury or disease of the mare or foal while housed at the stud farm. Responsibilities in this area should be clearly spelled out. It may be advisable to include some guarantee that the stallion is fertile and specifics as to what occurs if he cannot settle the mare.

The leasing of stallions, mares, and even performance geldings has become rather commonplace. Such a written lease (Figure 24–2) begins with identification of the horse involved and the two parties of the agreement. In cases of breeding stock it must be clear how the technicalities of registration will occur, i.e., how the necessary forms will be signed and received. Some registries do not recognize lease arrangements and require the registered owner to sign all foal registration applications. The lease should state the time of beginning and ending.

Often a stallion is leased with a certain number of breedings each year reserved. The written agreement should detail how this procedure will occur in terms of stallion management, mare care, and charges for such.

When a mare is leased in foal, provisions should be made for care and ownership of the foal. As is usually the case, when a mare lease is terminated with the mare in foal, the lease should provide for what will happen with the subsequent foal while it is still nursing the mare.

Standards for care of all animals are a necessary consideration in a lease. Provisions should be made for transportation at both ends of the contract. It is helpful to specify who will pay for veterinary and/or insurance costs. The price of the lease and the manner in which it is to be paid must not be forgotten. A court battle may be avoided if the contract outlines what happens if the breeding animal proves to be sterile and specifics are included as to how it is judged sterile.

A boarding contract (Figure 24–3) or agreement is commonly used by most horsemen. It defines the stable and owner, as well as identifying the horse to be boarded and its owner. The price of boarding and the method of payment are included. The contract should clarify who is responsible in case of injury or sickness and what are the limits of liability to the stable owner. It is important to specify the exact type of feed if the owner is not feeding. Misunderstandings often arise because the owner of the horse was expecting grain and all the horse got was poor hay. The length of time of the contract should be specified. Other items often covered are: provisions for water; identification of a specific stall or paddock; the use of stable facilities; the person who will clean the stalls; taxes, liens, and exercise.

A training contract (Figure 24–4) is common but often is neglected. Aside from the usual items of a contract, identifying the horse and parties, a training contract should include: a complete description of the training to be accomplished; what shows or races will be entered, or who will decide; the fee; what expenses there will be and who will pay them; feeding specifics; veterinary care; division of the prize money; and the beginning and termination dates of the contract.

BUYING

Buying a broodmare is probably the most critical task a horse breeder will ever undertake. So many factors, other than price, must be considered; yet, price usually determines which buys are good and which are poor.

Within each breed, differing factors determine value. Conformation is of greatest value in some, coat color in others, and the amount of money won in still others. Bloodlines are important in all breeds because, as explained earlier in this book, bloodlines can be a good indicator of prepotency, or the ability to pass on a likeness to the offspring.

At least one common denominator establishes value in both mares and stallions, and that is their production record. Those that have produced winners are extremely valuable regardless of their record or their looks. Generally, however, the producers of winners will be the pick of the lot regardless of other desirable traits.

At this point the reader is advised to review Part III, the Breeding Program; Chapters 14 and 15 are especially valuable in helping one to arrive at proper selection criteria. Also Chapter 4 should be reviewed if a breeder has not yet firmly established what type of horse he wants to breed.

The price range may drastically affect the type of bargain that can be found. In the range of $1500 or less, there are always sellers willing to take a loss just to get out of trouble. Many would-be breeders starting on a shoestring develop other interests and decide to sell at a loss. Sometimes quality animals can be found under such circumstances. At a price above $5000, there generally has been much more dedicated financial planning, resulting in less opportunity for bargain hunting.

The best place to begin looking and pricing is the classified section of the breed magazine of choice. Much information can be obtained on the telephone about price versus quality of these horses. Usually horses sold in this manner bring the top dollar, sometimes justifying the fact that the buyer has a much better opportunity to investigate every aspect of the horse before purchase. Often the breeder advertising in the breed magazine classifieds is somewhat of a barn-blind dreamer. He has deluded himself into thinking his horses are of the same quality as the highest selling horses at auction. Yet he has enough lingering doubt that he is afraid to put the horse in a good auction to prove it.

The real action in buying and selling occurs at the reputable auctions in all parts of the county. Every breed has one or more groups of outstanding breeders who support an annual auction of the "best of the breed."

Some auctions are not legitimate. It can happen that seller and buyer conspire before the sale to run up the price of certain horses. The conspiring buyer later returns the horse to the seller, or in some way is compensated for the excessive sale price paid. The purpose is to "sucker" an unknowing buyer into paying a higher price. This illegal activity is a blight on the reputation of horse auctions, and the reputable auction managers do everything in their power to eliminate this kind of practice at their sales. For this reason, an annual auction of national fame is often an honest one. Buyers that receive good quality for the dollar return year after year. Regular consigners at these reputable sales develop the attitude that this is "the" market. They view it much as someone selling on the stock exchange. The price brought is the "real" value at the time sold—a healthy attitude for both seller and buyer.

One looking to buy good foundation breeding stock should keep an eye open for dispersal sales of prominent breeders. Every year good breeding programs are dispersed at auction. At these times, mares and stallions that would never have been available in any other way can be purchased.

Some breeders are large enough to hold their own production sale annually. Many of these develop an honest reputation over the years, and serve as an excellent source of good horses. It is seldom that these breeders place one of their best mares up for sale. Were they to do this, they would soon deplete their breeding stock.

Often a busy breeder does not have time to attend all of the sales but yet wants to buy certain types of horses on a regular basis. In this case, an agent can be employed; this person regularly studies the market and attends the sales. This field has many good agents; some of them are veterinarians who know horses. Their fee is usually well worth the price when they have an old and impeccable reputation.

Thoroughbred and racing quarter horse buyers have an added source of purchase at the racetrack. The claiming race has become an established barometer of the value of a horse. A trainer or owner may put in a claim prior to the race for the amount of the claiming race. A $10,000 claiming race will often have some fancy fillies that have won considerable amounts of money.

SELLING

Selling is no problem for those breeders who consistently produce winners, whether it be in the field of showing, jumping, or racing. For this reason, many breeders enter the training and competing areas. Even though their principal concern is breeding and raising horses, they want to prove their produce in order to get good prices. Realizing the high cost of competing, some breeders offer special considerations to buyers who will take the horses to the show, or the races.

Breeders of long-standing reputation who have proven ability of their produce many times find it unnecessary to train and compete any longer. They have found it quite satisfactory to put their stock in the most reputable annual sale and reap a dependable annual income without a lot of advertising and promotion.

Some breeders like to merchandise. They develop extensive, dynamic sales programs which even include sending quarterly newsletters, or propaganda sheets, to a large list of prospective buyers. They find such a mailing serves a dual purpose. Acquainting prospects with the merits of their horses encourages breeding to their stallions, and increases sales of their produce. Taking display ads in the breed journals is essentially the same tactic. Some enthusiastic promoters conduct clinics and horse shows at their own farm to gain exposure to the buying public. Often these activities pay big dividends.

Jack Conner, manager of Rapidan River Farm, explains his philosophy about successful merchandising of Morgan horses:

> Horses leaving our barn have new halters and shank, a health chart showing their medical history, immunization record, internal parasite control schedule, record of farrier work, and in the case of breeding stock, their production history. Interstate health papers and Coggins test are furnished by us. In the case of bred mares, we have instituted the practice of calling in an outside vet for a

pregnancy check before the mares leave the farm. This avoids a possible conflict of interest with our regular vet. The same approach is used for soundness exams which we encourage a purchaser to have done.

We furnish a bale of hay and 50 pounds of grain with each horse. This helps to avoid digestive upsets by the horse and the upset owners that are a consequence of the upset horse.

Registration papers are transferred from us to the new owner and at our expense. Some breeds, such as the Arab, require that the breeder supply the registration or signed transfer but not necessarily to pay the cost of transfer. I feel that the new owner should be treated like a valued customer and to have him pay a $10 transfer fee on a horse doesn't give him a good feeling.

Yes, these extras cost money and they can be included in your pricing schedule. The extra costs will be returned many times in the form of higher net prices and the repeat business that is generated.

BIBLIOGRAPHY

Dyke, Bill, and Jones, Bill: The Horse Business: An Investor's Guide. Fort Collins, CO, Caballus Publishers, 1975.

Oppenheimer, H.L.: Cowboy Arithmetic. Danville, IL, Interstate Press, 1971.

Self, Margaret Cabell: How to Buy the Right Horse. Phoenix, Farnum Horse Library, 1973.

Taylor, Joyce: Horses in Suburbia. London, J.A. Allen, 1966.

Tyler, George: Making Money Raising Horses. Houston, TX, Cordovan Corp., 1970.

Vance, J.D.: Cost factors in commercial thoroughbred breeding. The Washington Horse 23(8):940–942, 1969.

Willis, Larryann C.: The Horse Breeding Farm. San Diego, CA, A.S. Barnes, 1973.

Wood, Kenneth A.: The Business of Horses. Chula Vista, CA, Wood Publications, 1973.

index

Pages in italics indicate a figure; pages followed by t indicate a table.

653